# DEVELOPMENTAL IMMUNOTOXICOLOGY

# DEVELOPMENTAL IMMUNOTOXICOLOGY

EDITED BY

Steven D. Holladay

CRC PRESS

Boca Raton London New York Washington, D.C.

**Library of Congress Cataloging-in-Publication Data**

Developmental immunotoxicology / edited by Steven D. Holladay.
p. ; cm.
Includes bibliographical references.
ISBN 0-415-28457-0 (alk. paper)
1. Immunotoxicology. 2. Developmental toxicology. I. Holladay, Steven D.
[DNLM: 1. Immunotoxins—adverse effects—Infant, Newborn. 2. Immune System—growth & development—Infant, Newborn. 3. Prenatal Exposure Delayed Effects. QW 630.5.I3 D489 2004]
RC582.17.D484 2004
616.97—dc22 2004054496



Direct all inquiries to CRC Press LLC, 2000 N.W. Corporate Blvd., Boca Raton, Florida 33431.


International Standard Book Number 0-415-28457-0
Library of Congress Card Number 2004054496
Printed in the United States of America 1 2 3 4 5 6 7 8 9 0
Printed on acid-free paper

# Preface

This textbook of developmental immunotoxicology was created to provide a nonexhaustive but reasonably extensive overview of available information concerning adverse postnatal consequences of perinatal immunotoxicant exposure. A general familiarity with development of the immune system greatly assists understanding of how early-life toxicant challenges may transiently or permanently change later immune responsiveness. For this reason we begin with two chapters that review immune system development in the most-studied species, rodents and humans. Industry- and government-sponsored symposia and workshops over the past 2 to 3 years considering developmental immunotoxicity issues have without exception expressed interest in effective risk assessment methodologies and research model availability for characterizing and detecting developmental immunotoxicity. The next five chapters of this book were solicited as our attempt to address these needs. The first of these chapters specifically addresses risk assessment perspectives for developmental immunotoxicology, while authors of the other four chapters present and discuss utility of mouse, rat, pig, and nonhuman primate research models for developmental immunotoxicology.

Five chapters of the textbook have been devoted to different classes of developmental immunotoxicants, including halogenated and polycyclic aromatic hydrocarbons (e.g., TCDD; benzo[a]pyrene), pesticides (chlordane), and heavy metals (lead). The apparent permanent nature of immune deficits reported in mice after developmental exposure to chlordane or benzo[a]pyrene is startling, and discussed as parts of these chapters. Related literature further suggests that early exposure to estrogenic compounds may permanently imprint immune cells with an altered response capacity (e.g., interferon-$\gamma$ production by T cells), and raises concern that such imprinting may increase risk of later-life exaggerated immune responses (hypersensitivity disorders; autoimmune diseases). Maternal neurotoxicity, offspring gender, and maternal nutritional status may further influence the spectrum of immune defects caused by immunotoxic compound exposure during development, and as such are considered in the final chapters of the textbook.

It will become apparent to readers that, while much is known, developmental immunotoxicology is nonetheless a discipline in its infancy. Many important gaps in the present knowledge base exist, especially regarding identification of heightened windows of immunotoxicant sensitivity during development, and potential long-term effects of immunotoxic agents (e.g., therapeutic immunosuppressants) on human immune development. It is our hope that information provided in this textbook will serve as an important resource to scientists devoted to filling these data gaps.

# Introduction

It is well established that diseases associated with abnormal immune function, including asthma and common infectious diseases, are considerably more prevalent in the very young when compared to the adult population. A number of factors are thought to account for these differences. The most obvious are age-related differences in the integrity of the host's anatomical and functional barrier and general immunocompetency. During the last decade, however, experimental and clinical studies have also suggested that exogenous factors can influence immune-mediated diseases in the young. These factors include, among others, obesity, nutrition and, as will be the focus of the following chapters, certain therapeutic and environmental exposures. The latter is supported, in part, by the alarming observation that the number of children who develop allergic asthma has increased markedly over the past four decades, particularly in urban areas (e.g., Timonen and Pekkanen 1997; Akinbami and Schoendorf 2002; Thurston et al. 1997; Smyth 2002). This is further supported by a plethora of laboratory studies conducted in rodents, guinea pigs, and nonhuman primates which demonstrate that, while not a causative agent for asthma, air pollutants, such as ozone and diesel exhaust, exacerbate asthma to common protein allergens such as house dust mites (i.e., shift the dose response curve to the left) (e.g., Schelegle et al. 2003). Interestingly, it has been found that children living in rural environments (i.e., farming communities) are less likely to develop allergic asthma (Ehrenstein et al. 2000). This is thought due to high levels of endotoxin in their environment, which skew Th1:Th2 cells toward a predominant Th1 phenotype. Less studied from a clinical perspective, but equally alarming, are results from studies demonstrating increases in infectious disease in children following perinatal exposure to certain therapeutic or environmental agents. An important study in this area was conducted by Weisglas-Kuperus et al. (2000), who demonstrated that exposure to levels of halogenated aromatic hydrocarbons normally found in highly industrialized countries are associated with increases in childhood infections and lower vaccination responses. Clinical undertakings in this area are rare but, as detailed in many of the following chapters, are supported by extensive laboratory studies.

While these studies have drawn some concern, it was not until the landmark report prepared in 1993 by the National Research Council entitled "Pesticides in the Diets of Infants and Children" (NRC 1993) that the regulatory agencies expressed interest regarding children's health protection from agents that may potentially damage the immune system. The scientific basis for this concern stems not only from likely pharmacokinetic differences between the adult and neonate, which would not be unique for immunotoxicity, but also from the fact that the developing immune system, which can be thought of as beginning at conception and completed at puberty, differs inherently from the adult system in its biological and physiological responses to injury. Thus, it has been suggested that the developing immune system, by its nature, may be more sensitive to injury or the toxic effects may be more persistent when compared to adult exposure. Clinical effects may also be expressed immediately or later in life and be manifested as either increased infectious or neoplastic diseases resulting from immunosuppression or increased incidences/severity of allergic or autoimmune disease resulting from immune dysregulation. The specific type(s) of immunotoxicity that might ensue would not only be dependent upon the toxicokinetic properties of the particular agent but also the time of exposure, referred to as critical windows of development. The NRC report discussed the potential unique vulnerability of children to pesticides and argued for a more thorough assessment of children's health in risk assessment. This report was followed in 1996 by the Food Quality Protection Act (FQPA), which required EPA and other regulatory agencies that deal with pesticides to specifically consider children's health risks. Amendments to the Safe Drinking Water Act (SDWA) in the same year (1996) also emphasized children's health in setting health advisories for drinking water.

Following publication of the NRC report and passage of the FQPA, more attention was focused on the development, characterization, and standardization of methods for developmental neurotoxicity and reproductive toxicity assessment (e.g., Garman et al. 1993; Levine and Butcher 1990), while at least initially, assessment of the developing immune system was not addressed. This was not due to any notion that the developing immune system is any more or less vulnerable than other developing organ systems to toxic injury, but rather reflects the fact that immunotoxicity assessment, in general, represents a relatively new effort in toxicology and efforts in this area have traditionally followed behind those in other areas. During the last 5 years, a number of workshops have been held where experts have convened to present their research on the role of environmental factors in childhood asthma. These conferences have been useful as they have allowed identification of additional research needs and have been instrumental in spurring interest in the area. Regarding immunosuppression, a workshop was convened by ILSI/HESI in June 2001, which not only provided an opportunity for experts to present their findings but also allowed discussions to address specific questions that relate to developmental immunotoxicity tests and their interpretation for risk assessment (Sandler, 2002). This workshop was followed by a "consensus" workshop sponsored by NIEHS/NIH and NIOSH/CDC on the most appropriate screening tests to evaluate developmental immunotoxicity (Luster et al. 2003). ILSI/HESI has recently held a panel discussion that discussed additional methodological details for conducting developmental immunotoxicology screening tests (Holsapple 2003; Holsapple, in preparation). This information should be useful to investigators in the area of developmental immunotoxicology.

The current volume represents the first textbook dealing exclusively with developmental immunotoxicology. It begins with overviews of immune system development in experimental animal models and human, areas that have been the focus of immunological research for a number of decades as well as the complex issues dealing with the evolution of developmental immunology and risk assessment. This is followed by sections describing different animal models used to study developmental immunotoxicity and examples of specific developmental immunotoxic agents, including therapeutics. Although the volume is focused on immunosuppression, chapters have been included describing the consequences of immune dysregulation (i.e., autoimmune and allergic diseases) following developmental exposures. The text concludes with several chapters that describe the role of neuroimmune interactions as it relates to developmental immunotoxicology.

## REFERENCES


Akinbami LJ and Schoendorf KC. Trends in childhood asthma: prevalence, health care utilization, and mortality. *Pediatrics* 110:315–322, 2002.

Ehrenstein OS, von Mutius E, et al. Reduced risk of hay fever and asthma among children of farmers. *Clin Exp Allergy* 30:187-193, 2000.

Garman RH, Fix AS, Jortner BS, Jensen KF, Hardisty JF, Claudio L, et al. Methods to identify and characterize developmental neurotoxicity for human health risk assessment. II Neuropathology. *Environ Health Perspect* 109(1):93-100, 2001.

Holsapple MP. Developmental immunotoxicity testing: a review. *Toxicol* 185:193-203, 2003.

Levine TE and Butcher RE. Workshop on the qualitative and quantitative comparability of human and animal developmental neurotoxicity. Work group IV report: triggers for developmental neurotoxicity testing. *Neurotoxicol Teratol* 12:281-284, 1990.

Luster MI, Dean JH, and Germolec DR. Consensus workshop on methods to evaluate developmental immunotoxicity. *Environ Health Perspect* 111(4):579-583, 2003.

NRC (National Research Council). Pesticides in the diets of infants and children. National Academy Press, Washington, D.C., 1993.

Sandler JD (Ed). Developmental immunotoxicology and risk assessment: special issue. *Human Exp Toxicol* 21(9-10):469-572, 2002.

Schelegle ES, Miller LA, Gershwin LJ, Fanucchi MV, Van Winkle LS, Gerriets JE, Walby WF, Mitchell V, Tarkington BK, Wong VJ, Baker GL, Pantle LM, Joad JP, Pinkerton KE, Wu R, Evans MJ, Hyde DM, and Plopper CG. Repeated episodes of ozone inhalation amplifies the effects of allergen sensitization and inhalation on airway immune and structural development in Rhesus monkeys. *Toxicol Appl Pharmacol* 191:74-85, 2003.

Smyth RL. Asthma: a major pediatric health issue. *Respir Res* 3(1):S3-S7, 2002.

Thurston GD, Lippmann M, et al. Summertime haze air pollution and children with asthma. *Am J Respir Crit Care Med* 155:654-600, 1997.

Timonen KL and Pekkanen J. Air pollution and respiratory health among children with asthmatic or cough symptoms. *Am J Respir Crit Care Med* 156:546-552, 1997.

Weiglas-Kuperus N, Sas TC, Koopman-Esseboom C, van der Zwan CW, De Ridder MA, Beishuizen A, et al. Immunologic effects of background prenatal and postnatal exposure to dioxins and polychlorinated biyphenyls in Dutch infants. *Pediatr Res* 28:404-410, 1995.

# The Editor

**Steven D. Holladay** obtained his Ph.D. in toxicology from North Carolina State University, in 1989. He then spent two years in postdoctoral training at the National Institute of Environmental Health Sciences (NIEHS), studying the developmental immunotoxicity of TCDD and DES. He has published 85 research manuscripts over the past 15 years, and served on the editorial review board for 5 journals and as an ad-hoc reviewer for 19 other journals over the past 10 years. He is presently a full professor at the College of Veterinary Medicine, Virginia Polytechnic Institute and State University, Blacksburg, Virginia.

Dr. Holladay was selected as the 1986 "Employee of the Year" by both the North Carolina State University College of Veterinary Medicine and by North Carolina State University. He received the North Carolina Governor's Award of Excellence for "Devotion to Duty" in 1987, presented by then Governor James Martin. He was selected by the Merck Foundation in both 1994 and 1995 to receive the AGVET Award for Creativity in Teaching, which recognizes innovative approaches to veterinary medical education. In 1995 he was also the recipient of the Virginia-Maryland Regional College of Veterinary Medicine Teaching Excellence Award and the Virginia Tech Academy Award for Teaching Excellence. Dr. Holladay received the 1996 Pfizer Animal Health Award for Research Excellence, and the Carl J. Norden Distinguished Teaching Award in 1998. The latter is considered to be the highest teaching award in veterinary medical education.

Dr. Holladay is a member of the following professional societies:

National Society of Toxicology
Immunotoxicology Specialty Section of the Society of Toxicology
Teratology Society
American College of Toxicology
International Congress of Vertebrate Morphology
American Association of Veterinary Anatomists
World Association of Veterinary Anatomists

# Contributors

**S. Ansar Ahmed**
College of Veterinary Medicine
Virginia Polytechnic Institute and State University
Blacksburg, Virginia

**John M. Armstrong**
Senior Clinical Scientist
Department of Medical Affairs
Centocor, Inc.

**John B. Barnett**
Department of Microbiology, Immunology and Cell Biology
Robert C. Byrd Health Sciences Center
West Virginia University
Morgantown, West Virginia

**Patricia V. Basta**
Research Triangle Institute
Research Triangle Park, North Carolina

**Benny L. Blaylock**
School of Pharmacy
College of Health Sciences
University of Louisiana at Monroe
Monroe, Louisiana

**Rodney R. Dietert**
Department of Microbiology and Immunology
College of Veterinary Medicine
Cornell University
Ithaca, New York

**Sarah V. M. Dodson**
Department of Microbiology, Immunology, and Cell Biology
and the Mary Babb Randolph Cancer Center
Robert C. Byrd Health Sciences Center
West Virginia University
Morgantown, West Virginia

**Rolf C. Gaillard**
Division of Endocrinology, Metabolism and Diabetology
University Hospital
Lausanne, Switzerland

**Laura F. Gibson**
Department of Microbiology and Cell Biology
Robert C. Byrd Health Sciences Center
West Virginia University
Morgantown, West Virginia

**Robert M. Gogal, Jr.**
Edward Via Virginia College of Osteopathic Medicine
Blacksburg, Virginia

**Andrew G. Hendrickx**
Center for Health and the Environment
University of California
Davis, California

**Steven D. Holladay**
College of Veterinary Medicine
Virginia Polytechnic Institute and State University
Blacksburg, Virginia

**Terry C. Hrubec**
Department of Biomedical Sciences
Virginia College of Osteopathic Medicine
Blacksburg, Virginia

**Carole A. Kimmel**
National Center for Environmental Assessment
Office of Research and Development
U.S. Environmental Protection Agency
Washington, D.C.

**Marquea D. King**
National Center for Environmental Assessment
Office of Research and Development
U.S. Environmental Protection Agency
Washington, D.C.

**Kenneth S. Landreth**
Department of Microbiology, Immunology, and Cell Biology
and the Mary Babb Randolph Cancer Center
Robert C. Byrd Health Sciences Center
West Virginia University
Morgantown, West Virginia

**Ji-Eun Lee**
Chevron Phillips Chemical Company, LP
The Woodlands, Texas

**Ramona Leibnitz**
Popular Immunology, Inc.
Aurora, Colorado

**Deborah Loer-Martin**
Environmental Health Scientist
Loer-Martin and Associates
Fort Myers, Florida

**N. Makori**
Center for Health and the Environment
University of California
Davis, California

**Susan L. Makris**
Office of Pesticide Programs, Office of Prevention, Pesticides and Toxic Substances
U.S. Environmental Protection Agency
Washington, D.C.

**Reinhard Pabst**
Center of Anatomy
Medical School
Hannover, Germany

**Néstor B. Pérez**
Head, Immunology Section
Children's Hospital
La Plata, Argentina

**Pamela E. Peterson**
Center for Health and the Environment
University of California
Davis, California

**M. Renee Prater**
Edward Via Virginia College of Osteopathic Medicine
Blacksburg, Virginia

**Hermann J. Rothkötter**
Institute of Anatomy
Medical Faculty, Otto von Guericke University
Magdeburg, Germany

**Ralph J. Smialowicz**
National Health and Environmental Effects
Research Laboratory
Office of Research and Development
U.S. Environmental Protection Agency
Research Triangle Park, North Carolina

**Eveline Sowa**
Center of Anatomy
Medical School
Hannover, Germany

**Eduardo Spinedi**
Neuroendocrine Unit
Multidisciplinary Institute on Cell Biology
La Plata, Argentina

# Contents

# PART I

# Development of the Immune System

CHAPTER 1

# Development of the Rodent Immune System

**Kenneth S. Landreth and Sarah V. M. Dodson**

## CONTENTS



## INTRODUCTION

Initial understanding of critical developmental events that shape the immune system of vertebrates came from seminal studies of the immune system in avian species (Glick et al. 1956). From these studies, it became obvious that the immune system could be divided into functional subsets based on cellular production in unique anatomic sites (Cooper et al. 1965). Innate immune responsive cells are produced in the bone marrow of birds; however, lymphocytes are derived from a set

0-415-28457-0/05/$0.00+$1.50

of gut-associated lymphoid tissues: the thymus located at the origin of the gut and the bursa of Fabricius, an outpocketing of the cloacal region. It was the unexpected finding that precursors of antibody-forming cells in the chicken were uniquely produced in the bursa of Fabricius that ultimately established the role of lymphocytes as primary effector cells of adaptive immunity, the division of lymphoid architecture into central lymphoid tissues in which cells develop and peripheral lymphoid tissues in which they function, and the identity of the unique lymphopoietic role of the thymus. The association of antibody-forming cell and cell-mediated immune function with respective bursal and thymus organ development resulted in the use of the terms B and T cell, respectively, and ushered in a period of intense investigation of developmental aspects of the immune system.

Although many laboratories pursued the identity of "mammalian bursal equivalent tissues," it soon became clear that no such exclusive organ for B cell development existed in mammals. For this reason, most laboratories moved quickly to establish a better mammalian animal model for continued study in this field. Rodents rapidly supplanted avian species as animals of choice for investigating the developmental origins of immune function, and remain, to this day, the best characterized animal model of developmental immunology for investigation (Yoder 2002). It is for this reason that most mechanistic studies designed to better understand effects of experimental manipulation or toxic insult on embryonic development of immune function have been conducted in rodent species.

In order to understand the potential for environmental agents to affect the development and eventual function of this complex biological system, it is necessary to understand the temporal sequence of events that occur during embryonic development and precede immune function. In this chapter, we review the current state of knowledge of embryonic and postnatal immune function in rodents. These data primarily derive from studies in mice and, to a considerably much lesser extent, rats, drawing on observations from both immunology and experimental hematology. Immunocompetent cells derive from the same set of hematopoietic stem cells that give rise to all remaining circulating blood cell forms, and developmental manipulation is known to impact the development of all blood cell types. For that reason, we feel that this broader approach to understanding the development of immunity is necessary in order to understand the overall effect of exposure to environmental agents on immune cell development in the context of overall damage to hematopoietic tissues.

## EMBRYOGENESIS OF HEMATOPOIETIC AND LYMPHOPOIETIC ORGANS

Establishment of hematolymphopoietic organs in the mouse initiates with the formation of pluripotential hemolymphopoietic stem cells (HSC) in intraembryonic splanchnopleura and is followed by a temporal migration of these HSC from intraembryonic mesenchyme in this region to the fetal liver, fetal spleen, and ultimately to final residence in bone marrow and in the thymus (Metcalf and Moore 1971; Dieterlen-Lievre 1975; Good 1995; Medvinsky and Dzierzak 1996). HSC are a unique

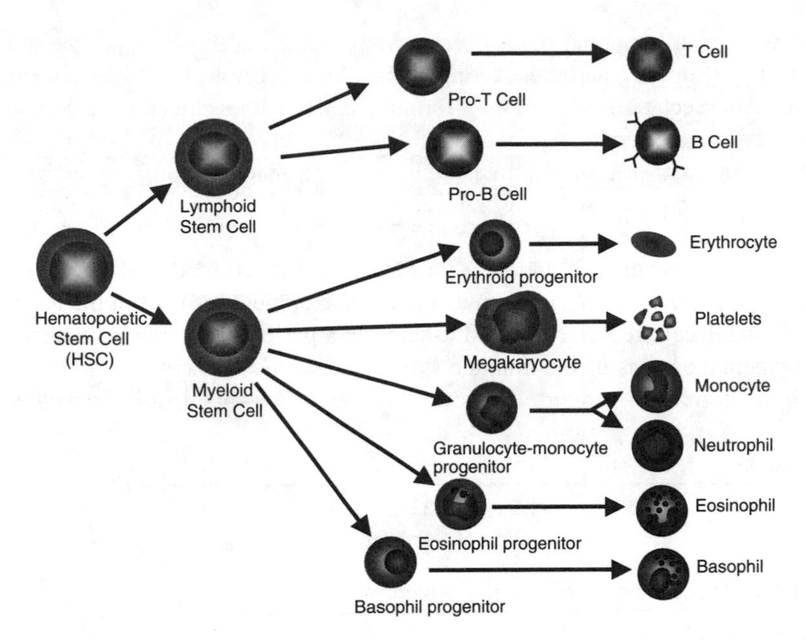


**Figure 1.1** Overview of the hematopoietic system detailing lineage relationships between hematopoietic stem cells (HSC) and lineage-restricted stem cells for hematopoietic and immunocompetent cells.

population of cells that maintain the capacity to self-renew to repopulate themselves when the system is depleted, proliferate to expand the numbers of cells that become committed to development down a specific blood cell lineage, and differentiate to form all of the morphologically identifiable leukocytes that participate in immune responses, as well as megakaryocytic and erythrocytic cells (Terskikh et al. 2003; Weissman 2000) (Figure 1.1).

HSC first appear in mice during the $8^{th}$ gestational day in intraembryonic splanchnopleuric mesenchyme surrounding the heart (Cumano and Godin 2001), a tissue also identified as the aorto-gonado-mesonephros (AGM) (Figure 1.2). HSC are found at essentially the same embryonic stage in extraembryonic blood islands of the yolk sac in rodents as well (Marcos et al. 1997). It remains unclear to what extent there is an exchange of cells from intraembryonic hematopoietic tissues to these extraembryonic tissues and whether this exchange is responsible for the appearance of HSC in those anatomic sites. Circulation in the embryo is not established until gestational day 8 in embryonic mice, making this exchange likely; however, it is also clear that HSC in these two tissues differ dramatically in developmental fate (McGrath et al. 2003). Recent studies have clearly shown that intraembryonic stem cells, but not those which appear in the yolk sac, ultimately contribute to sustained intraembryonic blood cell development in the embryo and to the emergence of functional immune responsive cells in rodents (Godin et al. 1999; Cumano et al. 2001). One interpretation of these results is that HSC in extraembryonic sites derive separately and have severely restricted developmental potential as compared to those that arise in intraembryonic tissues. It should be noted that both of these populations of HSC form spleen colonies (day 12 CFU-s) following *in vivo* transplantation of AGM or

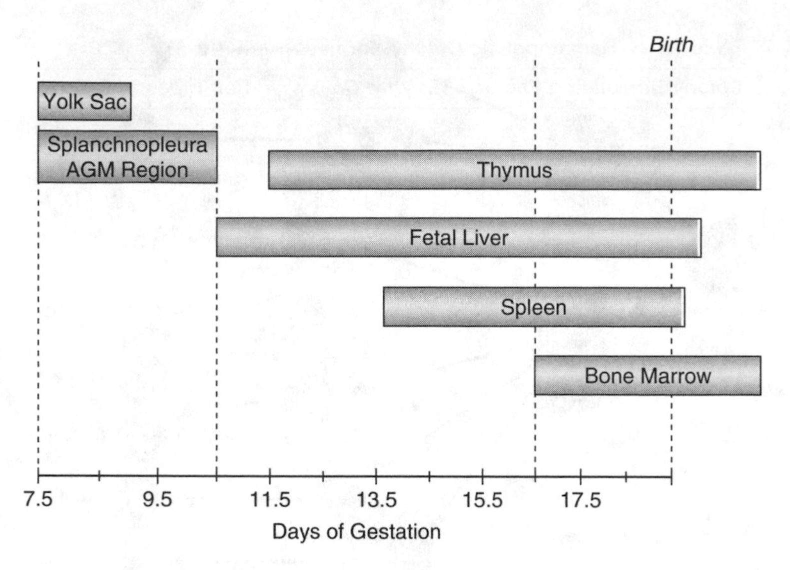


**Figure 1.2** Ontogeny of hematopoiesis in rodents.

yolk sac cells in rodents a standard assay for pluripotential hematopoietic cells (Till and McCulloch 1961; Magli et al. 1982).

This initial period of *de novo* development of hematolymphoid stem cells from uncommitted mesenchymal cells culminates as these cells migrate to other intraembryonic anatomic locations in the fetus, primarily the liver and spleen (Metcalf and Moore 1971; Tavassoli and Yoffey 1983). In these new tissues, lineage-restricted subpopulations of stem cells begin to emerge and expand. This second wave of hematolymphoid development, characterized by emergence of committed stem cells for lymphoid and myeloid cell production, initiates in the liver on gestational day 10 in mice (Godin et al. 1999).

## FETAL LIVER AS A HEMATOPOIETIC ORGAN

At approximately day 10 of gestation in mice, HSC relocate from the AGM to the developing fetal liver organ (Tavassoli 1991; Delassus and Cumano 1996; Godin et al. 1999; Cumano and Godin 2001). As HSC accumulate in the fetal liver, these cells differentiate to form the complex array of progressively more differentiated and lineage-restricted stem cells that characterize postnatal hematopoietic tissues (Mebius et al., 2001). This appearance and expansion of lineage-restricted stem cells is followed by differentiation of the same cells to form more mature progenitor cells that retain the ability to proliferate. These committed cells do not self-renew and are further restricted in their developmental options. Lineage-restricted cells are operationally defined by their proliferation and differentiation in response to a known set of colony stimulating factors (CSFs) and the formation of characteristic colonies *in vitro* (colony forming cells in culture or CFC) (Gwatkin et al. 1957; Lotem and Sachs 2002) (Table 1.1). As progenitor cells are stimulated to divide in response to

**Table 1.1 Hematopoietic Colony Formation *In Vitro***

| Colony Stimulating Factor | *In Vitro Colony* | Cell Types in Colony |
|---|---|---|
| IL-3 | CFU-multi | mixed |
| GM-CSF | CFU-GM | neutrophils<br>monocytes |
| G-CSF | CFU-G | neutrophils |
| M-CSF | CFU-M | monocytes |
| IL-5 | CFU-eo | eosinophils |
| IL-3 + TP | CFU-meg | megakaryocytes |
| IL-3 | BFU-e | erythrocytes |
| Epo | CFU-e | erythrocytes |
| IL-7 | CFU-IL-7 | lymphocytes |

specific cytokine stimulation, they differentiate, lose the capacity to proliferate, and eventually undergo apoptosis (Bradley and Metcalf 1966). Colonies enumerated in most of these *in vitro* assays, then, are primarily composed of nonproliferating end cells and contain few cells with repopulating potential. However, the availability of recombinant cytokines and the use of these assays have been invaluable in enumerating specific progenitor cells in embryonic tissues that are otherwise indistinguishable by morphologic analysis (Landreth and Dorshkind 1988). These assays are central to any study of the direct effects of toxic compounds on hematopoietic tissue in the developing embryo or in postnatal tissues.

The fetal liver continues to be the principal hematolymphopoietic organ until near the end of gestation (Owen et al. 1974; Owen et al. 1977). In this organ, hematopoietic cells expand rapidly; however, it is noteworthy that few morphologically or functionally identifiable mature leukocytes are found in the embryo until near the time of birth (Godin et al. 1999). In fact, mature lymphocytes are not found in the developing fetal liver until day 18 of gestation in the mouse, a time at which the liver is no longer the primary hematopoietic organ (Kincade 1981).

B lymphocyte production in the fetal liver has been well characterized and serves as a model of fetal blood cell production in that organ (Andrew and Owen 1978; Antoine et al., 1979; Velardi and Cooper 1984). Cells bearing immunoglobulin gene rearrangements are first found in the liver on gestational day 11 and increase rapidly to easily detectable numbers by day 13 (Velardi and Cooper 1984). This is mirrored in the expression of the cell surface antigen CD45R that is expressed on B lymphocyte progenitors in the fetal liver (Landreth et al. 1983). CD45R+ cells are first detected on day 11 of gestation in mice and are uniquely associated with the fetal liver at that time (Landreth et al. 1983). CD45+ cells rapidly expand in the liver until day 18 and then decline rapidly at birth as liver functions change (Landreth et al. 1983; Alonso-C et al. 2000). Cell surface immunoglobulin + lymphocytes (B cells) are not detected in the fetal liver until day 18 of gestation (Landreth et al. 1983).

Mechanisms that down-regulate the final maturation steps of leukocytes in the fetal liver and result in the failure of hematopoietic cells to mature in that site during embryogenesis are unknown. However, it seems reasonable to conclude that the process of blood cell development in the fetal liver vs. that in postnatal bone marrow is fundamentally different in this respect. For that reason, comparisons of the effect of toxic compounds on these two tissue sites should be made with that caveat in mind, and such studies may reveal quite different sensitivities to toxic insult. It should also be noted that the persistence and predominance of hematopoiesis in the fetal liver throughout much of gestation has made it the organ of choice for mechanistic studies of normal and disrupted rodent blood cell development and immune function development (Cumano et al. 2000).

At 18 days of gestation, numbers of hematopoietic cells decline in the fetal liver as the bone marrow assumes primary hematopoietic function (Kincade 1981). At birth, the function of the liver dramatically changes and hematopoietic function ceases. It is well known that hematopoietic cells are exquisitely sensitive to alterations in oxygen tension, and the dramatic change in the pattern of blood flow in the embryo as umbilical circulation is terminated may contribute to this change in hematopoietic function.

## SPLENIC HEMATOPOIESIS

HSC and lineage-restricted hematopoietic progenitor cells are found in the developing spleen on gestational day 13, at approximately the same time they are found in the fetal liver (Landreth 1993) (Figure 1.2). Although the expansion of hematolymphoid cells in the spleen never rivals that found in the developing liver, the spleen maintains this low level of hematopoietic activity well into postnatal life. For this reason, splenic hematopoiesis serves as a reserve of HSC following damage to the bone marrow organ following chemotherapy or radiation in postnatal animals. The persistence of the hematopoietic microenvironment in the spleen is revealed by the seeding of HSC to that organ following hematopoietic transplantation in rodents.

It is of some interest that lymphopoiesis has never been convincingly demonstrated in the spleen of postnatal mice under any experimental conditions tested (Paige et al. 1981). These observations led to the proposal of specific hematopoietic microenvironments in the bone marrow and spleen. The bone marrow hematopoietic microenvironment substantively differs from that found in the spleen such that it is uniquely required for lymphopoiesis in mice.

## EMBRYONIC EMERGENCE OF THE THYMUS

Organogenesis of the thymus anlagen initiates in mice on day 11 of gestation (Teh 1993; Shortman et al. 1998) and derives from the 3$^{rd}$ and 4$^{th}$ pharyngeal pouches (Manley and Blackburn 2003). The thymus architecture is composed of both epithelial and mesenchymal cell components, both of which contribute to the selection of functional thymocytes during development (Manley and Blackburn 2003). The thymus is immediately colonized by immigrant HSC, which are detectable by gestational day 11 (Auerbach 1961; Rodewald and Fehling 1998). Hematopoietic cells in the thymus on day 11 are ckit$^+$, Thy-1$^-$, CD3$^-$, CD4$^-$, CD8$^-$, CD25$^-$, a phenotype characteristic of uncommitted HSC (Peault et al. 1994). However, by day 13, the majority of thymic hematopoietic cells have begun to express Thy-1, CD3, CD4, and CD8, cell surface antigens more closely associated with T lymphocyte development (Rodewald and Fehling 1998). Hematopoiesis in the developing thymus is largely limited to lymphocyte production; however, HSC continue to be detectable in this organ throughout gestation and retain the capacity to form all blood cell lineages when transplanted.

Although less is known about the role of cytokines in fetal T cell development than about B lymphopoiesis, immature proliferating thymocytes form colonies *in vitro* when exposed to the proliferative cytokines IL-7 and ckit-ligand (CFU-IL-7) (Namen et al. 1998; Lee et al. 1989; Moore et al. 1996) and it is likely that both of these cytokines play a key role during *in utero* T cell development. Developing thymic lymphocytes initiate expression of the T-cell receptor (TcR) within the thymic cortex and then undergo sequential interactions with epithelial and mesodermal-derived stromal cells which result in selection to delete nonreactive and autoreactive cells (Cosgrove et al. 1992). Soon after birth in rodents, T lymphocytes with characteristic expression of the full complement of TcR-associated molecules are found in peripheral lymphoid organs, including the spleen (Rodewald and Fehling 1998).

## EMBRYONIC EMERGENCE OF SECONDARY LYMPHOID TISSUES

Lymph nodes are formed in the embryo by endothelial budding of the venous circulatory system, initiating on gestational day 10.5 in mice (Wigle et al. 2002). These primitive lymph sacs form the lymphatic vasculature by endothelial sprouting, a process that initiates on day 11 and is completed in mice by day 15.5 of gestation (Wigle and Oliver 1999; Mebius 2003). These lymph sacs are populated by IL-7-responsive cells (presumably lymphocytes) by day 13 (Yoshida et al. 2002). The fact that IL-7 receptor-deficient mice fail to form lymph nodes suggests that the interaction of immature hematopoietic cells with the lymph sac is necessary for maturation of the lymph node anatomical structure (Cao et al. 1995; von Freeden-Jeffry et al. 1995). Peyer's patches are formed from clusters of cells on the proximal end of the intestine in mice on gestational day 15.5. Nasopharyngeal lymphoid tissues are formed after birth in mice in a similar manner (Adachi et al. 1997).

## HEMATOPOIETIC DEVELOPMENT OF THE BONE MARROW

Near the termination of gestation in rodents, long bones of the embryo mineralize and the central marrow cavity is excavated to form a marrow cavity (Miller et al. 2002). This marrow cavity is formed on gestational day 17.5 in mice (Kincade 1981) and is immediately populated by hematopoietic cells, including HSC. This relationship between ossification of bone, formation of a marrow cavity, and emergence of hematopoietic function is a particularly intriguing one, and it has been proposed that the migration of hematopoietic cells into this tissue actually initiates the process of bone calcification and excavation of the marrow cavity. Osteoblastic cells are known to arise from the same mesenchymal stem cells that give rise to bone marrow stromal cells, but the origin of bone marrow stromal cells, their role as precursors of osteoblastic cells, and their relationship to hematopoietic stem cells remain controversial and a subject of intense debate (Mbalaviele et al. 1999). What is clear is that bone marrow is colonized embryonically by HSC derived from AGM (not yolk sac HSC), and this population of cells establishes hematopoietic tissue in the bone marrow which serves as a reserve of HSC, blood cell development, and production of immune responsive cells for the remainder of postnatal life (Cumano et al. 2001).

Considerable expansion of progenitor cells for both myeloid and lymphoid lineages takes place in the fetal liver prior to their appearance in embryonic bone marrow; however, it remains unknown whether committed progenitor cells circulate and populate newly formed bone marrow compartments or whether this process is replicated from HSC in the marrow microenvironment. It is also clear that bone marrow rapidly assumes primary hematopoietic function after embryonic day 18 in mice and persists as the primary organ of hematolymphopoiesis throughout postnatal life (Rolink et al. 1993).

Mature immunocompetent cells leave the bone marrow during postnatal life and migrate via the blood to the secondary immune organs, the spleen, lymph nodes, and mucosal lymphoid tissues (Stevens et al. 1982; Gray 1988). In addition, plasma cells, the antibody-producing end cells of the B lymphoid compartment, take up residence in the bone marrow and produce substantial quantities of antibody in that tissue (Manz et al. 1998). For that reason, the bone marrow is the primary anatomic location of both newly formed and fully mature B lymphoid cells.

## HEMATOPOIETIC MICROENVIRONMENTS

Bone marrow and thymus appear to be unique in providing microenvironment factors necessary for the development of the full array of functionally immunocompetent cells (Landreth 1993; Melchers 1997; Shortman et al. 1998). The microenvironment of the bone marrow, in particular, has been the focus of considerable attention in recent years (Lichtman 1984; Collins and Dorshkind 1987; Kierney and Dorshkind 1987; Landreth and Dorshkind 1988; Dorshkind and Landreth 1992). This microenvironment is functionally composed of fibroblastic stromal cells, endothelial cells, and macrophages. However, most unique hematopoietic functions have been assigned to the fibroblastic stromal cells.

Fibroblastic bone marrow stromal cells are the best characterized element of the hematopoietic microenvironment and are known to produce an array of cytokines that regulate the tempo of differentiation, maturation, and migration of hematopoietic cells. This same set of cytokines appears to be active in the thymus as well, suggesting that the origin of bone marrow and thymic mesenchymal stromal elements may be related (Chagraoui et al. 2003). Bone marrow stromal cells are required for establishing long-term bone marrow cultures *in vitro* (Dexter and Lajtha 1974; Whitlock and Witte 1987; Dorshkind and Landreth 1997) and the absence or failure of stromal cell function is known to result in failure of hematopoietic cells to survive *in vivo*. The regulatory role proposed for the bone marrow hematopoietic microenvironment in maintaining the continual production of leukocytes required for sustained immunocompetence has led to the proposal that exposure to toxic substances may affect production of hematolymphoid cells by damaging this stromal microenvironment (Mudry et al. 2000; Banfi et al. 2001). This form of indirect hematotoxicity is intriguing and currently under investigation in a number of laboratories.

The hematopoietic microenvironment is characterized by the expression of adhesion molecules and cytokines, which are both produced by stromal cells (Dorshkind, 1990; Dorshkind and Landreth 1992). Adhesion contacts with stromal cells are known to contribute to the regulation of hematopoiesis and cellular adhesion molecules fibronectin, CD44, and VCAM-1, which appear to be particularly important in this respect (Kincade et al. 1989; Hahn et al. 2000). This binding involves stimulation of cell signaling cascades in both hematopoietic cells and stromal cells (Yamada and Geiger 1997). Cytokines produced by stromal cells regulate hematopoietic cell proliferation and differentiation and include GM-CSF, G-CSF, M-CSF, IL-5 IL-7, c-kit-ligand, SDF, and IGF-1 (Namen et al. 1998; Kierney and Dorshkind 1987; Billips et al. 1992; Landreth et al. 1992; Hogan et al. 2000). Changes within the hematopoietic microenvironment resulting in alterations in cytokine production or adhesion molecule expression can alter the homeostasis of blood cell development.

## POSTNATAL DEVELOPMENT OF THE IMMUNE SYSTEM

It is during the immediate postnatal period that acquired immune function is first detectable in mammals (Ghia et al. 1998). Functional B and T lymphocytes are produced in the bone marrow and thymus, respectively, and migrate to the spleen, lymph nodes, and mucosal associated lymphoid tissues (MALT) as functional end cells (Osmond 1985). However, a mature pattern of immune response to antigens is not achieved until approximately one month of age in rodents. During the first month of postnatal life in mice, the immune system remains immature, fails to produce antibodies to carbohydrate antigens, and the humoral immune response to antigens is dominated by the production of IgM (Mosier et al. 1979; Dorshkind 1986). After this period of perinatal immunodeficiency, mice mount mature immune responses to protein and carbohydrate antigens and respond to intracellular pathogens, and the immune response can be assumed to accurately portray mature postnatal immune function (Raff et al. 1970).

## ESTABLISHMENT OF IMMUNE MEMORY AND IMMUNE SENESCENCE

During the first six months of life in rodents, there is enormous production of T and B lymphocytes from primary lymphoid tissues (Kincade 1981). Exposure to specific antigens during this period results in a rapidly expanding accumulation of lymphocyte specificities in the pool of memory B and T cells in secondary lymphoid tissues (Brooks and Feldbush 1983). As thymic function wanes at sexual maturity and thymocytes are no longer produced in that tissue, it is this pool of memory B and T cells that maintains immunocompetence for the life of the individual (Makinodan 1980). There is a similar, but less dramatic decline in the production of B lymphocytes with advancing age. This loss of B cell function serves to increase the dependence of older animals on memory cell function to maintain effective immunocompetence (Miller and Allman 2003).

Senescence of the immune system is well documented but poorly understood. Both innate and acquired immune responses to antigens are different in rodents in the last quartile of their life (McGlauchlen and Vogel 2003; Romanyukha and Yashin 2003). This failure of immune function is due, in part, to failure of production of newly formed cells and decreased survival of long-lived cells in lymphoid tissues.

## CRITICAL DEVELOPMENTAL WINDOWS FOR THE IMMUNE SYSTEM IN RODENTS

The developmental progression of cells, sequential migration of immature cells to a series of embryonic tissue sites, and differences in hematopoietic microenvironments in each of these tissue sites (Figure 1.3) strongly suggests that different periods of embryonic development may be differentially susceptible to immunotoxic insult. For that reason, a hypothetical sequence of windows of vulnerability for the developing immune system was recently proposed (Dietert et al. 2000). Details of this scheme as they apply to rodent development are presented in Figure 1.4 of this chapter and include five critical windows of potential vulnerability. Embryonic development does not proceed in lock-step sequence, and these windows of vulnerability are clearly not discrete. However, even though there is a considerable overlap of these developmental windows, they are based on a set of unique developmental events that can be followed in temporal sequence and describe the initial appearance of specific immunohematopoietic cell types or specific immune functions during embryonic and postnatal development in rodent species.

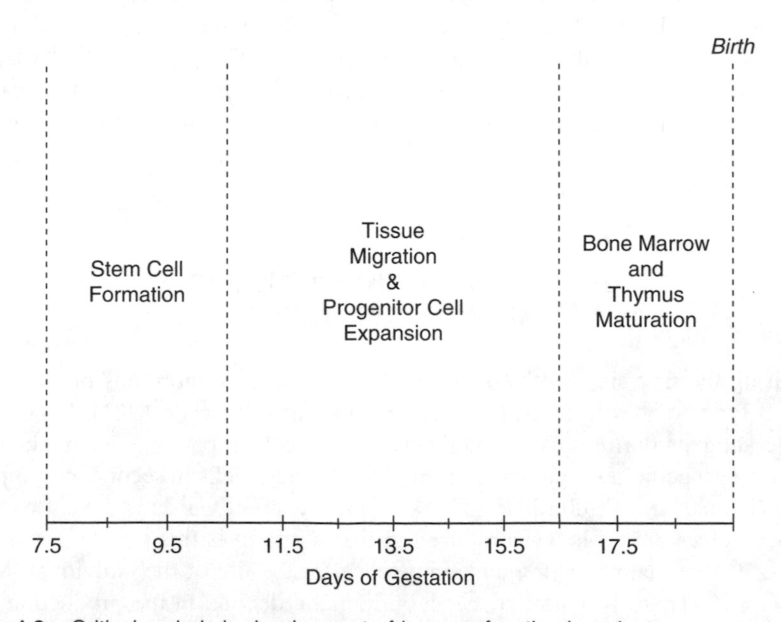


**Figure 1.3** Critical periods in development of immune function in rodents.

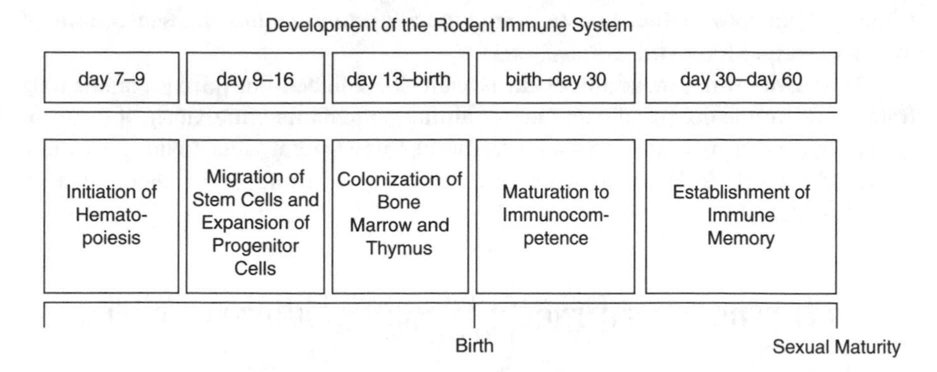


**Figure 1.4** Critical windows of vulnerability to environmental insult during development of the rodent immune system.

The first critical window (gestational days 7 to 9) is characterized by initial organogenesis of the hematolymphoid system from undifferentiated mesenchymal cells (Fehling et al. 2003) and generation of sufficient HSC to initiate both blood formation and immunity (Cumano et al. 2000). Damage to this process of initial stem cell formation is likely to have dramatic and persistent effects during the remainder of development and during postnatal life. The second critical window (gestational days 9 to16) represents a period of extensive migration of HSC to new tissue environments, generation and expansion of lineage-specific progenitor cells for leukocytes, and organogenesis of secondary lymphoid tissues (Godin and Cumano 2002). It is intriguing that the absence of lymphoid cell progenitors during this critical period results in the failure of secondary lymphoid tissues to develop (Godin and Cumano 2002). This suggests that secondary systemic failures of the immune system result from failure of hematopoietic cell development due to the requirement for these cells in induction of vascular budding and secondary lymphoid organ formation.

The third critical window described (gestational days 13 to birth) is dominated by ossification of bone, formation of bone marrow, and colonization of this marrow space by hematopoietic stem cells and the spectrum of hematopoietic progenitor cells (Dietert et al. 2000). Bone marrow is the hematopoeitic organ that persists throughout postnatal life, and immunotoxic damage during this critical period of transition for the hematopoietic system and establishment of the bone marrow has the potential for lasting effects on both blood cell development and immune function.

There are two critical periods of immune system development that follow birth in rodents. For the first month of life, the spleen persists as a hematopoietic organ and immune responses have a characteristic immature nature, characterized by the failure of rodents to respond to carbohydrate antigens (Tavassoli and Yoffey 1983). Rodents share this period of perinatal immunodeficiency with other mammals (including humans) and damage to lymphoid organs or cells during this period has the unique potential to result in lifelong damage to immune function. And finally, the initiation of normal mature immune responses to environmental pathogens and antigens during the first few months of life is essential to development of a set of memory lymphocytes that maintain immunocompetence following senescence of primary lymphoid organs.

These five critical windows of developmental vulnerability provide a working framework for the design of experiments aiming to test immunotoxicity of environmental agents on rodents. It will be essential to test both acute toxicity of these agents at each developmental stage and the potential for acute damage to persist into postnatal life.

## *IN VIVO* RECONSTITUTION OF IMMUNOCOMPETENT CELLS

Understanding the effect of immunotoxic compounds on hematopoietic and lymphopoietic cells during the critical windows of vulnerability and the persistence of that damage on postnatal immunity is complicated by at least two factors unique to immune responsive cells. The first is that cells at identical stages of lineage development may have differential susceptibility to toxic compounds based on their anatomic location and specific microenvironment. Lymphoid progenitor cells in the fetal liver may have different susceptibility to chemical exposure than cells with the same phenotype in the bone marrow. This is further complicated by the fact that resident cells in any of these tissues may have experienced toxic damage in other anatomic locations and then migrated to a different tissue site when they are assayed.

A second complication to understanding the effect of toxic compounds on hematolymphoid cells is that they have extensive renewal potential, and acute damage to hematopoietic cells may not be detectable later in life because of the unique ability of these cells to reconstitute losses (Bacigalupo et al. 2000). Numerous studies have demonstrated rapid recovery of immune responsive cells following cytotoxic drug therapy, and the ability of hematopoietic cells to recover following depletion is sufficiently robust to justify clinical utility of bone marrow transplantation (Banfi et al. 2001; Mudry et al. 2000). For that reason, relatively dramatic acute toxicity to either blood cell development or immune cell function may not lead to persistent detection of immunotoxicity. However, underlying damage to stem and progenitor cell populations, not detected by standard assays that enumerate numbers of cells responding to a single antigen, may in fact manifest itself when the hematopoietic system is stressed by myeloablative drug treatment (e.g., chemotherapy) or following bone marrow transplantation. For that reason, investigation of persistent damage to the blood-forming system of postnatal mice following prenatal exposure to toxic agents should address this issue.

## SUMMARY

Hematopoietic and immune system development in rodents can be segmented into five critical windows based on organogenesis, cell migration, and maturation of immune function. This well-documented series of developmental events has resulted in a working hypothesis that critical windows of vulnerability to chemical exposure exist for the developing immune system, and these critical windows may be differentially susceptible to chemical exposure (Dietert et al. 2000). This hypothesis is currently being tested and will need to be considered in designing experimental protocols to determine acute and persistent immunotoxic effects of environmentally applied chemicals. It is also important to realize that renewal of immunocompetent cells in postnatal animals continues to depend on differentiation of stem cells and expansion of progenitor cells in the bone marrow. This process is fundamentally different in neonatal, young, and older animals. For that reason, developmental windows of vulnerability to chemical exposure are not restricted to the period of prenatal development.

## ACKNOWLEDGMENT

This work was supported, in part, by the Department of Defense (DOD) Grant PR012230. S.V.M. Dodson is supported by National Institutes of Health–National Institute of Environmental Health Sciences (NIH–NIEHS) Training Grant ES010953.

## REFERENCES

Adachi S, Yoshida H, Kataoka H, and Nishikawa S. 1997. Three distinctive steps in Peyer's patch formation of murine embryo. *Int Immunol* 9:507–514.

Alonso-C LM, Munoz JJ, and Zapata AG. 2000. Early T cell development can be traced in rat fetal liver. *Eur J Immunol* 30:3604–3613.

Andrew TA and Owen JJT. 1978. Studies on the earliest sites of B cell differentiation in the mouse embryo. *Dev Comp Immunol* 2:339–346.

Antoine JC, Bleux C, Avrameas S, and Liacopoulos P. 1979. Murine embryonic B lymphocyte development in the placenta. *Nature* 277:219–221.

Auerbach R. 1961. Experimental analysis of the origin of cell types in the development of the mouse thymus. *Dev Biol* 3:336–354.

Bacigalupo A, Frassoni F, and Van Lint MT. 2000. Bone marrow or peripheral blood as a source of stem cells for allogeneic transplants. *Curr Opin Hematol* 7:343–347.

Banfi A, Bianchi G, Galotto M, Cancedda R, and Quarto R. 2001. Bone marrow stromal damage after chemo/radiotherapy: occurrence, consequences and possibilities of treatment. *Leuk Lymphoma* 42:863–870.

Billips L, Petitte D, Dorshkind K, Narayanan R, Chiu C, and Landreth K. 1992. Differential roles of stromal cells, interleukin-7, and kit-ligand in the regulation of B lymphopoiesis. *Blood* 79:1185–1192.

Bradley T and Metcalf D. 1966. The growth of bone marrow cells *in vitro*. *Aust J Exp Biol Med Sci* 44:287–299.

Brooks KH and Feldbush TL. 1983. The correlation between the activation state of B cells and their capacity for *in vitro* propagation of immunologic memory. *Cell Immunol* 76:213–223.

Burrows PD, Kearney J, Lawton AR, and Cooper MD. 1978. Pre-B cells: Bone marrow persistence in anti-μ suppressed mice, conversion to B lymphocytes, and recovery following destruction with cyclophosphamide. *J Immunol* 120:1526–1531.

Cao X, Shores EW, Hu-Li J, Anver MR, Kelsall BL, Russell SM, Drago J, Noguchi M, Grinberg A, Bloom ET, Paul WE, Katz SI, Love PE, and Leonard WJ. 1995. Defective lymphoid development in mice lacking expression of the common cytokine receptor gamma chain. *Immunity* 2:223–238.

Chagraoui J, Lepage-Noll A, Anjo A, Uzan G, and Charbord P. 2003. Fetal liver stroma consists of cells in epithelial-to-mesenchymal transition. *Blood* 101:2973–2983.

Collins L and Dorshkind K. 1987. A stromal cell line from myeloid long-term bone marrow cultures can support myelopoiesis and B lymphopoiesis. *J Immunol* 138:1082–1087.

Cooper MD, Peterson RDA, and Good RA. 1965. Definition of the thymic and bursal lymphoid systems in the chicken. *Nature* 205:143–145.

Cosgrove D, Chan SH, Waltzinger C, Benoist C, and Mathis D. 1992. The thymic compartment responsible for positive selection of CD4+ T cells. *Int Immunol* 4:707–710.

Cumano A, Ferraz C, Klaine M, Santo J, and Godin I. 2001. Intraembryonic, but not yolk sac hematopoietic precursors, isolated before circulation, provide long-term multilineage reconstitution. *Immunity* 15:477–485.

Cumano A, Dieterlen-Lievre F, and Godin I. 2000. The splanchnopleura/AGM region is the prime site for the generation of multipotent hemopoietic precursors, in the mouse embryo. *Vaccine* 18:1621–1623.

Cumano A and Godin I. 2001. Pluripotent hematopoietic stem cell development during embryogenesis. *Curr Opin Immunol* 13:166–171.

Delassus S and Cumano A. 1996. Circulation of hematopoietic progenitors in the mouse embryo. *Immunity* 4:97–106.

Dexter T and Lajtha L. 1974. Proliferation of haemopoietic stem cells *in vitro*. *Brit J Haematol* 28:525–530.

Dieterlen-Lievre F. 1975. On the origin of haemopoietic stem cells in the avian embryo: An experimental approach. *J Embryol Exp Morphol* 33:607–619.

Dietert RR, Etzel RA, Chen D, Halonen M, Holladay S, Jarabek AM, Landreth K, Pedan D, Pinkerton K, Smialowicz RJ, and Zoetis T. 2000. Workshop to Identify Critical Windows of Exposure for Children's Health: Immune and Respiratory Systems Work Group Summary. *Environ Health Perspect* 108, Suppl 3:483–490.

Dorshkind K. 1990. Regulation of hemopoiesis by bone marrow stromal cells and their products. *Annual Rev Immunol* 8:111–137.

Dorshkind KD. 1986. *In vitro* differentiation of B lymphocytes from primitive hemopoietic precursors present in long term bone marrow cultures. *J Immunol* 36: 422–429.

Dorshkind K and Landreth K. 1992. Regulation of B cell differentiation by bone marrow stromal cells. *Internat J Cell Clon* 10:12–17.

Dorshkind K and Landreth K. 1997. Use of long-term bone marrow cultures and cloned stromal cell lines to grow B lineage cells. In: Lefkovits I (Ed.), *The Immunology Methods Manual CD-ROM*. Academic Press Ltd, Orlando, FL, section 11.5.

Fehling HJ, Lacaud G, Kubo A, Kennedy M, Robertson S, Keller G, and Kouskoff V. 2003. Tracking mesoderm induction and its specification to the hemangioblast during embryonic stem cell differentiation. *Development* 130:4217–4227.

Fields KK. 1992. Autologous bone marrow transplantation and melanoma: a focused review of the literature. *Ann Plast Surg* 28:70–73.

Ghia P, Boekel E, Rolink A, and Melchers F. 1998. B-cell development: a comparison between mouse and man. *Immunol Today* 19:480–485.

Glick B, Chang TS, and Jaap RG. 1956. The bursa of Fabricius and antibody production in the domestic fowl. *Poultry Sci* 35:224–225.

Godin I and Cumano A. 2002. The hare and the tortoise: an embryonic hematopoietic race. *Nat Rev Immunol* 2:593–604.

Godin I, Garcia-Porrero J, Dieterlen-Lievre F, and Cumano A. 1999. Stem cell emergence and hemopoietic activity are incompatible in mouse intraembryonic sites. *J Exp Med* 190:43–52.

Good RA. 1995. Organization and development of the immune system. Relation to its reconstruction. *NY Acad Sci* 77:8–33.

Gray D. 1988. Recruitment of virgin B cells into an immune response is restricted to activation outside lymphoid follicles. *Immunol* 65:73–79.

Gwatkin RBL, Till JE, Whitmore GF, Siminovitch L, and Graham AF. 1957. Multiplication of animal cells in suspension measured by colony counts. *Proc Natl Acad Sci USA* 43:451–457.

Hahn BK, Piktel D, Gibson LF, and Landreth KS. 2000. The role of stromal integrin interactions in pro-B cell proliferation. *Hematol* 5:153–160.

Hogan MB, Piktel D, and Landreth KS. 2000. IL-5 production by bone marrow stromal cells: Implications for eosinophilia associated with asthma. *J. Allergy Clin Immunol* 106:329–336.

Kierney P, and Dorshkind K. 1987. B lymphocyte precursors survive in diffusion chamber cultures but B cell differentiation requires close association with stromal cells. *Blood* 70:1418–1424.

Kincade PW. 1981. Formation of lymphocytes in fetal and adult life. *Adv Immunol* 31:177–245.

Kincade PW, Lee G, Pietrangeli CE, Hayashi S, and Gimble JM. 1989 Cells and molecules that regulate B lymphopoiesis in bone marrow. *Ann Rev Immunol* 7: 111–143.

Landreth KS. 1993. B lymphocyte development as a developmental process. In: Cooper EL, Nisbet-Brown E (Eds.), *Developmental Immunology.* Oxford University Press, New York, NY, pp 238–273.

Landreth KS. 2002. Critical windows in development of the rodent immune system. *Hum Exp Toxicol* 21:493–498.

Landreth KS, and Dorshkind K. 1988. Pre-B cell generation potentiated by soluble factors from a bone marrow stromal cell line. *J Immunol* 140:845–852.

Landreth KS, Kincade PW, Lee G, and Medlock ES. 1983. Phenotypic and functional characterization of murine B lymphocyte precursors isolated from fetal and adult tissues. *J Immunol* 131:572–580.

Landreth KS, Narayanan R, and Dorshkind K. 1992. Insulin-like growth factor-1 regulates pro-B cell differentiation. *Blood* 80:1207–1212.

Lee G, Namen AE, Gillis S, Ellingsworth LR, and Kincade PW. 1989. Normal B cell precursors responsive to recombinant murine IL-7 and inhibition of IL-7 activity by transforming growth factor-beta. *J Immunol* 142:3875–3883.

Lichtman M. 1984. The relationship of stromal cells to hemopoietic cells in marrow. *Kroc Found Ser* 18:3–29.

Lotem J and Sachs L. 2002. Cytokine control of developmental programs in normal and leukemia. *Oncogene* 21:3284–3294.

Magli MC, Iscove NN, and Odartchenko N. 1982. Transient nature of early haematopoietic spleen colonies. *Nature* 295:527–529.

Makinodan T. 1980 Nature of the decline in antigen-induced humoral immunity with age. *Mech Ageing Dev* 14: 165–172.

Manley NR and Blackburn CC. 2003. A developmental look at thymus organogenesis: where do the non-hematopoietic cells in the thymus come from? *Curr Opin Immunol* 15:225–232.

Manz RA, Lohning M, Cassese G, Thiel A, and Radbruch A. 1998. Survival of long-lived plasma cells is independent of antigen. *Int Immunol* 10:1703–1711.

Marcos MA, Morales-Alcelay S, Godin I., Dieterlen-Lievre F, Copin SG, and Gaspar ML. 1997. Antigenic phenotype and gene expression pattern of lymphohemopoietic progenitors during early mouse ontogeny. *J Immunol* 158:2627–2637.

Mbalaviele G, Jaiswal N, Meng A, Cheng L, Van Den Bos C, and Thiede M. 1999. Human mesenchymal stem cells promote human osteoclast differentiation from CD34+ bone marrow hematopoietic progenitors. *Endocrinol* 140:3736–3743.

McGlauchlen KS and Vogel LA. 2003. Ineffective humoral immunity in the elderly. *Microbes Infect* 5:1279–1284.

McGrath KE, Koniski AD, Malik J, and Palis J. 2003. Circulation is established in a stepwise pattern in the mammalian embryo. *Blood* 101:1669–1676.

Mebius RE. 2003. Organogenesis of lymphoid tissues. *Nature Rev Immunol* 3:292–303.

Mebius RE, Miyamoto T, Christensen J, Domen J, Cupedo T, Weissman IL, and Akashi K. 2001. The fetal liver counterpart of adult common lymphoid progenitors gives rise to all lymphoid lineages, CD45+CD4+CD3- cells, as well as macrophages. *J Immunol* 166:6593–6601.

Medvinsky A and Dzierzak E. 1996. Definitive hematopoiesis is autonomously initiated by the AGM region. *Cell* 86:897–906.

Melchers F. 1997. The Carl Prausnitz Memorial Lecture: The development of lymphocytes. *Int Arch Allergy Immunology* 113:11–30.

Metcalf D and Moore MAS. 1971. Haemopoietic Cells. In: Neuberger A and Tatum EL (Eds.), *Frontiers of Biology*, Vol. 24. North-Holland, Amsterdam, pp. 172–271.

Miller J, Horner A, Stacy T, Lowrey C, Lian JB, Stein G, Nuckolls GH, and Speck NA. 2002. The core-binding factor beta subunit is required for bone formation and hematopoietic maturation. *Nat Genet* 32:545–649.

Miller JP and Allman D. 2003. The decline in B lymphopoiesis in aged mice reflects loss of very early B-lineage precursors. *J Immunol* 171:2326–2330.

Miller JF. 1961. Immunological function of the thymus. *Lancet* 30:748–749.

Moore TA, von Freeden-Jeffry U, Murray R, and Zlotnik A. 1996. Inhibition of gamma delta T cell development and early thymocyte maturation in IL-7 -/- mice. *J Immunol* 157:2366–2373.

Mosier DE, Goldings EA, and Bottomly K. 1979. Activation requirements of neonatal B lymphocytes *in vitro* and *in vivo*. In: Cooper et al. (Eds.), B Lymphocytes in the Immune Response. Elsevier/North-Holland New York, NY, pp. 91–95

Mudry RE, Fortney JE, York T, Hall BM, and Gibson LF. 2000. Stromal cells regulate survival of B-lineage leukemic cells during chemotherapy. *Blood* 96: 926–1932.

Namen AE, Lupton S, Hjerrild K, Wignall J, Mochizuki DY, Schmierer A, Mosley B, March CJ, Urdal D, Gullis S, Cosman D, and Goodwin RG. 1998. Stimulation of B-cell progenitors by cloned murine interleukin-7. *Nature* 333:571–573.

Osmond DG. 1985. The ontogeny and organization of the lymphoid system. *J Invest Dermatol* 85, Supp 1:2s–9s.

Owen JJT, Cooper MD, Raff MC. 1974. *In vitro* generation of B lymphocytes in mouse fetal liver, a mammalian bursal equivalent. *Nature* 249:361–363.

Owen JJT, Wright DE, Habu S, Raff MC, and Cooper MD. 1977. Studies on the generation of B lymphocytes in fetal liver and bone marrow. *J Immunol* 118:2067–2072.

Paige CJ, Kincade PW, Shinefield LA, and Satio VL. 1981. Precursors of murine B lymphocytes: physical and functional characterization and distinctions from myeloid stem cells. *J Exp Med* 153:154–165.

Peault B, Khazaal I, and Weissman IL. 1994. *In vitro* development of B cells and macrophages from early mouse fetal thymocytes. *Eur J Immunol* 24:781–784.

Raff MC, Sternberg M, and Taylor RB. 1970. Immunoglobulin determinants on the surface of mouse lymphoid cells. *Nature* 225:553–554.

Rodewald H and Fehling HJ. 1998. Molecular and cellular events in early thymocyte development. *Adv Immunol* 69:1–112.

Rolink A, Haasner D, Nishikawa S, and Melchers F. 1993. Changes in frequencies of clonable pre-B cells during life in different lymphoid organs of mice. *Blood* 81:2290–2300.

Romanyukha AA and Yashin AI. 2003. Age related changes in population of peripheral T cells: towards a model of immunosenescence. *Mech Ageing Dev* 124:433–443.

Shortman K, Vremec D, Cocoran IM, Georgopolos K, Lukas K, and Wu L. 1998. The linkage between T-cell and dendritic cell development in the mouse thymus. *Immuno Rev* 165:39–46.

Stevens SK, Weissman IL, and Butcher EC. 1982. Differences in the migration of B and T lymphocytes: organ-selective localization *in vivo* and the role of lymphocyte-endothelial cell recognition. *J Immunol* 128:844–851.

Tavassoli M. 1991. Embryonic and fetal hemopoiesis: An overview. *Blood Cells* 17:282–286.

Tavassoli M and Yoffey JM. 1983. *Bone Marrow: Structure and Function*. Alan R. Liss Inc., New York, NY.

Teh H. 1993. T cell development and repertoire selection. In: Cooper EL and Nisbet-Brown E (Eds.), *Developmental Immunology*. Oxford University Press, New York, pp. 217–237.

Terskikh AV, Miyamoto T, Chang C, Diatchenko L, and Weissman IL. 2003. Gene expression analysis of purified hematopoietic stem cells and committed progenitors. *Blood* 102:94–101.

Till JE and McCulloch EA. 1961. A direct measurement of the radiation sensitivity of normal mouse bone marrow cells. *Radiat Res* 14:213–222.

Velardi A and Cooper MD. 1984. An immunofluorescence analysis of the ontogeny of myeloid, T and B lineage cells in mouse hematopoietic tissues. *J Immunol* 133:672–677.

von Freeden-Jeffry U, Vieira P, Lucian LA, McNeil T, Burdach SE, and Murray R. 1995. Lymphopenia in interleukin (IL)-7 gene-deleted mice identifies IL-7 as a nonredundant cytokine. *J Exp Med* 181:1519–1526.

Weissman I. 2000. Stem cells: Units of development, units of regeneration and units in evolution. *Cell* 100:157–168.

Whitlock C and Witte O. 1987. Long-term culture of murine bone marrow precursors of B lymphocytes. *Meth Enzymol* 150:275-286.

Wigle JT, Harvey N, Detmar M, Lagutina I, Grosveld G, Gunn MD, Jackson DG, and Oliver G. 2002. An essential role for Prox1 in the induction of the lymphatic endothelial cell phenotype. *EMBO J* 21:1505–1513.

Wigle JT and Oliver G. 1999. Prox1 function is required for the development of the murine lymphatic system. *Cell* 98:769–778.

Yamada KM, and Geiger B. 1997. Molecular interactions in cell adhesion complexes. *Curr Opin Cell Biol* 9:76–85.

Yoder MC. 2002. Embryonic hematopoiesis in mice and humans. *Acta Paediatr Suppl* 438:5–8.

Yoshida H, Naito A, Inoue J, Satoh M, Santee-Cooper SM, Ware CF, Togawa A, Nishikawa S, and Nishikawa S. 2002. Different cytokines induce surface lymphotoxin-alphabeta on IL-7 receptor-alpha cells that differentially engender lymph nodes and Peyer's patches. *Immunity* 17:823–833.

CHAPTER 2

# Development of the Human Immune System

**Ramona Leibnitz**

## CONTENTS



0-415-28457-0/05/$0.00+$1.50

## INTRODUCTION

A review of the "ontogeny" of the human immune system carries the implication that this maturation process is genetically programmed and unalterable. In fact, environmental influences affect the development of the human immune system even during gestation, such that insult to, pathogenic exposure of, and immunologic responses by the mother can perturb the immune system of the fetus. This reactivity underscores the inherent responsiveness of the body's immune system to its environment even at the earliest stages. Therefore, this chapter's title refers to the *development* of the human immune system to incorporate the responsive nature of an ontologically emerging immune system.

A core of literature on developmental studies of the immune system dates back to the late 1970s and early 1980s. Much of the human data are related to fundamental observations using research animals—particularly rodents—for obvious ethical reasons. Information on human immune system development stems from either functionality studies performed using umbilical cord blood collected at birth after a full-term pregnancy (>37 weeks gestation) (Holt and Jones 2000) or the more clinically related studies analyzing susceptibility to disease due to immune system insult from genetic defects or environmental trauma. A picture emerges from these studies of an intricate developmental system of cells, their cell-surface receptors, and the soluble factors of a specific organ environment that enable the body's defense system to respond quickly, specifically, and with memory towards immunologic threats and not toward self-antigens. An awareness of how the immune system develops during ontogeny and the factors influencing maturation of the body's defense system is crucial to understanding the impact of environmental factors such as toxicants, pathogens, or even vaccines.

## DEVELOPMENT OF THE FETAL IMMUNE SYSTEM

Hematopoietic stem cells (HSC) are classically defined as those cells that: 1) give rise to the specialized cells of blood; and 2) have a capacity for extensive proliferation resulting in renewal of their own kind (Orkin and Zon 2002; Owen 1972). A bulk of information about hematopoiesis derives from the early 20$^{th}$ century (Tavassoli and Yoffey 1983), yet technological advances have since added to and called into question established developmental paradigms. Two modified paradigms are discussed here. First, historic studies concluded that HSC develop extraembryonically in the yolk sac (Metcalf and Moore 1971; Migliaccio et al. 1986; Owen 1972; Tavassoli and Yoffey 1983) and then migrate to the fetal liver, bone marrow, and other sites. Recent reports from avian and mouse studies now indicate that HSC arise in the aorta-gonad-mesonephros (AGM) region (Godin et al. 1993; Jaffredo et al. 2000; Muller et al. 1994) (reviewed in: Kincade et al. 2002; Nishikawa et al. 2001). Whether this type of HSC derives originally from yolk sac HSC is unknown (Nishikawa et al. 2001; Orkin and Zon 2002), but this population is able to repopulate the hematopoietic cells in an irradiated adult mouse. Therefore, HSC from the AGM have been termed *definitive HSC*, as opposed to *primitive HSC* from the yolk sac that appear to persist only during embryonic life (Orkin and Zon 2002). A modern paradigm for hematopoiesis is described later in this chapter.

A second recent paradigm shift involves lineage commitment of the pluripotent cells developing from the HSC. HSC have been assumed to give rise to a common myeloid precursor (CMP) with erythroid potential and a common lymphoid precursor (CLP) that produces both B and T lymphocytes (Cooper et al. 1993). Recent findings by Kondo et al. support this traditional scheme (1997), but research by Katsura and colleagues makes use of a multilineage progenitor (MLP) assay to refute this idea (reviewed in: Katsura 2002; Kincade et al. 2002). Katsura's group suggests that the idea that T and B cells derive from a common progenitor developed because both T and B cells recognize antigens specifically with clonally distributed receptors, use a similar molecular apparatus to develop antigen-receptor genes, and are deficient in patients and mice with severe combined immunodeficiency (SCID) disease (Katsura 2002). However, similar use of genes (coopted perhaps from an evolutionary precursor) cannot argue for a common progenitor. Moreover, a definite answer can only come from an assay system permissive for the development of all cell lineages and utilizing purified progenitor cells. Katsura and colleagues provide evidence that T- and B-cell progenitors arise from differentiation of a common myelolymphoid progenitor through a bipotential myeloid/T and myeloid/B stage, respectively (Kawamoto et al. 1997). Their research implies that T and B cells are related more to myeloid cells than to each other (Kincade et al. 2002). Further research should elucidate whether fetal progenitors conform to a stepwise lineage-restriction program of development or have an inherent degree of plasticity, enabling them to generate the spectrum of cell lineages in response to environmental cues.

## Hematopoiesis

Stages of immune system development appear to be fairly conserved within mammals, so for the purposes of this review, when fundamental human data are unavailable, mouse data will be provided. A thorough review of the development of the immune system in mice is given in Chapter 1.

In humans, the development of the immune system begins with HSC formation in the yolk sac from 15 to 18 days through 6 weeks of gestation, with the HSC becoming undetectable by the 10$^{th}$ week (Tavassoli 1991). (Definitive HSC in the mouse appear in the AGM region at 9 to 11 days postcoitus [Muller et al. 1994; Nishikawa et al. 2001; Orkin and Zon 2002]). Between the 5$^{th}$ and 8$^{th}$ weeks of human gestation, the fetal liver becomes the center of hematopoiesis (Bellanti et al. 2003; Cooper et al. 1993; Holt and Jones 2000; Tavassoli and Yoffey 1983; West 2002), with some activity persisting until shortly before birth (Cooper et al. 1993). The spleen also transiently participates in blood formation from 10 to 12 weeks post conception until hematopoietic activity shifts to the bone marrow beginning around 11 to 12 weeks. It is also at this time (variably reported as between 8 to 9 or 11 to 12 weeks of gestation [Bellanti et al. 2003; Holt and Jones 2000]) that the thymus is seeded with precursor cells. By 20 weeks, essentially all hematopoiesis in human embryos occurs in the bone marrow (Cooper et al. 1993).

These waves of blood cell production from different organs have long fueled the concept that hematopoiesis is a migratory phenomenon (Holt and Jones 2000; Metcalf and Moore 1971; Tavassoli and Yoffey 1983). Although not definitive, there is some evidence to support this notion, including experiments in mice demonstrating that HSC fail to colonize fetal liver in the absence of β1 integrin expression (Potocnik et al. 2000). Moreover, the tyrosine kinase receptor Flk-1 appears required for migration and expansion of HSC (Schuh et al. 1999).

## Microenvironments

Early morphological studies and, more recently, *in vitro* culture studies of murine embryonic stem cells support the existence of a bipotential hematopoietic-vascular precursor, i.e., the hemangioblast (reviewed in: Orkin and Zon 2002; Tavassoli and Yoffey 1983). The parallel development of endothelium and hematopoietic precursors foreshadows the significant future role played by the microenvironment in development of hematopoietic lineages.

In the bone marrow, stromal cells provide physical support as well as the growth factors and hormones important for the initiation of differentiation programs that produce the many cell lineages found there (reviewed in: Chaplin 2003). Particularly significant for immune function is the development of natural killer (NK) cells, B lymphocytes, dendritic cells, and macrophages (Muramatsu 1993) in the bone marrow compartment. Moreover, precursors of T lymphocytes originate in the bone marrow but emigrate to the thymus for further differentiation.

Like the bone marrow, the thymus also regulates the microenvironment ensuring proliferation and maturation of its resident hematopoietic cells. Unlike the bone marrow, the thymus also regulates the entry and maturation of hematopoietic cells to those that will function as T cells in the periphery (reviewed in: Chaplin 2003). The thymic stroma forms in the 6$^{th}$ week of gestation and divides into cortex and medulla by 10 to 12 weeks (Loke 1978). The medulla itself starts to develop around 14 weeks of gestation, and although thymocytes populate the thymus at approximately week 9 postconception (von Gaudecker 1991), the thymic medulla is not fully formed until week 17 (Kendall 1991).

The development of the thymus, while essential for maturation of most T cells, in turn requires population by T cells (Boyd et al. 1993; Kendall 1991) (reviewed in: Chaplin 2003). Moreover, it should be noted that the thymic microenvironment is under direct influences of the hypothalamus/pituitary axis. Furthermore, the central nervous system exerts control over the thymus through neural pathways (Boyd et al. 1993; Kendall 1991). The complexity of the T cell maturation environment underscores the contention that T cell development is not purely a genetically ordered event but can be significantly affected by other systemic regulatory systems.

The anatomic features of secondary lymphoid organs develop concomitantly with the central lymphoid organs in the fetus. Gut-associated lymphoid tissue (GALT) can be identified at week 8 of gestation; the *lamina propria* develops during weeks 8 to 10, followed by the occurrence of Peyer's patches and maturation of the appendix from 11-15 weeks postconception. Lymph nodes (LN) also appear from 8 to 12 weeks of gestation, followed by tonsils (and spleen, as mentioned previously) at 10 to 14 weeks (West 2002). The signals leading to the establishment and maintenance of the secondary lymph organs are not well defined, but IL-7 (interleukin-7), lymphotoxin (LT), and tumor necrosis factor (TNF) are known among these factors to play a critical role (reviewed in: Chaplin 2003; Mestecky et al. 2003). Other, less well-defined clusters of lymphoid and hematopoietic cells associated with the genitourinary, gastrointestinal, and respiratory tracts also contribute to host defense, but the origins and development of these systems are not clearly understood (Chaplin 2003).

## Interuterine Environment

Reproduction of a species represents a unique immunologic challenge. Despite allogeneic differences, the maternal immune response does not reject the fetus, and this situation is maintained in humans for 40 weeks of pregnancy. A body of research has demonstrated that many levels of regulation operate in a successful pregnancy. Most significantly, the placenta serves as a physical barrier, sequestering the fetus away from the mother's immune system. The trophoblast cells that form the outer layer of the placenta in contact with maternal tissues do not express classical MHC proteins, acting as a nonimmunogenic shield.

A second level of protection includes production by the placenta of immunomodulatory molecules to actively protect the fetus. Tryptophan metabolites and IL-10 inhibit T cell activation and proliferation, and expression of FasL (CD95L) eliminates activated T cells (reviewed in: Holt 2003). Moreover, the expression by placental trophoblasts of a range of molecules such as IL-10, prostaglandin E2, and progesterone promotes T-helper 2 (Th2) responses while dampening Th1 immunity. This Th2 skew is maintained both within the innate immune system by, for instance, down-regulation of the IL-12(p35) gene expressed in neonatal DC and, in the adaptive immune system, by CpG hypermethylation of the IFNγ promoter in neonatal $CD4^{+}CD45RA^{+}$ T cells (Holt and Sly 2002). The skewing of the lymphokine environment to Th2 dominance during pregnancy protects against toxicity to the placenta caused by Th1-type cytokines, particularly IFNγ.

A third level of protection of the fetal immune system from negative maternal influences involves regulation of sex steroid receptors on fetal cells. Lymphocyte precursors, particularly those of B lymphocytes, decline in the mother during pregnancy when sex steroids and glucocorticoid levels are elevated (Medina et al. 1993; Medina and Kincade 1994). Protection of fetal lymphoid precursors from glucocorticoids comes from placental enzymes, but estrogen in the fetal environment is not similarly neutralized (Kincade et al. 2002). Instead, recent studies indicate that receptors for estrogen and androgen on fetal cells do not appear until after birth (Igarashi et al. 2001; Kincade et al. 2002).

## Major Components of the Fetal Immune System

### *T Cells*

Various phenotypes have been reported for the earliest bone marrow-derived T cell precursors; however, strong evidence exists that some cells become committed to the T cell lineage prethymically (Rodewald et al. 2003; Rothenberg et al. 2003). Prothymocytes found in the fetal liver at 7 weeks postconception are positive for CD7, CD45, and cytoplasmic CD3. $CD7^+$ T cell precursors seed the thymus at 8 to 9 weeks of gestation, increase their expression of T cell receptor (TCR) γδ chain from 9.5 weeks to birth (Haynes et al. 1988; Haynes et al. 1989a; reviewed in: Holt and Jones 2000), and first express TCR αβ chain at 10 weeks (Haynes et al. 1989a). Prethymic T cell precursors have also been documented to express Thy-1 and low levels of CD4 (Chaplin 2003). Another report states that the most primitive precursors of T cells in the thymus are $CD34^+$, $CD38^{lo}$ (reviewed in: Rothenberg et al. 2003), whereas a fourth lists the expression of the T cell markers CD3, CD4, CD5, and CD8 on fetal thymocytes at 10 weeks of gestation (Lobach and Haynes 1987). What is clear even without reference to phenotype is that the majority of pre-T cells requires the thymic microenvironment for maturation, moving from the cortex of the newly formed thymus to the medulla during maturation (reviewed in: Rothenberg et al. 2003).

Thymic T cell precursors proceed along ordered pathways of development. Gene rearrangement in developing thymocytes occurs at approximately week 11, resulting first in γδTCR expression, followed by αβ TCR expression (reviewed in: West 2002). By 15 to 20 weeks of gestation, the number of T cell precursors in the thymus is relatively similar to those found postnatally (Wilson et al. 1992). Sequential expression of coreceptors CD3, CD4, and CD8 proceeds from TCR gene expression, and the expression of these molecules enables the thymocyte to respond to MHC molecules expressed on thymic epithelial and dendritic cells. The complex process of thymic education follows, culminating in an enormous reduction in the total lymphocyte population emerging from the thymus (Rothenberg et al. 2003).

$CD2^+$ $CD3^+$ $CD5^+$ T cells detectable in the fetal circulation at about 15 to 16 weeks of gestation (Haynes et al. 1989b) coincide with maturation of the thymus medulla (von Gaudecker 1991) and with the beginning of an intense expansion of T cells occurring between weeks 14 and 26 of gestation (West 2002). Interestingly, immature T cells express CD45RO until the final step of maturation in the thymus, when T cells begin expressing the "naïve" marker CD45RA. Therefore, the increased frequency of $CD45RO^+/RA^-$ T cells found in the spleen and blood of premature babies (Fujii et al. 1992; Hannet et al. 1992) may indicate very immature, rather than a previously activated, T cell population (Fujii et al. 1992; Holt and Jones 2000).

Functional maturation of T cells as tested by responsiveness to mitogens and alloantigens occurs at the end of the first trimester and the beginning of the second trimester (11 to 18 weeks of gestation) (Holladay and Smialowicz 2000; Holt and Jones 2000; West 2002). In particular, fetal thymus lymphocytes respond to PHA by 12 weeks and ConA by 13 to 14 weeks of gestation. Lymphocytes from spleen or peripheral blood respond to PHA and ConA by 16 weeks and 18 weeks of gestation, respectively. The ability of fetal lymphocytes to respond to and cause stimulation of allogeneic lymphocytes in a mixed-lymphocyte reaction (MLR) has been variably reported as occurring between weeks 11 to 19 of gestation (Holladay and Smialowicz 2000; West 2002). Overall, the T cell compartment of the human immune system has acquired a significant level of functional maturity in terms of responding to antigens by birth.

## *Natural Killer Cells*

Natural killler (NK) cells share a progenitor with T cells (Douagi et al. 2002; Ikawa et al. 1999; Raulet 2003; Rothenberg et al. 2003) but do not require the thymic microenvironment for maturation. Instead, strontium 89 and other agents that disrupt the bone marrow have been shown in mice to impair NK cell development (Kumar et al. 1979; Seaman et al. 1979). Moreover, a cytokine produced by bone marrow stromal cells, IL-15, plays a central role in NK cell development and survival (Kennedy et al. 2000; Lodolce et al. 1998). Administration of another factor produced by stromal cells, Flt3L, has been shown to cause expansion of NK cells (Brasel et al. 1996; Shaw et al. 1998). Very little is known about NK cell ontogeny in humans (Raulet 2003); however, Toivanen et al. (1981) reported that the development of functional NK cells in the human fetus occurs at 28 weeks of gestation. Moreover, levels of NK cells have been demonstrated to be significantly higher during fetal life than during the neonatal period, perhaps indicating a more important role for NK cells in fetal development than in adaptive immunity (West 2002).

## *B Cells*

The development of human B lymphocytes can be charted with the detection of pro-(CD24$^{+}$/surface IgM$^{-}$) and pre-B cells (cytoplasmic IgM$^{+}$/surface IgM$^{-}$) in the fetal liver and omentum as early as 8 weeks of gestation (Hardy 2003; Holladay and Smialowicz 2000; Holt and Jones 2000). These cells in the liver express CD20 but not CD21 or CD22. Assays for N region addition indicate that early B cells lack terminal deoxynucleotide transferase (TdT) until the 8$^{th}$ or 9$^{th}$ week of gestation (Hardy 2003). Surface IgM becomes apparent on liver B cells by 10 to 12 weeks, and surface IgD can be detected from 13 weeks of gestation. B cells can be found in the spleen between 13 to 23 weeks of gestation; these cells strongly express IgM on their cell surface. Likewise, B cells can be detected in the LN by IgM$^{+}$ staining from 16 to 17 weeks of gestation (Holt and Jones 2000). B cells positive for CD19, CD20, CD21, CD22, HLA-DR, IgM, and IgD can be detected in the peripheral circulation as early as 12 weeks of gestation (Bofill et al. 1985; Holt and Jones 2000). A higher percentage of fetal B cells express CD19 compared to adult B cells (Berry et al. 1992); CD20 expression by these CD19$^{+}$ B cells was not significantly different in the two populations (De Waele et al. 1988). Interestingly, the CD19$^{+}$ CD20$^{+}$ B cell population declines in the fetus during the period in which fetal circulation of maternally derived IgG dramatically increases. Maternal IgG at 22 weeks of gestation rises from 10 to 20% of adult levels to 100% of adult levels by 26 weeks of gestation (Gitlin and Biasucci 1969).

The cytokine requirements for human B cell maturation continue to be elucidated. Surprisingly, B-precursor growth in humans is not dependent on IL-7, although B cell formation in adult murine bone marrow is completely dependent on this cytokine (Carvalho et al. 2001; Pribyl and Lebien 1996). In addition, the lack of cytokine common gamma chain ($\gamma c$) has been shown to block B-cell development in mice but, in humans, the blockage is of T-cell development (Gougeon et al. 1990;

Noguchi et al. 1993). Clear differences exist in cytokine dependency between mouse and human lymphopoiesis (Hardy 2003).

$CD5^+$ B cells can be found in both the human peritoneal cavity and pleural cavity at 15 weeks of gestation (Namikawa et al. 1986). $CD5^+$ B cells are largely T-independent, and they produce polyreactive antibodies that may play an important role in the primary immune response of newborns. Fetal circulation has been reported to have a higher percentage of $CD5^+$ B cells than adult circulation (Bofill et al. 1985; Hao and Rajewsky 2001; Tucci et al. 1991). Indeed, the CD5 marker has been reported to be on the majority of B cells in fetal LN, spleen, and liver (Hannet et al. 1992).

T-dependent B cell responses lag behind $CD5^+$ B cell production. Fetal LN lack germinal center B cells, presumably due to a lack of antigen. Primary nodule development in fetal LN at week 17 precedes development of primary nodules in the spleen at 24 weeks of gestation. B cells do not become abundant in the bone marrow until 16-20 weeks of gestation (Holt and Jones 2000). Although the bone marrow is the predominant site for B lymphopoiesis in adults (Kane et al. 1999; Landreth 1993), the liver and spleen are major sites in the fetus (Cooper 1981; Hardy 2003).

B cells bearing surface immunoglobulin of all classes have been reported to reach adult levels by 14 to 15 weeks of gestation (Anderson et al. 1981). However, by parturition, no IgG- or IgA-producing cells can be identified (Holladay and Smialowicz 2000). Early IgG and IgM synthesis occurs in the spleen with large amounts of both antibodies being produced by the spleen as early as 10 weeks of gestation. Maximal levels are reached by 17 to 18 weeks of gestation (Holt and Jones 2000). IgE synthesis occurs at 11 weeks of gestation in fetal liver and lung, and by 21 weeks in the spleen (Miller et al. 1973).

### *Macrophages and Dendritic Cells*

The macrophage (mϕ) plays a critical role as an important effector cell of the innate immune system and as an activator of the adaptive response. This cell type plays a key tissue remodeling role during fetal organ development, scavenging the cells that are genetically programmed to die. In mice, a feature distinguishing fetal mϕ from adult cells is their high proliferative capacity, which may be due to their activation status (reviewed in: Muramatsu 1993).

In contrast to mϕ, dendritic cells (DC) lack phagocytic activity, but they have an extraordinary ability to stimulate naïve T cells and initiate primary immune responses. Importantly, DC induce different types of T cell immune responses depending on the type of original maturation signal; therefore, DC are key regulators of the immune response (Liu 2001). Very little is known about DC in the fetal period. Cells with a DC/mϕ morphology can be found in the yolk sac, prehematopoietic liver, and in the mesenchyme by 4 to 6 weeks of gestation (Holt and Jones 2000). Thereafter, these mostly MHC class II-negative mϕ appear in the thymic cortex, the marginal zones of lymphnodes (LN), splenic red pulp, and in the bone marrow (Holt and Jones 2000). DC have been cited as being present in the thymic rudiment by week 6 to 7 of gestation (Muramatsu 1993) and have subsequently been found in several tissues, including the intestine and epidermis. Monocytes, however, are the first cell type to appear in fetal circulation (Forestier et al. 1991).

Differentiation pathways leading to DC are complicated and controversial (Ardavin et al. 2001), yet the potent ability of DC to initiate immune responses has lead to extensive research in the past decade aimed at using DC as antigen presenting cells (APC) in vaccine trials or in immunotherapeutic anticancer treatments. This body of research indicates that GM-CSF, IL-4, IL-6, IL-7, TNF-α, LT, and LIF play important roles in DC differentiation and maturation, and that Flt3L is a required factor (Ardavin et al. 2001; Liu 2001). IL-4 has been shown to be a key cytokine for inducing DC from human monocytes, whereas IL-7 is essential in the differentiation of DC from early thymic lymphoid precursors (reviewed in: Ardavin et al. 2001). The growth and differentiation of monocytes and mϕ are tightly regulated by specific growth factors (e.g., IL-3, CSF-1, GM-CSF, IL-4, IL-13) and inhibitors (e.g., IFN-γ, TGF-β, LIF) (Gordon 2003). MHC class II-positive mϕ appear in the liver by 7 to 8 weeks of gestation, in the LN at 11 to 13 weeks of gestation, and in the T-cell areas of the developing thymic medulla by 16 weeks of gestation. A few hepatic sinusoidal mϕ (Kupffer cells) are seen by 17 weeks of gestation and reach nearly adult levels in the neonatal period. Langerhans cells can be detected in the skin by 6 to 7 weeks of gestation and develop the larger and more dendritic adult morphology by the second trimester (Holt and Jones 2000).

### *Complement*

A discussion of the ontogeny of the immune system is not complete without mention of the 35+ plasma or membrane proteins that make up complement and that serve as an auxiliary defense system (Prodinger et al. 2003). The major complement component C3 appears in fetal tissues as early as 6 weeks of gestation, whereas C2 and C4 are not detected until 8 weeks of gestation (Adinolfi 1997). Regulatory proteins that protect the fetus from maternal complement such as decay-accelerating factor (DAF), membrane co-factor protein (MCP), and CD59 are expressed in the liver from at least 6 weeks of gestation (Simpson et al. 1993). Complement receptors CR1 and CR3 have been detected on monocytes and neutrophils and on cells in the bone marrow, spleen, and thymus of 14-week-old fetuses (Adinolfi et al. 1988). Recently, Mastellos and Lambris have postulated that complement is crucial in ontogeny, not because of a role in immunity, but because of a role in modulating cellular responses and cell-cell interactions that are crucial to early development and cell differentiation (Mastellos and Lambris 2002).

## DEVELOPMENT OF THE NEWBORN IMMUNE SYSTEM

Upon parturition, the immune system of the neonate becomes abruptly released from the immunosuppressive intrauterine environment (Holt 2003; Janeway Jr. et al. 1997b; West 2002) and is concomitantly exposed to a world of antigens. Unlike the situation in rodents, the human neonatal immune system has achieved a significant level of maturity (Holt and Jones 2000; Peakman et al. 1992; West 2002).

However, in comparison with the human adult immune system and with respect to the antigenic challenges facing the neonate, the neonatal immune system is considered immature (Holt and Jones 2000; Miller 1978). The neonate's susceptibility to infection has sparked the term "immunodeficiency of immaturity" (Schelonka and Infante 1998). In general, the neonate has an insignificant inflammatory response (Holladay and Smialowicz 2000), low levels of antibody (Holt and Jones, 2000), and reduced T cell reactivity (Holt et al. 1992; Pirenne et al. 1992). Moreover, the percentages of γδ T cells (Peakman et al. 1992) and $CD5^+$ B cells (Hannet et al. 1992; Holt and Jones 2000; Peakman et al. 1992) are higher in the fetus than in the adult, and they decline postnatally.

A case has been made by Kincade et al. (2002) that with respect to antigenic exposure, the emerging immune system of fetuses is not immature, but represents a unique stage of development that undergoes replacement by more reactive "adult" components in time. These authors emphasize that a direct connection has yet to be established between fetal stem cells and ones that sustain adult blood cell formation.

The neonatal period bridges the fetal and adult immune systems. Some factors governing maturational aspects of this transient period have begun to be recognized. In particular, several groups provide evidence supporting the role of antigenic exposure on regulating adaptive immune cell function through the type of cytokine response (Bjorksten 1999a; Cunningham-Rundles et al. 2002; Holt 2003; Holt and Jones 2000; Wilson et al. 1992). Microbially derived molecules not normally encountered in fetal life signal APC via specific Toll-like receptors as well as through CD14, the high-affinity receptor for bacterial lipopolysaccharide. DC appear to be the crucial type of APC influencing the immune system of the newborn from the Th2-skewed state characteristic of the fetal compartment to the more balanced (by comparison, more Th1-polarized) adult-equivalent state (Holt and Sly 2002; Holt and Jones 2000). Epidemiologic evidence supports a heightened risk for neonates both for infectious disease and allergic sensitization, and in their commentary, Holt and Sly (2002) suggest that this susceptible state results from a "generalized maturational deficit in Th1 function."

As mentioned earlier, mechanisms maintaining the Th2-skew in fetal and early postnatal life include deficient production by DC of IL-12, which is required for stabilization of the IFNγ transcriptional machinery in T cells, and hypermethylation of the promoter of the IFNγ gene in neonatal $CD4^+$ T cells, which directly inhibits transcription. In addition, neonatal $CD4^+$ T cells require higher levels of costimulation to achieve maximum activation. Release of these inhibitory mechanisms primarily depends on microbial exposure. Interestingly, $CD8^+$ cytolytic T cells in neonates lack the hypermethylation of their IFNγ promoter characteristic of $CD4^+$ T cells. Moreover, the precursor frequency of $CD8^+$ cytotoxic T lymphocytes (CTL) in neonates is comparable to that in adults, indicating a functionally mature neonatal $CD8^+$ cytotoxic compartment. Although terminal differentiation of $CD8^+$ T cell precursors and reactivation of $CD8^+$ memory cells rely on cytokine "help" from $CD4^+$ T cells, at least the primary responses of $CD8^+$ effector cells can be utilized to provide protection during infancy (Hermann et al. 2002; Holt 2003). A description of the maturational state of the major cell types in the immune system during the neonatal period is detailed below.

## Major Components of the Newborn Immune System

### *T Cells*

The classic neonatal tolerance experiments of Billingham et al. (1953; 1956) have long championed the idea of a unique neonatal immunodeficiency period. Although this appears true in many regards, the inability of the neonatal T cells to "see" antigens has been refuted by Ridge, Fuchs, and Matzinger (Ridge et al. 1996). Neonatal mice primed with purified DC—as opposed to an innoculum containing many tolerogenic nonprofessional APC (B cells)—become activated, not tolerized (Ridge et al. 1996). The dose of antigen, type of adjuvant, and type of APC (Forsthuber et al. 1996; Pirenne et al. 1992; Ridge et al. 1996) impact the response of the mature virgin T cells in neonates.

Evidence also exists for the presence in the human newborn of functionally primed, as well as virgin, T cells. Umbilical cord mononuclear cells demonstrate antigen-specific responses to a number of allergens, parasite antigens, and autoantigens, including myelin basic protein (reviewed in: Holt and Jones 2000). Vaccine research indicates that maternal exposure to certain diseases or vaccines can prime the fetus or neonate (Glezen 2003; Goldman and Marchant 2003; Holt 2003; Zinkernagel 2003).

Despite examples of functional maturity, evidence abounds for reduced T cell appearance and reactivity in the neonate. Most notably, as discussed above, is a reduced capacity by neonatal T cells to produce Th1-type cytokines (particularly IFNγ) (Holt et al. 1992). The capacity to produce IFNγ in response to polyclonal stimuli rises after birth and reaches adult levels by 5 years of age (Miyawaki et al. 1985; Rowe et al. 2000); the deficiency in IFNγ production by neonates occurs at the T-cell clonal level (Holt et al. 1992). The proportion of lymphocytes represented by T cells is lowest at birth and increases over time (Berry et al. 1992; Hannet et al. 1992; West 2002). Almost all neonatal T cells express the naïve $CD45RA^+$ phenotype (Fujii et al. 1992; Hannet et al. 1992; Peakman et al. 1992; West, 2002). Little is known of the functional status of T cell subsets. The proportion of γδ T cells declines after birth (Peakman et al. 1992) and, as noted earlier, $CD4^+$ T cells have a reduced capacity to produce IFNγ (White et al. 2002). In limiting dilution assays (LDA), neonatal $CD4^+$ T cells have a lower precursor frequency than adult $CD4^+$ T cells, whereas the precursor frequency of $CD8^+$ T cells in neonates is comparable to the adult level of activity (Deacock et al. 1992). In spite of this, neonatal T cells exhibit lower levels of cytotoxicity in cell-mediated lympholysis assays than adult T cells (West 2002). Presumably, the reduced cytolytic potential, together with a sensitivity of neonatal T cells to chemical immunosuppression, provide an advantageous clinical environment for neonatal heart transplantation. Specifically, a 10-12% survival advantage has been noted through 10 years post-transplantation if heart transplantation is performed within the first 30 days of life compared to one to six months of age. This survival advantage has been ascribed to decreased immune related events in the immediate postnatal period (Morrow and Chinnock 2000).

### Natural Killer Cells

Natural killer cells have been reported to be both fewer at birth than later in life and of reduced functional capacity (Hannet et al. 1992; Kincade et al. 2002; Toivanen et al. 1981; West 2002). Acquisition of the recognition molecules that allow an NK cell to discriminate normal host cells from damaged or malignant counterparts is dependent on developmental age (Takei et al. 2001).

### B Cells

Human B lymphocytes are inherently immature at birth (Holladay and Smialowicz 2000). Despite significant production of IgM and IgG immunoglobulin in the fetus, production of these immunoglobulin isotypes at birth is impaired (Holt and Jones 2000). Neonates have very low levels of serum IgM, IgA, and IgE, and essentially all IgG is of maternal origin (Holladay and Smialowicz 2000; Holt and Jones 2000). IgM synthesis is reported to begin during the first week of extrauterine life regardless of the gestational age at birth (Vetro and Bellanti 1989). IgM levels rapidly reach 50% of the adult values by 6 months and 75 to 80% of the adult levels by 1 year of age (de Muralt 1978). IgA synthesis results in only 20% of the adult level by 12 months of age; adult levels are not reached until 10 years of age (de Muralt 1978; Vetro and Bellanti 1989). Maternally derived newborn IgG decreases to a low at 3 months of age, then synthesis by the newborn begins and adult levels are reached by 5 to 6 years of age (Holladay and Smialowicz 2000). The low IgE levels are the consequence of a lack of IL-4 produced by immature helper T cells and reduced CD40L expression on these T cells (Holt and Jones 2000). Approximately three-quarters of B cells in the neonate express the CD5 cell surface marker, whereas only approximately one-quarter of the B cells express this marker in adult cells (Hannet et al. 1992). It has been hypothesized that because $CD5^+$ B cells produce polyreactive antibody and are largely T-independent (Peakman et al. 1992), these cells may play a vital role in the infant's primary immune response (Holt and Jones 2000). Despite the presence of this type of B cell, the neonate's inability to respond to polysaccharide vaccines such as *Haemophilus influenzae* type b (Hib) (Ahmad and Chapnick 1999; Cadoz, 1998; Gathings et al. 1981) has revealed a weakness in the defense system of the infant that lasts until 1 to 2 years of age (Ahmad and Chapnick 1999; Cadoz 1998; Rijkers et al. 1998; West 2002). Polysaccharides activate B cells through a required costimulation of surface immunoglobulin and CD21, type 2 complement receptor. Reduced CD21 in neonates may explain their unresponsiveness to polysaccharides (Rijkers et al. 1998). The delays in B lymphocyte function have been attributed to a variety of different causes: deficiency in B cells (Anderson et al. 1981; Gathings et al. 1981), regulation by T suppressor cells (Miyawaki et al. 1981), and defects in accessory cell function (Bondada et al. 2000).

Similar to the requirement of antigenic exposure for maturation of the T cell compartment, B lymphocytes appear to be influenced by antigenic exposure *in utero* and during early postnatal life. Evidence from laboratory studies demonstrates that animals raised in germ-free environments have delayed lymphoid development (Thorbecke 1959). Moreover, a significant body of research addresses maternal influences on the baby during pregnancy (due to diet [Bjorksten1999a; Moore 1998], illness [Goldman and Marchant 2003; Zinkernagel 2003], allergens [Warner and Warner 2000], or vaccines [Glezen and Alpers 1999]) and after parturition (through breastfeeding [Bjorksten 1999b; Hanson 1998; Oddy 2002]), indicating that exposure to antigens through the mother can aid or impair maturation of the infant's immune system.

Finally, exposure of the infant to microorganisms, especially those of the gut, and to vaccines enhances serum immunoglobulin titers (de Muralt 1978). The response to encapsulated gut flora (particularly strains of *Escherichia coli*) (Fong et al. 1974) drives development of IgM isohemagglutinins that recognize the ABO blood group antigens. Clinically, a window of opportunity exists because the response to carbohydrates does not develop for 6 months or more (Fong et al. 1974). Heart transplantation in infants has now been successfully used despite ABO incompatibility of the donor and recipients (West et al. 2001).

### *Macrophages and Dendritic Cells*

The crucial role of professional APC in initiating immune responses against pathogens and not towards self has recently been given much attention (Janeway Jr. 1992; Matzinger 1994; Wright 1999). Surprisingly little, however, is known about the neonatal macrophage and DC. The only monocyte/mϕ populations that have been functionally assessed come from umbilical cord blood samples collected at birth (Holt and Jones 2000). Neonatal macrophages have been reported to have reduced chemotaxis compared to adult levels (Baron and Lafuro 1985; Miller 1978; Speer and Johnson Jr. 1984; West 2002; Weston et al. 1997) and a decreased production of a number of cytokines, including TNF-α (Serushago et al. 1996). Several reports indicate that antigen processing impairments in neonatal mϕ lead to defective phagocytosis (Baron and Lafuro 1985; Holladay and Smialowicz 2000; West 2002), while several other reports suggest that these cells can phagocytose (Clerici et al. 1993) and present antigen at adult levels (Weston et al. 1997). Bactericidal activity in neonatal macrophages, however, have been reported to be at adult levels (Miller 1978; Speer and Johnson Jr. 1984).

A key to the concept of neonatal immunodeficiency may be the hindered ability of APC to produce IL-12 (Prescott et al. 2003). Specifically, DC derived from neonatal monocytes appear deficient in IL-12(p35) gene expression (Goriely et al. 2001). Addition of IFNγ to the activated newborn DC restored expression of IL-12(p35) to adult levels, but because IL-12 is required for stabilization of the IFNγ transcriptional machinery in T cells (Yap et al. 2000), a positive feedback loop seems to be curtailed (Goriely et al. 2001). In addition, neonatal DC have poor accessory function (Hunt et al. 1994) and express lower levels of ICAM-1 and MHC class I and II than peripheral blood DC from adults (reviewed in: Holt and Jones 2000). These characteristics further contribute to a decreased ability by neonatal DC to stimulate an adaptive immune response.

### Complement

Many components of complement are decreased in newborns compared to adults (Holladay and Smialowicz 2000; Vetro and Bellanti 1989). In one study, the lytic activity associated with the classical pathway matured between 1–3 months and then exceeded that of adults between 7 and 24 months (Ferriani et al. 1990). The lytic activity associated with the alternative pathway reaches adult levels around the $13^{th}$ month (Ferriani et al. 1990). A subsequent study by the same researchers demonstrated age-dependent variations in C3 and factor B concentrations, but not in C4, in the sera of healthy children 3 to 14 years old. The highest values occurred in 5–6 year old children (Ferriani et al. 1999).

## THE ADULT IMMUNE SYSTEM

Once the cells and organs of the immune system have established "adult" levels or function, few studies ascertain maturation of these compartments beyond the formative years (Kincade et al. 2002). Several studies indicate that adult patterns of lymphopoiesis are remarkably stable even into the eighth decade of life (Hao et al. 2001; Nunez et al. 1996; Rossi et al. 2003). A notable exception is the thymus, which gradually involutes after puberty (Kendall 1991). Few new T cells are exported after the thymic cortex fills with adipose tissue, and immunity to new antigens depends on cross-reactions with previously encountered antigens (Nossal 2003). The thymus plays an important role in the development of sex organs, and infants with little or no thymus at birth (e.g., ataxia telangectesia and DiGeorge and cri du chat syndromes) fail to develop normal gonads (Kendall 1991). Although steroid hormones have been called "the gatekeepers" of entry into and progression within lymphoid lineages (Kincade et al. 2002) and obviously impact thymic development (Kendall 1991), few studies do more than state that final maturation of the immune system occurs during adolescence.

In general, aging beyond adolescence is associated with a reduced number of B cell progenitors in the bone marrow and a reduced proliferative capacity of the T cells. Surprisingly, the number of DC and NK cells increase in the periphery of older individuals, and DC seem unimpaired in their antigen-presenting functions (Longo 2003; Nossal 2003). The elderly seem more susceptible to infections, yet at the same time, they benefit significantly from immunization against influenza and pneumococci (Longo 2003; Nossal 2003). Knowledge of the relative senescence of various immune system components can aid in clinical manipulation to effect better health in older individuals.

## MANIPULATORS OF THE DEVELOPING IMMUNE SYSTEM

The immune system matures according to developmental pathways but also in response to the environment. Four manipulators impact many facets of the immune system and deserve attention. First and foremost, nutrition plays a vital role in the ability of the body to defend against pathogens (Bardare et al. 1993; Devereux et al. 2002). The majority of immunocompromised individuals in the world simply lack nourishment (Janeway Jr. et al. 1997a). Studies implicate maternal diet in development of atopic disease (Bardare et al. 1993; Devereux et al. 2002). Other maternal behaviors such as breastfeeding can decrease risk of allergy, asthma, and autoimmune disease (Bjorksten 1999a; Hanson 1998). The benefits of breastfeeding indicate that the neonate's diet plays a significant role in maturation of immune system function.

Second, maternal disease and vaccination can impact immune development in the fetus. This maternal influence can be harmful to the fetus (for example, the association of fetal and perinatal loss with malaria infection during pregnancy [Shulman and Dorman 2003]), but can also provide an important avenue for positive clinical manipulation (Gruber 2003). This area of research has much to offer as a critical window into the developing immune system.

Likewise, the third area, exposure to toxins, offers insight into vulnerable periods during development. Much work needs to be done on assessment of toxicant exposure in early life and the development of disease—particularly autoimmune disease—in later life. Popular opinion generates diffuse anxiety about scientific advances (such as offered through genetically modified foods) because of the unknown effect on people's immune systems and the subsequent ability to defend against diseases such as cancer.

Finally, medical advances have resulted in modern pharmacologic manipulation of the immune system. Effective immunosuppressive drugs allow transplantation to be a viable treatment for numerous abnormalities, but often at a cost of the body's effective inherent defense system. Likewise, medical treatments for cancers and other diseases manipulate the immune response, forfeiting future health in the interest of immediate survival. How these manipulators affect the developing immune system will be revealed with future research.

## SUMMARY

Ordered events outline the development of the immune response, particularly in fetal and early postnatal life. However, the intrinsic capacity of the immune system to respond to environmental cues impacts maturation. Manipulation of the internal environment that supports immune system development can provide insights into the interrelated network of cells and soluble mediators, potentially leading to clinically beneficial outcomes in many areas of immune dysfunction.

## REFERENCES

Adinolfi M. 1997. Ontogeny of human natural and acquired immunity. *Curr Top Microbiol Immunol* 222:67–102.

Adinolfi M, Cheetham M, Lee T, and Rodin A. 1988. Ontogeny of human complement receptors CR1 and CR3: expression of these molecules on monocytes and neutrophils from maternal, newborn and fetal samples. *Eur J Immunol* 18(4):565–569.

Ahmad H and Chapnick EK. 1999. Conjugated polysaccharide vaccines. *Infect Dis Clin North Am* 139(1):113–133.

Anderson U, Bird AG, Britton S, and Palacios R. 1981. Humoral and cellular immunity in humans studied at the cell level from birth to two years of age. *Immunol Rev* 57:1–38.

Ardavin C, del Hoyo GM, Martin P, Anjuere F, Arias CF, Marin AR, Ruiz S, Parrillas V, and Hernandez H. 2001. Origin and differentiation of dendritic cells. *Trends in Immunol* 22:691–700.

Bardare M, Vaccari A, Allievi E, Brunelli L, Coco F, de Gaspari GC, and Flauto U. 1993. Influence of dietary manipulation on incidence of atopic disease in infants at risk. *Ann Allergy* 71(4):366–371.

Baron M and Lafuro P. 1985. Symposium on infections in the compromised host. The extremes of age: The newborn and the elderly. *Nurs Clin North Am* 20:181–190.

Bellanti JA, Malka-Rais J, Castro HJ, de Inocencio JM, and Sabra A. 2003. Developmental immunology: clinical application to allergy-immunology. *Ann Allergy Asthma Immunol* 90 (6 Suppl 3):2–6.

Berry SM, Fine N, Bichalski JA, Cotton DB, Dombrowski MP, and Kaplan J. 1992. Circulating lymphocyte subsets in second- and third-trimester fetuses: comparison with newborns and adults. *Am J Obstet Gynecol* 167:895–900.

Billingham RE and Brent L. 1956. Acquired tolerance of foreign cells in newborn animals. *Proc R Soc London B Biol Sci* 146(922):78–90.

Billingham RE, Brent L, and Medawar PB. 1953. Activity acquired tolerance of foreign cells. *Nat* 172:603–606.

Bjorksten B. 1999a. Environment and infant immunity. *Proc Nutr Soc* 58(3):729–732.

Bjorksten B. 1999b. The intrauterine and postnatal environments. *J Allergy Clin Immunol* 104(6):1119–1127.

Bofill M, Janossy M, Burford GD, Seymour GJ, Wernet P, and Kelemen E. 1985. Human B cell development. II. Subpopulations in the human fetus. *J Immunol* 134:1531–1538.

Bondada S, Wu H, Robertson DA, and Chelvarajan RL. 2000. Accessory cell defect in unresponsiveness of neonates and aged to polysaccharide vaccines. *Vaccine* 19(4–5):557–565.

Boyd RL, Tucek CL, Godfrey DI, Izon DJ, Wilson TJ, Davidson NJ, Bean AGD, Ladyman HM, Ritter MA, and Hugo P. 1993. The thymic microenvironment. *Immuno Today* 14(9):445–459.

Brasel K, McKenna HJ, Morrissey PJ, Charrier K, Morris AE, Lee CC, Williams DE, and Lyman SD. 1996. Hematologic effects of flt3 ligand *in vivo* in mice. *Blood* 88:2004–2012.

Cadoz M. 1998. Potential and limitations of polysaccharide vaccines in infancy. *Vaccine* 16 (14/15):1391–1395.

Carvalho TL, Mota-Santos T, Cumano A, Demengeot J, and Vieira P. 2001. Arrested B lymphopoiesis and persistence of activated B cells in adult interleukin 7(-/-) mice. *J Exp Med* 194:1141–1150.

Chaplin DD. 2003. Lymphoid tissues and organs. In: Paul WE (Ed.), *Fundamental Immunology.* Lippincott, Williams, and Wilkins, Philadelphia, pp. 419–453.

Clerici M, Depalma L, Roilides E, Baker R, and Shearer GM. 1993. Analysis of T helper and antigen-presenting cell functions in cord blood and peripheral blood leukocytes from healthy children of different ages. *J Clin Invest* 91:2829–2836.

Cooper EL, Mueller-Sieburg CE, and Spangrude GJ. 1993. Stem Cells. In Developmental Immunology. In: Cooper EL and Nisbet-Brown E (Eds.), *Developmental Immunology.* Oxford University Press, New York, pp.177–197.

Cooper MD. 1981. Pre-B cells: Normal and abnormal development. *J Clin Immunol* 1:81–89.

Cunningham-Rundles S, Ahrn S, Abuav-Nussbaum R, and Dnistrian A. 2002. Development of immunocompetence: role of micronutrients and microorganisms. *Nutr Rev* 60(5 Pt 2):S68–S72.

de Muralt G. 1978. Maturation of cellular and humoral immunity. In: Stave U (Ed.), *Perinatal Physiology.* Plenum, New York, pp. 267.

De Waele M, Foulon W, Renmans W, Segers E, Smet L, Jochmans K, and Van Camp B. 1988. Hematologic values and lymphocyte subsets in fetal blood. *Am J Clin Pathol* 89(6):742–746.

Deacock SJ, Schwarer AP, Bridge J, Batchelor JR, Goldman JM, and Lechler RI. 1992. Evidence that umbilical cord blood contains a higher frequency of HLA class II-specific alloreactive T cells than adult peripheral blood. A limiting dilution analysis. *Transplantation* 53:1128–1134.

Devereux G, Barker RN, and Seaton A. 2002. Antenatal determinants of neonatal immune responses to allergens. *Clin Exp Allergy* 32(1):43–50.

Douagi I, Colucci F, DiSanto JP, and Cumano A. 2002. Identification of the earliest prethymic bipotent T/NK progenitor in murine fetal liver. *Blood* 99:463–471.

Ferriani VP, Barbosa JE, and de Carvalho IF. 1990. Serum haemolytic classical and alternative pathways of complement in infancy: age-related changes. *Acta Paediatr Scand* 79(3):322–327.

Ferriani VP, Barbosa JE, and de Carvalho IF. 1999. Complement haemolytic activity (classical and alternative pathways), C3, C4 and factor B titres in healthy children. *Acta Paediatr* 88(10):1062–1066.

Fong SW, Qaqundah BY, and Taylor WF. 1974. Developmental patterns of ABO isoagglutinins in normal children correlated with the effects of age, sex, and maternal isoagglutinins. *Transfus* 14:551–559.

Forestier F, Daffos F, Catherine N, Renard M, and Andreux JP. 1991. Developmental hematopoiesis in normal human fetal blood. *Blood* 77:2360–2363.

Forsthuber T, Yip HC, and Lehmann PV. 1996. Induction of $T_H1$ and $T_H2$ immunity in neonatal mice. *Science* 271:1728.

Fujii Y, Okumura M, Inada K, Nakahara D, and Matsuda H. 1992. CD45 isoform expression during T cell development in the thymus. *Eur J Immunol* 22:1843–1850.

Gathings WE, Kubagawa H, and Cooper MD. 1981. A distinctive pattern of B cell immaturity in perinatal humans. *Immunol Rev* 57:107–126.

Gitlin D and Biasucci A. 1969. Development of gamma G, gamma A, gamma M, beta IC-beta IA, C 1 esterase inhibitor, ceruloplasmin, transferrin, hemopexin, haptoglobin, fibrinogen, plasminogen, alpha 1-antitrypsin, orosomucoid, beta-lipoprotein, alpha 2-macroglobulin, and prealbumin in the human conceptus. *J Clin Invest* 48:1433–1445.

Glezen WP. 2003. Effect of maternal antibodies on the infant immune response. *Vaccine* 21:3389–3392.

Glezen WP and Alpers M. 1999. Maternal immunization. *Clin Infect Dis* 28(2):219–224.

Godin IE, Garcia-Porrero JA, Coutinho A, Dieterlen-Lievre F, and Marcos MA. 1993. Para-aortic splanchnopleura from early mouse embryos contains B1a cell progenitors. *Nat* 364(6432):67–70.

Goldman M and Marchant A. 2003. The impact of maternal infection or immunization on early-onset autoimmune diabetes. *Vaccine* 21(24):3422–3425.

Gordon S. 2003. Macrophages and the immune response. In: Paul, WE (Ed.), *Fundamental Immunology.* Lippincott, Williams, and Wilkins, Philadelphia, pp. 481–495.

Goriely S, Vincart B, Stordeur P, Vekemans J, Willems F, Goldman M, and De Wit D. 2001. Deficient L-12(p35) gene expression by dendritic cells derived from neonatal monocytes. *J Immunol* 166:2141–2146.

Gougeon ML, Drean G, Le Beist F, Cousseau M, Fevrier M, Diu A, Theze J, Griscelli C, and Fischer A. 1990. Human severe combined immunodeficiency disease: phenotypic and functional characteristics of peripheral B lymphocytes. *J Immunol* 145:2873–2879.

Gruber MF. 2003. Maternal immunization: US FDA regulatory considerations. *Vaccine* 21(24):3487–3491.

Hannet I, Erkeller-Yuksel F, Lydyard P, Deneys V, and DeBruyere M. 1992. Developmental and maturational changes in human blood lymphocyte subpopulations. *Immunol Today* 13(6):215–218.

Hanson LA. 1998. Breastfeeding provides passive and likely long-lasting active immunity. *Ann Allergy Asthma Immunol* 81(6):523–533.

Hao GL, Zhu J, Price MA, Payne KJ, Barsky IW, and Crooks GM. 2001. Identification of a novel, human multilymphoid progenitor in cord blood. *Blood* 97:3683–3690.

Hao Z and Rajewsky K. 2001. Homeostasis of peripheral B cells in the absence of B cell influx from the bone marrow. *J Exp Med* 194:1151–1164.

Hardy RR. 2003. B-Lymphocyte Development and Biology. In: Paul WE (Ed.), *Fundamental Immunology.* Lippincott, Williams, and Wilkins, Philadelphia, pp. 159–194.

Haynes BF, Denning SM, Singer KH, and Kurtzberg J. 1989a. Ontogeny of T-cell precursors: a model for the initial stages of human T-cell development. *Immunol Today* 10(3):87–91.

Haynes BF, Singer KH, Denning SM, and Martin ME. 1989b. Analysis of expression of CD2, CD3 and T cell antigen receptor molecule expression during early human thymic development. *J Immunol* 141:3776–3784.

Haynes BF, Singer KH, Denning SM, and Martin ME. 1988. Analysis of expression of CD2, CD3, and T cell antigen receptor molecules during early human fetal thymic development. *J Immunol* 141(11):3776–3784.

Hermann E, Truyens C, Alonso-Vega C, Even J, Rodriguez P, Berthe A, Gonzalez-Merino E, Torrico F, and Carlier Y. 2002. Human fetuses are able to mount an adultlike CD8 T-cell response. *Blood* 100:2153–2158.

Holladay SD and Smialowicz RJ. 2000. Development of the murine and human immune system: differential effects of immunotoxicants depend on time of exposure. *Environ Health Perspect* 108(Suppl3):463–473.

Holt PG. 2003. Functionally mature virus-specific $CD8^+$ T memory cells in congenitally infected newborns: proof of principle for neonatal vaccination? *J Clin Invest* 111:1645–1647.

Holt PG, Clough JB, Holt BJ, Baron-Hay MJ, Rose AH, Robinson BW, and Thomas WR. 1992. Genetic "risk" for atopy is associated with delayed postnatal maturation of T-cell competence. *Clin Exp Allergy* 22:1093–1099.

Holt PG and Sly PD. 2002. Interactions between RSV infection, asthma, and atopy: unraveling the complexities. *J Exp Med* 196:1271–1275.

Holt PG and Jones CA. 2000. The development of the immune system during pregnancy and early life. *Allergy* 55:688–697.

Hunt DW, Huppertz HI, Jiang HJ, and Petty RE. 1994. Studies of human cord blood dendritic cells: evidence for functional immaturity. *Blood* 12:4333–4343.

Igarashi H, Kouro T, Yokota T, Comp PC, and Kincade PW. 2001. Age and stage dependency of estrogen receptor expression by lymphocyte precursors. *Proc Natl Acad Sci USA* 98:15131–15136.

Ikawa T, Kawamoto H, Fujimoto S, and Katsura Y. 1999. Commitment of common T/Natural killer (NK) progenitors to unipotent T and NK progenitors in the murine fetal thymus revealed by a single progenitor assay. *J Exp Med* 190:1617–1626.

Jaffredo T, Gautier R, Brajeul V, and Dieterlen-Lievre F. 2000. Tracing the progeny of the aortic hemangioblast in the avian embryo. *Dev Biol* 224(2):204–214.

Janeway Jr. CA. 1992. The immune system evolved to discriminate infectious nonself from noninfectious self. *Immunol Today* 139(1):11–16.

Janeway Jr. CA, Travers P, Hunt S, and Walport M. 1997a. Failures of Host Defense Mechanisms. In *Immunobiology.* Current Biology Limited, London, San Francisco and New York, pp. 10:1–10:41.

Janeway Jr. CA, Travers P, Hunt S, and Walport M. 1997b. Immune responses in the absence of infection. In *Immunobiology.* Current Biology Limited, London, San Francisco and New York, pp. 12:1–12:42.

Kane MP, Jaen CR, Tumiel LM, Bearman GM, and O'Shea RM. 1999. Unlimited opportunities for environmental interventions with inner-city asthmatics. *J Asthma* 36:371–379.

Katsura Y. 2002. Redefinition of lymphoid progenitors. *Nat Rev Immunol* 2(2):127–132.

Kawamoto H, Ohmura K, and Katsura Y. 1997. Direct evidence for the commitment of hematopoietic stem cells to T, B and myeloid lineages in murine fetal liver. *Int Immunol* 9(7):1011–1019.

Kendall MD. 1991. Functional anatomy of the thymic microenvironment. *J Anat* 177:1–29.

Kennedy MK, Glaccum M, Brown SN, Butz EA, Viney JL, Embers M, Matsuki N, Charrier K, Sedger L, Willis CR, Brasel K, Morrissey PJ, Stocking K, Schuh JC, Joyce S, Peschon JJ. 2000. Reversible defects in natural killer and memory CD8 T cell lineages in interleukin 15-deficient mice. *J Exp Med* 191:771–780.

Kincade PW, Owen JJT, Igarashi H, Yokota T, and Rossi MID. 2002. Nature or nurture? Steady-state lymphocyte formation in adults does not recapitulate ontogeny. *Immunol Rev* 187:116–125.

Kondo M, Weissman IL, and Akashi K. 1997. Identification of clonogenic common lymphoid progenitors in mouse bone marrow. *Cell* 91(5):661–672.

Kumar V, Ben-Ezra J, Bennett M, and Sonnenfeld G. 1979. Natural killer cells in mice treated with 89Sr: normal target-binding cell numbers but inability to kill even after interferon administration. *J Immunol* 123:1832–1838.

Landreth KS. 1993. B lymphocyte generation as a developmental process. In: Cooper EL and Nisbet-Brown E (Eds.), *Developmental Immunology.* Oxford University Press, New York, NY, pp. 238–273.

Liu Y.-J. 2001. Dendritic cell subsets and lineages, and their functions in innate and adaptive immunity. *Cell* 106:259–262.

Lobach DF and Haynes BF. 1987. Ontogeny of the human thymus during fetal development. *J Clin Immunol* 7:81–97.

Lodolce JP, Boone DL, Chai S, Swain RE, Dassopoulos T, Trettin S, and Ma A. 1998. IL-15 receptor maintains lymphoid homeostasis by supporting lymphocyte homing and proliferation. *Immunity* 9:669–676.

Loke YW. 1978. Development of immunocompetence in the human foetus. In Immunology and immunopathology of the human foetal-maternal interaction. In: Loke YM (Ed.), *Immunology and Immunopathology of the Human Foetal-Maternal Interaction.* Elsevier North-Holland Biomedical Press, Amsterdam.

Longo DL. 2003. Immunology of aging. In: Paul WE (Ed.), *Fundamental Immunology.* Lippincott, Williams, and Wilkins, Philadelphia, pp. 1043–1075.

Mastellos D and Lambris JD. 2002. Complement: more than a 'guard' against invading pathogens? *Trends in Immunol* 23(10):485–491.

Matzinger P. 1994. Tolerance, danger, and the extended family. *Ann Rev Immunol* 12:991–1045.

Medina KL and Kincade PW. 1994. Pregnancy-related steroids are potential negative regulators of B lymphopoiesis. *Proc Natl Acad Sci USA* 91:5382–5386.

Medina KL, Smithson GM, and Kincade PW. 1993. Suppression of B lymphopoiesis during normal pregnancy. *J Exp Med* 178:1507–1515.

Mestecky J, Blumberg RS, Kiyono H, and McGhee JR. 2003. The Mucosal Immune System. In: Paul WE (Ed.), *Fundamental Immunology*. Lippincott, Williams, and Wilkins, Philadelphia, pp. 965–1020.

Metcalf D and Moore MAS. 1971. Haemopoietic cells: their origin, migration and differentiation. American Elsevier, New York.

Migliaccio G, Migliaccio AR, Petti S, Mavilio F, Russo G, Lazzaro D, Testa U, Marinucci M, and Peschle C. 1986. Human embryonic hemopoiesis. Kinetics of progenitors and precursors underlying the yolk sac-liver transition. *J Clin Invest* 78(1):51–60.

Miller DL, Hirvonen T, and Gitlin D. 1973. Synthesis of IgE by the human conceptus. *J Allergy Clin Immunol* 52:182–188.

Miller ME. 1978. *Host Defenses in the Human Neonate*. Grune and Stratton, New York.

Miyawaki T, Seki H, Taga K, Sato H, and Taniguchi N. 1985. Dissociated production of interleukin-2 and immune (gamma) interferon by phytohaemaglutinin stimulated lymphocytes in healthy infants. *Clin Exp Immunol* 59:505–511.

Miyawaki T, Moriya N, Nagaoki T., and Tonigachi N. 1981. Maturation of B-cell differentiation ability and T-cell regulatory function in infancy and childhood. *Immunol Rev* 57:69–87.

Moore SE. 1998. Nutrition, immunity and the fetal and infant origins of disease hypothesis in developing countries. *Proc Nutr Soc* 57(2):241–247.

Morrow WR and Chinnock RE. 2000. Survival after heart transplantation. In: Tejani AH, Harmon WE, and Fine RN (Eds.), *Pediatric Solid Organ Transplantation*. Munksgaard, Copenhagen, pp. 417–426.

Muller AM, Medvinsky A, Strouboulis J, Grosveld F, and Dzierzak E. 1994. Development of hematopoietic stem cell activity in the mouse embryo. *Immunity* 1(4):291–301.

Muramatsu S. 1993. Monocytes, macrophages, and accessory cells. In: Cooper EL and Nisbet-Brown E (Eds.), *Developmental Immunology*. Oxford University Press, New York, pp. 199–216.

Namikawa R, Mizuno T, Matsuoka H, Fukami H, Ueda R, Itoh G, Matsuyama M, and Takahashi T. 1986. Ontogenic development of T and B cells and non-lymphoid cells in the white pulp of human spleen. *Immunol* 57:61–69.

Nishikawa M, Tahara T, Hinohara A, Miyajima A, Nakahata T, and Shimosaka A. 2001. Role of the microenvironment of the embryonic aorta-gonad-mesonephros region in hematopoiesis. *Ann N Y Acad Sci* 938:109–116.

Noguchi M, Yi H, Rosenblatt HM, Filipovich AH, Adelstein S, Modi WS, McBride OW, and Lenard WJ. 1993. Interleukin-2 receptor gamma chain mutation results in X-linked severe combined immunodeficiency in humans. *Cell* 73:147–157.

Nossal GJV. 2003. Vaccines. In: Paul WE (Ed.), *Fundamental Immunology*. Lippincott Williams and Wilkins, Philadelphia, pp. 1319–1369.

Nunez C, Nishimoto N, Gartland GL, Billips LG, Burrows PD, Kubagawa H, and Cooper MD. 1996. B cells are generated throughout life in humans. *J Immunol* 156:866–872.

Oddy WH. 2002. The impact of breastmilk on infant and child health. *Breastfeed Rev* 10(3):5–18.

Orkin SH and Zon LI. 2002. Hematopoiesis and stem cells: plasticity *versus* developmental heterogeneity. *Nat Immunol* 3:323–328.

Owen JJT. 1972. The origins and development of lymphocyte populations. In: *Ontogeny of Acquired Immunity, A Ciba Foundation Symposium*. Elsevier, New York, pp. 35–54.

Peakman M, Buggins AG, Nicolaides KH, and et al. 1992. Analysis of lymphocyte phenotypes in cord blood from early gestation fetuses. *Clin Exp Immunol* 90:345–350.

Pirenne H, Aujard Y, Eljaafari A, Bourillon A, Oury JF, Le Gac S, Blot P, and Sterkers G. 1992. Comparison of T cell functional changes during childhood with the ontogeny of CDw29 and CD45RA expression on CD4+ T cells. *Pediatr Res* 32:81–86.

Potocnik AJ, Brakebusch C, and Fassler R. 2000. Fetal and adult hematopoietic stem cells require beta 1 integrin function for colonizing fetal liver, spleen, and bone marrow. *Immunity* 12:653–663.

Prescott SL, Taylor A, King B, Dunstan J, Upham JW, Thornton CA, and Holt PG. 2003. Neonatal interleukin-12 capacity is associated with variations in allergen-specific immune responses in the neonatal and postnatal periods. *Clin Exp Allergy* 33(5):566–572.

Pribyl JAR and Lebien TW. 1996. Interleukin 7 independent development of human B cells. *Proc Natl Acad Sci USA* 93:10348–10353.

Prodinger WM, Wuerzner R, Stoiber H, and Dierich MP. 2003. Complement. In: Paul WE (Ed.), *Fundamental Immunology*. Lippincott, Williams, and Wilkins, Philadelphia, pp. 1077–1103.

Raulet DH. 2003. Natural Killer Cells. In: Paul WE (Ed.), *Fundamental Immunology*. Lippincott, Williams, and Wilkins, Philadelphia, pp. 365–391.

Ridge JP, Fuchs EJ, and Matzinger P. 1996. Neonatal tolerance revisited: turning on newborn T cells with dendritic cells. *Science* 271:1723–1726.

Rijkers GT, Sanders EA, Breukels MA, and Zegers BJ. 1998. Infant B cell responses to polysaccharide determinants. *Vaccine* 16(14–15):1396–1400.

Rodewald HR, Kretzschmar K, Takeda S, Hohl C, and Dessing M. 2003. Identification of prothymocytes in murine fetal blood: T lineage commitment can precede thymus colonization. *EMBO J* 13(18):4229–4240.

Rossi MI, Yokota T, Medina KL, Garrett KP, Comp PC, Schipul AHJ, and Kincade PW. 2003. B lymphopoiesis is active throughout human life, but there are developmental age related changes. *Blood* 101(2):576–584.

Rothenberg EV, Yui MA, and Telfer JC. 2003. T-cell developmental biology. In: Paul WE (Ed.), *Fundamental Immunology*. Lippincott, Williams, and Wilkins, Philadelphia, pp. 259–301.

Rowe J, Macaubas C, Monger TM, Holt BJ, Harvey J, Poolman JT, Sly PD, and Holt PG. 2000. Antigen-specific responses to diphtheria-tetanus-acellular pertussis vaccine in human infants are initially Th2 polarized. *Infect Immun* 68(7):3873–3877.

Schelonka RL and Infante AJ. 1998. Neonatal immunology. *Semin Perinatol* 22(1):2–14.

Schuh AC, Faloon P, Hu Q.-L., Bhimani M, and Choi K. 1999. *In vitro* hematopoietic and endothelial potential of *flk-1*$^{-/-}$ embryonic stem cells and embryos. *Proc Natl Acad Sci USA* 96:2159–2164.

Seaman WE, Gindhart TD, Greenspan JS, Blackman MA, and Talal N. 1979. Natural killer cells, bone, and the bone marrow: studies in estrogen-treated mice and in congenitally osteopetrotic (mi/mi) mice. *J Immunol* 122:2541–2547.

Serushago B, Issekutz AC, Lee SH, Rajaraman K, and Bortolussi R. 1996. Deficient tumor necrosis factor secretion by cord blood mononuclear cells upon *in vitro* stimulation with Listeria monocytogenes. *J Interferon Cytokine Res* 16(5):381–387.

Shaw SG, Maung AA, Steptoe RJ, Thomson AW, and Vujanovic NL. 1998. Expansion of functional NK cells in multiple tissue compartments of mice treated with flt3-ligand; implications for anti-cancer and anti-viral therapy. *J Immunol* 161:2817–2824.

Shulman CE and Dorman EK. 2003. Importance and prevention of malaria in pregnancy. *Trans R Soc Trop Med Hyg* 97(1):30–35.

Simpson KL, Houlihan JM, and Holmes CH. 1993. Complement regulatory proteins in early human fetal life: CD59, membrane co-factor protein (MCP) and decay-accelerating factor (DAF) are differentially expressed in the developing liver. *Immunol* 80(2):183–190.

Speer CP and Johnson RB, Jr. 1984. Phagocytic function. Ogra PL (Ed.), *Neonatal Infections: Nutritional and Immunological Interactions.* Harcourt Brace Jovanovich, New York, pp. 21.

Takei F, McQueen KL, Maeda M, Wilhelm BT, Lohwasser S, Lian RH, and Mager DL. 2001. Ly49 and CD94/NKG2: developmentally regulated expression and evolution. *Immunol Rev* 181:90–103.

Tavassoli M. 1991. Embryonic and fetal hemopoiesis: an overview. *Blood Cells* 1:269–281.

Tavassoli M and Yoffey JM. 1983. Ontogeny. In: *Bone Marrow: Structure and Function.* AR Liss, New York, pp. 31–45.

Thorbecke GJ. 1959. Some histological and functional aspects of lymphoid tissue in germfree animals. I. Morphological Studies. *Ann N Y Acad Sci* 78:237–244.

Toivanen P, Uksila J, Lenino A, Lassila O, Hirvonen T, and Ruuskanen O. 1981. Development of mitogen responding T cells and natural killer cells in the human fetus. *Immunol Rev* 57:89–105.

Tucci A, Mouzaki A, James H, Bonnefoy JY, and Zubler RH. 1991. Are cord blood B cells functionally mature? *Clin Exp Immunol* 84:389–394.

Vetro SW and Bellanti JA. 1989. Fetal and neonatal immunoincompetence. *Fetal Ther* 4(Suppl 1):82–91.

von Gaudecker B. 1991. Functional histology of the human thymus. *Anat Embryol* 183:1–15.

Warner JA and Warner JO. 2000. Early life events in allergic sensitisation. *Med Bull* 56(4):883–893.

West L. 2002. Defining critical windows in the development of the human immune system. *Hum Exp Toxicol* 21:499–505.

West LJ, Pollock-BarZiv SM, Dipchand AI, Lee KJ, Cardella CJ, Benson LN, Rebeyka IM, and Coles JG. 2001. ABO-incompatible heart transplantation in infants. *N Engl J Med* 344:793–800.

Weston WL, Carson BS, Barkin RM, Slater GD, Dustin RD, and Hecht SK. 1997. Monocyte-macrophage function in the newborn. *Am J Dis Child* 131:1241–1242.

White GP, Watt PM, Holt BJ, and Holt PG. 2002. Differential patterns of methylation of the IFN-gamma promoter at CpG and non-CpG sites underlie differences in IFN-gamma gene expression between human neonatal and adult CD45RO- T-cells. *J Immunol* 168:2820–2827.

Wilson CB, Penix L, Weaver WM, Melvin A, and Lewis DB. 1992. Ontogeny of T lymphocyte function in the neonate. *Am J Reprod Immunol* 28(3–4):132–135.

Wright SD. 1999. Toll, a new piece in the puzzle of innate immunity. *J Exp Med* 189940:605–609.

Yap G, Pesin M, and Sher A. 2000. IL-12 is required for the maintenance of IFN-Gamma production in T cells mediating chronic resistance to the intracellular pathogen, *Toxoplasma gondii*. *J Immunol* 165:628–631.

Zinkernagel RM. 2003. On natural and artificial vaccinations. *Annu Rev Immunol* 21:515–546.

# Part II

# Developmental Immunotoxicology: Models

CHAPTER 3

# Risk Assessment Perspectives for Developmental Immunotoxicity

**Carole A. Kimmel, Marquea D. King, and Susan L. Makris**

## CONTENTS



## INTRODUCTION

Risk assessment is a process involving *hazard characterization* (the evaluation of data on the hazard potential of an agent or group of agents), *dose-response assessment* (the dose-response relationship and extrapolation of hazard data to

0-415-28457-0/05/$0.00+$1.50

typically low environmental exposure levels), and *exposure assessment* (the evaluation of sources, pathways, and conditions of exposure, as well as characterization of the populations potentially exposed). Information from these evaluations is integrated to characterize the risk for various outcomes as a result of exposure (*risk characterization*). The risk assessment paradigm was described originally in a landmark report by the National Research Council (NRC 1983), and has been refined in a number of subsequent reports and publications (e.g., NRC 1993a). Since the original NRC report, the U.S. Environmental Protection Agency (EPA) has published risk assessment guidelines in several health-related areas, e.g., developmental toxicity (USEPA 1986a; 1991), reproductive toxicity (USEPA 1996), neurotoxicity (USEPA 1998a), and carcinogenicity (USEPA 1986b; draft 2003a,b). To date, risk assessment guidelines have not been developed for adult or developmental immunotoxicity. The only testing guidelines available on immunotoxicity involve adult exposures, and no protocol has been established for developmental immunotoxicity testing.

In 1993, the NRC published another landmark report on children's environmental health that has spurred a number of legislative and regulatory activities. This report, titled "Pesticides in the Diets of Infants and Children" (NRC 1993b), pointed out the potential vulnerability of children to pesticide exposures, and called for more careful assessment of children's health in risk assessment. The Food Quality Protection Act (FQPA), passed in 1996, echoed many recommendations of the NRC report and required EPA and other agencies that regulate pesticides to specifically consider children's health risks and to add an additional 10-fold safety factor for concerns about children's health, unless data were available to assure protection by setting a different level. In addition, the FQPA called for the development of a battery of tests to screen for the potential estrogenicity of pesticides. The same year (1996), amendments to the Safe Drinking Water Act (SDWA) also emphasized the importance of children's health in setting health advisories for drinking water.

In response to these laws, the EPA established an Endocrine Disruptor Screening and Testing Advisory Committee (EDSTAC) to advise on the development of a screening battery for testing pesticides for endocrine activity. The endocrine disruptor screening battery is currently being developed and validated. The EPA also established an internal 10X Task Force to advise on the application of the additional 10-fold safety factor to protect children, as called for in FQPA (USEPA 1999). In 2002, EPA published its final guidance document on application of the 10-fold safety factor (USEPA 2002a). The guidance document clarified that the 10-fold factor ("FQPA factor") is in addition to the uncertainty factors (UFs) applied to account for interspecies extrapolation and interhuman variability. The FQPA factor overlaps with other UFs, but may sometimes be retained if there are residual uncertainties in the hazard characterization or exposure assessment.

As an outgrowth of the 10X Task Force activities and concerns over the application of these principles to agents other than pesticides, EPA conducted a review of the reference dose and reference concentration (RfD/RfC) process and made a number of recommendations for improving children's health risk assessment in the context of a life stage approach to risk assessment (USEPA 2002b). Recommendations from the 10X Task Force, as well as the RfD/RfC review, include the development of a testing protocol for developmental immunotoxicity and consideration of immunotoxicity data in hazard characterization and in setting RfDs and RfCs.

This chapter provides an overview of developmental immunotoxicity in the context of testing and risk assessment, and discusses possible approaches to incorporating developmental immunotoxicity testing into the regulatory process. The information in this chapter is discussed from the perspective of our experiences within EPA and may or may not be similar to that for other regulatory agencies.

## VULNERABILITY OF THE DEVELOPING IMMUNE SYSTEM

Development of the immune system begins in the early embryo and continues well into postnatal life, the greatest difference between humans and rodents being the stage of development at birth. Studies have demonstrated susceptibility of the immune system to environmental exposures at several developmental stages (Holladay and Luster 1994; Luster et al. 2002). Laboratory mice have been the most extensively used animal models for demonstrating postnatal immune effects following perinatal exposure to agents that affect the immune system. Fetal thymus development has been well characterized in mice, in which thymic colonization occurs in the latter portion of gestation. In humans, the first trimester is a period when both thymic colonization and early T-cell development are found. The limited data available indicate that early stages of T-lymphocyte development may be similar in the two species. Because of the early timing of thymic colonization in humans and involution by the time of birth, efforts to gain sufficient cells for molecular, phenotypic, or functional analysis of fresh cultures have been hindered; thus, experimental animal models have been useful in this regard (Holladay and Luster 1994).

Most of the studies conducted to investigate the effects of chemicals on the immune system have used animals and focused on immunosuppression or impaired host resistance following a subchronic exposure. Immunotoxicity has mainly been studied in adult animals following prenatal or early postnatal exposures and employing multiple immunological assays (van Loveren et al. 1999). Rats are often used because they are the best-defined models in terms of their reproduction and development (Sandler 2002). However, immune assays that were validated in an adult mouse exposure model by the National Toxicology Program (NTP) also have commonly been used to investigate immunotoxicity after developmental exposures (Holladay and Blaylock 2002).

## CHEMICAL SUPPRESSION AND SUSCEPTIBILITY

Chemicals often suppress the developing immune system and its functional capabilities while affecting the adult system with less severity. When given during the maturation of the immune system, several well-known chemicals (e.g., TCDD, lead) are found to be more persistent immunosuppressants than when exposure occurs in adult life. Studies conducted to examine the toxic effects of chemical exposures on the developing immune system are limited (Holladay and Luster 1994).

Environmental agents can affect the immune system in various ways. Effects tend to include immunosuppression, possibly resulting in altered host resistance against infections or neoplastic agents; hypersensitivity or autoimmunity; and uncontrolled proliferation of immune components, i.e., lymphoma or leukemia. Quantification of T cell-dependent antibody responses is known to be an extremely reliable indicator of immunotoxicity and could very well be used to predict changes in host resistance as they develop (Luster et al. 2002).

Cell-mediated immunity (CMI) is considered more sensitive than humoral immunity when assessing appropriate endpoints for developmental immunotoxicity (DIT), although different responses are being measured (Holsapple 2002). Effects on CMI after prenatal or postnatal exposure include suppression of T-cell mitogen responses, skin graft rejection times, graft-versus-host reactivity, cytotoxic T lymphocyte (CTL) activity, and delayed hypersensitivity. Long-lasting postnatal impairment of CMI has been found after chemical exposure during development. Exposure to low levels of TCDD given to pregnant rats at both prenatal and postnatal time points results in severe thymic atrophy and cellular depletion in offspring (Holladay and Luster 1994).

Conceptually, a number of dynamic changes associated with the developing immune system could characterize critical windows for unique susceptibility of the immune system during its ontological development. Major benchmarks in the developing immune system that could potentially aid in the identification of critical exposure windows include: initiation of hematopoiesis; stem cell migration; progenitor cell expansion; thymus and bone marrow colonization; and establishment of immune memory (Dietert et al. 2000). Determinants of susceptibility include the following: exposure age, test species, strain, gender, and assessment age. Species differences in maturation times are variable; therefore, study designs must accommodate these differences to allow for extrapolation of animal data to humans.

The lack of standardization of both *in vivo* and *in vitro* models in current immunotoxicity studies has plagued the scientific community and its ability to measure toxic responses consistently. For example, the *in vitro* CTL assay to alloantigen is thought to be useful to determine developmental immunotoxicity status between 4–6 days after birth, whereas between 11–20 days of age, adult levels are reached (Holladay and Blaylock 2002). These data are useful when attempting to extrapolate laboratory animal or human clinical data to predict human disease from exposure scenarios, but the timing of exposure relative to immune development must be carefully considered. There are several examples of endpoints where this construct has been used to successfully establish human homology for adults, but not for juveniles (Dietert et al. 2000). Whole animal studies at this time are the most appropriate model to use when determining developmental alterations. *In vitro* testing allows for the characterization of a mode of action (MOA), and is often verified and linked with the results of *in vivo* experimentation. These methods, in conjunction with whole animal testing, allow data to be extrapolated from animals to humans at different stages of development. The following chemicals have been found to induce effects developmentally in both *in vitro* and *in vivo* assays.

### Cyclosporin

Cyclosporin A (CsA), a fungal metabolite used in therapeutic immunosuppression for the prevention of human organ transplant rejection, has been found to produce autoimmunity in both rodents and humans (Holladay and Smialowicz 2000). Women who received CsA during their pregnancies often delivered prematurely or had babies that were small-for-gestational-age (SGA) and had birth defects (cataracts, cleft palate, dysmorphic facial appearance, and so on). CsA has been found to prevent activation of CD8+ cytotoxic T-cell precursors by interfering with selective events during critical periods in thymocyte differentiation and clonal deletion. CsA also interferes with deletion of autoreactive T-cells after bone marrow reconstitution which in turn induces a variety of autoimmune responses in adult rodents. After exposure to CsA, fetal thymic organ cultures demonstrated that mature T-cell generation was completely blocked. These studies support the concept that the fetal immune system could be targeted by maternal dosing with CsA (Holladay and Luster 1994). Newborn mice after prenatal and postnatal CsA treatments were also found to display impaired thymocyte differentiation along with an absence of mature T cells in lymph nodes and spleen, and a lack of T cell function. Concluding evidence also predicts that CsA given prenatally may affect the same target organs in the embryo as in the adult and may further cause autoimmune disease later in the life of gestationally exposed individuals (Holladay and Smialowicz 2000).

### TCDD

TCDD (2,3,7,8-tetrachloridebenzo-*p*-dioxin), the most studied chemical for suppression of the immune system, is also a persistent environmental contaminant (Holladay 1999). Although TCDD does not readily cross the placenta (1% of the maternal dose), a single dose or subacute exposure (1.5 or 3.0 μg/kg) causes a very significant involution of the fetal thymus, along with an inhibition of thymocyte differentiation at both prenatal and postnatal time points (Holladay and Smialowicz 2000). Prenatal exposure to TCDD in rodents leads to a reduction in the number of pluripotent stem cells in the bone marrow. It also alters normal maturation and development during ontogenesis of the immune system. TCDD, when given during immune system development, causes more persistent immunosuppression than when exposure occurs during adult life (Holladay and Smialowicz 2000). Long-lasting impairment of cell-mediated immunity (CMI) following TCDD exposure during immune system development causes an increase in susceptibility to challenge with tumor cells and infectious agents (Blaylock et al. 1992). Additional data suggest that TCDD has the potential to induce autoimmune disease in genetically predisposed animals, induce apoptosis in thymocytes, and affect thymic epithelium (Holladay and Luster 1994).

### Lead

Lead (Pb) is the most ubiquitous toxic metal in the environment (Goyer 1996) and has been found to alter T-lymphocyte function and macrophage activity as well as depress mitogenic responses in young rats (seven weeks of age) (Dietert et al. 2000). A shift in the T-helper cell functional balance causes a depression of Th1 and an elevation of Th2 after early exposure in both young rats and in chickens (Dietert et al. 2000; Holladay and Smialowicz 2000). Early developmental lead exposure in these two species caused effects on T-helper cells that altered humoral immune function (Holladay and Smialowicz 2000). In a dose-response study comparing pregnant rats and their female pups exposed *in utero* and assessed postpartum, the developing offspring were found to be susceptible to persistent immunotoxic effects, while, at the same doses, there was no effect on the pregnant dams (Dietert et al. 2000). Preliminary data on lead exposure to chickens also revealed that windows of differential vulnerability exist for lead-induced developmental immunotoxicity.

## CURRENT ASSESSMENT OF THE IMMUNE SYSTEM IN EPA GUIDELINE STUDIES

A number of guideline toxicity studies utilized in the regulation of pesticides, industrial chemicals, and environmental pollutants include assessments of immune system structure and/or function. The majority of these studies are performed in adult animals. Although they were not developed to assess the developing immune system, they can provide valuable information that could indicate the necessity for further testing to evaluate effects on the immune system following developmental exposure. Figure 3.1 demonstrates the immunotoxic endpoints assessed in several typical guideline toxicology studies, in relationship to the life stage of the test animals at the time of study conduct.

### Structural Assessments in Adult Animals

Macro- and/or microscopic structural anatomy of immune system organs and tissues are examined in a number of general guideline screening studies utilized by the U.S. EPA, including the acute inhalation toxicity study with histopathology (40 CFR 799.9135; USEPA 1997), the 90-day subchronic study (OPPTS 870.3100, 870.3150, 870.3250, 870.3465; USEPA 1998c-f), and the chronic/carcinogenicity studies (OPPTS 870.4100, 870.4200, 870.4300; USEPA 1998i-k).

In guideline subchronic and chronic/carcinogenicity studies, an evaluation of macroscopic structure and general qualitative histopathology is only conducted on a few immune system tissues. In subchronic studies that include young adult animals (e.g., rats 45 days to 5 months of age), the spleen, thymus, and lymph nodes from two locations (one near to and the other distal from the site of administration) are examined, and the spleen and thymus are weighed. In chronic and carcinogenicity study guidelines, there is no specification that the thymus be examined and/or weighed. For rodents (e.g., rats or mice 18 months to 2 years of age), it is reasonable to assume that the thymus would have undergone normal age-related atrophy by study termination. However, the thymus would be present at early interim sacrifices of rodents (e.g., at 6 months or 12 months of study) during a long-term study, and

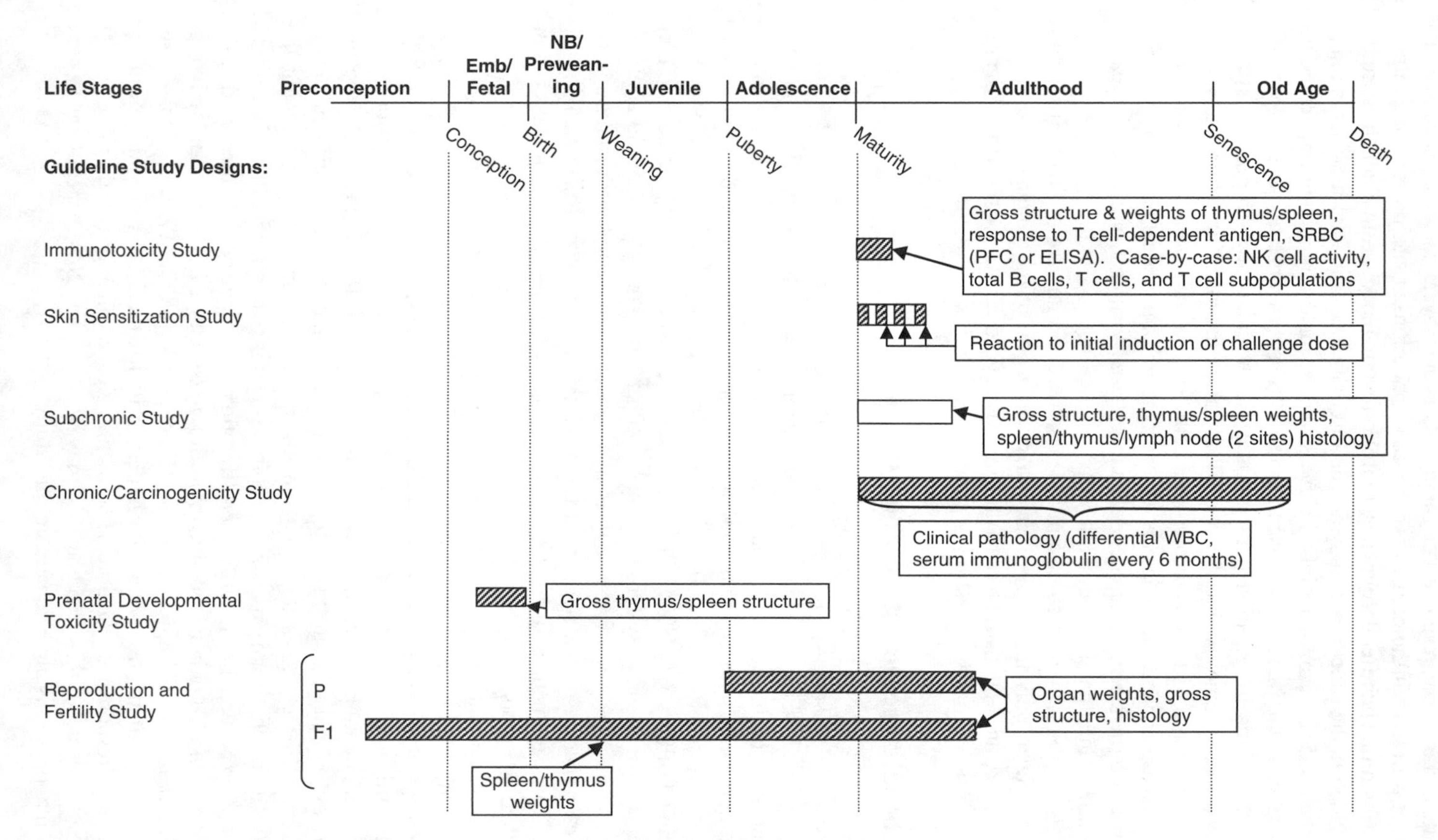


**Figure 3.1** Life stage at which immunotoxicity endpoints are assessed in guideline studies.

it would certainly be present at study termination in a 1-year canine chronic toxicity study (at which point the dogs are young adults of approximately 1.5 years of age).

Differential white cell counts in the circulating blood are examined at study termination in the subchronic study and at approximately 6-month intervals in long-term studies. Serum immunoglobulin levels may be measured at the same intervals. Perturbations may indicate increased immune system response to some unspecified initiator, but this information does not address the adequacy of immune system function. In the same manner, histopathological evaluation of other organ systems in the subchronic and chronic/carcinogenicity studies may identify cellular alterations that are nonspecific indicators of an effect on immune response, for example, the presence of increased numbers of macrophages in lung tissue or an increased incidence of inflammatory dermal lesions.

In the chronic toxicity studies in rodents, aged animals are available for evaluation; however, only indirect evidence of perturbation of the immune system may be observed through macroscopic and microscopic evaluation of various organs. Corollary functional assessment is not performed in this population.

### Functional Assessments in Adult Animals

Indirect assessment of immune system function is conducted in adult animals of various ages via the evaluation of peripheral blood cells and chemistry in subchronic and chronic/carcinogenicity studies. Specific direct functional assessments of the immune system in young adult animals are included only in the skin sensitization study (OPPTS 870.2600; USEPA 1998b) and the immunotoxicity testing guideline (OPPTS 870.7800; USEPA 1998L). These studies do not include a direct assessment of immune system function in either very young or very old animals. They incorporate only a few examples of potential immune system response (e.g., delayed hypersensitivity, humoral immunity, or nonspecific cell-mediated immunity). No assessment of autoimmune disease is conducted in any of the current guideline protocols.

The skin sensitization study has been generally conducted in guinea pigs as a Guinea Pig Maximization Test (GPMT) or a Buehler test. In the future, skin sensitization methods will preferentially include the local lymph node assay (LLNA), which utilizes young adult mice (USEPA, 2001). The skin sensitization study involves an initial intradermal (GPMT) and/or epidermal (Buehler; LLNA) exposure of the test animal to a substance, followed by challenge exposure approximately 1 week later. In the guinea pig tests, sensitization is determined by examining the reaction to the challenge exposure and comparing this reaction with that of the initial induction exposure. In the LLNA, proliferation of lymphocytes is measured (as a function of *in vivo* radioisotope incorporation into cellular DNA) in draining lymph nodes proximal to the application site. Hence, while the GPMT and Buehler tests result in a qualitative assessment of hypersensitivity, the LLNA provides a quantitative dose-response evaluation. Histopathological evaluation of the skin is not required with any of these methods, but may be conducted. No other immune system endpoints or organs are evaluated in this study.

In the guideline immunotoxicity study, young adult rats (6–8 weeks of age) are exposed to the test substance for 28 days, at which time they are terminated. The spleen and thymus are examined macroscopically, and organ weights are recorded; histopathological evaluation is not required. Assessments of immune system function include an evaluation of the response to T cell-dependent antigen, sheep red blood cells (SRBC). The SRBC antigen response assays can be conducted either by an antibody plaque-forming cell (PFC) assay or by immunoglobulin quantification by enzyme-linked immunosorbent assay (ELISA). In addition, an assessment of natural killer (NK) cell activity and/or enumeration of splenic or peripheral blood total B cells, total T cells, and T cell subpopulations may be performed on a case-by-case basis.

## Structural Assessments Following Developmental Exposures

Only two guideline studies include an evaluation of immune system structure following developmental exposures; these are the prenatal developmental toxicity study (OPPTS 870.3700; USEPA 1998g), and the two-generation reproduction study (OPPTS 870.3800; USEPA 1998h).

In the prenatal developmental toxicity study, an evaluation of the macroscopic structure of the thymus and spleen is conducted in at least half of the fetuses from each litter. In the two-generation reproduction study in rats, a macroscopic evaluation of all organ systems is conducted in a sample of offspring at weaning and in the mature adult parental animals at the termination of each generation. Additionally, the spleen and thymus are weighed from those pups that are necropsied at weaning. These measurements are intended to provide additional information to assess the need for further evaluation of the immunotoxic potential of a chemical to an immature animal.

There are a number of obvious gaps in the assessment of the immune system in immature animals in these two studies. The assessments of immune system organs conducted in immature animals in these studies consist entirely of the evaluation of macroscopic changes, with no microscopic examination. Pharmacokinetic data that characterize the exposure in the young (i.e., exposure of the fetus via the placenta or of the neonate via breast milk to the chemical or its metabolites) are not routinely required and are seldom available. Functional assessments are not addressed in these guideline studies. A developmental immunotoxicity guideline, i.e., one that evaluates potential perturbation of immune system function following early pre- and/or postnatal exposure, is in the early stages of development at the EPA. Comparisons of immune effects following exposure at various life stages (i.e., during *in utero* or postnatal development, adulthood, or old age), including data that analyze whether these effects are more severe or persistent in one age group, are not included in standard test batteries. Latent effects on immune function that result from early lifetime exposure are not assessed; these could include effects in aged animals that result from *in utero*, neonatal, or young adult exposure. Exacerbation of effects in relation to aging and response to subsequent immunological challenge are not routinely or systematically assessed. Although the two-generation reproduction study offers an opportunity to evaluate immunotoxic response in adulthood that resulted from prenatal or early postnatal exposure, there is little focus on the evaluation of the immune system in the current guidelines.

## PROPOSED CONTENT OF A DEVELOPMENTAL IMMUNOTOXICITY STUDY

Although the need for developmental immunotoxicity testing has long been recognized by the scientific and regulatory communities (NRC 1993b; USEPA 1999, 2002b), there is currently no formal EPA testing guideline protocol for this study. Studies have, however, been conducted in numerous laboratories to assess the developmental immunotoxic potential for a myriad of pharmaceuticals, pesticides, and environmental pollutants. Additionally, the National Toxicology Program (NTP) has devised an immunotoxicity testing battery (in adult mice) that has demonstrated the value of a combination of two or three immune system tests to achieve >90% predictability of immunotoxic compounds in rodents (Holladay and Luster 1994). The use of multiple tests, both structural and functional, is an approach that is also applicable in assessing immunotoxicity in juvenile animals following developmental exposures. Recent scientific workshops on immunotoxicity conducted in 2001 by the International Life Science Institute (ILSI) (Sandler 2002) and the National Institute of Environmental Health Sciences (NIEHS) (Luster et al. 2002) have discussed the most appropriate experimental design and tests for evaluating the potential developmental immunotoxic effect of chemicals in experimental models.

Certain general attributes have been identified as critical in an adequate developmental immunotoxicity testing battery:

- The study design must accommodate for temporal differences in immune system maturation between rodents and humans, since the level of immune system development and of immunocompetence varies between species during early postnatal life.
- The stage of immune system development at the time of exposure can affect the nature, magnitude, and persistence of a chemically induced immune response. While critical windows of vulnerability to immune system development differ across species, and may be chemically dependent, exposure throughout the prenatal and early postnatal periods should generally be employed in a screening level study to maximize the potential for eliciting a response.
- Endpoints that are assessed should be predictive and sensitive, and well understood biologically. They should be chosen to accommodate the normal temporal aspects of immune system development, and to assess the function and interactions of the major cellular components of the immune system.
- Pharmacokinetic data and information on the mode of action of the chemical should be used to design the study and select dose levels.

A number of specific aspects have been proposed for inclusion in a standardized developmental immunotoxicity study. These are illustrated in Figure 3.2 and listed below:

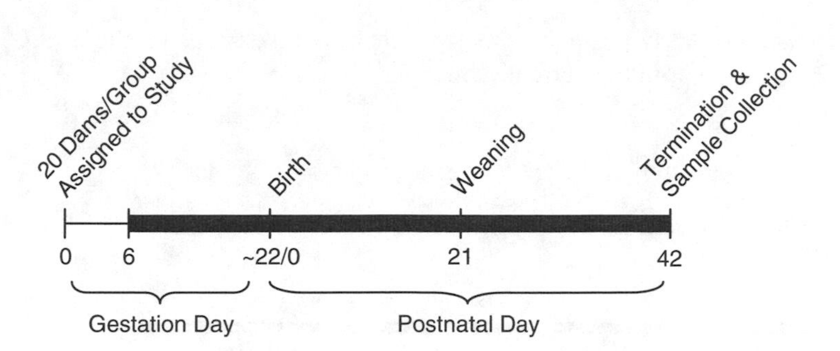


**Figure 3.2** Proposed developmental immunotoxicity study timetable.

*Species*—The rat is generally regarded as the species of choice for developmental immunotoxicity testing for the purpose of risk assessment, since there is extensive information available in the immune system of the rat, a large historical database of reproductive data, and a low background incidence of malformations and stress effects.

*Number of test subjects*—As with all guideline studies that examine developmental toxicity, the litter is considered the primary unit for evaluation. Each test group should contain at least 20 pregnant dams, and from the resulting litters, minimally one pup/sex per litter should be selected for immunotoxicological assessment.

*Exposure period*—The exposure period should cover the entire period of immunological ontogenesis, ranging in the rat approximately from the time of implantation through to the time of sexual maturation, e.g., gestation day (GD) 6 through postnatal day (PND) 42. Evaluation of the immune status in rat pups at a younger age (e.g., PND 21) has been considered as a possible alternative methodology (Holsapple 2002). However, further study is required to verify the sensitivity of this procedure.

*Routes of exposure*—As with most studies conducted for the purpose of risk assessment, the route of exposure should be predicated upon the primary known or potential route of human exposure. No matter what route is employed, the reproduction schedule of the rat will govern some aspects of exposure, while the pharmacokinetic profile of the chemical, including the likelihood for exposure via the maternal placenta or milk, may influence others. The extent of exposure duration through several early life stages in the rodent would require test substance administration to the maternal animal during gestation and in early lactation, direct administration to the offspring during late lactation and postweaning ages, and to the dam and/or directly to the pups during mid- to late lactation. Potential differences between gavage and dietary treatment are shown in Figure 3.3.

*Sampling schedule*—Blood and other tissue samples should be taken at necropsy, on approximately PND 42, when the antibody response in rats approximates that of a mature animal. An evaluation of the long-term consequences of developmental exposures on immune system function (e.g., in aged adult rats) is advisable but may not be practical in a stand-alone developmental immunotoxicity study, since it would require maintaining a subset of animals for an additional year or more.

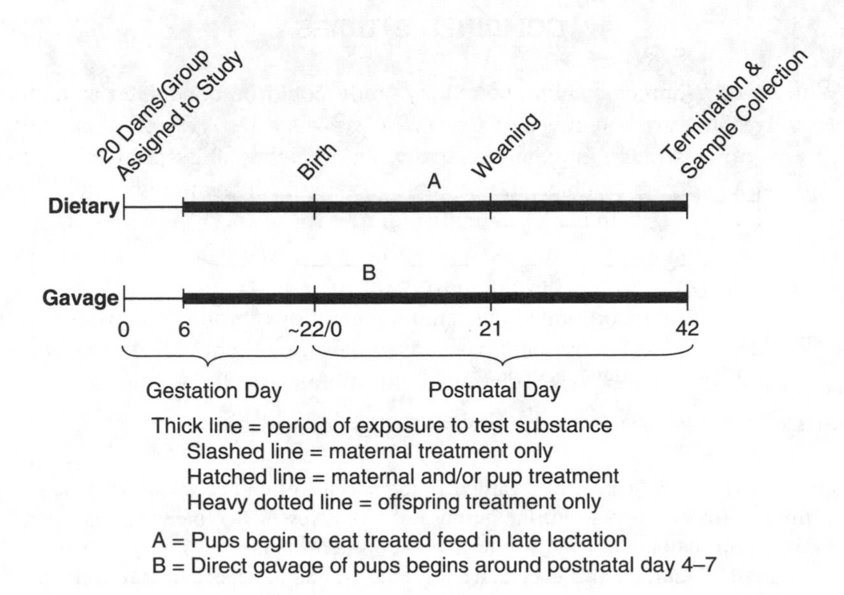


**Figure 3.3** Potential differences between dietary and gavage treatment in a developmental immunotoxicity study.

*Endpoints*—As with other studies that include developmental exposures, a developmental immunotoxicity study should include standard measurements of growth (body weight), survival, clinical response, and macroscopic pathology. Additionally, a number of research efforts in the field of immunotoxicology has demonstrated the sensitivity of the following morphological and functional endpoints in detecting effects on the immune system of young animals following developmental exposures:

- Complete total and differential blood cell count (CBC)
- Thymus, spleen, and lymph node weights
- Evaluation of primary antibody response to a T-dependent antigen (e.g., sheep red blood cell [SRBC])
- Functional test of Th1 immunity (e.g., cytotoxic T lymphocyte [CTL] or delayed hypersensitivity response [DHR/DTH]

Other assays that are commonly used to characterize immunotoxicity in adult animals, including macrophage function (NK cell activity), complement analysis, and cell surface marker expression, are believed to require further evaluation for use in developmental immunotoxicity testing (Luster et al. 2002). Flow cytometric analysis for the frequency and intensity of common surface cell markers is more applicable to mechanistic studies than to hazard characterization studies. Histopathology of lymphoid organs/tissues, including evaluation of lymphocyte subpopulations in the spleen and thymus, and other functional tests, such as host resistance assays, may have a place in second tier testing following an initial confirmation of an effect on the immune system following a developmental exposure.

## COMBINED STUDIES

While a developmental immunotoxicity study could be conducted as a stand-alone study, there are a number of reasons to consider a *combined study* design, which incorporates immunotoxicity endpoints into another study protocol. A combined study avoids the conduct of a separate developmental immunotoxicity study (thereby reducing the number of animals required for testing), provides an opportunity to more efficiently utilize offspring that are already in a study, and often maximizes exposure to the developing organism. For example, immunotoxicity assessments could be incorporated into a multigeneration reproductive toxicity study. This study includes test substance exposure to developing animals during *in utero* and postnatal life stages, coinciding with the ontogeny of the immune system. Publications by Chapin et al. (1997) and Smialowicz et al. (2001) have demonstrated success in conducting such combined studies. Protocol adjustments that must be addressed when attempting to incorporate immunotoxicity testing of offspring into a reproduction study are not extensive. A separate subgroup of either F1 or F2 offspring, other than those already designated as parental animals for the second generation or as subjects for other extensive postmortem or extended in-life evaluations, must be assigned to the developmental immunotoxicity study phase. In order to ensure an adequate population from which to select offspring, it may be necessary to retain five pups/sex/litter (rather than four pups/sex/litter) when standardizing litter size on day 4 of lactation. Developmental immunotoxicity testing requires termination of offspring at PND 42; therefore, pups that are designated for these procedures must be retained on study past the time of weaning. Care should be taken to ensure that exposure to the offspring is adequate throughout the pre- and postnatal dosing periods since lack of exposure during these critical periods of immune system development could limit the study. It is not sufficient to simply rely on an unsubstantiated assumption that test substance administration to the dam automatically results in transplacental exposure to the offspring during gestation and exposure in the milk during lactation (which is an approach that is often taken with multigeneration reproduction studies). The uses of pharmacokinetic data to more fully characterize the dose during each life stage and to refine the exposure methodology, if necessary, could define the difference between a developmental immunotoxicity study that is scientifically defensible and one that is not.

## DEVELOPMENTAL TOXICITY RISK ASSESSMENT

The EPA Guidelines for Developmental Toxicity Risk Assessment (1991) discuss how various types of developmental toxicity data should be considered in hazard characterization and dose-response analysis, and their integration with exposure information in risk characterization. In those guidelines, developmental toxicity is defined as:

> the study of adverse effects on the developing organism that may result from exposure prior to conception (either parent), during prenatal development, or postnatally to the

time of sexual maturation. Adverse developmental effects may be detected at any point in the lifespan of the organism. The major manifestations of developmental toxicity include: (1) death of the developing organism, (2) structural abnormality, (3) altered growth, and (4) functional deficiency.

Although not discussed in any detail in those guidelines, developmental immunotoxicity is considered part of developmental toxicity, as is developmental neurotoxicity or any other type of functional alteration resulting from an exposure during the developmental period (conception to sexual maturity). In the case of developmental neurotoxicity, further guidance is given in the Guidelines for Neurotoxicity Risk Assessment (USEPA 1998a), and the same approach might be considered for development of guidelines for immunotoxicity risk assessment that would include the assessment of potential effects on the immune system at multiple life stages.

Dose-response assessment for health effects that are not assumed to have a linear low-dose response relationship (i.e., most noncancer health effects and carcinogens that act via indirect mechanisms) is conducted by applying various uncertainty factors (UFs) to the no-observed-adverse-effect level (NOAEL) or a benchmark dose lower confidence limit (BMDL) if dose-response modeling can be done. A default 10-fold factor is used for the UFs, and various factors are applied to account for animal to human extrapolation ($UF_A$), within human variability ($UF_H$), the lack of a NOAEL and use of the lowest-observed-adverse-effect level, LOAEL ($UF_L$), use of a subchronic study to set a chronic reference value when no chronic study is available ($UF_S$), and a database factor ($UF_D$) to account for missing data that are considered essential in characterizing risk. If the exposure is by inhalation, the animal to human extrapolation UF is divided into pharmacokinetic and pharmacodynamic components. EPA has developed methodology for deriving appropriate conversions for inhalation exposures to account for the pharmacokinetic component of the extrapolation (USEPA 1994), so that the $UF_A$ applied as a default in the case of inhalation exposures is $10^{1/2}$. An additional 10-fold FQPA factor is applied for pesticides, but is acknowledged to overlap with several of the traditional UFs; thus, the FQPA factor may be reduced or removed, depending on the quality of the data on children's health and exposure, e.g., residual concerns about controlling exposures to children may result in retention of the FQPA factor (USEPA 2002a).

## THE USE OF DEVELOPMENTAL IMMUNOTOXICITY DATA IN RFD/RFC DERIVATION

Developmental immunotoxicity testing is expected to provide hazard and dose response information that can be utilized in formulating risk assessment decisions for any chemical. Adverse outcomes identified through developmental immunotoxicity testing, and their associated NOAELs or LOAELs, are considered appropriate for use in establishing endpoints and doses for risk assessment, particularly for any population group that includes fetuses, infants, or children. It is important to compare information on the immune response of adult animals with that of the immature individual, as obtained through developmental immunotoxicity testing. When the database contains evidence for concern regarding toxicity to adult immune system structure or function, and no data are available that assess developmental immunotoxic potential, the resulting uncertainties should be accounted for in the risk assessment. A database uncertainty factor applied to RfD/RfC derivation can be used to address this deficiency. The need for and the magnitude of such an uncertainty factor should be based upon an assessment of the chemical database, considering the weight-of-the-evidence that was used in arriving at the conclusion that a developmental immunotoxicity study would be required.

As mentioned earlier, the EPA has recently released a review of the RfD and RfC processes (USEPA 2002b) that calls for setting reference values not only for chronic exposure, but also for less-than-lifetime durations of exposure. These include acute (<24 hours), short-term (up to 30 days), and longer-term (up to 10% of the lifespan) exposures. The review also calls for a life-stage approach to the assessment of data, and particularly to the identification of potential data gaps that should be considered in applying uncertainty factors. Developmental immunotoxicity is one of many developmental endpoints that would be considered in setting reference values for various durations of exposure. In this process, various endpoints considered relevant and having sufficient data are carried through the process of sample reference value derivation. Then a final reference value is derived for each duration of exposure, based on a weight of evidence characterization of the database available, the degree of uncertainty in the database as a whole, and those endpoints and sample reference values considered to be protective of the exposed population including susceptible subgroups. Latent effects are important to consider in the derivation of these less-than-lifetime values, as the definition of these reference values includes the concept that they are to be protective of effects that may occur over a lifetime. Although these approaches have yet to be implemented, some offices within EPA have been setting various duration reference values for a number of years, e.g., the Office of Pesticide Programs (OPP) has established acute and chronic reference doses for pesticides since 1998, and the Office of Water (OW) has set acute, short-term, longer-term, and chronic health advisories for water contaminants for a number of years. EPA's Integrated Risk Information System (IRIS) program is implementing a pilot effort to evaluate the practicability of setting various duration reference values for inclusion in the IRIS database.

Once reference values have been derived, this information is integrated with the exposure assessment to characterize the overall risk. Risk characterization summarizes all of the information that has been evaluated, and develops an estimate of risk or a comparison of the toxicity data with estimated human exposure. This ratio of the NOAEL, LOAEL, or BMDL to the human exposure estimate is called the Margin of Exposure (MOE). Consideration of whether the size of the MOE is adequate is a risk management decision that includes economic, social, political and other regulatory factors, but the same issues that go into determining the size of the uncertainty factors applied in the reference value process should be considered in judging the adequacy of the MOE. A detailed description of the considerations in risk characterization can be found in the Handbook on Risk Characterization (USEPA 2000).

## SUMMARY

Several recent reviews of the adequacy of current risk assessment practices, particularly in regard to the protection of the fetus and child from adverse effects of exposure to pesticides and environmental pollutants, have resulted in recommendations for a more robust assessment of potential developmental immunotoxicity. The vulnerability of the developing immune system has been well established. Perturbations in immune system structure and function following exposure during critical windows of development have been demonstrated in animal models, thus supporting the need to evaluate the potential for developmental immunotoxicity. Current EPA testing guidelines primarily provide only indirect measures of immune system toxicity, focusing on structural alterations. Only a few guidelines address immune system function, and there is no guideline that directly assesses immunotoxicity following developmental exposures. Consequently, the stage is set for formulating a test guideline protocol for developmental immunotoxicity. Such a guideline could be used to conduct a more thorough assessment in cases where other data, derived for example from other available guideline studies or from the peer-reviewed literature, are suggestive of a potential effect on the developing immune system. Since situations of this nature have already arisen at EPA, the need for a standardized developmental immunotoxicity guideline has even more urgency.

The use of developmental immunotoxicity data in risk assessment is discussed. Consideration of developmental immunotoxic effects in testing and risk assessment is particularly important for determining whether children's health is adequately protected, or whether data deficiencies exist that can impact the size of the uncertainty factors applied to the RfD or RfC, or in the case of pesticides, whether the database is adequate to justify the retention or reduction of the FQPA safety factor. Not all chemicals can or should be tested in depth for every type of developmental exposure and effect, and new and innovative strategies are needed for determining when such testing should be done. But immunotoxic effects in children can severely impair quality of life and have been demonstrated to result from environmental exposures. Therefore, they cannot be ignored in the evaluation of children's environmental health risks.

## DISCLAIMER

The views expressed in this chapter are those of the authors and do not necessarily reflect the views or policies of the U.S. Environmental Protection Agency.

## REFERENCES

Blaylock BL, Holladay SD, Comment CE, Heindel JJ, and Luster MI. 1992. Exposure to tetrachlorodibenzo-p-dioxin (TCDD) alters fetal thymocyte maturation. *Toxicol Appl Pharmacol* 112:207–213.

Chapin RE, Harris MW, Davis BJ, Ward SM, Wilson RE, Mauney MA, Lockhart AC, Smialowicz RJ, Moser VC, Burka LT, and Collins BJ. 1997. The effects of perinatal/juvenile methoxychlor exposure on adult rat nervous, immune, and reproductive system function. *Fundam Appl Toxicol* 40:138–157.

Dietert RR, Etzel RA, Chen D, Halonen M, Holladay SD, Jarabek AM, Landreth K, Peden DB, Pinkerton K, Smialowicz RJ, and Zoetis T. 2000. Workshop to identify critical windows of exposure to children's health: immune and respiratory systems work group summary. *Environ Health Perspect* 108 (Suppl.3):483–490.

Dietert RR, Lee JE, and Bunn TL. (2002) Developmental immunotoxicology: emerging issues. *Hum & Experimental Toxicol* 21:479–485.

Goyer RA 1996. Toxic effects of metals. In: Klaassen, CD (Ed.), *Casarett & Doull's Toxicology: The Basic Science of Poisons*, 5th edition. McGraw-Hill, Health Professions Division, New York, NY.

Holladay SD. 1999. Prenatal immunotoxicant exposure and postnatal autoimmune disease. *Environ Health Perspect* 107:687–691.

Holladay SD and Luster MI. 1994. Developmental immunotoxicology. In: Kimmel CA and Buelke-Sam J, (Eds.), *Developmental Toxicology*, 2nd Ed. Raven Press, New York, pp. 93–118.

Holladay SD and Smialowicz RJ. (2000) Development of the murine and human immune system: differential effects of immunotoxicants depend on time of exposure. *Environ Health Perspect* 108: 463–473.

Holladay SD and Blaylock BL. 2002. The mouse as a model for developmental immunotoxicology. *Hum & Experimental Toxicol* 21:525–531.

Holsapple MP. 2003. Developmental immunotoxicity testing: a review. *Toxicology* 185:193–203. Available at http://www.sciencedirect.com.

Luster MI, Dean JH, and Germolec DR. 2003. Meeting report: Consensus workshop on methods to evaluate developmental immunotoxicity. *Environ Health Perspect* 111:579–583. doi:10.1289/ehp.5860 [Online 26 November 2002]. Available via http://dx.doi.org/.

NRC (National Research Council). 1983. *Risk Assessment in the Federal Government: Managing the Process*. National Academy Press, Washington, D.C.

NRC (National Research Council). 1993a. *Issues in Risk Assessment*. National Academy Press, Washington, D.C.

NRC (National Research Council). 1993b. *Pesticides in the Diets of Infants and Children*. National Academy Press, Washington, D.C.

Sandler D (Ed.). 2002. Developmental immunotoxicology and risk assessment; special issue. *Hum & Experimental Toxicol* 21(9–10):469–572.

Smialowicz RJ, Williams WC, Copeland CB, Harris MW, Overstreet D, Davis BJ, Chapin RE. 2001. The effects of perinatal/juvenile heptachlor exposure on adult immune and reproductive system function in rats. *Toxicol Sci* 61:164–175.

USEPA (U.S. Environmental Protection Agency). 1986a. Guidelines for the health assessment of suspect developmental toxicants. *Fed Regist* 51(185):34028–34040. Available at http://www.epa.gov/ncea/raf/.

USEPA (U.S. Environmental Protection Agency). 1986b. Guidelines for carcinogen risk assessment. *Fed Regist* 51(185):33992–34003. Available at http://www.epa.gov/ncea/raf/.

USEPA (U.S. Environmental Protection Agency). 1991. Guidelines for developmental toxicity risk assessment. *Fed Regist* 56(234):63798–63826. Available at http://www.epa.gov/ncea/raf/.

USEPA (U.S. Environmental Protection Agency). 1994. *Methods for Derivation of Inhalation Reference Concentrations and Application of Inhalation Dosimetry*, EPA/600/8-90/066F. Available from the U.S. EPA Risk Information Hotline at telephone 1-513-569-7254, fax 1-513-569-7159, email RIH.IRIS@epamail.epa.gov and/or the National Technical Information Service, Springfield, VA 22161, (703) 605-6000, http://www.ntis.gov.

USEPA (U.S. Environmental Protection Agency). 1996. Guidelines for reproductive toxicity risk assessment. *Fed Regist* 61(212):56274–56322. Available at http://www.epa.gov/ncea/raf/.

USEPA (U.S. Environmental Protection Agency). 1997. Toxic substances control act test guidelines. *Fed Regist* 62(158):43819–43864. Available at http://www.epa.gov/EPA_TOX/1997/August/Day_15/t21413.htm

USEPA (U.S. Environmental Protection Agency). 1998a. Guidelines for Neurotoxicity Risk Assessment. *Fed Regist* 63(93):26926–26954. Available at http://www.epa.gov/ncea/raf/.

USEPA (U.S. Environmental Protection Agency). 1998b. OPPTS 870.2600, *Skin Sensitization, Health Effects Test Guidelines*, EPA 712–C–98–197, Washington, DC. Available at http://www.epa.gov/oppts-frs/OPPTS_Harmonized/870_Health_Effects_Test_Guidelines.

USEPA (U.S. Environmental Protection Agency). 1998c. OPPTS 870.3100, *90-Day Oral Toxicity in Rodents, Health Effects Test Guidelines*, EPA 712–C–98–199, Washington, DC.

USEPA (U.S. Environmental Protection Agency). 1998d. OPPTS 870.3150, *90-Day Oral Toxicity in Nonrodents, Health Effects Test Guidelines*, EPA 712–C–98–200, Washington, DC.

USEPA (U.S. Environmental Protection Agency). 1998e. OPPTS 870.3250, *90-Day Dermal Toxicity, Health Effects Test Guidelines*, EPA 712–C–98–202, Washington, DC.

USEPA (U.S. Environmental Protection Agency). 1998f. OPPTS 870.3465, *90-Day Inhalation Toxicity, Health Effects Test Guidelines*, EPA 712–C–98–204, Washington, DC.

USEPA (U.S. Environmental Protection Agency). 1998g. OPPTS 870.3700, *Prenatal Developmental Toxicity Study, Health Effects Test Guidelines*, EPA 712–C–98–207, Washington, DC.

USEPA (U.S. Environmental Protection Agency). 1998h. OPPTS 870.3800, *Reproduction and Fertility Effects, Health Effects Test Guidelines*, EPA 712–C–98–208, Washington, D.C..

USEPA (U.S. Environmental Protection Agency). 1998i. OPPTS 870.4100, *Chronic Toxicity, Health Effects Test Guidelines*, EPA 712–C–98–210, Washington, DC.

USEPA (U.S. Environmental Protection Agency). 1998j. OPPTS 870.4200, *Carcinogenicity, Health Effects Test Guidelines*, EPA 712–C–98–211, Washington, DC.

USEPA (U.S. Environmental Protection Agency). 1998k. OPPTS 870.4300, *Combined Chronic Toxicity/Carcinogenicity, Health Effects Test Guidelines*, EPA 712–C–98–212, Washington, DC.

USEPA (U.S. Environmental Protection Agency). 1998L. OPPTS 870.7800, *Immunotoxicity, Health Effects Test Guidelines*, EPA 712–C–98–351, Washington, DC.

USEPA (U.S. Environmental Protection Agency). 1999b. *Toxicology Data Requirements for Assessing Risks of Pesticide Exposure to Children's Health - Report of the Toxicology Working Group of the 10X Task Force* (April 28, 1999 draft). Available at http://www.epa.gov/scipoly/sap/1999/index.htm#may.

USEPA (U.S. Environmental Protection Agency). 2000. *Science Policy Handbook on Risk Characterization*, EPA 100–B–00–002. Available at http://www.epa.gov/ORD/spc/2riskchr.htm.

USEPA (U.S. Environmental Protection Agency). 2001. *Scientific Applicability for Use of an Alternative Test Guideline for Dermal Sensitivity.* FIFRA Scientific Advisory Panel review, December 11, 2001. Available at http://www.epa.gov/scipoly/sap/2001/index.htm.

USEPA (U.S. Environmental Protection Agency). 2002a. *Determination of the Appropriate FQPA Safety Factor(s) for Use in the Tolerance-Setting Process.* Office of Pesticide Programs, Office of Prevention, Pesticides, and Toxic Substances. Available at http://www.epa.gov/oppfead1/trac/science/#10_fold.

USEPA (U.S. Environmental Protection Agency). 2002b. *A Review of the Reference Dose and Reference Concentration Processes.* External review draft, EPA/630/P–02/002A, May, 2002. Risk Assessment Forum, U.S. Environmental Protection Agency, Washington, D.C. National Center for Environmental Assessment, Office of Research and Development, Washington, D.C. Available at http://www.epa.gov/ncea/raf.

USEPA (U.S. Environmental Protection Agency). 2003a. *Draft Guidelines for Carcinogen Risk Assessment.* Available at http://www.epa.gov/ncea/raf/.

USEPA (U.S. Environmental Protection Agency). 2003b. *Draft Supplemental Guidance for Assessing Cancer Susceptibility from Early-Life Exposure to Carcinogens.* Available at http://www.epa.gov/ncea/raf/.

van Loveren H, Piersma AH, de Jong WH, and de Waal EJ. 1999. Developmental immunotoxicity of diazepam in prenatally treated weanling Wistar rats. Rijksinstituut voor Volkgesondheid en Millieu (RIVM), *Rapport* 640080001, pp.1–24. Available at http://www.rivm.nl/bibliotheek/rapporten/640080001.html.

CHAPTER 4

# The Mouse as a Model for Developmental Immunotoxicology

M. Renee Prater, Benny L. Blaylock, and Steven D. Holladay

## CONTENTS



## INTRODUCTION

The field of developmental immunotoxicology is relatively new. Nonetheless, many reports are available to show that prenatal exposure to immunotoxicants may cause transient to persistent postnatal immune suppression (reviewed in: Holladay and Smialowicz 2000); more recent studies suggest developmental immunotoxicant exposures may exacerbate postnatal aberrant immune responses including hypersensitivity disorders or autoimmune disease (see Chapters 13 and 14 of this book). Although important contributions to the literature have been made regarding such effects of prenatal chemical exposure on postnatal and adult immunocompetence, many significant information gaps remain, especially in the area of human immune health.

The laboratory mouse has been the most studied model for immunotoxicity research. The availability of several strains of inbred and outbred mice, strain crosses, gene deletion (knock-out) and gene insertion (transgenic) models has been of critical importance for the study of immune alterations following chemical exposures. This

0-415-28457-0/05/$0.00+$1.50

has particularly been true when specific mechanisms were examined by which immunotoxicants impair immune function, including how these relate to transient, long-lasting, or permanent effects in the mouse. It is also important to note that most of the common immune assays used for studying chemically induced immunotoxicity were developed in adult mice. It must be acknowledged, however, that limited information is available regarding predictive strength of these assays for immunotoxicity detection in subadult mice, or adult mice following perinatal immunotoxicant challenges. Further, it is not known how data collected thus far in adult mouse models, intended for risk assessment statements and immunotoxicity predictions, can be compared or extrapolated to other laboratory animals or to humans.

## NATIONAL TOXICOLOGY PROGRAM TESTS IN MICE

More than 10 years ago, a massive multilaboratory study was undertaken in the mouse to examine quantitative and qualitative indicators of immune suppression following chemical exposure. This National Toxicology Program (NTP)-sponsored effort had as its primary objective the development of a testing paradigm that could accurately detect chemical immunotoxicants (Luster et al. 1988) (Table 4.1). Approximately 50 chemicals were evaluated, usually at three dosage levels, in young adult female B6C3F1 mice. These mice are a cross between C57Bl/6N female and C3H/J male mice. The collective data were utilized to calculate the predictive value of eleven immune tests, and combinations of these tests, to aid in quantifying risk assessment for immunotoxicity. Results from this extensive study demonstrated that employment of two or three select immune tests can be sufficient to predict the immunotoxicity of chemicals in rodents (Luster et al. 1992). The highest associations for immunotoxicity detection included the ability to produce specific antibody after challenge with allogeneic (sheep) red blood cell antigen (78%) and splenic lympho-

**Table 4.1 National Toxicology Program Immune Tests**

| Test | |
|---|---|
| Splenic leukocyte enumeration | 0.83 |
| Plaque forming cell assay | 0.78 |
| Natural killer cell activity | 0.69 |
| Thymus/body weight ratio | 0.68 |
| Cytotoxic T lymphocyte activity | 0.67 |
| T cell mitogen response | 0.67 |
| Spleen/body weight ratio | 0.61 |
| Delayed type hypersensitivity response | 0.57 |
| Mixed lymphocyte response | 0.56 |
| Spleen cellularity | 0.56 |
| B cell mitogen (LPS) response | 0.50 |

*Source*: Modified from Luster MI, Portier C, Pait DG, White KL Jr, Gennings C, Munson AE, and Rosenthal GJ. 1992. Risk assessment in immunotoxicology. I. Sensitivity and predictability of immune tests. *Fundam Appl Toxicol* 18, 200–210.

cyte quantitation as determined by surface marker expression (83%). An association between immunotoxicity, carcinogenicity, and genotoxicity of the test chemicals further suggested relationships between immune tests and nonimmune, but immune-related, toxic effects of the chemical exposures.

A major contributor to the strength of the mouse as a model for immunotoxicity testing is the existence of the NTP risk assessment database. Admittedly, it is not currently known how closely the studied panel of immune tests (predictive values) relate to risk assessment in other strains of mice, different strain crosses, or different species. Other strains are nonetheless widely used for immunotoxicity testing, for example, inbred C57BL/6N (Johnson et al. 1990), LAF-1 (Goidl et al. 1976), and BALB/c (Barnett et al. 1985) mice and outbred Swiss albino (ICR) mice (Das et al. 1990). Data collected in these strains appear representative of those collected in B6C3F1 mice. However, comparison of chemically induced immunomodulation results between such different strains of mice is sometimes difficult, as variations in the genetic immunosensitivities amongst murine strains have not been clearly defined.

To minimize intersubject variability in data due to genetic diversity, inbred strains of adult mice have most commonly been used for immunotoxicity studies. Many of these inbred strains of mice are poor breeders and expensive as compared to outbred mice. In part for this reason, reproductive toxicity testing and developmental (teratology) studies have historically been conducted using outbred rodents. A good example of this is the NTP reproductive assessment by continuous breeding (RACB) paradigm, which uses outbred mice and rats to identify reproductive toxicants (Chapin and Sloan 1997). It should be considered that developmental immunotoxicity studies are teratology experiments by design, and for this reason outbred mice may have utility for studies aimed simply at identifying developmental immunotoxicants (Holladay and Blaylock 2002). Advantages to an outbred mouse model would include improved reproductive performance and considerably lower cost, the latter of which would extend model availability to more laboratories and, as such, presumably result in more rapid expansion of the developmental immunotoxicology literature database. There is precedent for use of timed-pregnant ICR mice in developmental immunotoxicity studies, with results suggesting excessive variability in data was not a major problem (Das et al. 1990; Holladay and Smith 1995). The latter studies suggest that a traditional teratology mouse may be useful for developmental immunotoxicity risk assessment. It is likely, but not verified, that NTP tests developed for immunotoxicity detection in adult female B6C3F1 mice will show similar predictive values in such other mouse strains. It remains to be demonstrated that predictive values calculated in adult mouse exposure models will be of approximately equal strength for detecting postnatal immune deficits in the offspring of maternally-exposed animals. Indeed, such may not be the case. As already mentioned, depression of the PFC assay in adult mice has an estimated predictive value of 0.78 for decreased resistance to immunologic challenge. Such immunologic challenge was in most cases administered shortly after the immunotoxicant exposure that depressed the PFC response. In contrast, if a developmental immunotoxicant exposure causes depression of the PFC assay in offspring mice at greater than 6 weeks of age (i.e., the time when this response reaches adult levels), such depression

**Table 4.2 Maturational Landmarks of Immune Development in the Mouse**

| Maturational Landmark | Time of Development | Ref. |
|---|---|---|
| Fetal liver hematopoiesis first detected | GD 10 | The 1993 |
| Pluripotent stem cells seed bone marrow | GD 11 | The 1993 |
| Lymphocyte precursors in thymus | GD 12 | The 1993 |
| SIg expression on B cells | GD 17 | Velardi and Cooper 1984 |
| Mitogen responsive thymocytes | GD 17 | Santoni et al. 1982 |
| T-independent Ab responses: | | |
| First detected | PNW 1 | Tyan 1981 |
| Adult level | PNW 2-3 | Tyan 1981 |
| T-dependent Ab responses: | | |
| First detected | PNW 2 | Tyan, 1981 |
| Adult level | PNW 6-8 | Tyan, 1981 |
| Functional NK cells | PNW 3? | Santoni et al. 1982 |

*Note:* Abbreviations: GD = gestation day; PNW = postnatal week.

*Source*: Modified from Holladay SD and Blaylock BL. 2002. The mouse as a model for developmental immunotoxicology. *Human Exper Toxicol* 21:525–532.

may be more likely to represent a permanent immunologic change, and may carry a different predictive value for immune suppression.

## WINDOWS OF SENSITIVITY

Several studies have identified maturational landmarks of immune development in the mouse (Table 4.2) (see Chapter 1). Major information gaps exist regarding the significance of maternal chemical exposure during these times of potential differential sensitivity within maturational stages. For example, offspring of pregnant mice dosed with 150 mg/kg benzo(a)pyrene from days 11 to 17 of gestation showed life-long suppression of antibody, graft-vs.-host, and mixed lymphocyte responses (Urso and Gengozian 1984). It is not known how these results may vary with different dosing schedules, e.g., lower doses given early in gestation or a single-dose exposure on gestation day 10, just prior to development of the thymic rudiment. Beyond chlordane (see Chapter 9) and, to some extent, lead (see Chapter 10), little has been determined about how such temporal changes in chemical exposures during gestation may differentially result in transient, prolonged, or permanent alterations of the progeny's immune responses.

## TIMING OF IMMUNE ASSAYS FOR DETECTING DEVELOPMENTAL IMMUNOTOXICITY

Development of the immune system has been more studied and better defined in the mouse than any other species (see Chapter 1). With this in mind, many immune function tests utilized in adult mouse models are not valid in perinatal mice simply because of the immaturity of their immune systems. Therefore, an adult mouse immunotoxicity risk assessment screening battery must be modified to be age-appropriate, if it is to be used for detecting postnatal immune deficits following prenatal chemical exposure. Several NTP assays are not functional assays, e.g., splenic cellularity, splenic and thymic organ to body weight ratios, leukocyte counts, and splenic lymphocyte surface antigen expression. These assays may have utility for detecting developmental immunotoxicity at very early ages, or even prior to birth (Holladay and Luster 1996). The considerable information available in the mouse as to when different arms of the immune system become functional at an adult level adds strength to the mouse for immunotoxicity testing, and will need to be considered when employing NTP functional assays for developmental immunotoxicity screening. These functional assays include antibody production, cytotoxic T lymphocyte activity, delayed type hypersensitivity response, natural killer cell activity, T and B lymphocyte mitogen assays, and the mixed lymphocyte response.

The individual assay that afforded the highest predictive value (0.78) for immunotoxicity detection in adult mice was the antibody plaque forming cell (PFC) assay (Luster et al. 1992). This assay measures ability to produce specific antibody following challenge with allogeneic (sheep) red blood cells. The PFC assay has been used to demonstrate developmental immunotoxicity, for instance mice exposed to the nonsteroidal estrogen diethylstilbestrol (DES) during the first five days of life displayed inhibited PFC response to sheep red blood cells into adulthood (Kalland 1980). Exposure of mice to benzo[a]pyrene during gestation similarly resulted in a profound and persistently depressed PFC response in the progeny that was still present 18 months after birth (Urso and Johnson 1987). However, it is not certain how early in life the PFC assay or a PFC-like assay may be useful in mice for detecting immunotoxicity. B lymphocytes collected from fetal mouse livers acquire the capacity to produce plaques against allogeneic antigens at about days 16-18 of gestation. These animals are then capable of producing a heterogeneous or "adult-type" response to normal immunization between one and two weeks after birth (Goidl et al. 1976). Performing the traditional PFC assay in two week old mice is therefore possible; however, an adult-level antibody response is not seen until mice reach about 6 weeks of age and is required for maximal predictive value of the assay (Holladay and Blaylock 2002). Nonetheless, the demonstration of altered capacity of late-gestation fetal liver B lymphocytic cells to produce plaques against allogeneic antigens could prove to be a useful predictor of developmental immunotoxicity.

The cytotoxic T lymphocyte (CTL) assay is a measure of cell-mediated immunity that has been widely used in adult mice. This test has an estimated individual predictive value for immune suppression of 0.67 in mice, and, when codepressed with the PFC assay, was found to have 100% predictability for immunosuppression (Luster et al. 1992). One study is available that showed diminished CTL activity in 8-week old mice that were exposed 2,3,7,8-tetrachlorodibenzo-*p*-dioxin (TCDD) from days 6-14 of gestation (Holladay et al. 1991). Beyond this report, there are limited additional data regarding the use of the CTL assay to detect developmental immunotoxicity, including how early this assay might first be useful. *In vitro* CTL responses to alloantigen can be detected as early as 4 to 6 days after birth in mice, but do not reach adult levels until between 11 to 20 days of age (Sarzotti et al. 1996). Thus, CTL assays that are performed prior to postnatal day 20 in mice may provide useful immunotoxicity testing information, but, at the least, will need to be compared to age-matched controls.

The delayed-type hypersensitivity (DTH) response is a second measure of cell mediated immunity that has proved useful for detecting long-lasting effects of developmental immunotoxicant exposure. For instance, depressed DTH was demonstrated in mice 101 days after birth following prenatal exposure to chlordane (Barnett et al. 1985). This assay again may be useful in prepubertal mice. Full expression of DTH requires antigen processing and presentation by bone marrow-derived $CD1a^{+}$ epidermal Langerhans cells (LC). These LC begin to populate the mouse skin on day 19 of gestation, but are not fully functional until one week after birth (Elbe et al. 1989). Das et al. (1990) reported decreased DTH responses in very young mice at postnatal day 10, following maternal hexachlorocyclohexane exposure. It is not known if agents like TCDD or chlordane may similarly affect the DTH response in mice at similar early ages, or how such early depression of DTH might correlate to persistent effects on DTH caused by these agents.

Natural killer (NK) cell ability to lyse target cells also has a high predictive value (0.69) for risk of immunotoxicity in adult mice. Prolonged depression of NK cell activity as an indicator of developmental immunotoxicity has been demonstrated in mice exposed to low or moderate doses of steroidal and nonsteroidal estrogens during gestation (Kalland and Forsberg 1981; Lanier 1998). NK cells express the necessary receptors during development that control both positive and negative effector functions, including LY-49, NKR-P1 and CD94/NKG2 (Lanier 1998; Sivakumar et al. 1999). Late in gestation, NK1.1+ lymphocytes collected from mouse fetal liver possess the ability to lyse MHC class I-deficient targets (Sivakumar et al. 1997, Toomey et al. 1998). These data suggest that NK cell activity in perinatal mice might be employed as an early indicator of developmental immunotoxicity.

Lymphocyte mitogenesis assays have a history of common use in immunotoxicologic studies and are part of the NTP testing battery. These include T lymphocyte proliferative ability in the presence of mitogens concanavalin A (ConA) and phytohemagglutinin A (PHA), and B lymphocyte mitogenesis in the presence of lipopolysaccharide (LPS). Lymphocyte proliferation after mitogen stimulation has also been applied to developmental immunotoxicology experiments, with altered proliferation reported in adult mice that were exposed *in utero* to immunotoxicants (Luster et al. 1979; Ways et al. 1980; Spyker-Cranmer et al. 1982). Similar use of lymphocyte proliferation assays in subadults is limited. In one study, offspring of pregnant mice exposed to hexachlorocyclohexane showed enhanced splenic T and B lymphocyte mitogenesis in response to ConA and LPS, respectively, at postnatal day 10 (Das et al. 1990).

The age at which lymphocyte proliferation assays can first be effectively utilized to detect immunotoxicity again has not been well characterized; however, several reports indicate that proliferative ability in mouse lymphocytes is acquired early in life. Fetal mouse thymocytes respond comparably to adult thymocytes, to the mitogen PHA or in an MLR, by day 17 of gestation; and to the mitogen ConA at about 2 to 3 weeks postnatally (Mosier 1977). Mouse fetal liver B lymphocytes respond similarly to adult B lymphocytes to LPS stimulation beginning at about 14 to 16 days gestation (Goidl et al. 1976). These results suggest that T and B cell mitogenesis

**Table 4.3 Immunotoxicants Producing Fetal Thymic Atrophy**

| Immunotoxicant | Exposure Regimen | Ref. |
|---|---|---|
| 2,3,7,8-tetrachlorodibenzo-*p*-dioxin (TCDD) | 10 µg/kg, GD 14 | Fine et al. 1989; Holladay et al. 1991 |
| 3,3',4,4'-tetrachlorobiphenyl (TCB) | 6–16 mg/kg, GD 12 | d'Argy et al. 1987 |
| diethylstilbestrol (DES) | 3–8 µg/kg, GD 10–16 | Holladay et al. 1993a |
| ethylene glycol monomethyl ether (EGME) | 100–200 mg/kg, GD 14–17 | Holladay et al. 1994 |
| benzo(a)pyrene | 50–150 mg/kg, GD 14–17 | Holladay and Smith 1994 |
| 7,12-dimethylbenzanthracene (DBMA) | 10–25 mg/kg, GD 14–17 | Holladay and Smith 1995 |
| T2 mycotoxin (T2 toxin) | 1.2–1.5 mg/kg, GD 14–17 | Holladay et al. 1993b |

*Source*: Modified from Holladay SD and Luster MI. 1996. Alterations in fetal thymic and liver hematopoietic cells as indicators of exposure to developmental immunotoxicants. *Environ Health Perspect* 104 Suppl 4, 809–813.

assays have potential for use as indicators of immunotoxicity beginning in late gestation, with data again compared to age-matched controls.

The most widely used and technically most straightforward measurements of developmental immunotoxicity include fetal thymus and liver weights, and cell counts in these organs. Organ weight and cellularity data have generally not been considered to be among the most sensitive indicators of immune competence, with the exception of the relatively useful measurement of thymus/body weight ratio, which has an estimated predictive value for immunosuppression in adult mice of 0.68 (Luster et al. 1992). Table 4.3 summarizes studies where fetal thymic weight or cell counts were reduced after immunotoxicant exposure, showing the utility of this measurement.

Developmental immunotoxicity studies have also made use of flow cytometry to examine alterations in marker expression in fetal thymus and liver leukocytes. Fine et al. (1989) observed reduced numbers of progenitor T cells in fetal mouse liver after maternal TCDD exposure, detected as reduced expression of the marker terminal deoxyneucleotydal transferase (TdT) in fetal liver. The reduced liver TdT levels corresponded to fetal thymic atrophy. Holladay et al. (1991) and Blaylock et al. (1992) similarly reported fetal thymic atrophy in TCDD-exposed mice. These authors also noted inhibition of thymocyte maturation, detected by CD4 and CD8 marker expression. These combined effects, thymic hypocellularity and inhibited maturation, appeared to contribute to a relatively persistent diminishment in postnatal immune function (CTL activity) lasting until 8 weeks of age (Holladay et al. 1991). It may be noteworthy that lymphocyte phenotyping in adult mouse spleen has a high predictive value for immunosuppression (0.83) (Luster et al. 1992); however, no comparable data are available regarding the predictive strength for immunotoxicity detection of similar phenotyping in fetal thymus or liver. Table 4.4 and Table 4.5 show summary fetal thymocyte and liver marker expression data after maternal exposure to a variety of immunotoxicants.

Exposure of pregnant mice to pharmacologic doses of DES produced fetal liver and thymic effects that in many ways paralleled those reported for TCDD. These included reduced fetal liver TdT+ progenitor T cells, thymic atrophy, and inhibited

**Table 4.4 Effect of Gestational Immunotoxicant Exposure on CD4+8+ (DP) and CD4-8- (DN) GD 18 Fetal Thymocytes**

| | | % Within Each Phenotype | | | | |
|---|---|---|---|---|---|---|
| | Exposure | Control | | Chemical | | |
| Agent | Regimen | DP | DN | DP | DN | Ref. |
| TCDD | 3 μg/kg, GD 6–14 | 69.1 | 21.1 | 43.2* | 37.3* | Holladay et al. 1991 |
| DES | 8 μg/kg, GD 10–16 | 71.4 | 22.0 | 63.6* | 29.2* | Holladay et al. 1993 |
| EGME | 200 mg/kg, GD 10–17 | 82.1 | 12.0 | 72.3* | 19.9* | Holladay et al. 1994 |
| B(a)P | 150 mg/kg, GD 14–17 | 78.0 | 18.1 | 33.4* | 62.0* | Holladay and Smith 1994 |
| T2 toxin | 1.5 mg/kg, GD 14–17 | 71.6 | 23.6 | 55.0* | 40.2* | Holladay et al. 1993b |

* Indicates numbers reported were different from control, $p < 0.05$.

*Source*: Modified from Holladay SD and Blaylock BL. 2002. The mouse as a model for developmental immunotoxicology. *Human Exper Toxicol* 21:525–532.

**Table 4.5 Effect of Gestational Immunotoxicant Exposure on Fetal Liver Markers**

| | Markers Examined | | | | | |
|---|---|---|---|---|---|---|
| Agent | TdT | CD44 | CD45 | CD45R | Mac-1 | Ref. |
| TCDD | + | nd | nd | nd | nd | Holladay et al. 1991 |
| DES | + | nd | nd | nd | nd | Holladay et al. 1993 |
| EGME | nd | + | + | + | – | Holladay et al. 1994 |
| B(a)P | + | + | nd | + | nd | Holladay and Smith 1994 |
| DMBA | nd | + | nd | + | + | Holladay and Smith 1995 |
| T2 toxin | nd | + | nd | + | – | Holladay et al. 1993b |

*Note:* + indicates the compound significantly changed expression of the marker; – indicates the compound did not affect the marker; *nd* indicates not determined.

*Source*: Modified from Holladay SD and Blaylock BL. 2002. The mouse as a model for developmental immunotoxicology. *Human Exper Toxicol* 21:525–532.

thymocyte maturation (Holladay et al. 1993a). Ways et al. (1980) had previously reported immunosuppression in female mice that were exposed neonatally to DES, in the form of prolonged diminishment of splenocyte immune responsiveness and inhibited graft-vs.-host responses. Kalland (1980) reported depressed antibody response in female mice following neonatal exposure to DES. More recently, Karpuzoglu-Sahin et al. (2001) reported impaired interferon-gamma secretion at 1 to 1.5 years of age in mice exposed to a single dose of DES during development. The developmental immune lesions in fetal liver and thymus caused by DES may in part explain these postnatal observations of immune modulation.

Additional studies have made use of fetal thymus and liver to detect effects of developmental immunotoxicants in mice, with varying results. Prenatal exposure to T2 mycotoxin caused profound fetal thymic atrophy; however, *in vitro* studies demonstrated that fetal thymocyte viability increased, rather than decreased, after exposure to T2 toxin (Holladay et al. 1993b). The increased thymocyte viability appeared to be related to inhibited synthesis of endonuclease by thymocytes, required for apoptosis, while thymic atrophy was related to both inhibited thymocyte proliferation and targeting of progenitor cells in fetal liver. Similar fetal thymic antiproliferative and fetal liver progenitor effects were seen in mice exposed *in utero* to the PAH, 7,12-dimethylbenz[a]anthracene (DMBA), and corresponded to dose-dependent fetal thymic and liver hypocellularity (Holladay and Smith 1995). Another PAH, B(a)P, also caused depletion of fetal liver prolymphocytes, thymic cellular depletion, and inhibition of normal thymocyte maturation (Holladay and Smith 1994). The industrial solvent ethylene glycol monomethyl ether (EGME) caused fetal thymic atrophy without affecting thymocyte proliferation. Similar to TCDD, DES, T2 toxin and the two studied PAHs, the EGME-induced thymic atrophy appeared to at least in part be related to targeting of prolymphoid cells in the fetal liver (Holladay et al. 1994). For these reasons, Holladay and Luster (1996) concluded that altered marker expression in both fetal thymus and liver appears to be a highly sensitive indicator of gestational immunotoxicant exposure.

## SUMMARY

The strength of the mouse as a model system for detecting developmental immunotoxicants includes the literature database produced in this species; probably greater than 95% of existing immunotoxicology information has been generated using mouse models. As part of this literature database, the most accepted immunotoxicity risk assessment methodology, the NTP paradigm, was also produced and validated in a mouse model (Luster et al. 1988; 1992). It has generally been presumed that NTP tests with high predictive value for immunosuppression in the mouse are likely to offer similar efficacy in other species. This presumption appears reasonable based on presently available information; however, variations may exist that have not yet been recognized. More antibody and other reagents have been developed and are available for use in the mouse than in any other species, which aids routine immunotoxicity testing as well as mechanistic studies. Finally, gene addition and deletion models exist or are produced almost exclusively in the mouse, which again can be of great value for understanding mechanisms by which immunotoxicants act.

## REFERENCES

Barnett JB, Soderberg LS, and Menna JH. 1985. The effect of prenatal chlordane exposure on the delayed hypersensitivity response of BALB/c mice. *Toxicol Lett* 25, 173–183.

Blaylock BL, Holladay SD, Comment CE, Heindel JJ, and Luster MI. 1992. Modulation of perinatal thymocyte surface antigen expression and inhibition of thymocyte maturation by exposure to tetrachlorodibenzo-p-dioxin (TCDD). *Toxicol Appl Pharmacol* 112:207–213.

Chapin RE and Sloane RA. 1997. Reproductive assessment by continuous breeding: evolving study design and summaries of ninety studies. *Environ Health Perspec* 105, Supl. 1:199–205.

d'Argy R, Dencker L, Klasson-Wehler E, Bergman A, Darnerud PO, and Brandt I. 1987. 3,3'4,4'-Tetrachlorobiphenyl in pregnant mice: embryotoxicity, teratogenicity, and toxic effects on the cultured embryonic thymus. *Pharmacol Toxicol* 61, 53–57.

Das SN, Paul BN, Saxena AK, and Ray PK. 1990. Effect of *in utero* exposure to hexachlorocyclohexane on the developing immune system of mice. *Immunopharmacol Immunotoxicol* 12, 293–310.

Elbe A, Tschachler E, Steiner G, Binder A, Wolff K, and Stingl G. 1989. Maturational steps of bone marrow-derived dendritic murine epidermal cells. Phenotypic and functional studies on Langerhans cells and Thy-1+ dendritic epidermal cells in the perinatal period. *J Immunol* 143, 2431–2438.

Fine JS, Gasiewicz TA, and Silverstone AE. 1989. Lymphocyte stem cell alterations following perinatal exposure to 2,3,7,8-tetrachlorodibenzo-p-dioxin. *Mol Pharmacol* 35, 18–25.

Goidl EA, Klass J, and Siskind GW. 1976. Ontogeny of B-lymphocyte function. II. Ability of endotoxin to increase the heterogeneity of affinity of the immune response of B lymphocytes from fetal mice. *J Exp Med* 143, 1503–1520.

Holladay SD, Lindstrom P, Blaylock BL, Comment CE, Germolec DR, Heindel JJ, and Luster MI. 1991. Perinatal thymocyte antigen expression and postnatal immune development altered by gestational exposure to tetrachlorodibenzo-p-dioxin (TCDD). *Teratology* 44, 385–393.

Holladay SD, Blaylock BL, Comment CE, Heindel JJ, Fox WM, Korach KS, and Luster MI. 1993a. Selective prothymocyte targeting by prenatal diethylstilbestrol exposure. *Cell Immunol* 152, 131–142.

Holladay SD, Blaylock BL, Comment CE, Heindel JJ, and Luster MI. 1993b. Fetal thymic atrophy after exposure to T-2 toxin: selectivity for lymphoid progenitor cells. *Toxicol Appl Pharmacol* 121, 8–14.

Holladay SD and Smith BJ. 1994. Fetal hematopoietic alterations after maternal exposure to benzo[a]pyrene: a cytometric evaluation. *J Toxicol Environ Health* 42, 259–273.

Holladay SD, Comment CE, Kwon J, and Luster MI. 1994. Fetal hematopoietic alterations after maternal exposure to ethylene glycol monomethyl ether: prolymphoid cell targeting. *Toxicol Appl Pharmacol* 129, 53–60.

Holladay SD and Smith BJ. 1995. Alterations in murine fetal thymus and liver hematopoietic cell populations following developmental exposure to 7,12-dimethylbenz[a]anthracene. *Environ Res* 68, 106–113.

Holladay SD and Luster MI. 1996. Alterations in fetal thymic and liver hematopoietic cells as indicators of exposure to developmental immunotoxicants. *Environ Health Perspect* 104 Suppl 4, 809–813.

Holladay SD and Smialowicz RJ. 2000. Development of the murine and human immune system: differential effects of immunotoxicants depend on time of exposure. *Environ Health Perspect* 108 Suppl 3, 463–473.

Holladay SD and Blaylock BL. 2002. The mouse as a model for developmental immunotoxicology. *Human Exper Toxicol* 21:525–532.

Johnson BE, Bell RG, and Dietert RR. 1990. 3-Methylcholanthrene-induced immunosuppression in mice to Trichinella spiralis antigens. *Immunopharmacol Immunotoxicol* 12:237–56.

Kalland T. 1980. Alterations of antibody response in female mice after neonatal exposure to diethylstilbestrol. *J Immunol* 124, 194–198.

Kalland T and Forsberg JG. 1981. Natural killer cell activity and tumor susceptibility in female mice treated neonatally with diethylstilbestrol. *Cancer Res* 4, 5134–5140.

Karpuzoglu-Sahin E, Hissong BD, and Ahmed AS. 2001. Interferon-γ levels are upregulated by 17-β-estradiol and diethylstilbestrol. *J Reprod Immunol* 52:113–127.

Lanier LL. 1998. NK cell receptors. *Annu Rev Immunol* 16, 359–393.

Luster MI, Faith RE, McLachlan JA, and Clark GC. 1979. Effect of *in utero* exposure to diethylstilbestrol on the immune response in mice. *Toxicol Appl Pharmacol* 47, 279–285.

Luster MI, Munson AE, Thomas PT, Holsapple MP, Fenters JD, White KL Jr, Lauer LD, Germolec DR, Rosenthal GJ, and Dean JH. 1988. Development of a testing battery to assess chemical-induced immunotoxicity: National Toxicology Program's guidelines for immunotoxicity evaluation in mice. *Fundam Appl Toxicol* 10, 2–19.

Luster MI, Portier C, Pait DG, White KL Jr, Gennings C, Munson AE, and Rosenthal GJ. 1992. Risk assessment in immunotoxicology. I. Sensitivity and predictability of immune tests. *Fundam Appl Toxicol* 18, 200–210.

Mosier DE. 1977. Ontogeny of T cell function in the neonatal mouse. In: Cooper MD and Dayton DH (Eds.), *Development of Host Defenses.* Raven Press, New York, NY, p. 115.

Santoni A, Riccardi C, Barlozzari T, and Herberman RB. 1982. Natural suppressor cells for murine NK activity. In: Herberman RB (Ed.), *NK Cells and Other Natural Effector Cells.* Academic Press, New York, NY, p. 443.

Sarzotti M, Robbins DS, and Hoffman PM. 1996. Induction of protective CTL responses in newborn mice by a murine retrovirus. *Science* 271, 1726–1728.

Sivakumar PV, Bennett M, and Kumar V. 1997. Fetal and neonatal NK1.1+ Ly-49- cells can distinguish between major histocompatibility complex class I(hi) and class I(lo) target cells: evidence for a Ly-49-independent negative signaling receptor. *Eur J Immunol* 27, 3100–3104.

Sivakumar PV, Gunturi A, Salcedo M, Schatzle JD, Lai WC, Kurepa Z, Pitcher L, Seaman MS, Lemonnier FA, Bennett M, Forman J, and Kumar V. 1999. Cutting edge: expression of functional CD94/NKG2A inhibitory receptors on fetal NK1.1+Ly-49- cells: a possible mechanism of tolerance during NK cell development. *J Immunol* 162, 6976–6980.

Smith BJ, Holladay SD, and Blaylock BL. 1994. T2 mycotoxin alters thymocyte maturation and produces changes in other hematopoietic compartments that may relate to impaired immunity. *Toxicon* 32, 1115–1123.

Spyker-Cranmer JM, Barnett JB, Avery DL, and Cranmer MF. 1982. Immunoteratology of chlordane: cell-mediated and humoral immune responses in adult mice exposed *in utero. Toxicol Appl Pharmacol* 62, 402–408.

The H-S. 1993. T cell development and repertoire selection. In: Cooper EL and Nisbet-Brown E (Eds.), *Developmental Immunology.* Oxford University Press, New York, NY, p. 217.

Toomey JA, Shrestha S, de la Rue SA, Gays F, Robinson JH, Chrzanowska-Lightowlers ZM, and Brooks CG. 1998. MHC class I expression protects target cells from lysis by Ly49-deficient fetal NK cells. *Eur J Immunol* 28, 47–56.

Tyan ML. 1981. Marrow stem cells during development and aging. In: Kay MB and Makinoday T. (Eds.), *Handbook of Immunology and Aging.* CRC Press, Boca Raton, FL, p. 87.

Urso P and Gengozian N. 1984. Subnormal expression of cell-mediated and humoral immune responses in progeny disposed toward a high incidence of tumors after *in utero* exposure to benzo(a)pyrene. *J Toxicol Environ Health* 14:569–584.

Urso P and Johnson RA. 1987. Early changes in T lymphocytes and subsets of mouse progeny defective as adults in controlling growth of a syngeneic tumor after *in utero* insult with benzo[a]pyrene. *Immunopharmacology* 14, 1–10.

Velardi A and Cooper MD. 1984. An immunofluorescence analysis of the ontogeny of myeloid, T, and B lineage cells in mouse hemopoietic tissues. *J Immunol* 133, 672–677.

Ways SC, Blair PB, Bern HA, and Staskawicz MO. 1980. Immune responsiveness of adult mice exposed neonatally to diethylstilbestrol, steroid hormones, or vitamin A. *J Environ Pathol Toxicol* 3, 207–220.

CHAPTER 5

# The Rat as a Model for Developmental Immunotoxicology

**Ralph J. Smialowicz**

## CONTENTS



0-415-28457-0/05/$0.00+$1.50

## INTRODUCTION

Animal testing for the identification and characterization of hazard(s) associated with exposure to potentially toxic chemicals is an accepted approach in determining the potential risk to humans. The rodent, in particular the rat, has been the most commonly used species for routine toxicity evaluations. While most standardized toxicology testing occurs in adult animals, there are a few protocols that specifically evaluate developmental toxicity in the rat. These test paradigms include prenatal developmental, multigenerational reproductive, and developmental neurotoxicity studies (EPA 1998 a-c; OECD 2001 a-b; FDA 2000). This safety testing in the developing rodent provides important information relevant to the design, performance, and interpretation of developmental toxicity studies. Since rats primarily have been used in developmental toxicity studies, abundant information is available regarding breeding, husbandry, and dosing of this rodent species. There are some additional advantages for using rats, rather than mice, in developmental toxicity testing. Rats are generally easier to breed than mice; rat dams have larger litters (depending on strain, from 10 to 18 pups/litter); rat pups are generally more robust and are relatively easy to cross foster; and rat pups can be gavaged with test chemicals as early as postnatal day (PND) 3. Another important consideration in the use of the rat as a test species is that developmental immunotoxicity testing can potentially be evaluated by integrating immune function assessment with currently required multigenerational reproductive toxicity testing (Chapin 2002, Chapin et al. 1997).

Over the past fifteen years, there has been an incremental growth in experimental immunotoxicity research, which has primarily focused on assessment of the mature immune system of experimental animals following exposure to chemicals. It is generally accepted that the developing organism is more vulnerable than the adult to agent-induced organ-system toxicity. Since there is limited information on developmental immunotoxicity, safety information based solely on adult immunotoxicity data is unlikely to result in effective protection of the potentially most at-risk populations. During the past few years interest in and emphasis on developmental toxicology has increased, particularly as it relates to children's risk. This interest is reflected in documents such as *Pesticides in the Diets of Infants and Children* (NRC 1993) and *Research Needs on Age-Related Differences in Susceptibility to Chemical Toxicants* (ILSI. 1996), in which the immune system has been identified routinely as an organ system of concern. In addition, two U.S. statutes require that the U.S. Environmental Protection Agency (EPA) perform "separate assessments for infants and children" for potential toxic susceptibility (FQPA 1996; SDWA 1996).

Establishment of the mammalian immune system involves a sequential series of carefully timed and coordinated developmental events that begin early in embryonic/fetal life and continue in the postnatal and juvenile periods and through puberty. Perturbation or abrogation of this developmental sequence of events can result in immune dysfunctions that may be life-threatening. Defects in the development of the immune system due to heritable changes in the lymphoid elements have provided clinical and experimental examples of the devastating consequences of impaired immune system development (Rosen et al. 1995). Potential functional defects, caused by exposure to certain chemical or physical agents during development, may range from life-threatening suppression of vital components of the immune system to altered or poorly regulated responses that can be debilitating. For example, studies in experimental laboratory rodents indicate that exposure to immunosuppressants (e.g., chemicals, drugs, ionizing radiation) during immune system development can result in alterations of immune system function which may persist into adult life (Holladay and Smialowicz 2000). In this chapter, evidence is provided implicating chemicals from different classes, including metals, polyhalogenated aromatic hydrocarbons, pesticides, and drugs, as being developmental immunotoxicants in the rat.

## METALS

### Lead

A variety of environmental and occupational metals have been identified that alter immune function in humans and laboratory rodents (Zelikoff and Thomas 1995). One such metal is lead, a naturally occurring element found in the earth's crust as well as throughout the biosphere. Present-day concentrations of lead in the atmosphere are due primarily to anthropogenic sources, with insignificant contribution from natural sources. Lead exposure to both humans and animals at sufficiently toxic doses results in renal, hematopoietic, and central nervous system dysfunction. Lead has also been demonstrated to be a potent immunotoxicant in mice and rats as well as a developmental immunotoxicant in rats.

Early work by Luster et al. (1978) and Faith et al. (1979) examined the developmental immunotoxic effects of chronic pre- and postnatal exposure of rats to low levels of lead acetate in drinking water. Weanling (21-day-old) female Sprague-Dawley (SD) rats were exposed to lead acetate at concentrations of 25 and 50 ppm in their drinking water until 10 weeks of age. These female rats were then mated with untreated males and continued on the same lead acetate concentrations throughout gestation and lactation. Offspring of these females were weaned at 21 days of age and continued on the same lead exposure through immune function testing between 35 and 45 days of age. These lead exposures resulted in decreased thymus weights, compared to controls, but did not alter body weight nor result in overt signs of toxicity. The IgG antibody response to sheep red blood cells (SRBCs), as well as the IgM antibody plaque-forming cell (PFC) response to SRBCs, were both suppressed at 25 and 50 ppm lead acetate (Luster et al. 1978). The mitogen-stimulated lymphoproliferative (LP) responses of splenic lymphocytes to phytohaemagglutinin (PHA) and Concanavalin A (ConA) were suppressed at both lead acetate concentrations (Faith et al. 1979). Blood lead determinations in these studies indicated that the observed levels were comparable to blood levels found in many children in urban areas during the early 1970s (Caprio et al. 1974; Committee on Biological Effects 1972).

More recently, Miller et al. (1998), Chen et al. (1999), and Bunn et al. (2001a; 2001b) examined the developmental immunotoxic effects of lead acetate in the drinking water of female rats at different periods of development during pre- and postbreeding: during breeding and pregnancy; two weeks prior to mating and through parturition; throughout pregnancy; or on gestation day (GD) 3 to 9 or GD 15 to 21, respectively, for each study. In 13-week-old Fischer 344 female rat offspring (male offspring were not evaluated), the total blood leukocyte count, the delay-type hypersensitivity (DTH) response to keyhole limpet hemocyanin (KLH), and splenic interferon-γ (IFNγ) production were depressed at 250 ppm (Miller et al. 1998; Chen et al. 1999). In contrast, macrophage tumor necrosis factor-α (TNFα) and nitric oxide production, and splenic interleukin-4 (IL-4) production were elevated at 250 ppm lead acetate (Miller et al. 1998; Chen et al. 1999).

A comparison of the effect that lead acetate in the drinking water of rat dams has on the developing immune system of both male and female offspring was determined in studies by Bunn et al. (2001a; 2001b). Female offspring of F344 rat dams exposed to 250 ppm lead acetate in drinking water from GD 2 through pregnancy displayed persistent suppression of the DTH responses to KLH at both 5 and 13 weeks of age, while male offspring were not affected at either age (Bunn et al. 2001a). The DTH response to KLH was also suppressed in 12-week-old female offspring of pregnant SD rats given 500 ppm lead acetate in their drinking water during late gestation (i.e., access to lead acetate only during GD 15 to 21), but not during early gestation (GD 3 to 9). These same female offspring exhibited a persistent increase in IL-10 production. In contrast, 13-week-old male offspring did not display a decrease in the DTH response following either early or late gestational lead exposure (Bunn et al. 2001b). Taken together, these data indicate that prenatal lead exposure of rats leads to persistent alterations in immune function of adult offspring, with females being more sensitive than males.

## ORGANOTINS

Organotins have a wide variety of industrial and commercial applications including use as wood preservatives, agricultural pesticides, chemical catalysts, plastic heat stabilizers, marine antifoulants, and curing agents (Piver 1973). Certain dialkyl- and trialkyl-substituted organotins (e.g., di-*n*-octyltin dichloride [DOTC] and di-*n*-butyltin dichloride [DBTC], and tributyltin oxide [TBTO],) alter the structure and function of the thymus and consequently affect primarily T cell-dependent immune function in rodents. There is evidence that rat natural killer (NK) cell activity is also affected by organotins (Pennicks and Pieters1995).

Early work by Seinen et al. (1977; 1979) demonstrated that pre- and/or postnatal exposure of Wistar-derived WAG rats to either DOTC or TBTO results in suppression of the DTH response, antibody response to SRBCs, and graft-vs.-host response. Female rats were dosed by gavage with 0, 5, or 15 mg DOTC/kg body wt or with 0, 1, 3, 5, or 15 mg DBTC/kg body wt starting on day 2 of pregnancy and through gestation, followed by direct dosing of the pups on PND 2 and then three times per week until immune function testing. The antibody response to SRBCs, as determined by the PFC assay and serum antibody titers, was suppressed in both 6-week-old male and female offspring dosed at 15 mg/kg DBTC (Seinen et al. 1977). In addition, the graft-vs.-host skin graft rejection response was delayed, compared to control, in 7-week-old male rats dosed at 1 and 3 mg DBTC/kg (Seinen et al. 1977). Six-week-old female rats exposed pre- and postnatally to DOTC at 5 and 15 mg DOTC/kg had delayed skin graft rejection times in the graft-vs.-host reaction (Seinen et al. 1979). The authors of these papers concluded that it appears that DOTC and DBTC cause immune suppression in rats by a selective inhibition of T lymphocyte function, and that immune suppression is more pronounced in rats exposed to these organotins during immune system development (Seinen et al. 1977). No attempt was made in these studies to determine if the immune function deficits observed immediately following the last organotin exposure of the rats, resulted in a long-lived or persistent immunosuppression.

More recently, two studies were conducted to determine if exposure to either DOTC or TBTO during immune system development affected immune function in adult male Fischer 344 rats. In the first study (Smialowicz et al. 1988), dams were exposed to DOTC by gavage during the prenatal, both pre- and postnatal (i.e., lactation), or postnatal period(s). The offspring of these dams displayed no consistent alteration in immune function, as determined by the *in vitro* LP response to T and B cell mitogens or by the *in vitro* NK cell $^{51}$Cr-release assay. In contrast, direct gavage dosing of pups with 5, 10, or 15 mg DOTC/kg/d on PND 3 and then 3 times a week up to PND 24, for a total of 10 doses, resulted in suppression of the LP responses at up to10 weeks of age (i.e., 7 weeks after the last exposure to DOTC). LP responses returned to control levels after 10 weeks of age. In comparison, young adult (i.e., 8-week-old) rats dosed by gavage with 10 or 20 mg DOTC/kg/body wt/day, using the identical dosing schedule (i.e., 3 times per week for a total of 10 doses) showed no alteration of the LP responses one week after the last exposure (Smialowicz et al. 1988).

In the second study (Smialowicz et al. 1989), rat pups were dosed by gavage with TBTO, as described above (i.e., PND 3 through PND 24, 3 days per week, for a total of 10 doses). Rats were assessed for immune competency as determined by the *in vitro* LP response and NK assay, as well as mixed leukocyte response (MLR) and cytotoxic T lymphocyte (CTL) response, and the T cell-dependent antibody response to SRBCs as determined by the PFC assay. Young adult (i.e., 9-week-old) rats were similarly dosed (i.e., 3 times per week for a total of 10 times) with TBTO. LP responses were suppressed in pups dosed at 5 and 10 mg/kg/d, while in adults suppression was observed at doses of 10 and 20 mg/kg/d. No other immune function endpoints were affected. Within 3 weeks following the last exposure of adult rats, the LP responses returned to control levels. However, suppression of the LP responses in exposed pups persisted for up to 13 weeks of age (i.e., 10 weeks after the last exposure to TBTO), after which they returned to control levels.

These data (Smialowicz et al. 1988, 1989) indicate that the direct exposure of young male rat pups to DOTC or TBTO resulted in immune suppression at doses lower than those required to suppress immune function in the adult rat. Furthermore, immune suppression persisted for 7 to 10 weeks longer when pups, rather than adults, were exposed to DOTC or TBTO, respectively. It is interesting that direct early postnatal, but not perinatal (i.e., *in utero* and lactational) exposure to DOTC resulted in immune suppression that persisted for 7 weeks. This suggests that DOTC may not be available to the fetus via the placenta and the pup via milk at sufficient concentrations to alter immune system development. This observation underscores the importance of information relative to the pharmacokinetics of the test chemical. Such information would be extremely helpful in designing the dosing protocol for a chemical that might be considered for developmental immunotoxicity testing. Since only males were evaluated in these studies, important information relative to potential gender susceptibility to early postnatal organotin exposure was not available.

## POLYHALOGENATED AROMATIC HYDROCARBONS

### 2,3,7,8-Tetrachlorodibenzo-*p*-Dioxin (TCDD)

Exposure of experimental animals to TCDD results in a variety of toxic responses which differ in intensity among various species and strains (Poland and Knutson 1989). Of the many organs/systems affected by TCDD, one of the most sensitive is the immune system (Holsapple 1995). TCDD is a highly toxic compound that produces more severe effects when exposure occurs during development. In the rat and the mouse, perinatal exposure to TCDD has been associated with teratogenicity (Couture et al. 1990), reproductive toxicity (Murray et al. 1979; Mably et al. 1992; Gray et al. 1995), neurobehavioral dysfunction (Thiel et al. 1994), and immunotoxicity (Vos and Moore 1974; Faith and Moore 1977).

The results of TCDD studies in experimental animals are of concern because humans potentially have their greatest intakes of TCDD and related polyhalogenated aromatic hydrocarbons during development. Breastfeeding infants have been estimated to consume 35–53 pg TCDD toxic equivalents/kg body weight/day during the first year of life (Schecter et al. 1994). In comparison, various studies have calculated the adult intake to be in the range of 1 or 2 pg TCDD toxic equivalents/kg/day in Germany (Beck et al. 1989; Fürst et al. 1990), the Netherlands (Theelen et al. 1993), and the United States (Schecter et al. 1994). In one study, a nursing infant was found to absorb 96% of the consumed TCDD, since nearly all of ingested TCDD is absorbed (McLachlan 1993). While nursing represents the predominant source of perinatal exposure to TCDD, transplacental exposure also occurs (Korte et al. 1990; Li et al. 1995).

Perinatal TCDD exposure has been shown to alter a variety of immune functions in rodents. In rats, perinatal TCDD exposure suppresses T cell mitogen responsiveness (Vos and Moore 1974; Faith and Moore1977); skin graft rejection time (Vos and Moore 1974); graft- vs.- host activity (Vos and Moore 1974); and DTH responsiveness (Faith and Moore 1977). Perinatal TCDD exposure has also been shown to produce alterations in rat thymocyte subpopulations (Gehrs and Smialowicz 1997), which are similar to those observed in mice perinatally exposed to TCDD (Holladay et al. 1991; Fine et al.1989).

Early work with perinatal exposure of rats to TCDD (Vos and Moore 1974; Faith and Moore 1977) established that immune responses in the offspring associated with cell-mediated immunity were depressed while those of humoral immunity were not. Vos and Moore (1974) show that TCDD exposure studies were performed (i.e., a pre- and postnatal and postnatal-only TCDD exposure) using female Fischer 344 rats. In the first study, pregnant rats were dosed by gavage with 1.0 or 5.0μg

TCDD/kg on GD 11 and 18, and then postnatally on PND 4, 11, and 18. Most of the neonates in the 5.0μg TCDD/kg treatment group died. In the postnatal-only exposure study, nursing dams were gavaged with 5μg TCDD/kg on PND 0, 7, and 14. The PHA LP response in 25-day-old male offspring was reduced in the pre-and postnatally exposed and postnatal-only exposed males at 1.0 and 5.0μg TCDD/kg, respectively. The graft-vs.-host response was also suppressed in 25-day-old male offspring exposed to 5.0μg TCDD/kg postnatally only. Histopathological examination of the thymuses from 25-day-old male and female rats pre-and postnatally exposed to 1.0μg TCDD/kg, and male and female rats postnatally only exposed to 1.0 and 5.0μg TCDD/kg, had atrophy of the thymic cortex. In addition, these offspring did not display any adrenal hypertrophy, indicating that a stress-induced release of glucocorticoids was not considered responsible for the observed immunosuppression (Vos and Moore 1974).

In a follow-up to the Vos and Moore (1974) study, Faith and Moore (1977) employed the same two TCDD exposure regimens with the exception that in the pre- and postnatal TCDD exposure experiment, pregnant Fischer 344 rats were dosed by gavage with 5.0μg TCDD/kg on GD 18, and postnatally on PND 0, 7, and 14. 39-day-old male and female rats had reduced thymus and spleen weights in both TCDD-exposed groups. At 29 and 59 days of age, both TCDD-exposed groups had suppressed spleen cell LP responses to PHA and ConA. The DTH response to oxazolone was also suppressed in both TCDD-exposed groups at 145 days old. In contrast, there was no difference in the antibody titers to bovine gamma globulin (BGG) between control or TCDD-exposed rats. With the exception of the organ weight data, for the immune function results presented in this study, male and female offspring were grouped together rather than separated by gender (Faith and Moore 1977). Consequently, information relative to potential gender susceptibility to perinatal TCDD exposure was not available.

In a TCDD postnatal lactational exposure study, Badesha et al. (1995) exposed lactating Leeds female rats to TCDD in their diet starting on PND 1 through PND 18. The total administered doses over the 18 days of TCDD dosing were 0.2, 1.0, or 5.0μg TCDD/kg dam body weight. At PND 130, body and organ weights were determined and immune function tests were performed. Body weights were reduced in both male and female offspring at 5.0μg TCDD/kg, while liver and spleen, but not thymus, weights were reduced in male offspring, and only spleen weights were reduced in females at 5.0μg TCDD/kg. Offspring displayed suppressed *in vivo* T cell-dependent and -independent responses and mitogen induced *in vitro* production of IL-1 and IL-2. Antibody responses to the T cell-dependent antigen SRBCs, and to the T cell-independent antigens dinitrophenyl (DNP)-Ficoll, trinitrophenyl-lipopolysaccharide (TNP-LPS), and LPS were suppressed by 50% of the maximum (i.e., estimates in the range of 0.3–1.0μg TCDD/kg for the SRBC, DNP-Ficoll and TNP-LPS antigens and 3.5–3.9μg TCDD/kg for LPS antigen). *In vitro* production of IL-1 and IL-2, by rat splenic macrophages and lymphocytes from TCDD-exposed offspring in the presence of PHA or Con A, respectively, revealed that suppression of IL-1 production occurred at 0.2μg TCDD/kg while that of IL-2 occurred at 1.0μg TCDD/kg. As in Faith and Moore's (1977) study, male and female offspring were combined for immune function assays in the Badesha et al. (1995) study. Consequently, information relative to potential gender-specific differences in immune system susceptibility to lactational TCDD exposure was not available.

Three studies were performed in which timed-bred pregnant Fischer 344 rat dams were given a single gavage dose of TCDD on GD 14 (i.e., GD 0 = day of vaginal plug). Depending on the study, doses ranged from 0.1 to 3.0μg TCDD/kg. In the first study (Gehrs and Smialowicz 1997), GD 19 fetuses from the 3.0μg TCDD/kg maternal exposure group exhibited decreased relative thymus weight and thymic cellularity compared to control rats. There were decreased percentages of immature $CD3^-/CD4^+CD8^+$ thymocytes and increased percentage of $CD3^-/CD4^-CD8^+$ thymocytes in these fetuses. Development of thymocytes in the fetal thymus is a sequential and highly synchronized process. Kampinga and Aspinall (1990) found that all rat thymocytes are $CD3^-/CD4^-CD8^-$ on GD 14; by GD 19 most rat thymocytes are still $CD3^-$, but some have differentiated through the $CD3^+/CD4^+CD8^+$ stage. Fine et al. (1989) and Holladay et al. (1991) observed increased percentages of $CD4^-$ $CD8^-$ and $CD4^-$ $CD8^+$ mouse thymocytes on GD 18 as well as a decreased percentage of $CD4^+CD8^+$ thymocytes. Blaylock et al. (1992) used the J11d antigen to determine that the $CD4^-$ $CD8^+$ population represented immature thymocytes. Taken together with the decrease in thymic cellularity, these results suggest that TCDD mediates its effects in the fetal rat and the fetal mouse either by blocking maturation from $CD3^-/CD4^-CD8^+$ to $CD3^-/CD4^+CD8^+$, by selectively eliminating $CD3^-/CD4^+CD8^+$ cells, or by accelerating the differentiation of the $CD3^-/CD4^-CD8^+$ cells. The lack of an increase in $CD3^+$ cells in rats so exposed to TCDD rules out the latter of these possibilities. Whether these alterations are responsible for the immune function effects described below remains to be determined.

The second study (Gehrs et al. 1997) was designed to determine if adult immune function was compromised by GD 14 TCDD exposure and to determine whether transplacental and lactational transfer of TCDD to the pups was critical for immunosuppression. Immune function was assessed in pups born to dams exposed to 3μg TCDD/kg on GD 14 at 14 to 17 weeks of age. The DTH response to bovine serum albumin (BSA) was suppressed in both the TCDD-exposed males and females compared to controls. The LP responses to T-cell and B-cell mitogens and the PFC antibody response to SRBCs, while reduced compared to controls, were not significantly lowered. The latter result is the same as that observed by Faith and Moore (1977) in that the antibody response to BGG was not affected by perinatal TCCD exposure.

A second experiment in this study (Gehrs et al. 1997) involved different groups of timed-pregnant Fischer 344 rats dosed with 1.0μg TCDD/kg on GD 14. One day after birth, litters were cross-fostered to produce control, placental-only, lactational-only, and placental/lactational exposure groups. The DTH response to BSA was assessed in 5-month-old males. In these rats the severity of the suppression of the DTH response was related to the route of TCDD exposure (i.e., placental/lactational > lactational > placental), with suppression occurring only in the males receiving both placental and lactational exposure. These results indicated that the immunosuppressive effect of perinatal (i.e., placental and lactational) TCDD exposure of rats persisted into adulthood and that suppression of the DTH response represents the most sensitive biomarker for TCDD-induced developmental immunotoxicity in this species.

The third study (Gehrs and Smialowicz 1999) was designed to better characterize the suppression of the DTH response by perinatal TCDD exposure. "Persistence" of the DTH suppression was determined by measuring the DTH response to BSA in the offspring (at 4, 8, 12, and 19 months of age) of dams dosed orally with 3.0μg TCDD/kg on GD 14. TCDD significantly suppressed the DTH response in males through 19 months of age. While the DTH response of females was reduced at 8, 12, and 19 months, significant suppression was observed only at 4 months of age.

In a second experiment of this study (Gehrs and Smialowicz 1999), the lowest maternal dose of TCDD that produced DTH suppression was determined by measuring the DTH response to BSA in the 4- and 14-month-old offspring of dams dosed with 0.1, 0.3, or 1.0μg TCDD/kg on GD 14. In the males, suppression was observed at a maternal dose as low as 0.1μg TCDD/kg at 14 months of age, while a maternal dose of 0.3μg TCDD/kg was required to cause suppression in the 14-month-old females. Both males and females were more sensitive to the suppression of the DTH response to BSA at 14 months of age than at 4 months of age.

In the third experiment of this study (Gehrs and Smialowicz 1999), a comparison was made between the DTH response to BSA and that of keyhole limpet hemocyanin (KLH) in 5-month-old male offspring of dams dosed with 3.0μg TCDD/kg on GD 14. The DTH response to both antigens was equally suppressed by GD 14 TCDD exposure compared to controls. The contact hypersensitivity (CHS) response to dinitrofluorobenzene (DNFB) was also suppressed at 3.0μg TCDD/kg in a more recent study (unpublished observation). Phenotypic analysis was performed on thymus and lymph node suspensions from the DTH/BSA offspring. Significant effects in the thymus included an increased percentage of (γα $TCR^+$ cells and a decreased percentage of (αβ $TCR^+/CD4^-CD8^-$ and $MHCI^-$ $MHCII^-$ cells. In the popliteal lymph node draining the BSA-injected footpad, there was a decreased percentage of (αβ $TCR^+$ and $MHCI^-MHCII^-$ cells and an increased percentage of $MHCI^+$ cells. While phenotypic analysis identified differences in subsets of thymocytes and lymph node cells between control and TCDD exposed offspring, no clear correlations were established between altered subpopulations and suppressed DTH responses.

The results of these three studies indicate that suppression of the DTH response associated with perinatal TCDD exposure is "persistent" throughout adulthood, occurs at a low dose (i.e., 0.1μg TCDD/kg), and is more pronounced in males than females. Both placental and lactational TCDD exposure of the pups is required for suppression of this immune response. While exposure of adult male rats leads to suppression of the DTH response to KLH, a dose of 90μg TCDD/kg is required to produce this immunosuppression (Fan et al., 1996). These data indicate that suppression of the DTH response by TCDD administration during immune system development of the rat requires a TCDD dose that is almost three orders of magnitude lower than that required to suppress this response in adult rats, and that administration of this TCDD dose during this period of development results in "persistent" suppression. These findings provide evidence of two critical criteria (i.e., low dose and "persistent" effect) for labeling the developing immune system of the rat as a very sensitive period for the initiation of immune function impairment by TCDD.

# PESTICIDES

## Hexachlorobenzene

Hexachlorobenzene (HCB) was a widely used pesticide until 1965 and has not been commercially produced in the United States since the late 1970s (Beyer 1996). HCB is a by-product or impurity in the production of several chlorinated solvents (e.g., trichloroethylene, carbon tetrachloride), as well as in the production of several pesticides such as chlorothalonil, picloram, atrazine, propazine, simazine, and mirex (Tobin 1985). It also has been detected in the flue gas and the fly ash of municipal incinerators and other thermal processes (IARC 1979).

The effects that HCB has on immune system development in the rat were reported in studies by Vos et al. (1979a; 1983). Pregnant Wistar rats received diets containing 50 or 150 mg HCB/kg starting on day 1 to 3 of pregnancy and through gestation and lactation. Pups were weaned after PND 21 and continued on the same HCB dose as their dams for an additional 2 weeks prior to testing at 5 weeks of age (Vos et al. 1979a). Either male or female offspring, but not both, were evaluated for alterations in immune endpoints. The weight of the adrenal glands and livers of female pups exposed to 150 mg HCB/kg was increased, as were total serum IgM and IgG concentrations. Body, spleen, and thymus weights of females were not different from controls. In both dosage groups, males infected with *Listeria monocytogenes* displayed reduced resistance to this bacterial infection, as did a separate set of males infected with *Trichinella spiralis*. In both dosage groups, females had increased IgM and IgG antibody responses to tetanus toxoid and in the high dose group increased IgG response to *T. spiralis*. Mitogen-induced LP responses of spleen and thymus cells, and IgM response to LPS in females, were not affected by perinatal exposure to HCB (Vos et al. 1979a).

In a follow-up study, Wistar dams were fed 4, 20, or 100 mg HCB/kg from early pregnancy through lactation, and then the pups were fed the same doses for up to 7 months of age (Vos et al. 1983). Feeding at 4 mg HCB/kg resulted in increases in the IgM and IgG antibody responses to tetanus toxoid as well as increased DTH responses to ovalbumin (OVA). At 20 and 100 mg HCB/kg popliteal lymph node weights were increased which upon histopathological evaluation indicated increased cellular proliferation. Increased spleen, lung, and mesenteric lymph node weights were detected in rats exposed to 100 mg HCB/kg. At this dose, total serum IgM concentrations were increased. In contrast to the earlier study (Vos et al. 1979a), there was no effect found on the resistance to *T. spiralis* infection. These results indicate that pre- and postnatal exposure to HCB results in immunostimulation of both humoral and cell-mediated immunity in the rat.

Adult rats exposed to HCB display a similar enhanced immune stimulation to that of perinatally exposed female rats. Both display increased spleen and lymph node weights; increased splenic and serum levels of total IgM; and increased IgM against tetanus toxoid, SRBCs, and BSA; induced cellular proliferation in spleen marginal zones; and endothelial activation in high endothelial venules of mesenteric, popliteal lymph nodes, gut-associated lymphoid tissue, and small pulmonary venules (Vos et al. 1979b; Schielen et al. 1993; Schulte et al. 2002).

While perinatal (Vos et al.1979a; 1983) and adult (Vos et al. 1979b; Schielen et al. 1993; Schulte et al. 2002) HCB exposure of rats results in a generalized stimulation of the immune system, exposure of mice from GD 1 to 18 (Barnett et al. 1987) or of adult mice for 6 to 18 weeks to HCB (Loose et al. 1977; 1978) results in suppression of immune function. The DTH response to oxazolone was suppressed at 0.5 and 5.0 mg HCB/kg, as was the mixed lymphocyte response (MLR) at 5.0 mg HCB/kg in 45-day-old mice exposed perinatally (Barnett et al. 1987). However, there was no change in the LP response to PHA or LPS, nor was there any change in the PFC response to SRBC. The difference in the increased DTH response in HBC-exposed rats during the perinatal period vs. the decreased DTH response in mice perinatally exposed to HBC appears to be species-dependent. This difference may also be attributed to the fact that while the DTH response to OVA is a classic DTH reaction, oxazolone is a contact sensitizer (Barnett 1995). As such, these responses also involve different antigen-processing and -presenting cells, as well as different effector cells and cytokine profiles (Grabbe and Schwarz 1998).

### Dimethoate and Methylparathion

Dimethoate (DM) and methylparathion (MPT) are organophosphate insecticides and acaricides used to control many biting or sucking insect pests of agricultural crops. They kill insects systemically and on contact by stomach and respiratory action. They are used against a wide range of insects, including house flies as well as aphids, thrips, planthoppers, and whiteflies on ornamental plants, cotton, grains, vegetables, nuts, fruits, and tobacco. The mode of action of these organophosphate insecticides is through inhibition of acetyl cholinesterase at cholinergic synapses (Senanayake and Karalliedde 1987).

Institoris et al. (1995) examined the developmental immunotoxic effects that exposure to 7, 9, or 14 mg DM/kg or to 0.2, 0.3, or 0.4 mg MPT/kg has on Wistar rats over 3 generations. Exposure to either DM or MPT in drinking water of the first generation (G1) began in 4-week-old males and females. Parental males were dosed until separation of females, and after mating the females were treated until separation, at 4 weeks old, from their G2 offspring. The G3 rats were produced in the same manner as the parental G2 rats. Selected 4-week-old males from each generation were also exposed to DM or MPT for 4 weeks. Only male offspring were evaluated for immune function. G1 males exposed to 14 mg DM/kg had suppressed PFC response to SRBCs, as did G3 males exposed to 0.3 and 0.4 mg MPT/kg. Peripheral blood leukocyte counts were decreased in G1 males exposed to DM, while MPT exposed G3 males had reduced thymus weights at 0.2, 0.3, and 0.4 mg MPT/kg. There was no effect on the DTH response to KLH for either DM- or MPT-exposed rats. These results suggest that MPT, but not DM, may be a developmental immunotoxicant. Further studies are needed to confirm and expand these preliminary findings.

### Cypermethrin

Cypermethrin is a pyrethroid insecticide which is effective against a wide range of insects including ants, fire ants, roaches, spiders, termites, and crickets. It acts as a stomach and contact insecticide. Pyrethroids are potent neurotoxicants in both vertebrates and invertebrates. The principal target for pyrethroids is the voltage-dependent sodium channel in the neuronal membrane. Both Type I and Type II pyrethroids act potently and stereo selectively on sodium channels by slowing the kinetics of both opening and closing of individual channels. Inhibition of the (γ-aminobutyric acid (GABA) receptor is an additional mechanism for Type II pyrethroids, of which cypermethrin is a member (Aldridge 1990).

Santoni et al. (1997; 1998; 1999) examined the effects that cypermethrin, administered by gavage in corn oil during gestation, has on the developing immune system of rats. Pregnant Wistar rats were exposed to 50 mg cypermethrin for 10 days from GD 7 to GD 16. Pups were euthanized on PND 15, 30, 60, 90, or 120 and evaluated for immune dysfunction. There was no evidence of toxicity, as demonstrated by normal pregnancies, no mortality, and no loss of body weight compared to controls in the offspring. Male and female offspring results were grouped together; consequently, information relative to potential gender susceptibility to prenatal cypermethrin exposure is not available.

In the first study (Santoni et al. 1997), increased numbers of total peripheral blood lymphocytes (PBL) and bone marrow cells and decreased spleen and thymus cells were observed in PND 30 and PND 90 offspring. Peripheral blood NK cell activity against YAC-1 cells was increased compared to controls through PND 120, as was bone marrow NK cell activity. In contrast, NK activity was not altered in PBL nor in spleen cells from dams. Antibody-dependent cytotoxic cell (ADCC) activity against antibody-coated P815 target cells mirrored that of the NK cell activity (i.e., increased ADDC in PBL and decrease in spleen cells at PND 120). Regression analysis indicated that the kinetics of change in the percentages of NK-RP1$^+$ cells directly correlated with the increase of NK and ADCC cytotoxic activity observed at the same PNDs in PBL.

In the second study, Santoni et al. (1998) found that the absolute number of all thymocyte subsets decreased during the first 30 days after birth, with CD4$^-$CD8$^-$, CD4$^+$CD8$^-$, and CD4$^-$CD8$^+$ thymocytes being preferentially affected. By PND 60 to PND 90 the CD4$^+$CD8$^+$, and the CD4$^+$CD8$^-$ and CD4$^-$CD8$^+$ thymocytes gradually recovered, while the total number of CD4$^-$CD8$^-$ thymocytes increased. Thymocytes from PND 15, 30, 60 and 90 cypermethrin-exposed rats had suppressed ConA- and human recombinant IL-2 (hrIL-2)- induced LP responses, as well as decreased ability to produce and/or release IL-2. These results indicate that prenatal cypermethrin exposure affects several important steps in thymocyte differentiation pathways as demonstrated by the altered thymocyte subset distribution and altered thymocyte functions.

A correlation between cypermethrin-induced changes in adrenalin and noradrenaline plasma concentrations, and changes in T cell populations and immunomodulation in prenatally exposed rats was reported in the third study by Santoni et al. (1999). Adrenaline and noradrenaline plasma concentrations were increased compared to controls in PND 60 and 90, and PND 15, 30, and 90 offspring, respectively. These increased plasma catecholamine levels were accompanied by a marked increase of CD5$^+$, CD4$^+$, and CD8$^+$ total peripheral blood T cells, while CD5$^+$, CD4$^+$, and CD8$^+$ total spleen cells were reduced at PND 15 through PND 90. PBL from

PND 15 through PND 90 rats dosed prenatally with cypermethrin had enhanced LP responses to Con A and hrIL-2, whereas reduced LP responses to Con A and hrIL-2 were observed for splenic lymphocytes. The data demonstrated that the percentage increase in noradrenaline, but not adrenaline plasma concentrations, paralleled the immunomodulatory effects, induced by prenatal cypermethrin exposure, to T cell distribution and LP responses in blood and spleen. These results suggest that changes in the LP responses in PBL and splenic lymphocytes may be due to cypermethrin-induced catecholamine release leading to increased output of T lymphocytes from the spleen to peripheral blood (Santoni et al. 1999).

## Methoxychlor

The 1993 National Research Council (NRC) report, *Pesticides in the Diets of Infants and Children* identified the immune, reproductive, and nervous systems as potential targets of pesticide exposure. Alterations in the development of these systems, which continues through puberty, make them potential targets of toxicity due to pesticide exposure through food consumption, possibly resulting in long-term functional deficits. A collaborative research project between the National Institute of Environmental Health Sciences (NIEHS) and the U.S. EPA was initiated to address these scientific and regulatory concerns by exploring the long-term effects of pesticide exposure on these organ systems in rats (Chapin et al. 1997).

The exposure period of concern for humans was identified in the NRC report (1993) as encompassing the last trimester of pregnancy through 18 years of age. Consequently, studies were designed to ensure exposure to pesticides during development in the rat comparable to that in humans. To that end, pregnant rats were exposed to pesticides from mid-gestation through the first week postpartum, followed by direct dosing the pups from PND 8 to PND 42, the approximate end of puberty in rats. This exposure regimen was employed in order to ensure pesticide exposure during the windows of developmental vulnerability for the immune, reproductive, and nervous systems (Chapin et al. 1997).

The first pesticide examined was methoxychlor (MXC) (Chapin et al. 1997) an organochlorine pesticide which is more readily metabolized and excreted than is dichlorodiphenyltrichloroethane (DDT), and which is widely used for insect and larval control. MXC is one of only four remaining chlorinated pesticides approved for use in the U.S. because of these qualities. Timed-bred SD pregnant dams were dosed by gavage with 5, 50, or 150 mg MXC/kg/d starting on GD 14 through PND 8, and then their pups, normalized to four of each sex/litter, were gavaged until PND 42. Immune function evaluation of the pups so exposed, which consisted of splenocyte mitogen LP responses, NK cell activity, PFC antibody response to SRBCs, and flow cytometry phenotypic analysis of splenic lymphocytes, occurred at 8 to 9 weeks of age. Of the immune system parameters examined only the PFC response to SRBCs was suppressed in males but not females. A 35 and 42% suppression of the PFC response was observed at 5 and 50 mg MXC/kg/d, respectively, in 9-week-old males. Unfortunately, due to an unanticipated reduction in live births in the 150 mg MXC/kg/d group, there were an insufficient number of males for evaluation of this endpoint. Nevertheless, the male data indicate that the T cell-dependent antibody response to SRBCs was suppressed 3 weeks after the final exposure to MXC. These data suggest that not only does suppression of this immune endpoint "persist" for at least 3 weeks, but also that a dose lower than 5 mg MXC/kg may also result in immunosuppression.

## Heptachlor

The second organochlorine pesticide evaluated under the above dosing schedule, with the exception that dosing began on GD 12 rather than GD 14, was heptachlor (HEP) (Smialowicz et al. 2001). HEP is a chlorinated cyclodiene pesticide that was used primarily as an agricultural and domestic insecticide from the mid 1960s to the early 1980s. In 1976, the U.S. EPA canceled HEP registration for all uses except for subterranean termite control and fire ant control, primarily on pineapple crops. Because of its efficacy in the control of fire ants, which influence the survival of the mealybug, it was considered to be essential for protecting pineapples in Hawaii from lethal mealybug wilt. For 15 months, from 1981 to 1982, the commercial milk supply of Oahu, Hawaii, was contaminated with heptachlor epoxide (HEPX), the major metabolite of HEP, which is more toxic and more stable in biological systems than is HEP (Fendick et al. 1990). The source of HEPX in the milk on Oahu was HEP-tainted chopped pineapple leaves mixed into dairy cattle feed (Baker et al. 1991). The amount of HEPX in the dairy milk was found to be more than nine times FDA's "action level" of 0.3 ppm (Smith 1982). In the HEP study summarized here (Smialowicz et al. 2001), the doses (i.e., 0.03, 0.3 or 3.0 mg HEP/kg/d) were adjusted so that the low dose gave rat dam milk values of HEPX that approximated the 95th percentile of human milk HEPX values in Oahu, Hawaii, in 1981 (Baker et al. 1991; Siegel 1988), thereby establishing an environmentally relevant level of exposure to HEP.

Rats were evaluated using a battery of immune function tests, including splenic LP responses to mitogens; splenic lymphocyte flow cytometry phenotype analysis; MLR assay; NK cell activity; DTH response to BSA; CHS response to DNFB; and the antibody response to SRBCs, as determined by an enzyme-linked immunosorbent assay (ELISA).

The results indicated that the amount of HEP and HEPX found in milk, blood, fat and tissues was proportional to the dose of HEP administered. There were no effects on splenic NK cell activity, LP responses to mitogens, or the MLR. While HEP exposure of males resulted in a decrease in the DTH response to BSA at all doses, these decreases were not significant. In contrast, the CHS response to DNFB was suppressed at all doses in 17-week-old males but not females (Smialowicz 2002). The primary IgM antibody response to SRBCs was suppressed at all doses in males but not females at 8 weeks of age. The percentage of splenic B lymphocytes and plasma cells (i.e., $OX12^{+}OX19^{-}$ lymphocytes) was also reduced in the high-dose males at 8 weeks of age. At 26 weeks of age, the secondary IgG antibody response to SRBCs was suppressed in all of the HEP-exposed males, but not females. Taken together, these data indicate that perinatal exposure of male rats to HEP results in suppression of the primary IgM and secondary IgG anti-SRBC responses, as well as the CHS response to DNFB. Suppression of the antibody responses to SRBCs persisted for up to 20 weeks after the last exposure to HEP, at all of the doses employed, including the lowest dose that resulted in HEPX concentrations in the dam's milk comparable to those detected in human milk on Oahu, Hawaii, in 1981 (Baker et al. 1991; Siegel 1988). The "persistent" suppression of the anti-SRBC response occurred consequent to a total exposure of approximately 1.5 mg HEP/kg/rat. Evaluation of the antibody response to SRBCs in HEP-exposed dams, 6 to 8 weeks after parturition, indicated no alteration in this immune response (Smialowicz 2002). Taken together, these results indicate that perinatal, but not adult, exposure to HEP results in "persistent" suppression of immune system function in males but not females.

## DRUGS

### Diazepam (Valium)

Diazepam (DZP) is a benzodiazepine (BDZ) with central nervous system (CNS)-depressant properties. It has anxiolytic, sedative/hypnotic, muscle relaxant, and anticonvulsant effects. DZP is used in humans for the short-term relief of symptoms related to anxiety disorders; the treatment of agitation, tremors, delirium, seizures, and hallucinations as a result of alcohol withdrawal; the relief of muscle spasms in certain neurological diseases; and the control of active seizures (Costa and Guidotti 1979). GABA, the predominant inhibitory neurotransmitter in the brain, affects the activation of GABA receptors, which are ligand-gated chloride ion channels that mediate inhibitory synaptic transmission in the vertebrate CNS. BDZs potentiate GABA-mediated inhibition via the $GABA_A$ receptor in the CNS (Morrow 1995). In addition, the peripheral BDZ receptor (PBR) is expressed by several fetal and adult tissues and cells (Burgi et al. 1999; Krueger 1991). DZP has equal affinities for both central and peripheral type PBRs (Mohler and Okada 1977).

A study by Livezey et al. (1986) demonstrated that the offspring of SD pregnant rats, given daily subcutaneous (s.c.) injections of 6 mg DZP/kg on GD 15-20, had lower plasma total IgG concentrations, higher numbers of leukocytes, and higher incidence of uterine, lung, and skin infections as well as tumors, compared to controls. These DZP-induced adverse effects on the developing immune system were confirmed and extensively expanded upon by Schlumpf et al. (1995). Pregnant Long Evans (LE) rats were given DZP at 1.25 mg/kg in sterile olive oil, from GD 14 to GD 20, via s.c. injection. This dose resulted in drug tissue concentrations in the nanomolar range, comparable to the upper human treatment range. At birth no drug was detectable in dams or pups (Schlumpf et al. 1992).

In early studies, Schlumpf et al. (1989a, b) demonstrated that exposure to DZP during late (i.e., GD 14 to 20) rather than during early (i.e. GD 12 to 16) gestation resulted in suppression, of 50% or more, of the T lymphocyte LP responses to the mitogen ConA and in the MLR. These T cell responses were suppressed for up to two months of age in both male and female offspring, after which they returned to normal. A link was made between the fact that central BZD receptors are not present in all brain regions of the fetal rat brain during the early fetal period (GD 12 to 16), when T cell responses were not suppressed, while BZD receptors are present during the late period (GD 14 to 20) when suppression was observed (Schlumpf et al. 1989b). Subsequently, studies were performed to determine the physiological changes underlying DZP-induced immunosuppression.

The effects that perinatal DZP exposure has on splenocyte production of TNF-α, IL-6, IL-1 and IL-2 were examined in three different studies (Schreiber et al. 1993a; 1993b; Schlumpf et al. 1993a). Splenocytes from male offspring of LE dams exposed to DZP at 1.25 mg/kg on GD 14 to 20 were stimulated *in vitro* with LPS or ConA. IL-6 production from splenocytes of 2- and 8-week-old rats stimulated with LPS had lower IL-6 production than did controls. Splenocytes of 2-week-old male rats stimulated with ConA also had reduced IL-6 production compared to controls. In addition, in 8-week-old rats, IL-6 production by ConA stimulation of spleen cells and spleen macrophages was lower than controls (Schreiber et al. 1993a).

TNF-α production by spleen cells following stimulation by LPS, but not ConA, was reduced compared to control in 2- and 4-week-old males from dams exposed to DZP on GD 14 to 20. When splenic lymphocytes and macrophages were "purified," TNF-α production was found only in macrophage cultures stimulated with LPS. These results indicate that the TNF-α produced by LPS-stimulated splenocytes was due to splenic macrophages. The data suggest that an alteration in TNF-α production by macrophages is involved in the suppression of T cell-mediated immune responses in prenatally DZP-exposed rats (Schreiber et al. 1993b).

IL-1 production, following LPS *in vitro* stimulation of splenocytes from prenatally DZP-exposed males and females, was decreased compared to controls at 5 to 12 weeks of age and 9 to12 weeks of age, respectively. IL-2 production was reduced in ConA-stimulated splenocytes from males and females at 5 to 9 weeks of age. While IL-2 production was decreased in DZP-exposed male and female offspring, the percentage of IL-2 receptor expressing $CD45^+$ splenocytes was not different from that of control rats (Schlumpf et al. 1993a). These data implicate T lymphocytes, along with macrophages, as targets for DZP-induced immunosuppression in offspring of dams exposed on GD 14 to 20 (Schlumpf et al. 1993a; Schreiber et al. 1993a; 1993b).

DZP is an agonist not only at the central $GABA_A$ receptor complex, but also at the peripheral type T3 BDZ receptor. The peripheral T3 BDZ receptor is expressed during the early fetal period in rat placenta and fetuses (Schlumpf et al. 1990). It is also found on macrophages (Zavalac et al. 1984), rat spleen cells (Schlumpf et al. 1990), and polymorphonuclear leukocytes (Zavala et al. 1990). In light of this, the effects that DZP has on cytokine production and altered immune system development in the rat may be due to alterations in the binding efficiency of T3 on splenic macrophages and lymphocytes. To address this possibility, Schlumpf et al. (1993b) measured binding of [$^3$H] PK 11195, a T3 receptor ligand, to splenic macrophage and lymphocyte membranes from rats prenatally exposed to DZP. A marked decrease in maximum binding capacity of splenic macrophages membranes from both males and females, compared to controls, was observed at 8 weeks of age. Splenic lymphocytes showed a decreased binding affinity at both 2 and 8 weeks of age. As indicated above, depression of cellular immune responses and altered cytokine production were observed during these same postnatal periods.

Evidence of a link between suppressed macrophage and lymphocyte function and reduced cytokine production, in offspring of dams exposed to DZP on GD 14 to 20, and increased overall immunosuppression, was demonstrated using the PFC antibody response to SRBCs and the *T. spiralis* host-resistance model (Butikofer et al. 1993; Schlumpf et al. 1994). Four-week-old male and female offspring had significantly reduced (i.e., approximately 50% suppression) PFC responses compared with controls (Butikofer et al. 1993; Schlumpf et al. 1995). This immunosuppression coincides with the observed maximum inhibition of T lymphocyte proliferative responses (Schlumpf et al. 1989b). Male offspring of dams exposed to DZP were infected with *T. spiralis* at 8 weeks of age (Schlumpf et al. 1994). The number of total *T. spiralis* larvae per carcass and number of muscle larvae was increased at 12 weeks of age. IgG antibody titers to *T. spiralis* antigen were decreased, while IgA titers were elevated in 25-week-old rats. These data demonstrate that alterations in the cell biology of lymphocytes and macrophages in prenatally DZP-exposed rats results in significant suppression of the antibody response to SRBCs and decreased resistance to infection.

Postnatal exposure to DZP, given as either a single s.c. injection of 10 mg/kg on PND7 or three repeated injections of 5 mg/kg on PND 5 to 7 in male SD rats, was employed by Dostal et al. (1995) to determine if early postnatal exposure results in persistent immunosuppression in 6-,7-, 12-, and/or 24-month-old rats. The DTH response to BSA and the LP response of splenic lymphocytes to ConA were suppressed compared to controls in 6-month-old males given a single injection of 10 mg DZP/kg on PND7. In addition, the IgM and IgG antibody responses to SRBCs and OVA were suppressed in 12- and 24-month-old rats, dosed at 10 mg DZP/kg on PND7. Seven-month-old rats exposed to 5 mg DZP/kg on PND5-7 had suppressed IgM antibody responses to SRBCs and suppressed anti-OVA IgM and IgG antibody titers. These results, along with those of Schlumpf et al. (1995), demonstrate that both pre- and postnatal exposure of rats to DZP results in persistent immunosuppression.

## Acyclovir

Acyclovir is an effective antiviral drug that selectively inhibits herpes virus replication. It is phosphorylated by viral thymidine kinase to a triphosphate, which inhibits the DNA polymerase and thus the formation of viral DNA. Host cells are not significantly affected by acyclovir. Herpes simplex viruses are more sensitive to acyclovir than are varicella-zoster viruses. Resistant herpes viruses, that lack the thymidine kinase, have been observed almost exclusively in immunocompromised subjects (Collins et al. 1989).

The effect that prenatal exposure of rats to acyclovir has on immune system development was examined by Stahlmann et al. (1992). Pregnant Wistar rats were exposed to 100 mg acyclovir/kg body wt either with one or three (i.e., 1x100 and 3x100, respectively) s.c. injections on GD 10. There was considerable mortality during the first week after birth of pups from dams given 3x100 mg acyclovir/kg. Body weights of 12-week-old offspring born to dams exposed to 3x100 were reduced by 12.5 to 18.8% compared to controls for males and females respectively. Thymus and liver weights were also reduced by 22.1% to 35.5% and 8.8% to 11.6%, respectively, in both male and female offspring compared with controls.

Offspring were infected with *T. spiralis* at 6 weeks of age and assessed for host resistance to this parasitic nematode. Host resistance was evaluated by titration of Ig isotype antibody responses to *T. spiralis* antigen in offspring as well as determination of muscle *T. spiralis* larvae burdens. IgG, IgA and IgE anti-*T. spiralis* antibody titers were suppressed in 8- to 12-week-old 3x100 prenatally acyclovir-exposed offspring. The number of *T. spiralis* larvae in tongue muscle was significantly increased compared to that of controls. These results indicate that prenatal acyclovir exposure results in a long lived or "persistent" decrease in *T. spiralis*-specific Ig isotype antibodies and decreased resistance to *T. spiralis* infection (Stahlmann et al. 1992).

## Dexamethasone

Dexamethasone (DEX) is a synthetic adrenocortical steroid that suppresses inflammation and normal immune responses. It is used systemically and locally to treat chronic inflammatory disorders, severe allergies, lymphomas, shock, CNS trauma, and to reduce high blood calcium levels. DEX has also been used in preterm infants to decrease morbidity and mortality associated with bronchopulmonary dysplasia and to prevent chronic lung disease consequent to neonatal respiratory distress syndrome (Mammel et al. 1983; Cummings et al. 1989). DEX therapy improved pulmonary and neurodevelopmental outcomes in very-low-birth weight infants at high risk for bronchopulmonary dysplasia (Cummings et al. 1989). DEX treatment also hastened weaning infants from mechanical ventilation; however infections occurred in a substantial proportion of patients studied by Mammel et al. (1983).

Bakker et al. (2000) investigated the effects that DEX treatment of neonatal rats has on susceptibility to and severity of experimental autoimmune encephalomyelitis (EAE). Newborn female Wistar rats were injected intraperitoneally (i.p.) with DEX on PND1, 2 and 3 with 0.5, 0.3, and 0.1μg DEX/g/body wt, respectively. At 8 weeks of age, females exposed to DEX had decreased body weights compared with controls. These rats were injected s.c. in one hind footpad with 1.5 mg of guinea pig myelin basic protein (MBP) in incomplete Freunds adjuvant. Neurological aberrations were examined daily and graded 0 to 5 as follows: 0, no EAE; 0.5, loss of tip tail tonus; 1, loss of tail tonus; 2, partial tail paralysis; 3, complete tail paralysis; 4, hind limb paresis; and 5, hind limb paralysis. Disease severity was scored by observers in a blinded fashion. One week following MBP injection the first clinical signs of EAE appeared. The severity of EAE peaked at 13 days postimmunization and was significantly greater in DEX-treated rats compared to both saline and untreated cage control rats. Nine-week-old MBP-treated rats were injected with LPS and plasma corticosterone (CORT) levels were determined. Peak CORT levels were reduced in neonatal DEX-exposed versus control rats. These results are interesting in that an enhanced susceptibility to autoimmune disease has been correlated with reduction of the CORT response (Wick et al. 1993).

Peritoneal macrophages from 9- to 11-week-old neonatally exposed DEX female rats produced decreased levels of TNF-$\alpha$ and IL-1$\alpha$ upon stimulation with LPS. The reduced cytokine responses by macrophages suggest that these macrophages may be less able to mount an adequate immune response to infections. Spleen cells, from these same rats, cultured with ConA expressed increased mRNA levels of the proinflammatory cytokines TNF-$\alpha$ and IFN-$\gamma$. Taken together, these data suggest that neonatal exposure to DEX may be a risk factor for subsequent development of autoimmune disease and/or increased susceptibility to bacterial infections later in life (Bakker et al. 2000).

## SUMMARY

The evidence presented demonstrates that the rat is a robust yet sensitive rodent species for developmental immunotoxicity testing. The immune function assays performed in prepubertal and adult rats following pre- or postnatal exposure (i.e., during gestational, lactational, and/or juvenile development) to different classes of environmental chemicals or drugs proved to be predictive of altered immune function. The chemicals that have been examined for developmental immunotoxicity in the rat encompass a broad range of classes including metals, pesticides, drugs, and aromatic hydrocarbons. Suppression of immune function was observed in adult rats exposed to each of these chemicals during immune system development. The duration of immune function suppression in the rats so exposed ranged from 3 weeks (i.e., DOTC and MXC) to 19 months (i.e., TCDD) after the last exposure to the chemical.

Despite the limited data that exists, the rat appears to be a relatively sensitive species for identifying chemicals that alter the developing immune system in such a way as to compromise not only postweaning but also adult immune function. The alterations initiated during rat immune system development occur at lower doses than those required to suppress adult immune function, as well as "persist" for several weeks to several months after cessation of dosing, depending on the chemical. An interesting finding was the fact that male rats were more profoundly affected by developmental exposure to certain chemicals (i.e., TCDD, MXC, HEP, DZP) than were females. Whether these gender differences are authentic, and as such can be associated with a consequent perturbation of the neuro-endocrine-immune network, remains to be determined. The choice of the period of exposure to the chemical during immune system development is dependent on the pharmacokinetics of the chemical relative to placental and lactational transport of the parent compound or its metabolites. Based on the data review presented, for screening potential developmental immunotoxicants, it is recommended that dosing of the rat dam should occur during gestation and the first week of lactation, followed by direct dosing of the pups through 42 days of age. This exposure paradigm, as described by Chapin et al. (1997) and modeled after the multigenerational reproductive test guideline (EPA 1998b), could reasonably be employed to perform a comprehensive evaluation of the developmental vulnerability of not only the reproductive system but also the immune system.

## DISCLAIMER

This chapter has been reviewed by the Environmental Protection Agency's Office of Research and Development and approved for publication. Approval does not signify that the contents necessarily reflect the views and policies of the Agency, nor does mention of trade names or commercial products constitute endorsement or recommendation of use.

## REFERENCES

Aldridge WN. 1990. An assessment of the toxicological properties of pyrethroids and their neurotoxicity. *CRC Crit Rev Toxicol* 21:89–104.

Badesa JS, Maliji G, and Flaks B. 1995. Immunotoxic effects of exposure of rats to xenobiotics via maternal lactation. Part I 2,3,7,8-tetrachlorodibenzo-*p*-dioxin. *Int J Exp Path* 76:425–439.

Baker DB, Loo S, and Barker J. 1991. Evaluation of human exposure to the heptachlor epoxide contamination of milk in Hawaii. *Hawaii Med J* 50:108–118.

Bakker JM, Kavelaars A, Kamphuis PJGH, Cobelens PM, van Vugt HH, van Bel F, and Heijnen CB. 2000. Neonatal dexamethasone treatment increases susceptibility to experimental autoimmune disease in adult rats. *J Immunol* 165:5932–5937.

Barnett JB, Barfield L. Walls R, Joyner R, Owens R, and Soderberg LSF. 1987. The effect of *in utero* exposure to hexachlorobenzene on the developing immune response of BALB/C mice. *Toxicol Lett* 39:263–274.

Barnett JB. 1995. Developmental immunotoxicology. In: Smialowicz RJ and Holsapple MP (Eds.), *Experimental Immunotoxicology.* CRC Press, Boca Raton, FL, pp. 47–62.

Beck H, Eckart K, Mathar W, Wittkowski R. 1989. PCDD and PCDF body burden from food intake in the Federal Republic of Germany. *Chemosphere* 18:417–424.

Beyer WN. 1996. Accumulation of chlorinated benzenes in earthworms. *Bull Environ Contam Toxicol* 57:729–736.

Blaylock BL, Holladay SD, Comment CE, Heindel JJ, and Luster MI. 1992. Exposure to tetrachlorodibenzo-*p*-dioxin (TCDD) alters fetal thymocyte maturation. *Toxicol Appl Pharmacol* 112:207–213.

Bunn TL, Parsons PJ, Kao E, and Dietert RR. 2001a. Gender-based profiles of developmental immunotoxicity to lead in the rat: assessment in juveniles and adults. *J Toxicol Environ Health* 64 (Part A): 223–240.

Bunn TL, Parsons PJ, Kao E, and Dietert RR. 2001b. Exposure to lead during critical windows of embryonic development: differential immunotoxic outcome based on stage of exposure and gender. *Toxicol Sci* 64:57–66.

Burgi B, Lichtensteiger W, Lauber ME, and Schlumpf M. 1999. Ontogeny of diazepam binding inhibitor/acyl-CoA binding protein mRNA and peripheral benzodiazepine receptor mRNA expression in the rat. *J Neuroendocrinol* 11:85–100.

Burgi B, Lichtensteiger W, and Schlumpf M. 2000. Diazepam-binding inhibitor/acyl-CoA-binding protein mRNA and peripheral benzodiazepine receptor mRNA in endocrine and immune tissues after prenatal diazepam exposure of male and female rats. *J Endocrinol* 166:163–171.

Butikofer EE, Lichtensteiger W, and Schlumpf M. 1993. Prenatal exposure to diazepam causes sex-dependent changes of the sympathetic control of rat spleen. *Neurotoxicol Teratol* 15:377–382.

Caprio RJ, Margulis HL, and Joselow MM. 1974. Lead absorption in children and its relationship to urban traffic densities. *Arch Environ Health* 28:195–197.

Chapin RE. 2002. The use of the rat in developmental immunotoxicology studies. *Hum Exp Toxicol* 21:521–523.

Chapin RE, Harris MW, Davis BJ, Ward SM, Wilson RE, Mauney MA, Lockhart AC, Smialowicz RJ, Moser VC, Barone S, Padilla S, Burka LT, and Collins BJ. 1997. The effects of perinatal/juvenile methoxychlor exposure on adult rat nervous, immune and reproductive system function. *Fundam Appl Toxicol* 40:138–157.

Chen S, Golemboski KA, Sanders FS, and Dietert RR. 1999. Persistent effect of *in utero* meso-2,3-dimercaptosuccinic acid (DMSA) on immune function and lead-induced immunotoxicity. *Toxicol* 132:67–79.

Collins P, Larder BA, Oliver N M, Kemp S, Smith IW, and Darby G. 1989. Characterization of a DNA polymerase mutant of herpes simplex virus from a severely immunocompromised patient receiving acyclovir. *J Gen Virol* 70:375–382.

Costa E and Guidotti A. Molecular mechanisms in receptor action of benzodiazepines. 1979. *Ann Rev Pharmacol Toxicol* 19:531–545.

Committee on Biological Effects of Atmospheric Pollutants, Division of Medical Sciences. National Research Council. 1972. *Lead: airborne lead in perspective*, pp. 71–77, National Academy of Sciences, Washington, D.C.

Couture LA, Abbott BD, and Birnbaum LS. 1990. A critical review of the developmental toxicity and teratogenicity of 2,3,7,8-tetrachlorodibenzo-*p*-dioxin: recent advances toward understanding the mechanism. *Teratol* 42:619–627.

Cummings JJ, D'Eugenio DB, and Gross SJ. 1989. A controlled trial of dexamethasone in preterm infants at high risk for bronchopulmonary dysplasia. *N Engl J Med* 320:1505–1510.

Dostal M, Benesova O, Tejkalova H, and Soukupova D. 1995. Immune response of adult rats is altered by administration of diazepam in the first postnatal week. *Reprod Toxicol* 9:115–121.

EPA. 1998a. *U.S. Environmental Protection Agency, Prenatal Developmental Toxicity Study, Health Effects Test Guidelines*, OPPTS 870.3700, EPA 712-C-98-207. http://www.epa.gov/opptsfrs/OPPTS_Harmonized/870_Health_Effects_Test_Guidelines/Series/870-6300.pdf

EPA. 1998b. *U.S. Environmental Protection Agency, Reproductive and Fertility Effects, Health Effects Test Guidelines*, OPPTS 870.3800, EPA 712-C-98-208. http://www.epa.gov/opptsfrs/OPPTS_Harmonized/870_Health_Effects_Test_Guidelines/Series/870-6300.pdf

EPA. 1998c. *U.S. Environmental Protection Agency, Developmental Neurotoxicity Study, Health Effects Test Guidelines*, OPPTS 870.6300, EPA 712-C-98-239. http://www.epa.gov/opptsfrs/OPPTS_Harmonized/870_Health_Effects_Test_Guidelines/Series/870-6300.pdf

Faith RE and Moore JA. 1977. Impairment of thymus-dependent immune functions by exposure of the developing immune system to 2,3,7,8-tetrachlorodibenzo-*p*-dioxin (TCDD). *J Toxicol Environ Health* 3:451–464.

Faith RE, Luster MI, Kimmel CA. 1979. Effect of chronic developmental lead exposure on cell-mediated immune function. *Clin Exp Immunol* 35:413–420.

Fan F, Wierda D, and Rozman KK. 1996. Effects of 2,3,7,8-tetrachlorodibenzo-*p*-dioxin on humoral and cell-mediated immunity in Sprague-Dawley rats. *Toxicol* 106:221–228.

FDA. 2000. *U.S. Food and Drug Administration, Toxicological Principles for the Safety of Food Ingredients, Redbook 2000*. http://www.cfsan.fda.gov/~redbook/red-toca.html

Fendick EA, Mather-Mihaich E, Houck KA, St. Clair MB, Faust JB, Rockwell CH, Owens M. 1990. Ecological toxicology and human health effects of heptachlor. *Rev Environ Contam Toxicol*, 111:61–142.

Fine JS, Gasiewicz TA, and Silverstone AE. 1989. Lymphocyte stem cell alterations following perinatal exposure to 2,3,7,8-tetrachlorodibenzo-*p*-dioxin. *Mol Pharmacol* 35:18–25.

FQPA. 1996. Food Quality Protection Act of 1996, U.S. Public Law 104–170.

Fürst P, Fürst C, and Groebel W. 1990. Levels of PCDDs and PCDFs in food-stuffs from the Federal Republic of Germany. *Chemosphere* 20:787–792.

Gehrs BC and Smialowicz RJ. 1997. Alterations in the developing immune system of the F344 rat after perinatal exposure to 2,3,7,8-tetrachlorodibenzo-*p*-dioxin. I. Effects on the fetus and the neonate. *Toxicol* 122:219–228.

Gehrs BC, Riddle MM, Williams WC, and Smialowicz RJ. 1997. Alterations in the developing immune system of the F344 rat after perinatal exposure to 2,3,7,8-tetrachlorodibenzo-*p*-dioxin. II. Effects on the pup and the adult. *Toxicol* 122:229–240.

Gehrs BC and Smialowicz RJ. 1999. Persistent suppression of delayed-type hypersensitivity in adult F344 rats after perinatal exposure to 2,3,7,8-tetrachlorodibenzo-*p*-dioxin. *Toxicol* 134:79–88.

Grabbe S and Schwarz T. 1998. Immunoregulatory mechanisms involved in elicitation of allergic contact hypersensitivity. *Immunol Today* 19: 37–44.

Gray LE, Kelce WR, Monosson E, Ostby JS, and Birnbaum LS. 1995. Exposure to TCDD during development permanently alters reproductive function in male Long-Evans rats and hamsters: reduced ejaculated and epididymal sperm numbers and sex accessory gland weights in offspring with normal androgenic status. *Toxicol Appl Pharmacol* 131:108–118.

Holladay SD, Lindstrom P, Blaylock BL, Comment CE, Germolec DR, Heindel JJ, and Luster MI. 1991. Perinatal thymocyte antigen expression and postnatal immune development altered by gestational exposure to tetrachlorodibenzo-*p*-dioxin (TCDD). *Teratol* 44:385–393.

Holladay SD and Smialowicz RJ. 2000. Development of the murine and human immune system: differential effects of immunotoxicants depend on time of exposure. *Environ Health Perspect* 108 (Suppl. 3):463–474.

Holsapple MP. 1995. Immunotoxicity of Halogenated Aromatic Hydrocarbons. In: Smialowicz RJ and Holsapple MP (Eds.), *Experimental Immunotoxicology.* CRC Press, Boca Raton, FL, pp. 265–305.

IARC. 1979. *International Agency for Research of Cancer (IARC) monograph on the evaluation of the carcinogenic risk of chemicals to humans: some halogenated hydrocarbons.* Vol. 20. International Agency for Research in Cancer, World Health Organization, Lyon, France, pp. 155–178.

ILSI. 1996. *International Life Sciences Institute, Research Needs on Age-related Differences in Susceptibility to Chemical Toxicants. Report of an ILSI Risk Science Institute Working Group.* Washington, D.C.: ILSI Risk Science Institute.

Institoris L, Siroki O, and Desi I. 1995. Immunotoxicity study of repeated small doses of dimethoate and methylparathion administered to rats over three generations. *Hum Exp Toxicol* 14:879–883.

Kampinga J and Aspinall R. 1990. Thymocyte differentiation and thymic microenvironment development in the fetal rat thymus: an immunohistological approach. In: Kendall MD and Ritter MA. (Eds.), *Thymus Update*, Vol. 3. Harwood Academic Publishers, London, pp. 149–186.

Korte M, Stahlmann R, and Neubert D. 1990. Induction of hepatic monooxygenases in female rats and offspring in correlation with TCDD concentrations after single treatment during pregnancy. *Chemosphere* 20:1193–1198.

Krueger KE. 1991. Peripheral-type benzodiazepine receptor: A second site of action for benzodiazepine. *Neuropsychopharmacology* 4:237–244.

Li, X, Weber LWD, and Rozman KK. 1995. Toxicokinetics of 2,3,7,8-tetrachlorodibenzo-*p*-dioxin in female Sprague-Dawley rats including placental and lactational transfer to fetuses and neonates. *Fund Appl Toxicol* 27:70–76.

Livesey GT, Marczynski TJ, McGrew EA, and Beluhan FZ. 1986. Prenatal exposure to diazepam: late postnatal teratogenic effects. *Neurobehav Toxicol Teratol* 8:433–440.

Loose LD, Pittman KA, Benitz KF, and Silkworth JB. 1977. Polychlorinated biphenyl and hexachlorobenzene induced humoral immunosuppression. *J Reticuloendothel Soc* 22: 253–267.

Loose LD, Silkworth JB, Pittman K, Benitz KF, and Mueller W. 1978. Impaired host resistance to endotoxin and malaria in polychlorinated biphenyl- and hexachlorobenzene-treated mice. *Infect Immun* 20: 30–35.

Luster MI, Faith RE, and Kimmel CA. 1978. Depression of humoral immunity in rats following chronic developmental lead exposure. *J Environ Path Toxicol* 1:397–402.

Mably TA, Bjerke DL, Moore RW, Gendron-Fitzpatrick, and Peterson RE. 1992. *In utero* and lactational exposure of male rats to 2,3,7,8-tetrachlorodibenzo-*p*-dioxin. 3. Effects on spermatogenesis and reproductive capability. *Toxicol Appl Pharmacol* 114:118–126.

Mammel MC, Green TP, Johnson DE, and Thompson TR. 1983. Controlled trial of dexamethasone therapy in infants with bronchopulmonary dysplasia. *Lancet* 1:1356–1358.

McLachlan MS. 1993. Digestive tract absorption of polychlorinated dibenzo-*p*-dioxins, dibenzofurans, and biphenyls in a nursing infant. *Toxicol Appl Pharmacol* 123:68–72.

Miller TE, Golemboski KA, Ha RS, Bunn T, Sanders FS, and Dietert RR. 1998. Developmental exposure to lead causes persistent immunotoxicity in Fischer 344 rats. *Toxicol Sci* 42:129–135.

Mohler H and Okada T. 1977. Demonstration of benzodiazepine receptors in the central nervous system. *Science* 198:849–851.

Morrow AL. 1995. Regulation of $GABA_A$ receptor function and gene expression in the central nervous system. In: Bradley RI and Harris RA (Eds.), *International Review of Neurobiology*, Vol. 38. Academic Press, San Diego, pp. 1–41.

Murray FJ, Smith FA, Nitschke KD, Humiston C.G., Kociba RJ, and Schwetz BA. 1979. Three-generation reproduction study of rats given 2,3,7,8-tetrachlorodibenzo-*p*-dioxin in the diet. *Toxicol Appl Pharmacol* 50:241–252.

NRC, 1993. National Research Council, *Pesticides in the Diets of Infants and Children*, National Academy Press, Washington, D.C.

OECD. 2001a. Organization for Economic Cooperation and Development, Guideline 414, Prenatal Developmental Toxicity Study.

OECD. 2001b. Organization for Economic Cooperation and Development, Guideline 416, Two Generation Reproduction Toxicity Study.

Pennicks AH and Pieters RHH. 1995. Immunotoxicity of organotins. In: Smialowicz RJ and Holsapple MP (Eds.), *Experimental Immunotoxicology*. CRC Press, Boca Raton, FL, pp. 229–243.

Piver WT. 1973. Organotin compounds: industrial applications and biological investigations. *Environ Health Perspect* 4:61–79.

Poland A and Knutson JC. 1989. 2,3,7,8-Tetrachlorodibenzo-*p*-dioxin and related halogenated aromatic hydrocarbons: examination of the mechanism of toxicity. *Ann Rev Pharm Toxicol* 22:517–554.

Rosen FS, Cooper MD, and Wedgwood RJP. 1995. The primary immunodeficiencies. *N Engl J Med* 333:431–440.

Santoni G, Cantalamessa F, Mazzucca L, Romagnoli S, and Piccoli M. 1997. Prenatal exposure to cypermethrin modulates rat NK cell cytotoxic functions. *Toxicol* 120:231–242.

Santoni G, Cantalamessa F, Cavagna R, Romagnoli S, Spreghini E, and Piccoli M. 1998. Cypermethrin-induced alterations of thymocyte distribution and functions in prenatally-exposed rats. *Toxicol* 125:67–78.

Santoni G, Cantalamessa F, Spreghini E, Sagretti O, Staffolani M, and Piccoli M. 1999. Alterations of T cell distribution and function in prenatally cypermethrin-exposed rats: possible involvement of catecholamines. *Toxicol* 138:175–187.

Schecter A, Startin J, Wright C, Kelly M, Päpke O, Lis A, Ball M, and Olson JR. 1994. Congener-specific levels of dioxins and dibenzofurans in U.S. food and estimated daily dioxin toxic equivalent intake. *Environ Health Perspect* 102:962–966.

Schielen P, Schoo W, Tekstra J, Oostermeijer HH, Seinen W, and Bloksma N. 1993. Autoimmune effects of hexachlorobenzene in the rat. *Toxicol Appl Pharmacol* 122: 233–243.

Schlumpf M, Ramseier H, and Lichtensteiger W. 1989a. Prenatal diazepam induced persisting depression of cellular immune responses. *Life Sci* 44:493–501.

Schlumpf M, Ramseier H, Abriel H, Youmbi M, Baumann, and Lichtensteiger W. 1989b. Diazepam effects on the fetus. *Neurotoxicol* 10:501–516.

Schlumpf M, Parmar R, Ramseier HR, Lichtensteiger W. 1990. Prenatal benzodiazepine immunosuppression: possible involvement of peripheral benzodiazepine site. *Dev Pharmacol Ther* 15:178–85.

Schlumpf M, Parmar R, Schreiber A, Ramseier H, Butikofer E, Abriel H, Barth M, Rhyner T, and Lichtensteiger W. 1992. Nervous and immune system as targets for developmental effects of benzodiazepine. *Dev Pharmacol Ther* 18:145–158.

Schlumpf M, Lichtensteiger W, and Ramseier H. 1993a. Diazepam treatment of pregnant rats differentially affects interleukin-1 and interleukin-2 secretion in their offspring during different phases of postnatal development. *Pharmacol Toxicol* 73:335–340.

Schlumpf M, Parmar R, and Lichtensteiger W. 1993b. Prenatal diazepam induced persisting down regulation of peripheral (T3) benzodiazepine receptors on rat splenic macrophages. *Life Sci* 52:927–934.

Schlumpf M, Lichtensteiger W, and van Loveren H. 1994. Impaired host resistance to *Trichinella spiralis* as a consequence of prenatal treatment of rats with diazepam. *Toxicol* 94:223–230.

Schlumpf M, Parmar R, Butikofer E, Inderbitzin S, Salili AR, Schreiber AA, Ramseier HR, van Loveren H, and Lichtensteiger W. 1995. Delayed developmental neuro- and immunotoxicity of benzodiazepines. *Arch Toxicol* (suppl.) 17:261–287.

Schreiber AA, Frei K, Lichtensteiger W, and Schlumpf M. 1993a. Alterations in interleukin-6 production by LPS- and ConA-stimulated mixed splenocytes, spleen macrophages and lymphocytes in prenatally diazepam-exposed rats. *Agent Actions* 39:166–173.

Schreiber AA, Frei K, Lichtensteiger W, and Schlumpf M. 1993b. The effect of prenatal diazepam exposure on TNF-$\alpha$ production by rat splenocytes. *Agent Actions* 38:265–272.

Schulte A, Althoff J, Ewe S, and Richter-Reichhelm HB. 2002. Two immunotoxicity ring studies according to OECD TG 407. Comparison of data on cyclosporin A and hexachlorobenzene. *Reg Toxicol Pharmacol* 36:12–21.

SDWA. 1996. Safe Drinking Water Act Amendment of 1996, U.S. Public Law 104–182.

Seinen W, Vos JG, van Krieken R, Pennicks A, Brands R, and Hooykaas H. 1977. Toxicity of organotin compounds. III. Suppression of thymus-dependent immunity in rats by di-*n*-butyltindichloride and di-*n*-octyltindichloride. *Toxicol Appl Pharmacol* 42:213–224.

Seinen W, Vos JG, Brands R, and Hooykaas H. 1979. Lymphocytotoxicity and immunosuppression by organotin compounds. Suppression of graft-versus-host reactivity, blast transformation, and E-rosette formation by di-*n*-butyltindichloride and di-*n*-octyltindichloride. *Immunopharmacol* 1:343–355.

Senanayake N and Karalliedde, L. 1987. Neurotoxic effects of organophosphorus insecticides: an intermediate syndrome. *New Engl J Med* 316:761–763.

Siegel BZ. 1988. *Heptachlor Epoxide in Mothers' Milk in Hawaii 1981–1984 and Its Relationship to the Recall of Contaminated Dairy Products on Oahu in 1982. Completion Report.* Honolulu Pesticide Hazard Assessment Project, Pacific Biomedical Center, University of Hawaii, Honolulu, Hawaii.

Smialowicz RJ. 2002. The rat as a model in developmental immunotoxicology. *Hum Exp Toxicol* 21:513–519.

Smialowicz RJ, Riddle MM, Rogers RR, Rowe DG, Luebke RW, Fogelson LD, and Copeland CB. 1988. Immunologic effects of perinatal exposure of rats to dioctyltin dichloride. *Journal of Toxicol Environ Health* 25:403–422.

Smialowicz RJ, Riddle MM, Rogers RR, Luebke RW, and Copeland CB. 1989. Immunotoxicity of tributyltin oxide in rats exposed as adults or pre-weanlings. *Toxicology* 57:97–111.

Smialowicz RJ, Williams WC, Copeland CB, Harris MW, Overstreet D, Davis, BJ, and Chapin RE. 2001. Effect of perinatal/juvenile heptachlor exposure on adult immune and reproductive system function in rats. *Toxicol Sci* 61:164–175.

Smith RJ. 1982. Hawaiian milk contamination creates alarm. *Science* 217:137–140.

Stahlmann R, Korte M, Van Loveren H, Vos JG, Theil R, and Neubert D. 1992. Abnormal thymus development and impaired function of the immune system in rats after prenatal exposure to aciclovyr. *Arch Toxicol* 66:551–559.

Theelen RMC, Liem AKD, Slob W, and van Wijnen JH. 1993. Intake of 2,3,7,8 chlorine substituted dioxins, furans, and planar PCBs from food in the Netherlands: median and distribution. *Chemosphere* 27:1625–1635.

Thiel R, Koch E, Ulbrich B, and Chahoud I. 1994. Peri- and postnatal exposure to 2,3,7,8-tetrachlorodibenzo-*p*-dioxin: effects on physiological development, reflexes, locomotor activity and learning behavior in Wistar rats. *Arch Toxicol* 69:79–86.

Tobin P. 1985. Known and potential sources of hexachlorobenzene. In: Morris CR and Cabral JRP (Eds.), *Hexachlorobenzene: Proceedings of an International Symposium.* Lyon, France, IARC Scientific Publications, pp. 3–11.

Vos JG and Moore JA. 1974. Suppression of cellular immunity in rats and mice by maternal treatment with 2,3,7,8-tetrachlorodibenzo-*p*-dioxin. *Int Arch Allergy* 47:777–794.

Vos JG, van Logten MJ, Kreegtenberg JG, and Kruizinga W. 1979a. Effects of hexachlorobenzene on the immune system of rats following combined pre- and postnatal exposure. *Drug Chem Toxicol* 2:61–76.

Vos JG, van Logten MJ, Kreegtenberg JG, Steerenberg PA, and Kruizinga W. 1979b. Hexachlorobenzene-induced stimulation of the humoral immune response in rats. *Ann NY Acad Sci* 320:535–550.

Vos JG, Brouwer GMJ, van Leeuwen FXR, and Wagenaar SJ. 1983. Toxicity of hexachlorobenzene in the rat following combined pre- and post-natal exposure: comparison of effects on immune system, liver and lung. In: Gibson GG, Hubbard R, and Parke DV (Eds.), *Immunotoxicity.* Academic Press, New York, NY, pp. 219–235.

Wick G, Hu Y, Schwartz S, and Kroemer G. 1993. Immunoendocrine communication via the hypothalamo-pituitary-adrenal axis in autoimmune diseases. *Endor Rev* 14:539–563.

Zavala F, Haumont J, and Lenfant M. 1984. Interaction of benzodiazepines with mouse macrophages. *Eur J Pharmacol* 106:561–566.

Zavala, F, Veber F, and Descamps-Latsch, B. 1990. Altered expression of neutrophil peripheral benzodiazepine receptor in X-linked chronic granulomatous disease. *Blood* 76:184–188.

Zelikoff JT and Cohen MD. 1995. Immunotoxicology of inorganic metal compounds. In: Smialowicz RJ and Holsapple MP (Eds.), *Experimental Immunotoxicology.* CRC Press, Boca Raton, FL, pp. 189–228.

CHAPTER 6

# The Pig as a Model for Developmental Immunotoxicology

**Hermann J. Rothkötter, Eveline Sowa, and Reinhard Pabst**

## CONTENTS



## INTRODUCTION

The examination of the developing immune system faces so many limitations that experimental animal models are necessary to analyze environmental influences during the ontogeny of the immune system. Rodent models alone cannot be the basis for risk assessment because the intrauterine period of the rodent's development is much shorter than that of humans. The pig provides a variety of experimental approaches for developmental immunotoxicology in addition to studies in dogs and monkeys. The pig's gestation period is 115 days, and the appearance of T and B lymphocytes in the blood and organs of the porcine embryo and fetus has been well examined. B cells in the pig fetus represent a naïve population developing without maternal idiotypic-antiidiotypic influences because of the six-layered porcine placenta. Sufficient uptake of colostrum during the first 48 hours after birth is

0-415-28457-0/05/$0.00+$1.50

essential for a healthy postnatal development. The different strains of standard pigs and miniature pigs and the increasing number of immunological reagents available provide the basis for the pig as important experimental model with which to analyze the risk of environmental hazards. Many immunotoxicological experiments have been performed on pigs; although often documented on public databases, they are not often documented in the form of original publications in international biomedical journals.

There is an urgent need to study the immunotoxicity of various compounds, e.g., environmental substances, drugs, nutritional chemicals, and new food proteins (Dean et al. 1998; Dean and Thurmond 1987; Hinton 1992; Hinton 2000; Worden and Roberts 1972), as well as the effects of therapeutical procedures such as the effects of gene transfer carriers (Morrissey et al. 2002). The developing immune system is particularly at risk from various environmental hazards, especially in the pre- and postnatal periods. Only undisturbed intrauterine development and early postnatal growth result in a healthy immune system. Recently, it became obvious that the early postnatal period is critical with respect to the development of tolerance and immunity, and also to the direction of T lymphocytes toward the T helper 1 and T helper 2 type of immune reactions, important reactions that can influence asthma development, for example (Holt and Sly 2002). Animal models reflecting the situation in humans are necessary for risk assessment and for understanding the intrauterine and early postnatal development of the human immune system. For these studies, using species close to humans is important. Rodents, often used for biomedical research, are essential for the overall understanding of the developing immune system. However, the gestation period of rodents and the state of maturation of the immune system at birth are completely different from those of humans; therefore, different animal models are necessary. Great debate exists over experiments in primates and nonhuman primates; only under clearly defined conditions can experiments in monkeys be tolerated (Goodman and Check 2002; Hinton 1992; Hinton 2000; Smith et al. 2001). In this situation, the pig is a suitable model positioned between studies in rodents, experiments in primates, and observations in humans.

## PIG RACES AND BREEDS

In experimental studies, pigs obtained from local breeders often lead to the use of regionally available pig strains. These pigs are mostly hybrid and are crossbreeds of different land races. The standardization of the experiments and the basis for comparisons between studies in pigs is far less developed in comparison to experiments in rodents. Many efforts have been made to analyze the specific genetic peculiarities of the numerous pig races and breeds (see URLs at the end of this chapter). Future immunotoxicological studies in pigs should therefore include statements on the peculiarities of the strain used. So far, typical immunological regulatory differences for pigs are not known. Pig samples taken at the slaughterhouse show a wide variety of immunological results. For example, the lymphocyte subset distribution described may not only depend on intraindividual differences, but also on differences between the herds (Saalmüller et al. 1987).

Certain minipig breeds are used for biomedical research, e.g., the Hormel-Hanford minipig, the Goettingen minipig (www.minipigs.dk), the Minnesota minipig (Travnicek et al. 1989), and the Yucatan minipig. Gut development and intestinal immunology have been studied in Goettingen minipigs (Rothkötter et al. 1999a; Rothkötter et al. 1999b; Thielke et al. 1999). For use in immunology and transplantation research, MHC-homozygous pigs have been bred (Sachs et al. 1976). The typical microbial flora of minipigs has been determined, including rearing techniques to maintain healthy flora (Hansen et al. 1997; Rinke 1997). The effects of air pollutants on the pigs in their pens has been analyzed (Urbain et al. 1993), and chemical methods such as liquid chromatographic-mass spectrometry (LC-MS) have been used for the identification and quantification of chemical compounds (Maurer 1998).

## THE PIG'S IMMUNE SYSTEM DURING PRENATAL DEVELOPMENT

In pigs, the gestation period lasts 115 days (for review about pigs in general, see Straw 1999). As a six-layered placenta separates the embryo/fetus from the mother's blood supply during intrauterine development, no maternal cells and immunoglobulins can pass into the developing piglet (Pescovitz et al. 1998; Straw 1999). The development of the human placenta is different, which should be taken into consideration in developmental studies (Bright and Ockleford 1995). So far it is not known whether the porcine six-layered placenta is also impassable for DNA or whether it is a portal for its entry into the embryo/fetus, as has been reported in mice (Doerfler and Schubbert 1998). The development of the lymphoid cell populations in the pig embryo and fetus has been examined. Blood islands are observed on about day 16, rudiments of the thymus are present on day 21, and on day 22 the developing spleen can be observed. On about day 30 CD3+ lymphoid cells are detected, and $\gamma/\delta$ T cells are present both in and outside the thymus (Sinkora et al. 1998). It has not yet been examined whether the $\gamma/\delta$ T cell development in pigs is comparable to that in other species (Kagnoff 1998), but in the later embryonic and fetal period $\gamma/\delta$ T cells are observed in the peripheral blood and in the intestinal wall (Trebichavsky et al. 1995). Extrathymic $\gamma/\delta$ T cells are present in pigs later in life; their prenatal origin is not known so far (Licence and Binns 1995; Thielke et al. 2003). The specific clones of $\gamma/\delta$ T cells observed in adult pigs may well have their origin in the embryonic and fetal periods (Holtmeier et al. 2002). The CD4 and CD8 lymphocyte subpopulations are found in the thymus on about day 44 of gestation (Tlaskalova-Hogenova et al. 1994). A great number of T cells are neither CD4+ nor CD8+ at week 6 of the intrauterine period.

The fetal B cells represent a naïve population developing without idiotypic-antiidiotypic influences from the maternal site as the porcine fetus develops without the presence of immunoglobulins from the mother. The genomic basis of the B cell repertoire is different from that of rodents and humans, indicating the need for a porcine specific examination of this part of the developing piglet (Butler et al. 1996). The first B cells ($\mu$-chain+) occur in the liver on about day 40 of gestation, and surface IgM+ cells are found in the spleen (day 50) and bone marrow (day 60). A small amount of immunoglobulin is secreted by B cells of the spleen and liver beginning with day 50 of gestation, spontaneous isotype switching from IgM to IgG occurring in the thymus (Cukrowska et al. 1996). Porcine fetal B cell areas react with various conserved molecules and antigens (Butler et al. 2000; Tlaskalova-Hogenova et al. 1994).

## POSTNATAL ADAPTATION

Directly after birth, the piglet takes up macromolecules from the intestinal lumen in a nonselective way. IgG from the sow's colostrum is transported into the organism of the newborn via enterocytes (Komuves et al. 1993; Komuves and Heath 1992). A rapid "closure" of the gut for the macromolecule uptake occurs within 24 and 48 hours after birth (Leece 1973). Colostral intake is essential for efficient immune reactions, and the germ-free colostrum-deprived "virgin" piglet can serve as model for the assessment of the development of defense against toxins (Lee et al. 1998). Lymphocytes migrating into the lactating mammary gland obviously provide the immunological information necessary for the production of secretory IgA that is released into the sow's milk for maintenance of humoral immunity in the offspring (Salmon 2000).

Ovalbumin-fed animals were used to follow antigen uptake by the sow and transfer of antigen and specific antibodies to the offspring via colostrum (Telemo et al. 1991). After antigen feeding during gestation and lactation, the antigen was tolerated by the piglets; however, antigen feeding during lactation alone resulted in induction of IgG antibody production and diarrhea (Bailey et al. 1994b). These experiments demonstrate the immunological competence of the porcine fetus and newborn against environmental antigens. The immune response may be tolerance or defense against the antigen. In lymph duct cannulation experiments in young pigs, the nutritional uptake of antigens was described (Kiriyama et al. 1988).

## THE PIG'S INTESTINAL IMMUNE SYSTEM DURING POSTNATAL DEVELOPMENT

The early postnatal development of the gut immune system is important for effective immunological responses towards nutritional and microbial antigens in the intestinal lumen. Adaptation to the commensal flora, defense against pathogens, and the development of oral tolerance to food components are essential immunological processes. Depending on the food composition and on the presence of oral food components, the morphology of both the intestinal mucosa and the intestinal immune cells changes (Ganessunker et al. 1999; Park et al. 1998; Shulman et al. 1988; Shulman 1988). In humans, it has been reported that the maturity of the child at birth and the feeding pattern in the early postnatal period influences health later on (Saarinen and Savilahti 2000; Siltanen et al. 2001). The pig may well represent an animal model suitable for studying these observations in more detail, as, for example, the development of the B lymphocyte subsets in the intestinal mucosa is comparable in humans and pigs (Perkkiö and Savilahti 1980; Rothkötter et al. 1991).

In pigs, the structure of the Peyer's patches under various antigenic stimuli has been described (Barman et al. 1997; Makala et al. 2000; Pabst et al. 1988), and the appearance of *lamina propria* and intraepithelial lymphocyte subpopulations have been analyzed (Rothkötter et al. 1991; Rothkötter et al. 1999b; Vega-López et al. 1993). Various methods to separate lymphocytes and dendritic cells from the intestine have been developed (Bailey et al. 1994a; Haverson et al. 1999; Haverson et al. 2000; Rothkötter et al. 1994; Solano-Aguilar et al. 2000). Nonradioactive *in situ* hybridization for major T cell cytokines has been performed (Sowa et al. 2002). In pigs, the techniques for the analysis of the intestinal immune system have been established; thus, test systems for the effects of antigens or putative hazardous environmental substances can be developed on a solid basis.

## THE PIG AS AN ANIMAL MODEL

The pig as an omnivorous species is of major interest for studying the development of the immune system. In comparison to rodents, the piglet is almost completely developed at birth. In terms of digestion, the pig's gastrointestinal tract can be compared in several structural aspects to that of humans. So far, the pig has been used for several experimental approaches; especially in transplantation research is the pig an essential animal model. This is reflected by nearly 6000 bibliographic citations between 1997 and 2002 (between 1991 and 1996, there were approximately 1530 citations). The possible use of pigs as donors for xenotransplantation has inspired much research in this area (Davis et al. 1996; Robinson et al. 1998; Tucker et al. 2002).

As the skin of humans resembles that of the pig in many morphological and functional aspects (including sunburn), porcine skin is often used as model for humans (de Lange et al. 1992; Monteiro-Riviere and Riviere 1996), and about 1150 studies on such topics have been published between 1997 and 2002. This area covers burns of the skin (Rice et al. 2000), models for wound healing (Liu et al. 1999), cutaneous vasculature (Rogers and Riviere 1994), percutaneous persorption (Rohatagi et al. 1997; Williams et al. 1994), and the regulation of integrin expression (Zhang and Monteiro-Riviere 1997).

There is a long history of using pigs for testing drugs, toxic compounds, and trace elements including radiochemicals (Buck and Ewan 1973; Dean 1978; Funaki 1972; Hallesy et al. 1973; Hansen and Olsen 1978; Hashimoto 1972; Hildebrand 1994; Jacob 1969; Klocking 1991; Leddicotte 1971; Liu et al. 1999; Osuna et al. 1982; Osuna and Edds 1982b; Osuna and Edds 1982a; Stannard 1973; van Genderen and van Esch 1968; Van Ryzin and Trapold 1980; von Keutz and Schluter 1998; Wilber 1980). The effects of mycotoxins derived from contaminated food have been studied using several approaches (Kuiper-Goodman 1991; Pedersen and Miller 1999; Rotter et al. 1996). A combination of biochemical analysis and surgical placement of venous cannulas in the portal vein and a peripheral vein resulted in the description of the metabolic pathways of vitamin A (Arnhold et al. 2002). The capacity of the porcine organ systems to cope with environmental substances has been analyzed in terms of enzymes and cellular functions (Freudenthal et al. 1976; Gillette and Stripp

1975; Schneider et al. 1977; Szabo et al. 1974; Warner 1977). Many aspects of the use of pigs for these studies have been reviewed (Boisseau 1993; Bollen and Ellegaard 1997; Dean and Thurmond 1987; Jones et al. 1999; Pallardy et al. 1998; Stannard 1973; Sterzl-Eckert and Greim 1996; Worden and Roberts 1972).

Furthermore, pigs are used in studying environmental effects, including electric fields (Sikov et al. 1987), and in gene therapy experiments (about 200 citations during the last few years) such as the follow-up of DNA carriers (Cunningham et al. 2002; Morrissey et al. 2002). In neurophysiological research, pigs have been used to analyze the continuous intraventricular infusion of phenytoin and valproic acid. The analysis combined the measurement of drug levels, as well as animal behavior and morphological studies of the brain (Martinez et al. 1991). The pig can serve as organ donor for the development of primary cell cultures, e.g., for primary hepatocytes (Koebe et al. 1999; Koebe et al. 2000). In porcine kidney cell cultures, nephrotoxicity tests are performed (L'Eplattenier et al. 1990).

This overview can only cover some of the many studies on toxicology in the pig, and few original papers have concentrated so far on the developmental aspects of immunotoxicology. Therefore, when planning studies in pig developmental immunotoxicology, it may be of importance to look beyond the standard bibliographic databases, e.g., to check the database of the National Agricultural Library, U.S. Department of Agriculture (see URLs at the end of this chapter). This database contains many studies using pigs for toxicological studies. Furthermore, an overview of the broad spectrum of experiments carried out in pigs in the past was in part summarized during a symposium a few years ago (Tumbleson and Schook 1996).

Important for the design of immunotoxicological studies is the option to establish intravenous or intralymphatic cannulations in pigs and to take blood or lymph samples for several days without the stress effect that is present in restrained rodents or sheep, for example (Sudo et al. 2001; Zhang et al. 1998). This is of particular interest after increasing numbers of studies have shown major lymphocyte subset-specific stress effects in laboratory animals and in humans (Hennig et al. 2000; Hennig et al. 2001; Ottaway and Husband 1992).

The lack of suitable immunological reagents in pigs was often a difficulty in designing experimental studies in the past. The increased interest in pig immunology from the field of xenotransplantation, where the pig is the donor of choice, has increased the numbers of immunological markers, e.g., monoclonal antibodies (Saalmüller et al. 1998) and cytokine detection (Murtaugh 1994). Antibody secretion can be studied using ELISA-spot assays, a method established in pigs (Baltes et al. 2001).

## SUMMARY

The pig is an important model of developmental immunology. Morphological and descriptive studies, as well as functional experiments, provide a basis to design further experimental models in this species. With a growing number of reagents of different kinds, the pig may now be a cost-effective experimental model between studies in rodents and primates and an important addition to observations on the developing immune system in humans.

## REFERENCES

Arnhold T, Nau H, Meyer S, Rothkötter HJ, and Lampen AD. 2002. Porcine intestinal metabolism of excess vitamin A differs following vitamin A supplementation and liver consumption. *J Nutr* 132: 197–203.

Bailey M, Hall L, Bland PW, and Stokes CR. 1994a. Production of cytokines by lymphocytes from spleen, mesenteric lymph node and intestinal lamina propria of pigs. *Immunol* 82: 577–583.

Bailey M, Miller BG, Telemo E, Stokes CR, and Bourne FJ. 1994b. Altered immune response to proteins fed after neonatal exposure of piglets to the antigen. *Int Arch Allergy Immunol* 103: 183 –187.

Baltes N, Tonpitak W, Gerlach GF, Hennig-Pauka I, Hoffmann-Moujahid A, Ganter M, and Rothkötter HJ. 2001. Actinobacillus pleuropneumonia iron transport and urease activity: effects on bacterial virulence and host immune response. *Infect Immun* 69: 472–478.

Barman NN, Bianchi ATJ, Zwart RJ, Pabst R, and Rothkötter HJ. 1997. Jejunal and ileal Peyer's patches in pigs differ in their postnatal development. *Anat Embryol* 195: 41–50.

Boisseau J. 1993. Basis for the evaluation of the microbiological risks due to veterinary drug residues in food. *Vet Microbiol* 35: 187–192.

Bollen P and Ellegaard L. 1997. The Gottingen minipig in pharmacology and toxicology. *Pharmacol Toxicol* 80 Suppl 2: 3–4.

Bright NA and Ockleford CD. 1995. Cytotrophoblast cells: a barrier to maternofetal transmission of passive immunity. *J Histochem Cytochem* 43: 933–944.

Buck WB and Ewan RC. 1973. Toxicology and adverse effects of mineral imbalance. *Clin Toxicol* 6: 459–485.

Butler JE, Sun J, Kacskovics I, Brown WR, and Navarro P. 1996. The VH and CH immunoglobulin genes of swine: implications for repertoire development. *Vet Immunol Immunopathol* 54: 7–17.

Butler JE, Sun J, Weber P, Navarro P, and Francis D. 2000. Antibody repertoire development in fetal and newborn piglets. III. Colonization of the gastrointestinal tract selectively diversifies the preimmune repertoire in mucosal lymphoid tissues. *Immunol* 100: 119–130.

Cukrowska B, Sinkora J, Mandel L, Splichal I, Bianchi ATJ, and Kovaru F. 1996. Thymic B cells of pig fetuses and germ-free pigs spontaneously produce IgM, IgG and IgA: detection by ELISPOT method. *Immunol* 87: 487–492.

Cunningham S, Meng QH, Klein N, McAnulty RJ, and Hart SL. 2002. Evaluation of a porcine model for pulmonary gene transfer using a novel synthetic vector. *J Gene Med* 4: 438–446.

Davis EA, Pruitt SK, Greene PS, Ibrahim S, Lam TT, Levin JL, Baldwin WM, III, and Sanfilippo F. 1996. Inhibition of complement, evoked antibody, and cellular response prevents rejection of pig-to-primate cardiac xenografts. *Transplantation* 62: 1018–1023.

de Lange J, van Eck P, Elliott GR, de Kort WL, and Wolthuis OL. 1992. The isolated blood-perfused pig ear: an inexpensive and animal-saving model for skin penetration studies. *J Pharmacol Toxicol Methods* 27: 71–77.

Dean BJ. 1978. Genetic toxicology of benzene, toluene, xylenes and phenols. *Mutat Res* 47: 75–97.

Dean JH and Thurmond LM. 1987. Immunotoxicology: an overview. *Toxicol Pathol* 15: 265–271.

Dean JH, Hincks JR, and Remandet B. 1998. Immunotoxicology assessment in the pharmaceutical industry. *Toxicol Lett* 102–103: 247–255.

Doerfler W and Schubbert R. 1998. Uptake of foreign DNA from the environment: the gastrointestinal tract and the placenta as portals of entry. *Wien Klin Wochenschr* 110: 40–44.

Freudenthal RI, Leber P, Emmerling D, Kerchner G, and Campbell D. 1976. Characterization of the hepatic microsomal mixed-function oxidase enzyme system in miniature pigs. *Drug Metab Dispos* 4: 25–27.

Funaki H. 1972. [Fate of experimentally administered drugs. 3. Several problems concerning animal experimentals]. *Jikken Dobutsu* 21: 86–88.

Ganessunker D, Gaskins HR, Zuckermann FA, and Donovan SM. 1999. Total parenteral nutrition alters molecular and cellular indices of intestinal inflammation in neonatal piglets. *JPEN-Parenter Enter* 23: 337–344.

Gillette JR and Stripp B. 1975. Pre- and postnatal enzyme capacity for drug metabolite production. *Fed Proc* 34: 172–178.

Goodman S and Check E. 2002. The great primate debate. *Nat* 417: 684–687.

Hallesy DW, Shott LD, and Hill R. 1973. Comparative toxicology of naproxen. *Scand J Rheumatol* Suppl 2:20–28.

Hansen E and Olsen P. 1978. Peroral toxicity of Orange RN in pigs. Early haemotological changes. *Arch Toxicol Suppl* 313–315.

Hansen AK, Farlov H, and Bollen P. 1997. Microbiological monitoring of laboratory pigs. *Lab Anim* 31: 193–200.

Hashimoto Y. 1972. [Fate of experimentally administered drugs. 4. Drug absorption and species specificity]. *Jikken Dobutsu* 21: 89–95.

Haverson K, Bailey M, and Stokes CR. 1999. T-cell populations in the pig intestinal lamina propria: memory cells with unusual phenotypic characteristics. *Immunol* 96: 66–73.

Haverson K, Singha S, Stokes CR, and Bailey M. 2000. Professional and non-professional antigen-presenting cells in the porcine small intestine. *Immunol* 101: 492–500.

Hennig J, Netter P, and Voigt K. 2000. Mechanisms of changes in lymphocyte numbers after psychological stress. *Z Rheumatol* 59 Suppl 2: II/43–II/48.

Hennig J, Netter P, and Voigt KH. 2001. Cortisol mediates redistribution of CD8+ but not of CD56+ cells after the psychological stress of public speaking. *Psychoneuroendocrinol* 26: 673–687.

Hildebrand M. 1994. Inter-species extrapolation of pharmacokinetic data of three prostacyclin-mimetics. *Prostaglandins* 48: 297–312.

Hinton DM. 1992. Testing guidelines for evaluation of the immunotoxic potential of direct food additives. *Crit Rev Food Sci Nutr* 32: 173–190.

Hinton DM. 2000. US FDA "Redbook II" immunotoxicity testing guidelines and research in immunotoxicity evaluations of food chemicals and new food proteins. *Toxicol Pathol* 28: 467–478.

Holt PG and Sly PD. 2002. Interactions between respiratory tract infections and atopy in the aetiology of asthma. *Eur Respir J* 19: 538–545.

Holtmeier W, Käller J, Geisel W, Pabst R, Caspary W, and Rothkötter HJ. 2002. The development and compartmentalisation of the TCR δ repertoire at mucosal and extraintestinal sites: the pig as a model for analyzing the effects of age and microbial factors. *J Immunol* 169: 1993–2002.

Jacob SW. 1969. Dimethyl sulfoxide (DMSO)—current concepts in toxicology, pharmacology, and clinical usefulness in surgery. *Am Surg* 35: 564–573.

Jones RD, Stuart BP, Greufe NP, and Landes AM. 1999. Electrophysiology and pathology evaluation of the Yucatan pig as a non-rodent animal model for regulatory and mechanistic toxicology studies. *Lab Anim* 33: 356–365.

Kagnoff MF. 1998. Current concepts in mucosal immunity. III. Ontogeny and function of γ/δ T cells in the intestine. *Am J Physiol* 274: G455–G458.

Kiriyama H, Harada E, and Syuto B. 1988. Continual collection of the thoracic duct lymph for investigation of the protein absorption in conscious newborn pigs. *Nutr Reps Int* 37: 779–784.

Klocking HP. 1991. Toxicology of hirudin. *Semin Thromb Hemost* 17: 126–129.

Koebe HG, Deglmann CJ, Metzger R, Hoerrlein S, and Schildberg FW. 2000. *In vitro* toxicology in hepatocyte bioreactors-extracellular acidification rate (EAR) in a target cell line indicates hepato-activated transformation of substrates. *Toxicol* 154: 31–44.

Koebe HG, Muhling B, Deglmann CJ, and Schildberg FW. 1999. Cryopreserved porcine hepatocyte cultures. *Chem Biol Interact* 121: 99–115.

Komuves LG and Heath JP. 1992. Uptake of maternal immunoglobulins in the enterocytes of suckling piglets: improved detection with a streptavidin-biotin bridge gold technique. *J Histochem Cytochem* 40: 1637–1646.

Komuves LG, Nicols BL, Hutchens TW, and Heath JP. 1993. Formation of crystalloid inclusions in the small intestine of neonatal pigs: an immunocytochemical study using colloidal gold. *Histochem J* 25: 19–29.

Kuiper-Goodman T. 1991. Risk assessment of ochratoxin A residues in food. *IARC Sci Publ* 307–320.

L'Eplattenier HF, Zhao JM, Pfannkuch F, Scholtysik G, and Wuthrich A. 1990. Cell culture in nephrotoxicity testing. *Toxicol Lett* 53: 227–229.

Leddicotte GW. 1971. Activation analysis of the biological trace elements. *Methods Biochem Anal* 19: 345–434.

Lee WJ, Farmer JL, Hilty M, and Kim YB. 1998. The protective effects of lactoferrin feeding against endotoxin lethal shock in germfree piglets. *Infect Immun* 66: 1421–1426.

Leece JG. 1973. Effect of dietary regimen on cessation of uptake of macromolecules by piglet intestinal epithelium (closure) and transport to the blood. *J Nutr* 103: 751–756.

Licence ST and Binns RM. 1995. Major long-term changes in γ/δ T-cell receptor-positive and CD2+ T-cell subsets after neonatal thymectomy in the pig: a longitudinal study lasting nearly 2 years. *Immunol* 85: 276–284.

Liu L, Teng G, Zhang D, Song J, He S, Guo J, and Fang W. 1999. Toxicology of intrahepatic arterial administration of interventional phosphorus-32 glass microspheres to domestic pigs. *Chin Med J* (Engl) 112: 632–636.

Makala LH, Kamada T, Nishikawa Y, Nagasawa H, Igarashi I, Fujisaki K, Suzuki N, Mikami T, Haverson K, Bailey M, Stokes CR, and Bland PW. 2000. Ontogeny of pig discrete Peyer's patches: distribution and morphometric analysis. *Pathobiol* 68: 275–282.

Martinez R, Vaquero J, De La Morena LV, Tendillo F, and Aragones P. 1991. Toxicology and kinetics of long-term intraventricular infusion of phenytoin and valproic acid in pigs: experimental study. *Acta Neurochir Suppl* (Wien) 52: 3–4.

Maurer HH. 1998. Liquid chromatography-mass spectrometry in forensic and clinical toxicology. *J Chromatogr B Biomed Sci Appl* 713: 3–25.

Monteiro-Riviere NA and Riviere J. 1996. The pig as a model for cutaneous pharmacology and toxicology research. In: Tumbleson, ME and Schook, LB (Eds.), *Advances in swine in biomedical research.* Plenum Press, New York, pp. 425–458.

Morrissey RE, Horvath C, Snyder EA, Patrick J, Collins N, Evans E, and MacDonald JS. 2002. Porcine toxicology studies of SCH 58500, an adenoviral vector for the p53 gene. *Toxicol Sci* 65: 256–265.

Murtaugh MP. 1994. Porcine cytokines. *Vet Immunol Immunopathol* 43: 37–44.

Osuna O and Edds GT. 1982a. Toxicology of aflatoxin B1, warfarin, and cadmium in young pigs: clinical chemistry and blood coagulation. *Am J Vet Res* 43: 1387–1394.

Osuna O and Edds GT. 1982b. Toxicology of aflatoxin B1, warfarin, and cadmium in young pigs: performance and hematology. *Am J Vet Res* 43: 1380–1386.

Osuna O, Edds GT, and Simpson CF. 1982. Toxicology of aflatoxin B1, warfarin, and cadmium in young pigs: metal residues and pathology. *Am J Vet Res* 43: 1395–1400.

Ottaway CA and Husband AJ. 1992. Central nervous system influences on lymphocyte migration. *Brain Behav Immun* 6: 97–116.

Pabst R, Geist M, Rothkötter HJ, and Fritz FJ. 1988. Postnatal development and lymphocyte production of Jejunal and ileal Peyer's patches in normal and gnotobiotic pigs. *Immunol* 64: 539–544.

Pallardy M, Kerdine S, and Lebrec H. 1998. Testing strategies in immunotoxicology. *Toxicol Lett* 102–103: 257–260.

Park YK, Monaco MM, and Donovan SM. 1998. Delivery of total parenteral nutrition (TPN) via umbilical catheterization: development of a piglet model to investigate therapies to improve gastrointestinal structure and enzyme activity during TPN. *Biol Neonate* 73: 295–305.

Pedersen PB and Miller JD. 1999. The fungal metabolite culmorin and related compounds. *Nat Toxins* 7: 305–309.

Perkkiö M and Savilahti E. 1980. Time of appearance of immunoglobulin-containing cells in the mucosa of the neonatal intestine. *Pediatr Res* 14: 953–955.

Pescovitz MD, Pabst R, Rothkötter HJ, Murtaugh MP, Foss DL, Butler JE, Lunney JK, Vogeli P, Llanes D, Trebichavsky I, Tlaskalova-Hogenova H, Cukrowska B, Sinkora J, Stokes C, Bailey M, Hogasen K, and Morgan BP. 1998. Immunology of the pig. In: Pastoret P-P et al. (Eds.), *Handbook of Vertebrate Immunology.* Academic Press, London, pp. 373–419.

Rice P, Brown RF, Lam DG, Chilcott RP, and Bennett NJ. 2000. Dermabrasion—a novel concept in the surgical management of sulfur mustard injuries. *Burns* 26: 34–40.

Rinke M. 1997. How clean is a mini-pig?—Impressions and suggestions of a pathologist working in the field of toxicology. *Pharmacol Toxicol* 80 Suppl 2: 16–22.

Robinson LA, Tu L, Steeber DA, Preis O, Platt JL, and Tedder TF. 1998. The role of adhesion molecules in human leukocyte attachment to porcine vascular endothelium: implications for xenotransplantation. *J Immunol* 161: 6931–6938.

Rogers RA and Riviere JE. 1994. Pharmacologic modulation of the cutaneous vasculature in the isolated perfused porcine skin flap. *J Pharm Sci* 83: 1682–1689.

Rohatagi S, Barrett JS, McDonald LJ, Morris EM, Darnow J, and DiSanto AR. 1997. Selegiline percutaneous absorption in various species and metabolism by human skin. *Pharm Res* 14: 50–55.

Rothkötter HJ, Hriesik C, Barman NN, and Pabst R. 1999a. B and also T lymphocytes migrate via gut lymph to all lymphoid organs and the gut wall, but only IgA$^+$ cells accumulate in the lamina propria of the intestinal mucosa. *Eur J Immunol* 29: 327–333.

Rothkötter HJ, Kirchhoff T, and Pabst R. 1994. Lymphoid and non-lymphoid cells in the epithelium and lamina propria of intestinal mucosa of pigs. *Gut* 35: 1582–1589.

Rothkötter HJ, Möllhoff S, and Pabst R. 1999b. The influence of age and breeding conditions on the number and proliferation of intraepithelial lymphocytes (IEL). *Scand J Immunol* 50: 31–38.

Rothkötter HJ, Ulbrich H, and Pabst R. 1991. The postnatal development of gut lamina propria lymphocytes: number, proliferation, and T and B cell subsets in conventional and germ-free pigs. *Pediatr Res* 29: 237–242.

Rotter BA, Prelusky DB, and Pestka JJ. 1996. Toxicology of deoxynivalenol (vomitoxin). *J Toxicol Environ Health* 48: 1–34.

Saalmüller A, Pauly T, Lunney JK, Boyd P, Aasted B, Sachs DH, Arn S, Bianchi A, Binns RM, Licence S, Whyte A, Blecha F, Chen Z, Chu RM, Davis WC, Denham S, Yang H, Whittall T, Parkhouse RM, Dominguez J, Ezquerra A, Alonso F, Horstick G, Howard C, and Zuckermann F. 1998. Overview of the Second International Workshop to define swine cluster of differentiation (CD) antigens. *Vet Immunol Immunopathol* 60: 207–228.

Saalmüller A, Reddehase MJ, Bühring HJ, Jonjic S, and Koszinowski UH. 1987. Simultaneous expression of CD4 and CD8 antigens by a substantial proportion of resting porcine T lymphocytes. *Eur J Immunol* 17: 1297–1301.

Saarinen KM and Savilahti E. 2000. Infant feeding patterns affect the subsequent immunological features in cow's milk allergy. *Clin Exp Allergy* 30: 400–406.

Sachs DH, Leight G, Cone J, Schwarz S, Stuart L, and Rosenberg S. 1976. Transplantation in miniature swine. I. Fixation of the major histocompatibility complex. *Transplantation* 22: 559–567.

Salmon H. 2000. Mammary gland immunology and neonate protection in pigs - Homing of lymphocytes into the MG. *Biol Mammary Gland* 480: 279–286.

Schneider NR, Bradley SL, and Andersen ME. 1977. Toxicology of cyclotrimethylenetrinitramine: distribution and metabolism in the rat and the miniature swine. *Toxicol Appl Pharmacol* 39: 531–541.

Shulman RJ. 1988. Effect of different total parenteral nutrition fuel mixes on small intestinal growth and differentiation in the infant miniature pig. *Gastroenterol* 95: 85–92.

Shulman RJ, Henning SJ, and Nichols BL. 1988. The miniature pig as an animal model for the study of intestinal enzyme development. *Pediatr Res* 23: 311–315.

Sikov MR, Rommereim DN, Beamer JL, Buschbom RL, Kaune WT, and Phillips RD. 1987. Developmental studies of Hanford miniature swine exposed to 60-Hz electric fields. *Bioelectromagnetics* 8: 229–242.

Siltanen M, Kajosaari M, Pohjavuori M, and Savilahti E. 2001. Prematurity at birth reduces the long-term risk of atopy. *J Allergy Clin Immunol* 107: 229–234.

Sinkora M, Sinkora J, Rehakova Z, Splichal I, Yang H, Parkhouse RM, and Trebichavsky I. 1998. Prenatal ontogeny of lymphocyte subpopulations in pigs. *Immunology* 95: 595–603.

Smith D, Trennery P, Farningham D, and Klapwijk J. 2001. The selection of marmoset monkeys (Callithrix jacchus) in pharmaceutical toxicology. *Lab Anim* 35: 117–130.

Solano-Aguilar GI, Vengroski KG, Beshah E, and Lunney JK. 2000. Isolation and purification of lymphocyte subsets from gut- associated lymphoid tissue in neonatal swine. *J Immunol Methods* 241: 185–199.

Sowa E, Merkel S, Pabst R, Whiting C, Bailey M, and Rothkötter HJ. 2002. Expression of cytokine mRNA in intestinal lymphocytes. *Ann Anat* 184 (Suppl) 138–139.

Stannard JN. 1973. Toxicology of radionuclides. *Annu Rev Pharmacol* 13: 325–357.

Sterzl-Eckert H and Greim H. 1996. Occupational exposure. *Food Chem Toxicol* 34: 1177–1178.

Straw BE. 1999. *Diseases of swine*. Iowa State University Press, Ames, Iowa.

Sudo N, Oyama N, Yu XN, and Kubo C. 2001. Restraint stress-induced elevation of endogenous glucocorticoids decreases Peyer's patch cell numbers via mechanisms that are either dependent or independent on apoptotic cell death. *Neuroimmunomodulation* 9: 333–339.

Szabo S, Selye H, Kourounakis P, and Tache Y. 1974. Comparative studies on the effect of adrenocorticotrophic hormone (ACTH) and pregnenolone-16alpha-carbonitrile (PCN) upon drug response and distribution in rats. *Biochem Pharmacol* 23: 2083–2094.

Telemo E, Bailey M, Miller BG, Stokes CR, and Bourne FJ. 1991. Dietary antigen handling by mother and offspring. *Scand J Immunol* 34: 689–696.

Thielke KH, Hoffmann-Moujahid A, Weisser C, Waldkirch E, Pabst R, Holtmeier W, and Rothkötter HJ. 2003. Proliferating intestinal γ/δ T cells recirculate rapidly and are a major source of the γ/δ T cell pool in the peripheral blood. *Eur J Immunol* 33: 1649–1656.

Thielke KH, Pabst R, and Rothkötter HJ. 1999. Quantification of proliferating lymphocyte subsets appearing in the intestinal lymph and the blood. *Clin Exp Immunol* 117: 277–284.

Tlaskalova-Hogenova H, Mandel L, Trebichavsky I, Kovaru F, Barot R, and Sterzl J. 1994. Development of immune responses in early pig ontogeny. *Vet Immunol Immunopathol* 43: 135–142.

Travnicek J, Mandel L, Trebichavsky I, and Talafantova M. 1989. Immunological state of adult germfree miniature Minnesota pigs. *Folia Microbiol* (Praha) 34: 157–164.

Trebichavsky I, Sinkora J, Rehakova Z, Splichal I, Whyte A, Binns R, Pospisil R, and Tuckova L. 1995. Distribution of gamma delta T cells in the pig foetus. *Folia Biol* (Praha) 41: 227–237.

Tucker A, Belcher C, Moloo B, Bell J, Mazzulli T, Humar A, Hughes A, McArdle P, and Talbot A. 2002. The production of transgenic pigs for potential use in clinical xenotransplantation: microbiological evaluation. *Xenotransplantation* 9: 191–202.

Tumbleson ME and Schook LB (Eds.). 1996. *Advances in Swine in Biomedical Research.* Plenum, New York.

Urbain B, Gustin P, Prouvost JF, Ansay M, Michel O, and Nicks B. 1993. Microclimate and air composition in a closed chamber meant for the study of the toxicity of atmospheric pollutants for the piglet. *Vet Res* 24: 503–514.

van Genderen H and van Esch GJ. 1968. Toxicology of the herbicide dichlobenil (2,6-dichlorobenzonitrile) and its main metabolites. *Food Cosmet Toxicol* 6: 261–269.

Van Ryzin RJ and Trapold JH. 1980. The toxicology profile of the anti-inflammatory drug proquazone in animals. *Drug Chem Toxicol* 3: 361–379.

Vega-López MA, Telemo E, Bailey M, Stevens K, and Stokes CR. 1993. Immune cell distribution in the small intestine of the pig: immunohistological evidence for an organized compartmentalization in the lamina propria. *Vet Immunol Immunopathol* 37: 49–60.

von Keutz E and Schluter G. 1998. Preclinical safety evaluation of cerivastatin, a novel HMG-CoA reductase inhibitor. *Am J Cardiol* 82: 11J–17J.

Warner WL. 1977. Toxicology and pharmacology of adenine in animals and man. *Transfusion* 17: 326–332.

Wilber CG. 1980. Toxicology of selenium: a review. *Clin Toxicol* 17: 171–230.

Williams PL, Brooks JD, Inman AO, Monteiro-Riviere NA, and Riviere JE. 1994. Determination of physicochemical properties of phenol, p-nitrophenol, acetone and ethanol relevant to quantitating their percutaneous absorption in porcine skin. *Res Commun Chem Pathol Pharmacol* 83: 61–75.

Worden AN and Roberts CN. 1972. Species choice in safety evaluation. *Jpn J Pharmacol* 22: Suppl 22.

Zhang D, Kishihara K, Wang B, Mizobe K, Kubo C, and Nomoto K. 1998. Restraint stress-induced immunosuppression by inhibiting leukocyte migration and Th1 cytokine expression during the intraperitoneal infection of Listeria monocytogenes. *J Neuroimmunol* 92: 139–151.

Zhang Z and Monteiro-Riviere NA. 1997. Comparison of integrins in human skin, pig skin, and perfused skin: an *in vitro* skin toxicology model. *J Appl Toxicol* 17: 247–253.

## URL WEB SITES

http://www.lib.iastate.edu/
http://www.genome.iastate.edu/pig/html
http://www.toulouse.inra.fr/lgc/pig/RH/IMpRH.htm
http://www.minipigs.dk
http://netvet.wustl.edu/species/pigs/srb94-01.htm#toxicology

CHAPTER 7

# The Nonhuman Primate as a Model of Developmental Immunotoxicity

**Andrew G. Hendrickx, Pamela E. Peterson, and Norbert M. Makori**

## CONTENTS



0-415-28457-0/05/$0.00+$1.50

## INTRODUCTION

Developmental immunotoxicity testing and effects of test compounds on the immune system, both prenatally and postnatally, have become topics of increasing importance to the pharmaceutical industry and to the regulatory agencies. This is due to the increased use of drugs with a specific effect on the immune system, including immunosuppressive monoclonal antibodies. Although potentially useful specificities of these compounds can be made in rodents, human-specific chimeric monoclonal antibodies and other biotechnology test articles can only be evaluated in humans or a closely related species. In light of ethical constraints in human studies, it becomes apparent that nonhuman primates are the only available species for reproductive immunotoxicity studies, including general efficacy and safety studies. There is, therefore, a need to identify some of the critical features and developmental timelines of the lymphoid organs and lymphocytes in the macaque, the most common nonhuman primate species presently used in these studies. This review focuses on morphology as well as the phenotypic and functional expression of the developing immune system of rhesus (*Macaca mulatta*) and cynomolgus (*Macaca fascicularis*) macaques.

## COMPARATIVE PRE- AND POSTNATAL DEVELOPMENT

### Implantation

In humans and other nonhuman primates, including the baboon and rhesus macaque, the conceptus reaches the uterus on days 3 to 4 following fertilization. Just before implantation, the conceptus produces chorionic gonadotrophin (CG), which rescues the corpus luteum from regression. The trophoblast undergoes rapid differentiation, and following invasion of the endometrial epithelium, establishes vascular connections that provide a suitable cellular and nutritional environment for the embryo (Hearn et al. 1994). Human embryos undergo interstitial implantation (trophoblast sinks under the endometrial epithelium) and a massive decidual reaction, while in macaques, implantation is superficial with limited decidual development (Enders and Schlafke 1986). In both humans and monkeys, embryo-maternal tissue apposition and exchange are similar despite differences in the types of implantation. In addition, they both eventually exhibit extensive decidua formation by mid-pregnancy. The decidual reaction likely provides an immunologically safe site for the developing embryo, and concurrent with decidualization, leukocytes that have infiltrated the endometrial stroma during the late progestational phase of the menstrual cycle secrete interleukin-2 (IL-2) (Carlson 1994). IL-2 prevents maternal recognition of the embryo as a foreign body during the early stages of pregnancy.

### Organogenesis

Morphological changes following fertilization and implantation include cleavage and formation of the inner cell mass. The inner cell mass undergoes further changes

| Human | Cynologus | Rheasus | |
|---|---|---|---|
| Menstrual Cycle | + | + | + |
| Hormone Patterns | + | + | + |
| Length (days) | 25–29 | 27–28 | 26–29 |
| Early Pregnancy | + | + | + |
| Hormone Patterns | 10–40 | 10–40 | 7–term |
| CG Secretion (days) | | | |
| Organogenesis (days) | 20–50 | 20–50 | 18–60 |
| Spermatogenesis (days) | ~60 | ~60 | 70 |

**Figure 7.1** Comparative reproductive features in the cynomolgus macaque, rhesus macaque, and human. + = presence; CD = chorionic gonadotropin. (Hendrickx AG et al. *Hum Exp Toxicol* 19:1–7, 2000. With permission.)

with establishment of the primitive streak. Formation of the primitive streak is closely associated with the beginning of the period of early organ formation, which lasts until the time of palate closure and, possibly, ossification of the humerus. It is during this period of major organ formation (organogenesis) that the developmental timetables of rhesus and cynomolgus macaques closely resemble that of humans. Comparative embryonic studies in human embryos (O'Rahilly and Müller 1992) and macaques (Gribnau and Geijsberts 1981; Makori et al. 1996) also demonstrate overall morphological similarities during development of these primates. Figure 7.1 and Figure 7.2 summarize the timetable for major events encompassing

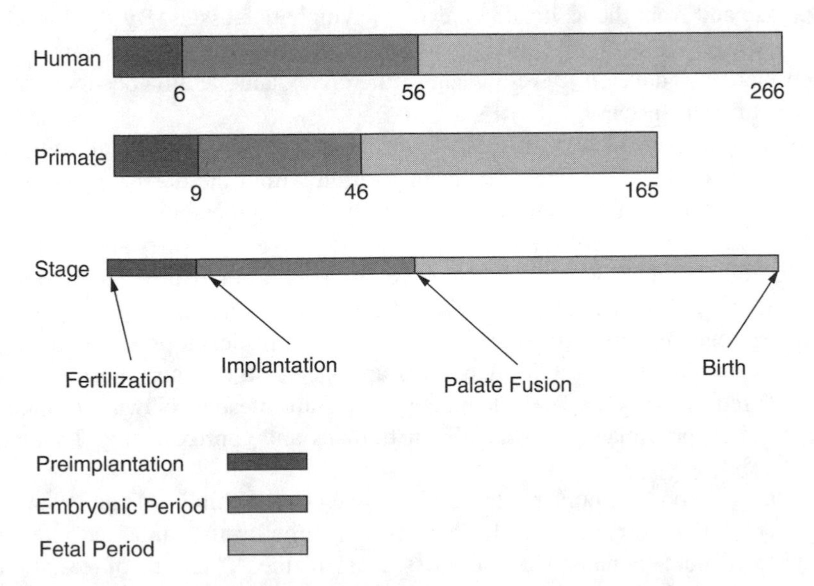


**Figure 7.2** Schematic summary of the major gestational periods (preimplantation, embryonic, and fetal) in humans and nonhuman primates. (Hendrickx AG et al. *Hum Exp Toxicol* 19:1–7, 2000. With permission.)

preimplantation, the embryonic period, and the fetal period in humans and macaques. During the fetal period, maximum growth and further development of the organ systems takes place.

## Embryology of the Lymphatic System

The lymphatic system includes lymph vessels and trunks, spleen, thymus, lymph nodes, bone marrow, intestines, tonsils, and lymphoid follicles in the mucous membranes of the alimentary canal and respiratory organs. Only the spleen, thymus, and lymph nodes will be discussed in this section. Since there is limited published information on this topic in the monkey, extrapolation has been done from human data, as needed, to provide a framework for the macaques.

In the human, development of the first lymphatics is apparent at approximately 5 weeks near the juncture of the precardinal and postcardinal veins (Nishimura 1983; O'Rahilly and Müller 1992). These bilateral jugular lymph sacs communicate with the adjacent veins. Axillary lymph sacs soon develop and fuse with jugular sacs; other primordia soon appear. By the end of the embryonic period, jugulo-axillary, mesenteric, and ilio-inguinal lymphatic sacs are present. The thoracic duct is formed between 6 and 8 weeks in the human by fusion of several separate primordia. The thoracic duct extends cranially and communicates with the left jugulo-axillary sac. The caudal extension of the thoracic duct links with the cisterna chyli, a dilatation of the lymphatic vessel.

Lymph nodes first appear in the fetal period as aggregations of lymphoblasts in the axillary and inguinal regions. Connective tissue invaginations subdivide a lymph sac, converting it to a plexus. The cervical nodes are reported to arise from the jugular sac and from the associated plexus of lymphatic vessels. By the end of the 1$^{st}$ trimester (3$^{rd}$ month in humans and 2$^{nd}$ month in macaque), a capsule and subcapsular sinus develop. Development of the cortex and medulla occurs in the 2$^{nd}$ trimester of each species.

The spleen forms as an intraepithelial organ and is first detected at 4 weeks in humans as a localized mesenchymal condensation within the dorsal mesogastrium (dorsal mesentery of the stomach). In its early development, the spleen consists mostly of mesenchymal cells, and the splenic pulp serves as a filter and modifier of blood (O'Rahilly and Müller 1992). Blood vessels soon form in the primordial tissue. The white pulp develops first, consisting of periarterial sheaths with lymphocytes, monocytes, and numerous macrophages. The red pulp develops early in the 2$^{nd}$ trimester and consists of arteries and veins arranged in a reticular meshwork with numerous red blood cells. The splenic corpuscles are present midway through the fetal period (approximately 4.5 months in humans and approximately 2.5 months in macaques).

By the 4$^{th}$ month in humans, trabeculae appear in the hilum of the developing spleen. The lobular architecture also becomes apparent at this time and is evident until birth in both humans and macaques. Late in the 4$^{th}$ month of gestation in humans, lymphoid colonization occurs in the fetal spleen and T-lymphocyte precursors appear. Reticular fibers are developed and erythropoiesis decreases after the 5$^{th}$ month. Hematopoiesis ceases near the end of pregnancy.

Thymus development is unique among the lymphoid organs because it receives major contributions from all three embryonic primitive germ layers—the endoderm of the 3rd pharyngeal pouch, the ectoderm of the 3rd pharyngeal pouch, and mesenchymal elements that include a large component of neural crest cells (Bockman and Kirby 1984). At the end of the 4th week in both humans and monkeys, the thymic primordia develop as paired ventral elongations arising from the endoderm of the 3rd pharyngeal pouches. Initially, the endodermal proliferations form hollow tubes that invade the underlying mesoderm and gradually transform into solid, branching cords (Larsen 1993). The boundaries of the pharyngeal arches are demarcated externally by intervening grooves. Between the 5th and 7th weeks in humans and macaques, the thymus separates from the pharynx and "descends" to a position inferior and ventral to the developing thyroid. Although the right and left thymic primordia become apposed at this time, fusion of the parenchyma does not occur until later in embryonic development.

In the late embryonic period, transient narrow diaphragms, the pharyngeal membranes, fill the interval between the pharyngeal arches. These pharyngeal membranes are lined externally by ectoderm and internally by endoderm. Before and during neural tube closure, neural crest cells migrate from its dorsal part in three distinct streams in the macaque and contribute to the pharyngeal arch mesenchyme (Peterson et al. 1996). These cells migrate to the arch region and become ectomesenchymal cells (through epithelial-mesenchymal transformation). In the 3rd pharyngeal arch, the crest cells surround the expanding epithelial mass (epithelial primordium) representing the primordium of the thymus (Bockman 1997; Bockman and Kirby 1984). Lymphoid precursor cells from neighboring blood vessels migrate into this epithelial primordium. The developing thymic epithelium forms a supporting network for the lymphoid cells and interacts with them to permit maturation of these cells.

The thymus becomes lobulated in the human in the 3rd month and the monkey in the early 2nd month. Differentiation into cortex and medullary regions also occurs early in the fetal period. Hassall's corpuscles, which can be identified in the medulla in the 2nd trimester, probably arise from the ectodermal cells of the 3rd pharyngeal cleft (membrane). However, the loosely organized epithelial reticulum of the thymus is of endodermal origin (O'Rahilly and Müller 1992). In these early stages of development, the thymus is infiltrated by lymphocytes derived from stem cells in the yolk sac, omentum, and liver. The homing of these lymphocytes on the thymus is probably a chemotactic mechanism. The human thymus remains highly active during the perinatal period and continues to grow throughout childhood. At puberty, it reaches its maximum size and then the gland involutes rapidly and is represented by fatty vestiges in the adult.

## THE DEVELOPING IMMUNE SYSTEM IN PRIMATES

### Major Lymphoid Organ Morphology and Immune Cell Development

Macaques have been established as excellent models for the study of abnormal fetal immune system development (Makori et al. 2002), including the pathogenesis and prevention of pediatric acquired immunodeficiency syndrome (Tarantal et al. 1995; Otsyula et al. 1996). This has necessitated adaptations of methods more routinely used in characterizing human and murine lymphoid system cells including immunohistochemistry, flow cytometry, ELISA, and ELISPOT. In nonhuman primates, the fetal lymphoid organs, like in the adult (Westerman and Pabst 1990), harbor the majority of the total lymphocyte pool. In recent years, it has been well established that a large number of monoclonal antibodies raised for human lymphocyte surface markers can cross-react with high specificity to antigen markers on monkey leukocytes (Sopper et al. 1997; Dykhuizen et al. 2000).

Embryo-fetal development in rhesus and cynomolgus macaques, like in humans, is divided into trimesters. The 1$^{st}$ trimester encompasses the period of organogenesis (gestation day [GD] 0 to 55), which is equivalent to 0 to 90 days in humans. During this period, organ progenitor modeling and migration of lymphocyte precursors to the thymus takes place. By the 2$^{nd}$ (GD 56 to 110 in macaque, 91 to 180 in human) and 3$^{rd}$ (GD 111 to 165 in macaque, 181 to 270 in human) trimesters, the lymphoid organs have acquired their definitive anatomical form even though there may be differences in specific microanatomical and lymphoid cell phenotypes compared to adult organs. For example, there are no secondary follicles in fetal lymph nodes or spleen, with the lymph nodes lacking a well-formed cortex and medulla (MacDonald and Spencer 1990).

In spleens from GD 75 macaque fetus, the immature white pulp (WP) areas appear as distinct clusters densely populated by both T (CD3$^+$) and B (CD20$^+$) cells (Makori et al. 2001). Most of the lymphocytes within the red pulp (RP) are immunoreactive to the B-cell marker CD20. About 80% of the T cells are organized around arterioles in 2 to 3 layers of cells, whereas the B subset forms 3 to 4 layers. By GD 80, most of the B cells are relocated from the ends of the arterioles to the outer margins of the rapidly expanding immature periarterial lymphoid sheath (PALS) T cell zone. The developing B-cell primary follicles and the associated primitive PALS T cell zone in most of the lymphoid aggregates at GD 80 are in a ratio of 1:1 (Makori et al. 2001). Between GD 100 to 145, the PALS T cell areas increase relatively rapidly in size compared to the B cell follicles to occupy a larger area of the WP. This is followed by an accelerated increase in the frequency of B-cell primary follicles appearing as outgrowths of the T cell PALS between GD 100 and GD 145. A single layer of CD20$^+$ B cells forms around most PALS areas by GD 100. The developing WP area expands more rapidly than the RP such that the ratio of WP:RP changes from 1:2 at GD 80 to 1:1 by GD 145.

The developmental timetable and appearance of antigen-presenting cells (APC) in the spleen is also remarkable. Between GD 75 and 145, macrophages (CD68$^+$) are scattered mostly in the RP with only a small number appearing in the WP. Fascin-positive (p55$^+$) dendritic cells (DCs) are present in moderate numbers in the outer WP in spleens from GD 100 fetuses (Makori et al. 2001). By GD 145, the DCs are numerous and exclusively found in T cell zones of the WP. These cells have extensive cytoplasmic extensions with intense immunostaining. The major histocompatability class II (MHC-II) expressing cells (human leukocyte antigen D-related positive cells [HLA-DR$^+$]) at GD 75 to 80 (including B cells, macrophages, and DCs) are mostly found in or around the developing PALS. The increase in number and widespread distribution of HLA-DR-expressing cells is coincident with the increase in number of CD3+ T cells. The presence and increase in HLA-DR+ B cells correlates with an increase in T cells colonizing the spleen between GD 75 and GD 145.

Overall lymph node ontogeny shows a similar progressive appearance of the various cell subsets as the spleen, with specific differences noted below. Macrophages (CD68+) at GD 75 to 80 are present mostly in areas with loose clusters of B cells. By GD 100, macrophages are also scattered in the medulla, medullary cords, and the outer rims of the deep T cell cortex. Between GD 75 to 80, loose clusters of CD20+ B and CD3+ T cells are formed with no definite aggregation into specific zones or B-cell primary follicles. The number of T cells is relatively higher than the B cells, a situation opposite to that seen in spleens from the same fetuses. By GD 100, the definite demarcation between cortex and medulla becomes apparent with the CD3+ T cell areas constituting the immature deep cortex and occupying a large portion of the developing lymph nodes (Makori et al. 2001). The CD20+ B cells form a thin layer of cells on the outer margins of the cortex subjacent to the subcapsular sinus. Immature B-cell primary follicles appear as aggregates with more cells in some areas along the thin layer of CD20+ B cells. The medulla and medullary cord areas are populated by scattered lymphocyte subsets. By GD 145, the primary B-cell follicles in the outer cortex are well defined as round entities with a thin layer of B cells between the follicles.

Development of gut-associated lymphoid tissue (GALT) is definitive by GD 80 (Makori et al. 2001). The appearance and subsequent ontogeny of B- and T-lymphocytes and their accessory cells follows a similar trend in the ileum, jejunum, and colon. Therefore, the descriptions that follow will refer to these components combined as GALT. At GD 75, CD3+ T cells are scattered throughout the intestinal villi, with most of these lymphocytes being intraepithelial (intraepithelial lymphocytes, i.e., L). The small number of B cells (CD20+) in intestinal tissue are found in the lamina propria (LP) in about 2 to 3 villi for every cross-section (approximately 6μm) of intestine examined. In the LP, a few cells are immunoreactive to HLA-DR (MHC-II). Macrophages (CD68+) are scattered in the LP and muscularis mucosa. DCs (Fascin+) are numerous within the LP. By GD 80 to 100, about a four-fold increase in lymphocyte numbers is evident, with most being of the T cell subset. The DCs are present in all villi and the HLA-DR+ cells are more widespread in the lamina propria. HLA-DR immunoreactive cells are found in the T cell zone at GD 145 in the immature lymphoid follicles. Macrophages are scattered in 50 to 70% of the villi. In the immature lymphoid aggregates at GD 100, the T- and B-cell zones are clearly distinguishable. Macrophages within the lymphoid aggregates are concentrated in developing B-cell areas, whereas HLA-DR+ staining cells are more abundant in T cell zones and in the lamina propria of the intestinal villi.

In the thymus, definitive thymic lobules at GD 75 with a characteristic medulla (central) and cortex (outer) are present; there is a higher lymphocyte density in the cortex (Makori et al. 2001). Hassall's corpuscles are evident at GD 75. Most of the thymocytes are immunoreactive to the pan-T cell marker, CD3. In the medulla and cortico-medullary junction (CMJ), a subset of the lymphocytes is of the B-cell phenotype (CD20+). These large thymic B-lymphocytes (TBL) are closely associated with HLA-DR+ cells. Some of these B cells are also HLA-DR. The CMJ and medulla

**Table 7.1 Frequency of IgG-Secreting Cells/10⁶ MNC in Normal Rhesus Macaque Fetal Tissues at Different Gestational Ages**

| Fetal Age[a] (days) | Thymus | Spleen | Liver | Colon | Small Intestine | Axillary LN | Bone Marrow | PBMC |
|---|---|---|---|---|---|---|---|---|
| 75 | 11 | 0 | 0 | 11500 | 3000 | NA | 769 | 0 |
| 80 | 0 | 106 | 0 | 11765 | 1063 | 0 | 1612 | 0 |
| 100 | 0 | 0 | 0 | 0 | 2456 | 0 | 0 | 0 |
| 100 | 1 | 0 | 0 | 5426 | 6929 | 0 | 0 | 0 |
| 145 | 3 | 14 | 0 | 38181 | 753 | 607 | 0 | 0 |
| 145 | 1 | 13 | 0 | 34375 | 895 | 10 | 0 | 0 |

*Note:* MNC: mononuclear cells; LN: lymph node; PBMC: peripheral blood mononuclear cells; NA: sample not available.

[a] Values represent one fetus at each time point (n=6 fetuses total).

also contain many large irregularly shaped DCs with their cytoplasmic protrusions extending between the lymphocytes. The DCs are scattered in the medulla, and more closely packed in the CMJ. At GD 80 and later stages, there are macrophages within the medulla and CMJ compared to the cortex. With increasing fetal age, the medullary region of the thymus appears to increase in size at a faster rate than the cortex. This is accompanied by increased spatial complexity of the various cell types in the medulla and CMJ, especially the B-cell subset, DCs, and expression patterns of the MHC-II (HLA-DR).

## Immunoglobulin and Cytokine-Secreting Cells

Phenotypic analysis of the frequency of immunoglobulin-secreting cells (Ig-SCs: IgM, IgG, IgA) and cytokines (IL6, IFNγ) shows unique patterns in different organs during development (Makori et al. 2001). In the fetal thymus, the Ig-SCs at all gestational days produce mainly IgM, but also small numbers of IgG and IgA (Table 7.1 to Table 7.3). Analysis of isotype expression by Ig-SCs during fetal rhesus macaque ontogeny shows marked differences in all organs examined. For example,

**Table 7.2 Frequency of IgA-Secreting Cells/10⁶ MNC in Normal Rhesus Macaque Fetal Tissues at Different Gestational Ages**

| Fetal Age[a] (days) | Thymus | Spleen | Liver | Colon | Small Intestine | Axillary LN | Bone Marrow | PBMC |
|---|---|---|---|---|---|---|---|---|
| 75 | 6 | 0 | 0 | 5500 | 1500 | NA | 31 | 0 |
| 80 | 0 | 266 | 0 | 10000 | 0 | 0 | 0 | 0 |
| 100 | 43 | 269 | 13 | 0 | 2280 | 0 | 0 | 0 |
| 100 | 5 | 71 | 0 | 0 | 0 | 0 | 0 | 0 |
| 145 | 5 | 16 | 0 | 3750 | 59 | 0 | 0 | 0 |
| 145 | 2 | 1 | 0 | 88 | 138 | 0 | 0 | 0 |

*Note:* MNC: mononuclear cells; LN: lymph node; PBMC: peripheral blood mononuclear cells; NA: sample not available.

[a] Values represent one fetus at each time point (n=6 fetuses total).

**Table 7.3 Frequency of IgM-Secreting Cells/10⁶ MNC in Normal Rhesus Macaque Fetal Tissues at Different Gestational Ages**

| Fetal Age[a] (days) | Thymus | Spleen | Liver | Colon | Small Intestine | Axillary LN | Bone Marrow | PBMC |
|---|---|---|---|---|---|---|---|---|
| 75 | 17 | 0 | 0 | 0 | 0 | NA | 0 | 0 |
| 80 | 19 | 851 | 0 | 0 | 0 | 0 | 0 | 0 |
| 100 | 117 | 1923 | 38 | 0 | 350 | 0 | 52 | 26 |
| 100 | 8 | 316 | 0 | 0 | 0 | 0 | 18 | 10 |
| 145 | 59 | 295 | 0 | 429 | 0 | 0 | 0 | 0 |
| 145 | 29 | 139 | 0 | 250 | 0 | 8 | 54 | 11 |

*Note:* MNC: mononuclear cells; LN: lymph node; PBMC: peripheral blood mononuclear cells; NA: sample not available.

[a] Values represent one fetus at each time point (n=6 fetuses total).

**Table 7.4 Frequency of IFNγ-Secreting Cells/10⁶ MNC in Normal Rhesus Macaque Fetal Tissues at Different Gestational Ages**

| Fetal Age[a] (days) | Thymus | Spleen | Liver | Colon | Small Intestine | Axillary LN | Bone Marrow | PBMC |
|---|---|---|---|---|---|---|---|---|
| 75 | ND | ND | ND | ND | ND | NA | ND | ND |
| 80 | 31 | 0 | 23 | 1470 | 0 | 0 | 0 | 0 |
| 100 | ND | ND | ND | ND | ND | ND | ND | ND |
| 100 | 3 | 1000 | 3063 | 1489 | 6143 | 0 | 279 | 471 |
| 145 | 4 | 625 | 2083 | 13961 | 1000 | 0 | 134 | 381 |
| 145 | 18 | 1111 | 4792 | 71875 | 1936 | 316 | 129 | 258 |

*Note:* MNC: mononuclear cells; LN: lymph node; PBMC: peripheral blood mononuclear cells; NA: sample not available; ND: not done.

[a] Values represent one fetus at each time point (n=6 fetuses total).

IgG- and IgA-secreting cells are numerous in the colon and small intestines but present in only small numbers in the spleen, lymph nodes, thymus, and cord blood of the same gestation age fetuses. There is spontaneous presence of IgG-SC in the colon, which increases to achieve maximum count at GD 145 without substantial IgM detection. This is accompanied by a proportional increase in the number of IFNγ-secreting cells to achieve high numbers at GD 145. On the other hand, IL6-secreting cells in the spleen increase only slightly. Table 7.4 and Table 7.5 provide some normal values of cytokines in macaque lymphoid organs, bone marrow, and cord blood as analyzed by ELISPOT.

## Differential Expression of Lymphocyte Markers

The surface marker expression on the individual T cell (CD3ε, CD4, and CD8) and B-cell (CD20, IgM) subsets in the fetal rhesus monkey is summarized in Table 7.6.

Approximating the proportions of CD4⁺ and CD8⁺ expressing lymphocytes with the use of CD3ε coincident staining shows significant differences in the various

**Table 7.5 Frequency of IL6-Secreting Cells/10⁶ MNC in Normal Rhesus Macaque Fetal Tissues at Different Gestational Ages**

| Fetal Age[a] (days) | Thymus | Spleen | Liver | Colon | Small Intestine | Axillary LN | Bone Marrow | PBMC |
|---|---|---|---|---|---|---|---|---|
| 75 | ND | ND | ND | ND | ND | NA | ND | ND |
| 80 | 0 | 0 | 2184 | 0 | 0 | 0 | 3467 | 0 |
| 100 | ND | ND | ND | ND | ND | ND | ND | ND |
| 100 | 100 | 1150 | 1458 | 319 | 0 | 15789 | 10214 | 6857 |
| 145 | 197 | 3909 | 2875 | 0 | 0 | 0 | 25568 | 9619 |
| 145 | 27 | 3597 | 11875 | 0 | 0 | 227 | 19286 | 7790 |

*Note:* MNC: mononuclear cells; LN: lymph node; PBMC: peripheral blood mononuclear cells; NA: sample not available; ND: not done.

[a] Values represent one fetus at each time point (n=6 fetuses total).

**Table 7.6 Various Surface Marker Phenotypes of Fetal Rhesus Monkey Lymphocyte Subsets**

| Subset | CD3ε | CD4 | CD8 | CD45RA | CD45RO | HLA-DR | IgM | IgD |
|---|---|---|---|---|---|---|---|---|
| CD4+CD8- T | + | + | – | + | + | + | – | – |
| CD4-CD8+ T | + | + | + | + | + | + | – | – |
| CD4+CD8+ T | + | + | + | + | – | – | – | – |
| CD20+IgM+ B | – | – | – | ? | + | ? | + | + |
| CD20-IgM+ B | – | – | – | + | – | ? | + | – |
| CD20+IgD+ B | – | – | – | + | + | ? | + | + |
| CD20-IgD+ B | – | – | – | ? | – | ? | – | + |

*Note:* (–) negative; (+) positive; (?) not determined.

subsets between the thymus and the spleen. When the values for all fetal stages are pooled for these organs, it is clear that the T cell $CD3^+CD4^+$: $CD3^+CD8^+$ ratio is approximately 3:1 in the spleen and 1:1 in the thymus (Figure 7.3).

The $CD4^+CD8^+$ subset comprises 80% of the thymocyte population and only 4% of the splenic T cells. A further comparison of T cell ontogeny reveals a 50 to 60% increase in the $CD3^+CD4^+$, $CD3^+CD8^+$, and $CD4^+CD8^+$ subsets between GD 75 to 100 and GD 145. An age-related increase is also noted in the single-positive category with the highest increases noted in the $CD4^-CD8^+$ category in the thymus (38%) and the $CD4^+CD8^-$ subset in the spleen (48%) (Figure 7.4 and Figure 7.5). A significant proportion of the B cells in the spleen and thymus express IgM (20 to 28%) in agreement with results obtained by the ELISPOT assay for Ig-SCs.

## Prenatal Immune System

The earliest lymphocyte subsets in the rhesus macaque to be detected in large numbers at GD 75 are $CD20^+$B cells in the spleen and $CD3^+$ T cells and $CD20^+$ B-cells in the thymus. In the lymph nodes, colonization by the B and T cells, in approximately equal proportions, is evident at GD 75. The early appearance of T cells in the thymus and then in the periphery in the macaque follows a similar trend as described in mice (Pardoll et al. 1987), and pigs (Sinkora et al. 1998). Unlike B

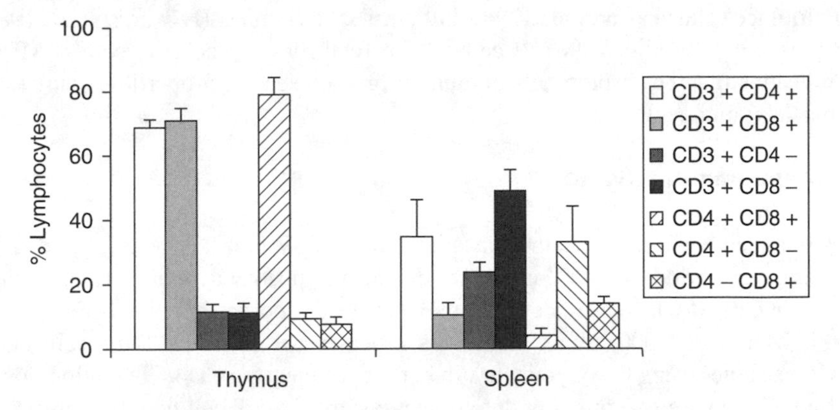


**Figure 7.3** Mean ± SD of CD3/CD4, CD3/CD8, CD4/CD8 ratios in the thymus and spleen pooled from 8 animals as follows: GD 75/80 ($n$=4), GD 100 ($n$=2), and GD 145 ($n$=2).

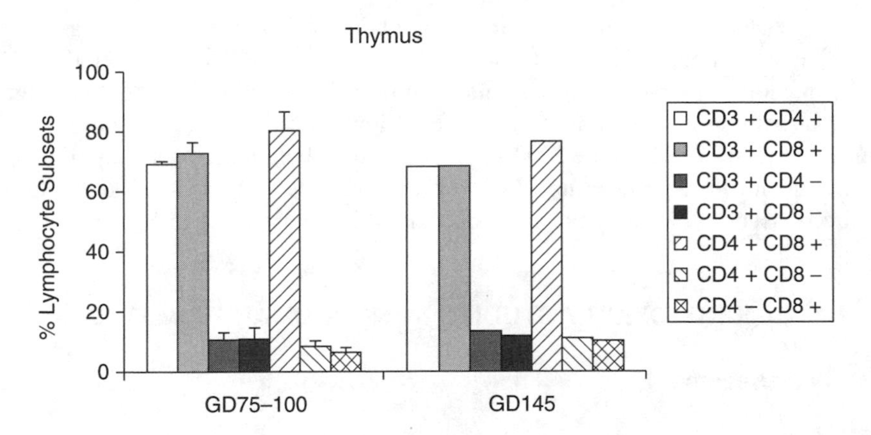


**Figure 7.4** Bar chart showing percentages of lymphocyte maturational stages as revealed by CD3, CD4, and CD8 surface marker staining in thymus lymphocytes.

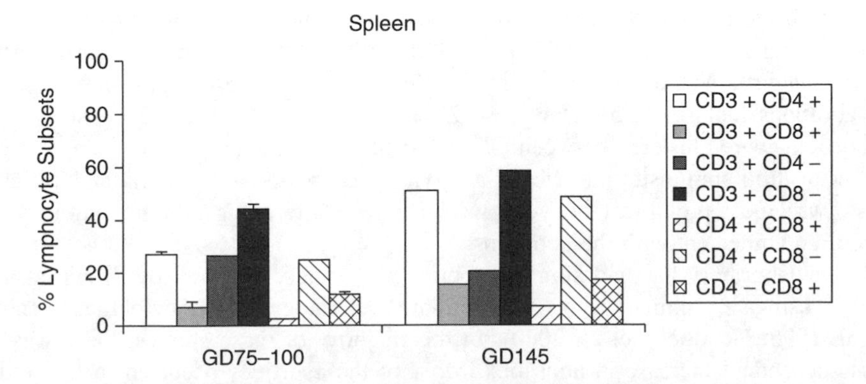


**Figure 7.5** Bar chart showing percentages of lymphocyte maturational stages as revealed by CD3, CD4, and CD8 surface marker staining in spleen lymphocytes.

cells in mice (Hayakawa et al. 1994), but similar to humans (Edwards et al. 1986) and pigs (Sinkora et al. 1998), B cells in the fetal rhesus macaque express MHC-II, indicating that the development of their antigen-presenting properties occurs early during fetal ontogeny.

### Postnatal Immune System

Continued maturation of the macaque immune system has been provided by evaluation of various serum parameters during the postnatal period. Longitudinal measurements of CD4/CD8 T cell ratios have been reported in colony-bred rhesus (Dykhuizen et al. 2000) and cynomolgus (Bleavins et al. 1993; Baroncelli et al. 1997) macaques using flow cytometry. In both macaques, the CD4/CD8 ratios slowly decline with age, predominantly due to decreasing $CD4^+$ T cell counts (Table 7.7). The CD4/CD8 values between 6 months and 1 year of age in these macaques (approximately 1.4 to 1.8) are in the same range as reported for humans between 4 to 5 years of age (Heldrup et al. 1992).

Age-related changes in serum immunoglobulin levels have also been assessed as indicators of immunologic competence in rhesus (Voormolen-Kalova et al. 1974) and cynomolgus (Terao 1981) macaques. Gradual increases in IgA, IgG, and IgM occur in both species with age (Table 7.6). Similar trends have been observed in humans (Siberry and Iannone 2000). For example, in both macaques ranging in age from 6 months to 1 year, serum IgA levels approximate 43 to 45% of adult values. A similar level (40%) of serum IgA is observed in humans at 4 to 5 years.

## IMMUNOTOXICITY STUDIES IN NONHUMAN PRIMATES

### Safety Assessment

A variety of human-derived cytokines have been studied in pregnant rhesus and cynomolgus monkeys according to specific protocols, which involve drug administration from GD 20 to 80 (Henck et al. 1996; Hendrickx et al. 2000). An example of a prototype teratology study is provided by Trown and colleagues (1986) in which intramuscular administration of interferon alpha-2$\alpha$ (Roferon R-A) intramuscularly to pregnant rhesus monkeys from GD 20 to 70 resulted in a dose-dependent increase in abortions (control 12.5%, low dose 25%, mid dose 37%, and high dose 66%). Abortions were clustered between GD 37 to 68, that is, about 17 to 48 days after onset of drug administration. Maternal toxicity, expressed as anorexia and weight loss, was most prominent between weeks 3 to 5 after drug administration, and occurred coincident with the abortions.

Our laboratory has had similar experiences to those reported by Trown et al. (1986) using very similar protocols to evaluate the teratogenicity of cytokines (Henck et al. 1996; Hendrickx et al. 2000). In the majority of these studies, there was a dose-dependent increase in abortions. Most of the abortions occurred in the mid-portion of the treatment period, beginning 6 to 12 days after onset of drug administration, or between GD 25 to 40. The well-defined clustering of the abortions may

**Table 7.7 Serum Immunological Parameters in Macaques**

| Rhesus Monkey | | Cynomolgus Monkey | | Humans | | |
|---|---|---|---|---|---|---|
| CD4 (%, mean from graph)[a] | | CD4 (%, mean ± SEM)[c] | | CD4 (%, median, range)[f] | | |
| 2 mo (n=10–12) | ~42 | Males (n=69) | 28.1 ± 0.8 | Newborn (n=13) | 49.0 (16–58) | |
| 6 mo (n=10–12) | ~39 | Females (n=77) | 30.4 ± 0.8 | 6 mo – 3yr (n=14) | 46.0 (22–87) | |
| 22–33 mo (n=10–12) | ~31 | | | 3.9–4.8 yr (n=10) | 37.5 (29–51) | |
| 36–60 mo (n=10–12) | ~31 | | | 7.7–10.6 yr (n=10) | 43.5 (28–55) | |
| 70–94 mo (n=10–12) | ~34 | | | Adult (n=10) | 40.5 (29–62) | |
| CD8 (%, mean from graph)[a] | | CD8 (%, mean ± SEM)[c] | | CD8 (%, median, range)[f] | | |
| 2 mo (n=10–12) | ~22 | Males (n=69) | 55.8 ± 1.3 | Newborn (n=13) | 19.0 (13–29) | |
| 6 mo (n=10–12) | ~23 | Females (n=77) | 53.8 ± 1.0 | 6 mo – 3yr (n=14) | 18.0 (12–52) | |
| 22–33 mo (n=10–12) | ~20 | | | 3.9–4.8 yr (n=10) | 22.0 (11–33) | |
| 36–60 mo (n=10–12) | ~26 | | | 7.7–10.6 yr (n=10) | 21.5 (17–31) | |
| 70–94 mo (n=10–12) | ~28 | | | Adult (n=10) | 25.5 (20–43) | |
| %CD4/%CD8 (mean from graph)[a] | | %CD4/%CD8 (mean ± SEM)[c] | | %CD4/%CD8 (median, range)[f] | | |
| 2 mo (n=10–12) | ~2.1 | Males (n=69) | 0.53 ± 0.03 | Newborn (n=13) | 2.5 (0.8–4.0) | |
| 6 mo (n=10–12) | ~1.7 | Females (n=77) | 0.59 ± 0.02 | 6 mo – 3yr (n=14) | 2.6 (0.4–5.4) | |
| 22–33 mo (n=10–12) | ~1.6 | | | 3.9–4.8 yr (n=10) | 1.8 (1.2–4.6) | |
| 36–60 mo (n=10–12) | ~1.2 | %CD4/CD8 | | 7.7–10.6 yr (n=10) | 2.1 (1.0–3.2) | |
| 70–94 mo (n=10–12) | ~1.3 | (mean from graph or mean ± SE)[d] | | Adult (n=10) | 1.6 (0.9–3.1) | |
| | | 1 wk | ~3.3 | | | |
| | | 1 yr | 1.4 ± 0.4 | | | |
| | | Adult | 0.8 ± 0.4 | | | |
| Serum IgA [% adult, mean (range)][b] | | Serum IgA [% adult, mean ± SD (range)][e] | | Serum IgA [mg/dl, mean (95% CL),% adult][g] | | |
| 6 mo–1yr (n=23) | 43 (19–98) | Newborn (n=11) | 0 | Birth | 2.3 (1.4–3.6) | 1% |
| 1–2 yr (n=21) | 67 (31–148) | 1 mo (n=23) | 5 ± 3 (1–10) | 1 mo | 13 (1.3–53) | 8% |
| 2–3 yr (n=22) | 86 (45–166) | 3–5 mo (n=13) | 12 ± 5 (8–23) | 6 mo | 25 (8.1–68) | 15% |
| 3–4 yr (n=17) | 67 (33–137) | 6–11 mo (n=11) | 45 ± 3 (17–93) | 1 yr | 44 (14–106) | 26% |
| 4–5 yr (n=17) | 96 (45–207) | 1–2 yr (n=14) | 63 ± 19 (27–93) | 4–5 yr | 68 (25–154) | 40% |
| 5–6 yr (n=14) | 112 (43–291) | 3–4 yr (n=8) | 48 ± 9 (41–63) | 6–8 yr | 90 (33–202) | 53% |
| 6–7 yr (n=9) | 112 (56–224) | 5–9 yr (n=7) | 68 ± 17 (41–86) | 9–10 yr | 113 (45–236) | 66% |
| > 7 yr (n=27) | 108 (52–221) | >10 yr (n=10) | 100 ± 20 (65–120) | Adult | 171 (70–312) | 100% |
| Serum IgG [% adult, mean (range)][b] | | Serum IgG [% adult, mean ± SD (range)][e] | | Serum IgG [mg/dl, mean (95% CL),% adult][g] | | |
| 6 mo–1yr (n=23) | 58 (37–93) | Newborn (n=18) | 88 ± 15 (78–111) | Birth | 1121 (636–1606) | 113% |
| 1–2 yr (n=21) | 51 (33–79) | 1 mo (n=22) | 52 ± 9 (36–66) | 1 mo | 503 (251–906) | 51% |
| 2–3 yr (n=22) | 56 (33–96) | 3–5 mo (n=14) | 50 ± 14 (29–72) | 6 mo | 407 (215–704) | 41% |
| 3–4 yr (n=17) | 60 (34–104) | 6–11 mo (n=12) | 76 ± 19 (40–109) | 1 yr | 679 (345–1213) | 68% |
| 4–5 yr (n=17) | 68 (45–101) | 1–2 yr (n=7) | 69 ± 14 (53–83) | 4–5 yr | 780 (463–1236) | 78% |
| 5–6 yr (n=14) | 67 (39–115) | 3–4 yr (n=10) | 109 ± 13 (89–130) | 6–8 yr | 915 (633–1280) | 92% |
| 6–7 yr (n=9) | 72 (49–106) | 5–9 yr (n=12) | 95 ± 17 (65–136) | 9–10 yr | 1007 (608–1572) | 101% |
| > 7 yr (n=27) | 91 (53–157) | >10 yr (n=16) | 100 ± 16 (65–128) | Adult | 994 (639–1349) | 100% |
| Serum IgM [% adult, mean (range)][b] | | Serum IgM [% adult, mean ± SD (range)][e] | | Serum IgM [mg/dl, mean (95% CL),% adult][g] | | |
| 6 mo–1yr (n=23) | 66 (38–112) | Newborn (n=13) | 5 ± 2 (2–9) | Birth | 13 (6.3–25) | 8% |
| 1–2 yr (n=21) | 80 (51–124) | 1 mo (n=22) | 25 ± 10 (11–47) | 1 mo | 45 (20–87) | 29% |
| 2–3 yr (n=22) | 78 (40–149) | 3–5 mo (n=13) | 33 ± 12 (11–56) | 6 mo | 62 (35–102) | 40% |
| 3–4 yr (n=17) | 58 (35–96) | 6–11 mo (n=10) | 66 ± 24 (34–112) | 1 yr | 93 (43–173) | 60% |
| 4–5 yr (n=17) | 67 (38–117) | 1–2 yr (n=7) | 54 ± 8 (39–68) | 4–5 yr | 99 (43–196) | 63% |
| 5–6 yr (n=14) | 82 (43–156) | 3–4 yr (n=10) | 76 ± 21 (50–104) | 6–8 yr | 107 (48–207) | 68% |
| 6–7 yr (n=9) | 84 (45–156) | 5–9 yr (n=14) | 102 ± 40 (58–197) | 9–10 yr | 121 (52–242) | 78% |
| > 7 yr (n=27) | 88 (38–202) | >10 yr (n=15) | 100 ± 38 (63–197) | Adult | 156 (56–352) | 100% |

[a] Laboratory born and housed rhesus monkeys (Dykhuizen et al. 2000).
[b] Male and female rhesus monkeys; values given as % of values in 30 normal adults (Voormolen-Kalova et al. 1974).
[c] Clinically normal wild-caught cynomolgus monkeys, 4-10 yr (Bleavins et al. 1993).
[d] Healthy newborn cynomolgus monkeys (Baroncelli et al. 1997).
[e] Male and female cynomolgus monkeys; values given as% of average adult level (Terao 1981).
[f] Healthy children and adults (Heldrup et al. 1992).
[g]% adult values calculated as follows: [(age value – adult value)/adult value] x100 (Siberry and Iannone 2000).

From Hendrickx AG et al. *Hum Exp Toxicol* 19:1–7, 2000. With permission.

indicate a pharmacological effect of the cytokine; alternatively, they may be associated with an immune response to the human-derived protein. Hormone analysis in one of these studies indicated a reduction in serum progesterone levels concurrent with abortions that occurred within the first week of treatment. This period coincides with the luteal-placental shift, a particularly fragile time during pregnancy in primates. These data suggested that an immediate failure in the function of the maternal corpus luteum or the early placenta may be associated with cytokine treatment. These data are consistent with the observation that early gestation is highly vulnerable to disruption because of endocrine changes that are necessary in establishing pregnancy in primates. Maternal toxicity, manifested as weight loss, anorexia, alterations in hematologic values, and skin lesions, was commonly seen in these studies.

## Immune System Teratogenicity

Several well-known teratogenic agents have been shown to have an adverse effect on the developing immune system, including the early thymus in macaques and baboons.

Triamcinolone acetonide (TAC), a potent corticosteroid, induced moderate to severe thymic hypoplasia and lymphocyte deficiency following exposure at doses of 3 to 28 mg/kg/day during early, mid-, and late gestation (Hendrickx et al. 1975). The treatment regimen encompassed four consecutive days between GD 37 to 48 (early proliferation of thymic anlage and separation from pharyngeal pouches), GD 50 to 73 (differentiation of lymphoid elements within the thymus) and GD 100 to 133 (maturation of fetal lymphoid system). Of the fetuses, 76% demonstrated a marked hypoplasia of the thymus, a severe depletion of the thymic lymphocytes, and a marked reduction in the epithelial component. A depletion of the lymphocytes from the thymic-dependent areas of the spleen and lymph nodes was also observed.

Zinc is an essential nutrient critical for the normal development and function of the immune system. Studies in our Primate Center have been directed toward developing the rhesus macaque model to define the influence of zinc deficiency on reproductive outcome and early postnatal development (Golub et al. 1984; Haynes et al. 1987), and to demonstrate that cell-mediated immunity is affected in these animals (Haynes et al. 1987). Experimentally induced zinc deficiency (4μg Zn/g in experimental cases compared to 100μg Zn/g in controls) from conception until 1 year postnatal in rhesus macaques results in severe alterations in immunocompetence (Haynes et al. 1985; Vruwink et al. 1991). The adverse effects include depressed peripheral blood lymphocyte mitogenesis, reduced polymorphonuclear leukocyte function, alterations in IgM levels, and presence of hypochromic microcyte anemia at 1 month of age.

Prenatal exposure to 13-*cis*-retinoic acid (cRA, Accutane®) in the cynomolgus macaque model results in a spectrum of abnormalities resembling retinoic acid embryopathy observed in human fetuses/newborns following inadvertent maternal use of this drug during the first trimester of pregnancy. Defects of the thymus (hypoplasia or aplasia) comprise this syndrome in both human (Lammer et al. 1985) and nonhuman primates (Hummler et al. 1990; Korte et al. 1993). Etiology has mainly been attributed to changes in both the cranial neural crest and pharyngeal arch endoderm that contribute to normal thymus development (Makori et al. 1999).

The thymus is the primary site for generation of T-lymphocytes, which then populate and expand extensively in the peripheral (secondary) lymphoid organs (Blackman et al. 1990; Mosley and Klein 1992).

Thymic aplasia due to 13-*cis*-retinoic acid could potentially induce immune deficiency in the macaque neonate. In a recent publication (Makori et al. 2002), we evaluated several morphological and cellular endpoints of immune ontogeny in the macaque fetus to characterize the adverse effects on this system due to embryonic exposure to RA. Assessment of T cells ($CD3^+$), B cells ($CD20^+$), DCs ($p55^+$), and MHC-II ($HLA\text{-}DR^+$) showed a clear reduction in size and proportion of T cells in the splenic WP in specimens with perturbed thymic development. Similar application of these endpoints could be useful in assessing perturbation of the developing immune system by immunomodulatory agents in the primate model.

## CURRENT AND FUTURE DEVELOPMENTAL TOXICITY PROTOCOLS

### Fertility Studies

Rhesus and cynomolgus macaques have served as test animals in studies of female and male reproductive toxicity of biologics and pharmaceuticals (Henck et al. 1996). The focus here will be on female fertility. As noted above, both macaque species closely resemble the human female in menstrual cycle length and other aspects of reproductive biology and endocrinology. Monitoring hormonal profiles is particularly important in evaluating reproductive toxicity. The hormones commonly monitored include serum progesterone, 17-beta-estradiol, and their urinary metabolites. Luteinizing hormone (LH) and follicle-stimulating hormone (FSH) may also be monitored in specific studies. Although all of these hormones may be informative in detecting potential toxicity, progesterone provides the most critical information regarding alterations in the cycle and in the ovulatory process. For example, progesterone values below 2 ng/ml in the luteal phase are indicative of anovulation, and an abbreviated profile is indicative of a shortened luteal phase, both of which are indicative of toxicity.

Our experience in developing an effective experimental design (Figure 7.6) has involved selection of sexually mature females that have experienced at least three normal menstrual cycles before assignment to the study. In most cases, it is advisable to begin the study with about 25% more animals than are required for the dosing component of the study in the event that a few animals may display cycle irregularity and are then readily replaced. Since rhesus macaques are often used, a thorough knowledge of their breeding season is essential in order to avoid seasonable effects at the beginning or at the end of study. (e.g., in California, the breeding season typically begins in early November and ends in May). Each female may serve as its own control, providing that normal cyclicity is established before initiation of the study; this can be verified by monitoring one or two cycles for progesterone before test article administration. This phase also provides the necessary pre-trial information if a recovery phase becomes necessary. Test article administration is usually done for one or more cycles, or in increments of 30 days to simulate a normal cycle

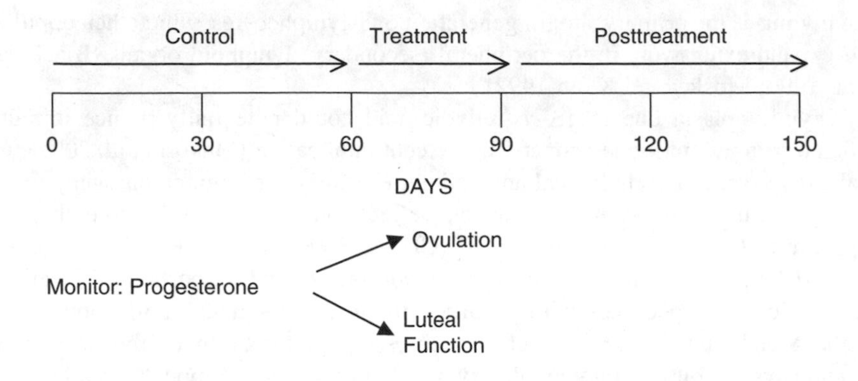


**Figure 7.6** Sample protocol for female fertility study in nonhuman primates includes a pretreatment (control) phase of one or two cycles, a treatment phase of one or more cycles, and a posttreatment (recovery) phase of one or two cycles. Serum progesterone is monitored at regular intervals throughout each phase to monitor ovulation and luteal function. Optional reproductive hormones that may also be monitored are LH, FSH, and estradiol. (Hendrickx AG et al. *Hum Exp Toxicol* 19:1–7, 2000. With permission.)

in the event that anovulation or luteal dysfunction occurs. A recovery phase of one to two cycles is sufficient to monitor for a return to normality. This approach was used in one of the initial studies in monitoring a cytokine (Trown et al. 1986). Other studies, also involving cytokines, have reported suppression of ovarian function and deficient luteal phases, based on decreased progesterone levels (Henck et al. 1996). The potential mechanisms of toxicity include direct actions on the ovary, or indirect actions on the hypothalamic–pituitary axis.

## Teratology Studies

Safety assessment studies of pharmaceuticals have typically involved administration of the test compound during the embryonic period, when the organ systems form, which encompasses GD 20 to 50 in the commonly used species, the rhesus and cynomolgus macaques. With the emergence of biopharmaceuticals, the safety assessment protocols have been altered to provide for longer exposure of the conceptus to include a significant portion of the fetal period. This period is characterized by rapid growth of the fetus and differentiation of organ systems, including the fetal immune system. The fetal period in the macaque extends from GD 50 to term (GD 155 and 165 in the cynomolgus and rhesus, respectively). Many protocols now include administration of the test compound until GD 70 to 100 (Figure 7.7). Future protocols involving immunomodulatory agents will probably include exposure throughout pregnancy (i.e., GD 20 to term); in certain cases, exposure may include the early postnatal period (i.e., birth to postnatal day 30).

Protocols in the near future may also require exposure from the time of conception or implantation to adequately cover early differentiation of the thymus and other components of the fetal immune system. Exposure from the time of conception will

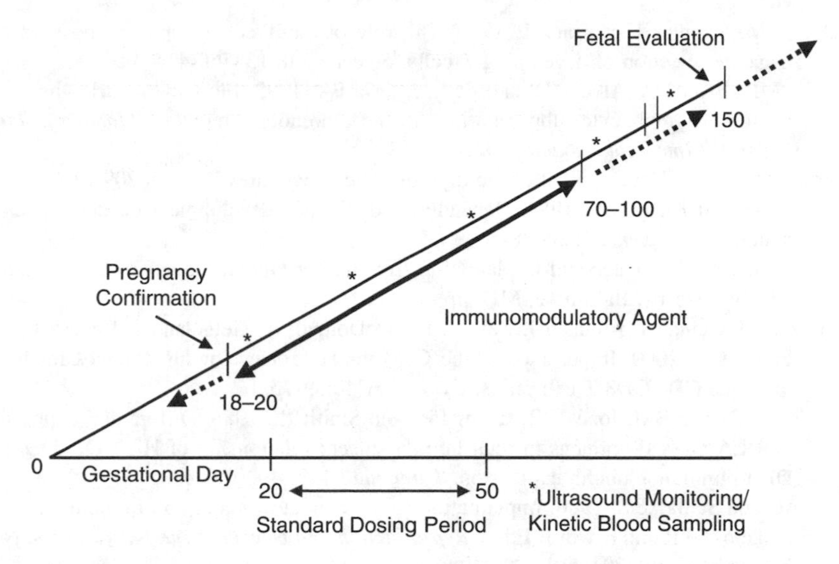


**Figure 7.7** The standard protocol for a teratology study in nonhuman primates includes exposure to a test agent during the major period of organ formation (GD 20-50) following pregnancy confirmation at GD 18 to 20. In the case of immunomodulatory agents, the test agent would be administered for a longer period (until GD 70 or 100) to cover growth and differentiation of the fetal immune system. Future protocols could extend the period of treatment to include earlier (before GD 20) or later (until term or during the postnatal period) exposure to an immunomodulatory drug. (Hendrickx AG et al. *Hum Exp Toxicol* 19:1–7, 2000. With permission.)

require administration of the test compound to approximately four times more females due to the average conception per mating success of about 25%. With the current availability of ultrasound technology confirmation of pregnancy is not feasible until 3 to 4 days after implantation (i.e., 2 to 4 days after embryo attachment in the macaque, approximately GD 13 to 14) (Guo et al. 1999).

In addition to standard fetal weights and measurements, useful endpoints to measure possible adverse effects on the developing immune system include proportions of T and B lymphocytes, as well as presence and numbers of antigen-presenting cells, and the cells secreting immunoglobulins and cytokines.

## ACKNOWLEDGMENT

This research was supported by NIH grant RR00169.

## REFERENCES

Baroncelli S, Panzini G, Geraci A, Pardini S, Corrias F, Iale E, Varano F, Turillazzi PG, Titti F, and Verani P. 1997. Longitudinal characterization of CD4, CD8 T cell subsets and of haemotological parameters in healthy newborns of cynomolgus monkeys. *Vet Immunol Immunopath* 59: 141-150.

Blackman M, Kappler J, Marrack P. 1990. The role of the T cell receptor in positive and negative selection of developing T cells. *Science* 248:1335-1341.

Bleavins MR, Brott DA, Alvey JD, and de la Iglesia FA. 1993. Flow cytometric characterization of lymphocyte subpopulations in the cynomolgus monkey (*Macaca fascicularis*). *Vet Immunol Immunopath* 37: 1-13.

Bockman DE. 1997. Development of the thymus. *Microscopy Res Tech* 38:209-215.

Bockman DE and Kirby ML. 1984. Dependence of thymus development on derivatives of neural crest. *Science* 223:498-500.

Carlson BM. 1994. Cleavage and Implantation. In: *Human Embryology and Developmental Biology*, Mosby, Baltimore, MD, pp. 33-50.

Dykhuizen M, Ceman J, Mitchen J, Zayas M, MacDougall A, Helgeland J, Rakasz E, and Pauza CD. 2000. Importance of the CD3 marker for evaluating changes in rhesus macaque CD4/CD8 T cell ratios. *Cytometry* 40:69-75.

Edwards JA, Durant BM, Jones DB, Evans PR, and Smith JL. 1986. Differential expression of HLA class II antigens in fetal human spleen: relationship of HLA-DP, DQ, and DR to immunoglobulin expression. *J Immunol* 137: 490-497.

Enders AC and Schlafke S. 1986. Implantation in nonhuman primates and in the human. In: Dukelow WR and Erwin J (Eds.), *Reproduction and Development*. Alan R. Liss, New York, Vol. 3, pp. 291-310.

Golub MS, Gershwin ME, Hurley LS, Saito WY, and Hendrickx AG. 1984. Studies of marginal zinc deprivation in rhesus monkeys. IV. Growth of infants in the first year. Am J Clin Nutr 40:1192-1202.

Gribnau AA and Geijsberts LG. 1981. Developmental stages in the rhesus monkey (*Macaca mulatta*). *Advances Anat Embryol Cell Biol* 69:1-84.

Guo Y, Hendrickx AG, Overstreet JW, Dieter J, Stewart D, Tarantal AF, Laughlin L, and Lasley BL. 1999. Endocrine biomarkers of early fetal loss in cynomolgus macaques (*Macaca fascicularis*) following exposure to dioxin. *Biol Reprod* 60:707-13.

Hayakawa K, Tarlinton D, and Hardy RR. 1994. Absence of MHC class II expression distinguishes fetal from adult B lymphopoiesis in mice. *J Immunol* 152:4801-4807.

Haynes DC, Golub MS, Gershwin ME, Hurley LS, and Hendrickx AG. 1987. Long-term marginal zinc deprivation in rhesus monkeys. II. Effects on maternal health and fetal growth at midgestation. *Am J Clin Nutr* 45:1503-1513.

Haynes DC, Gershwin ME, Golub MS, Cheung AT, Hurley LS, and Hendrickx AG. 1985. Studies of marginal zinc deprivation in rhesus monkeys: VI. Influence on the immunohematology of infants in the first year. *Am J Clin Nutr* 42:252-262.

Hearn JP, Hendrickx AG, Webley GE, and Peterson PE. 1994. Normal and abnormal embryofetal development in mammals. In: Lamming GE (Ed.), *Marshall's Physiology of Reproduction*, Vol. 3, Chapman and Hall: London, pp. 535-676.

Heldrup J, Kalm O, and Pellner K. 1992. Blood T and B lymphocyte subpopulations in healthy infants and children. *Acta Paediatrica* 81: 125-132.

Henck JW, Hilbish KG, Serabian MA, Cavagnaro JA, Hendrickx AG, Agnish ND, Kung AHC, and Mordenti J. 1996. Reproductive toxicity testing of therapeutic biotechnology agents. *Teratol* 53:185-195.

Hendrickx AG, Makori N, and Peterson P. 2000. Nonhuman primates: their role in assessing developmental effects of immunomodulatory agents. *Hum Exp Toxicol* 19:1-7.

Hendrickx AG, Sawyer RH, Terrell TG, Osburn BI, Henrickson RV, and Steffek AJ. 1975. Teratogenic effects of triamcinolone on the skeletal and lymphoid systems in nonhuman primates. *Fed Proc* 34:1661-1667.

Hummler H, Korte R, and Hendrickx AG. 1990. Induction of malformations in the cynomolgus monkey with 13-*cis*-retinoic acid. *Teratol* 42:263-272.

Korte R, Hummler H, and Hendrickx AG. 1993. Importance of early exposure to 13-*cis*-retinoic acid to induce teratogenicity in the cynomolgus monkey. *Teratol* 47:37-45.

Lammer EJ, Chen DT, Hoar RM, Agnish ND, Benke PJ, Braun JT, Curry CJ, Fernhoff PM, Grix, AW, Lott, IT, Richard JM, and Sun SC. 1985. Retinoic acid embryopathy. *New Engl J Med* 313:837-841.

Larsen WJ. 1993. Development of the head, the neck, and the eyes and ears. In: *Human Embryology*, Churchill Livingstone, New York, pp. 309-374.

MacDonald TT and Spencer J. 1990. Ontogeny of the mucosal immune response. *Springer Sem Immunopathol* 12:129-37.

Makori N, Rodriguez CG, Cukierski MA, and Hendrickx AG. 1996. Development of the brain in staged embryos of the long-tailed monkey (*Macaca fascicularis*). *Primates* 37:351-361.

Makori N, Peterson PE, Wei X, Hummler H, and Hendrickx AG. 1999. 13-*cis*-retinoic acid alters neural crest cells expressing Krox-20 and Pax-2 in macaque embryos. *Anat Rec* 255:142-154.

Makori N, Tarantal AF, Fabien XL, McChesney M, Marthas ML, AG Hendrickx, and Miller CJ. 2001. Lymphocytes and antigen presenting cells ontogeny in fetal rhesus monkeys (*Macaca mulatta*). *FASEB* J p.75 (Abstract).

Makori N, Peterson PE, Lantz K, and Hendrickx AG. 2002. Exposure of cynomolgus monkey embryos to retinoic acid causes thymic defects: effects on peripheral lymphoid organ development. *J Med Primatol* 31:91-97.

Mosley RL and Klein JR. Peripheral engraftment of fetal intestine into athymic mice sponsors T cell development: direct evidence for thymopoietic function of murine small intestine. 1992. *J Exp Med* 176:1365-1373.

Nishimura H. 1983. *Atlas of Human Prenatal Histology*, Igaku-Shoin, Tokyo.

O'Rahilly R, and Müller F. 1992. *Human Embryology and Teratology*, Wiley-Liss, Inc., New York, NY.

Otsyula MG, Miller CJ, Tarantal AF, Marthas ML, Greene TP, Collins JR, van Rompay KK, and McChesney MB. 1996. Fetal or neonatal infection with attenuated simian immunodeficiency virus results in protective immunity against oral challenge with pathogenic SIVmac251. *Virology* 222:275-8.

Pardoll DM, Fowlkes BJ, Bluestone JA, Kruisbeek A, Maloy WL, Coligan JE, and Schwartz RH. 1987. Differential expression of two distinct T cell receptors during thymocyte development. *Nat* 326:79-81.

Peterson PE, Blankenship TN, Wilson DB, and Hendrickx AG. 1996. Analysis of hindbrain neural crest migration in the long-tailed monkey (*Macaca fascicularis*). *Anat Embryol* 194:235-246.

Siberry GK and Iannone R. 2000. The Harriet Lane Handbook, Mosby, Inc., St. Louis.

Sinkora M, Sinkora J, Rehakova Z, Splichal I, Yang H, Parkhouse RM, and Trebichavsk I. 1998. Prenatal ontogeny of lymphocyte subpopulations in pigs. *Immunol* 95:595-603.

Sopper S, Stahl-Hennig C, Demuth M, Johnston IC, Dorries R, and ter Meulen V. 1997. Lymphocyte subsets and expression of differentiation markers in blood and lymphoid organs of rhesus monkeys. *Cytometry* 29:351-62.

Tarantal AF, Marthas ML, Gargosky SE, Otysula M, McChesney MB, Miller CJ, and Hendrickx AG. 1995. Effects of viral virulence on intrauterine growth in SIV-infected fetal rhesus macaques *(Macaca mulatta)*. *J Acquir Imm Def Syn Hum Retrovirol* 10:129-38.

Terao K. 1981. Serum immunoglobulin levels in relation to age in the cynomolgus monkey. *Jap J Med Sci Biol* 34: 246-249.

Trown PW, Wills RJ, and Kamm JJ. 1986. The preclinical development of Roferon R-A. *Cancer* 57:1648-1656.

Voormolen-Kalova M, Van den Berg P, and Radl J. 1974. Immunoglobulin levels as related to age in nonhuman primates in captivity. II. Rhesus monkeys. *J Med Primatol* 3: 343-350.

Vruwink KG, Fletcher MP, Keen CL, Golub MS, Hendrickx AG, and Gershwin ME. 1991. Moderate zinc deficiency in rhesus monkeys. An intrinsic defect of neutrophil chemotaxis corrected by zinc repletion. *J Immunol* 146:244-249.

Westermann J and Pabst R. 1990. Lymphocyte subsets in the blood: a diagnostic window on the lymphoid system? *Immunol Today* 11:406-10.

# Part III

# Developmental Immunotoxicant Exposure and Postnatal Immunosuppression

CHAPTER 8

# Developmental Immunotoxicity of Halogenated Aromatic Hydrocarbons and Polycyclic Aromatic Hydrocarbons

**Terry C. Hrubec, Benny L. Blaylock, and Steven D. Holladay**

## CONTENTS



## HALOGENATED AROMATIC HYDROCARBONS

The halogenated aromatic hydrocarbons comprise a large group of both man-made and naturally occurring environmental pollutants. This group includes 2,3,7,8-tetrachlorodibenzo-*p*-dioxin (TCDD) and many structurally related compounds, including other polychlorinated dibenzodioxins (PCDD), chlorinated dibenzofurans (PCDF), polychlorinated biphenyls (PCB), and polybrominated biphenyls (PBB). Of these, TCDD has been demonstrated to cause the most profound developmental immunotoxicity (Faith and Moore 1977; Vos and Luster 1989; Holladay and Luster 1994) and is probably the most studied of the chemicals in this group. Much of the toxicity, including immunomodulatory effects, produced by TCDD exposure (as well as the other HAHs) is directly related to the binding affinity of the HAH to the aryl hydrocarbon receptor (Vecchi et al. 1983; Dencker et al. 1985; Silkworth and Grabstein 1982; Silkworth et al. 1984, 1986; Davis and Safe 1988; Kerkvliet et al. 1985, 1990a,b; Vorderstrasse et al. 2001). The *Ah* receptor (*AhR*) is a cytoplasmic receptor and is bound to other cytoplasmic proteins when not bound to its ligand (Chen and

0-415-28457-0/05/$0.00+$1.50

Perdew 1994; Ma and Whitlock 1997). Once a ligand binds the *Ah* receptor, the receptor-ligand complex translocates to the nucleus, forms a heterodimer with the *AhR* nuclear translocator, and binds to dioxin response elements (DREs) found in the regulator region of many genes including P450s and cytokines (Okey et al. 1994; Hankinson 1995; Lai et al. 1996; Rowlands and Gustafsson 1997). TCDD has been shown to produce a profound atrophy of the thymus in all species tested (Vos and Luster 1989) and the effect is greater on the thymus when the exposure is perinatal compared to adult exposure (Holladay and Luster 1994). Studies have indicated that both thymic epithelial cells and thymocytes are sensitive to TCDD and are potential targets (Greenlee et al. 1985). Although the exact molecular mechanisms in which TCDD produces thymic atrophy have not been fully described, effects on thymic stromal cells could significantly alter thymocyte "education" and maturation, producing altered T-dependent immune responses. Direct effects on the thymocytes include apoptosis (McConkey et al. 1988) and alterations in maturation.

Greenlee et al. (1985) presented *in vitro* evidence that TCDD produces thymic atrophy and suppression of cell-mediated immune responses by acting directly on epithelial cells in the thymus. Treatment of thymic epithelial monolayers with TCDD ranging in concentration from 0.1 nM to 10 nM resulted in a concentration-dependent suppression of mitogenic stimulation of thymocytes with either concanavalin A (con A) or phytohemagglutinin (PHA) when the thymocytes were cocultured with the thymic epithelial monolayers. However, when thymocytes were cultured in TCDD-treated thymic epithelial cell-conditioned medium, the previously-observed suppression of mitogenic stimulation to con A or PHA was not observed. This indicated that TCDD was not inducing suppression by production of a soluble factor(s) by thymic epithelial cells but was altering the cell-cell interaction between treated thymic epithelial cells and thymocytes.

Fine et al. (1988) reported thymic atrophy and an alteration in the lymphocyte stem cell population in the fetus and neonate after treating pregnant BALB/c mice with a single dose of TCDD (10 μg/kg body weight) by oral gavage on gestational day (GD) 14. Thymic atrophy was present by GD 18, was maximal by postnatal days 4 to 7, and had essentially returned to normal by postnatal day 18. They observed a greater than 50% reduction in fetal liver stem cell specific enzyme terminal deoxynucleotidyl transferase (TdT) while, on prenatal days 4 and 11, bone marrow TdT was inhibited from 70 to 90% compared to age-matched controls. Flow cytometric analysis revealed slight decreases in the percent of $CD4^+CD8^+$ (DP) cells in TCDD-exposed fetuses on GD 18. Small increases in $CD8^+$ and $CD4^-CD8^-$ (DN) populations were also observed. Percentages of the four major subsets ($CD4^+$, $CD8^+$, DP and DN) were returning to normal by postnatal day 11. Perinatal TCDD exposure did not produce an observable effect on thymocyte expression of the IL-2 receptor. Taken together, these results indicate that perinatal TCDD exposure produces thymic atrophy by altering the pre-thymocyte population's ability to seed the thymus.

Further studies support the suggestion that TCDD, at least in part, is producing immunosuppression and thymic atrophy through alteration in the maturation of thymocytes that migrate from the fetal liver or bone marrow. Holladay et al. (1991) dosed pregnant C57Bl/6 mice for 9 consecutive days (GDs 6 to 14) with 0, 1.5, or 3.0 μg TCDD/kg body weight by oral gavage. Fetal thymic atrophy was pronounced

**Table 8.1 Effect of Gestational Exposure to TCDD on Perinatal Thymocyte, Populations Defined by CD4 and CD8 Antigen Expression**

| Population | Age | Control | 1.5 μg | 3.0 μg |
|---|---|---|---|---|
| CD4-8- | GD 18 | 21.1 ± 0.7 | 30.3 ± 1.8[a] | 37.3 ± 3.7[a] |
| | D 6 | 7.3 ± 0.7 | 7.5 ± 0.5 | 15.9 ± 4.2* |
| | D 14 | 4.3 ± 0.5 | 4.3 ± 0.4 | 5.4 ± 0.6 |
| CD4+8+ | GD 18 | 69.1 ± 1.2 | 52.6 ± 2.5[a] | 43.2 ± 4.5[a] |
| | D 6 | 83.0 ± 1.1 | 83.5 ± 0.4 | 75.5 ± 5.2 |
| | D 14 | 88.2 ± 0.6 | 87.6 ± 0.6 | 86.9 ± 0.6 |
| CD4+8- | GD 18 | 1.8 ± 0.2 | 1.5 ± 0.1 | 2.0 ± 0.2 |
| | D 6 | 6.4 ± 0.5 | 5.5 ± 0.3 | 3.2 ± 0.3[a] |
| | D 14 | 5.9 ± 0.2 | 6.1 ± 0.2 | 5.6 ± 0.6 |
| CD4-8+ | GD 18 | 8.1 ± 0.7 | 15.5 ± 0.9[a] | 17.5 ± 0.9[a] |
| | D 6 | 3.3 ± 0.3 | 3.5 ± 0.3 | 5.4 ± 1.2 |
| | D 14 | 1.5 ± 0.1 | 2.0 ± 0.1 | 2.1 ± 0.2 |

*Note:* Numbers are percentages (means ± SE) of cells within each phenotype.

[a] Significantly different from control, $p < 0.05$.

*Source*: Modified from Holladay et al. 1991.

with the high dose producing thymic cellularity of 26% of control. By postnatal day 14, thymic cellularity had recovered to control values. Flow cytometric analysis of GD 18 thymocytes showed significantly increased DN and $CD8^+$ populations and a significantly decreased DP population compared to control. Again, by postnatal day 14, these differences were no longer apparent (Table 8.1). Although cellularity and thymic phenotype changes induced by TCDD gestational exposure disappeared by postnatal day 14, functional suppression of cell-mediated immunity was present at 8 weeks of age as measured by depressed cytotoxic T lymphocyte (CTL) activity. Normal CTL function was measured at 10 weeks of age (Table 8.2). This indicates

**Table 8.2 Postnatal Immune Function in Mice after Maternal Exposure to Corn Oil, Vehicle, or 3.0 μg/kg TCDD from Days 6 to 14 of Gestation**

| Parameter | Control | TCDD |
|---|---|---|
| Mitogen Response (cpm × $10^3$) | | |
| PHA 2 μg/ml | 33.5 ± 2.4 | 26.4 ± 2.4 |
| Con A 2 μg/ml | 52.0 ± 2.3 | 52.9 ± 2.1 |
| LPS 40 μg/ml | 48.9 ± 3.1 | 41.1 ± 2.6 |
| Antibody Plaque Response | | |
| Spleen cellularity (× $10^6$) | 101.1 ± 5.8 | 108.2 ± 7.9 |
| PFC/$10^6$ spleen cells | 867 ± 93 | 1031 ± 175 |
| Total PFC/spleen (× $10^3$) | 84.4 ± 7.1 | 109.4 ± 20.1 |
| Cytotoxic T Lymphocyte Assay (% cytotoxicity at E:T 25:1) | 69.1 ± 4.7 | 51.5 ± 6.1[a] |

*Note:* Immune assays were performed in 7 to 8 week old mice, N = 12.

[a] Significantly different from control, $p < 0.05$.

*Source*: Modified from Holladay et al. 1991.

that the phenotypic changes observed earlier were associated with persistent postnatal alteration of immune function.

Blaylock et al. (1992) extended these observations of altered fetal thymocyte maturation by gestational TCDD exposure. Using flow cytometric analysis, gestational day 18 fetuses exhibited severe thymic atrophy with increased DN and $CD8^+$ populations and a significantly decreased DP population compared to control. Peanut agglutinin (PNA) binding was used as a means of determining the maturational stage of thymocytes (Scollay et al. 1984). There was a significant decrease in $PNA^+$ small thymocytes corresponding to the DP phenotype while there was a significant increase in $PNA^-$ thymocytes which represents an increase in the DN cell population. J11d expression was used to confirm that the increased $CD8^+$ population was intermediate between DN and DP and not the more mature medullary $CD8^+$ population. After complement-mediated lysis of $CD4^+$ cells (DP and $CD4^+$ single positive), it was found that the $CD8^+$ cells were $J11d^+$ indicating immature cortical thymocytes and not more mature medullary $CD8^+$ cells. These data further indicate that TCDD is inducing a significant phenotype and maturational alteration in the GD-18 thymus.

That TCDD causes thymic atrophy independent of significant apoptosis in double-positive thymocytes is supported by several studies. McConkey et al. (1988) showed that TCDD initiates apoptosis in young rat thymocytes by a $Ca^{++}$-dependent endonuclease. However, subsequent studies by Silverstone et al. (1994) failed to detect apoptosis in TCDD-treated BALB/cJ mouse thymocytes. Additionally, Comment et al. (1992) did not observe mouse thymocyte apoptosis after *in vitro* treatment. Staples et al. (1998a,b) reported that overexpression of the bcl-2 gene in transgenic mice did not prevent TCDD from inducing thymic atrophy. They showed that, in mice lacking the bcl-2 transgene, double positive thymocytes were reduced by only 10% in TCDD-treated animals compared to a 96% decrease in animals treated with dexamethasone, which is known to induce apoptosis in DP thymocytes. This study was subsequently verified using fetal thymus organ culture (FTOC) (Lai et al. 2000). Again, the presence or absence of the bcl-2 transgene did not affect the percentage reduction of cell numbers induced by TCDD. Also, TCDD did not induce detectable apoptosis in FTOC. TCDD treatment was observed to inhibit developing thymocytes from entry into S phase of the cell cycle. This inhibition was most pronounced in the most immature subpopulation, the double negative thymocytes, after continuous exposure.

Gehrs et al. (1997), using a F344 rat model, observed that prenatal TCDD exposure altered several immune parameters. Timed-pregnant F344 rats received a dose of 3.0 μg/kg by oral gavage on GD 14. Offspring were examined at 14 to 17 weeks of age. Organ/body weight ratios were altered in both sexes. The spleen/body weight ratio was increased while the thymus/body weight ratio was decreased. Blastogenesis in B cells and T cells induced by mitogen stimulation was generally not altered with the exception of suppression of pokeweed mitogen blastogenesis in female offspring. The splenic T cell phenotype was altered with a decrease in the percentage of $CD3^+/CD4^-CD8^-$ T cells in both sexes. In addition, the delayed-type hypersensitivity (DTH) response to bovine serum albumin (BSA) was suppressed in both sexes of offspring.

Gehrs and Smialowicz (1999) extended the observation above to show a significant suppression of the DTH to bovine serum albumin (BSA), after GD 14 exposure of F344 rats to TCDD. In this study, suppression in males was more pronounced than females. The maternal TCDD dose on GD 14 required to induce suppression in the male offspring at 14 months of age was 0.1 μg/kg. A maternal dose of 0.3 μg/kg TCDD was required for suppression of female offspring at the same age. The observation was also made that the duration of suppression was greater in males than females exposed to the same TCDD dose. After a 3 μg/kg TCDD maternal dose on GD 14, male offspring displayed significant DTH suppression through 19 months of age. Female offspring from dams receiving the same maternal dose showed significant suppression of the DTH response to BSA only at 4 months of age.

There is also evidence that prenatal TCDD exposure in mice may increase risk of postnatal autoimmune disease. These studies are discussed in Chapter 13 of this book.

## POLYCYCLIC AROMATIC HYDROCARBONS

Polycyclic aromatic hydrocarbons (PAHs) are a group of unsaturated cyclic hydrocarbons containing two or more rings, derived from the incomplete burning of organic materials. They are common environmental contaminants produced by diverse industrial (Ball and Dawson 1969; Ball et al. 1966; Gricuite 1979) and domestic (Huhti 1981; Aggarwal et al. 1982; Grimmer and Pott 1983) processes. Contamination and exposure to PAHs can occur in industrial, rural, and urban settings, as PAHs can be found in plastics, dyes, medicines, charbroiled meat, tobacco, pesticides, and industrial materials such as asphalt, petroleum, and coal tar products, creosote, and roofing tar. Environmental PAH contamination occurs mainly by release into the air from residential wood burning and automobile exhaust (Autrup et al. 1995). Occupational exposure occurs most prominently in industries utilizing petroleum products, while exposure in the home occurs mainly from cigarette smoke and charbroiled foods.

Naturally produced PAHs are generally found as a mixture of a variety of PAH compounds; thus, exposure to only one specific PAH is rare and only occurs in the laboratory. There are over 100 described natural and synthetic PAHs, but benzo(a)pyrene (B(a)P), 3-methylcholanthrene, 7,12-dimethyl-benz(a)anthracene, benzo(b)fluoranthene, chrysene, and phenanthrene are some of the more common and better characterized forms. B(a)P is the most studied PAH compound and is found in combustion pollutants including car exhaust and tobacco smoke; thus, it is often used as a model for other PAH exposure.

When present in the environment, PAHs are found in the air as vapors or adhered to small particulate matter. Some PAHs have limited solubility in water but most remain adhered to particulate matter and settle into soils or the bottoms of lakes and rivers. Human and animal exposure can then occur from inhalation, dermal contact, or ingestion. Once in the body, the lipophilic nature of PAHs causes them to preferentially partition to tissues with a high adipose content, such as the liver kidneys and fat stores. PAHs are metabolized rapidly by members of the cytochrome

P450 family and glutathione-S-transferase into compounds that may be more or less harmful than the original PAH. The metabolized compounds are more water-soluble, however, and are eliminated more rapidly than the original PAH. Elimination rates of PAHs and their resulting metabolites appear to be species-specific, e.g., transformation and elimination of B(a)P is rapid in the rat but slower in dogs and monkeys (U.S. DHHS 1995).

Placental transfer of PAHs occurs, but at a reduced rate compared to distribution throughout the body of the mother (Salhab et al. 1987; Autrup et al. 1995; U.S. DHHS 1995). Autrup et al. (1995) investigated the transplacental transfer of PAHs from cigarette smoke or ambient air by comparing the PAH-albumin adducts present in maternal and fetal umbilical cord blood. A positive association was observed between adduct levels in the maternal blood and umbilical cord blood, indicating that fetal exposure will parallel maternal exposure; however, the levels of adducts in the cord blood were lower than in the maternal blood in all cases. Pregnant women may be exposed to B(a)P levels as high as 150 mg/day in highly polluted areas and up to 3 μg/day due to heavy cigarette smoking (Rodriguez et al. 1999).

Exposure to these PAHs can cause cancer, and B(a)P is generally recognized as the most potent carcinogenic PAH (Urso and Gengozian 1984). In addition to being carcinogenic, PAHs can also induce immune suppression in humans and laboratory animals (Luster and Blank 1987). In coke oven workers, inhalation exposure to a mixture of PAHs lowered immunoglobulin levels, particularly IgG and IgA (Szczeklik et al. 1994). Suppression of cellular and humoral immune function has been demonstrated in adult mice. Adult female B6C3F1 mice were exposed to 10 different PAHs at subchronic exposures and immune response measured by antibody production to sheep red blood cells (White et al. 1985). Anthracene, chrysene, benzo(e)pyrene and perylene did not significantly suppress antibody production; however, benz(a)anthracene, benzo(a)pyrene, dibenz(a,c)anthracene and dibenz(a,h)anthracene suppressed the antibody forming cell response by 55 to 91%. In another study, B6C3F1 mice demonstrated reduced antibody production to both T-dependent and T-independent antigens after exposure to B(a)P (Dean et al. 1983).

PAH exposure during gestation can cause developmental abnormalities in animals, including stillbirths, reabsorbed fetuses, decreased birth weight, birth defects, and infertility or sterility of the offspring (U.S. DHHS 1995). PAHs are also toxic to the developing fetal immune system. Exposure of experimental animals to B(a)P during immune system ontogenesis appears to result in persistent alterations in immune function. Mouse progeny of dams exposed to B(a)P during mid-pregnancy had abnormalities in their cell mediated and humoral immune response. For example, offspring of pregnant mice receiving 150 mg/kg B(a)P from days 11 to 17 of gestation showed suppression in plaque-forming cells, graft-vs.-host, and mixed lymphocyte responses. Suppression of immune responses was detected during gestation, one week after birth, and still demonstrable at 18 months of age (Urso and Gengozian 1984). These animals further exhibited an 8 to 10 fold higher tumor incidence than vehicle-exposed controls. Other studies have also demonstrated an increase in tumor rate associated with suppression of cell and humoral immune function after *in utero* PAH exposure (Urso and Gengozian, 1980 1982). The authors concluded that *in utero* exposure to B(a)P alters development of immunity, and that such exposure during development can lead to severe and sustained postnatal immunosuppression (Urso and Gengozian 1984).

The toxicity of B(a)P is in large part due to production of a highly reactive epoxide metabolite by microsomal mixed-function oxidase enzymes such as cytochrome P450 isozymes, or prostaglandin H synthase and lipooxygenase (Rodriguez et al. 2002). Cytochrome P450 isozymes are generally considered to be the main enzymes responsible for B(a)P metabolism. As summarized by Rodriguez et al. (2002), maternal tissues contain high concentrations of the P4501A1 and P4501A2 isozymes to convert B(a)P, although concentrations of these isozymes are low in placental and fetal tissues. Fetal tissues metabolize B(a)P mainly with the fetal isozyme P4503A7; however, prostaglandin H synthase and lipooxygenase are present in the fetus also. The main placental isozyme is P4501A1, and while it is found in low concentrations in nonsmokers, it is highly inducible in women who smoke (Rodriguez et al. 2002). B(a)P is metabolized to B(a)P-7,8-diol-9,10-epoxide (BPDE) which covalently binds DNA and other nucleophilic intracellular macromolecules. Thus, rapidly proliferating cells, such as those composing the immune system, are targeted (Holbrook 1980). Interestingly, a study evaluating the covalent binding of B(a)P in fetal mouse tissues found liver hematopoietic cells to be the most active (Salhab et al. 1987). Further, the extent of transplacental enzyme induction compared to control was greatest in hematopoietic cells (18-fold), followed closely by whole fetal liver (16-fold) (Salhab et al. 1987; Urso and Johnson 1987). Such results indicate that fetal liver, the primary hematopoietic organ in the fetus, and its hematopoietic cells are specific targets of B(a)P, and may explain, at least in part, the significant toxicity to the developing immune system resulting from exposure to this compound.

The exact mechanisms of PAH immunotoxicity in the developing fetus are not known but it is clear that B(a)P disrupts T cell differentiation (Holladay and Smith 1994; Lummus and Henningsen 1995; Rodriguez et al. 1999). *In utero* exposure to B(a)P was found to alter expression of murine thymocyte and liver fetal cell-surface markers (Holladay and Smith 1994). Pregnant mice were treated orally with 0, 50, 100, or 150 mg B(a)P/kg/d on GDs 13 to 17, and offspring were examined on GD 18. Severe thymic atrophy and cellular depletion were found in B(a)P exposed fetal mice. Flow cytometric analysis indicated that the B(a)P treatment resulted in a significant decrease in the percentage of CD4+8+ fetal thymocytes, as well as significantly increased CD4-8- and CD4-8+ thymocytes (Table 8.3). The B(a)P treatment was also found to decrease total fetal liver cellularity, including numbers of cells within resident hematopoietic subpopulations. In particular, prolymphocytic cells, identified by CD45R antigen expression and by presence of nuclear terminal deoxynucleotidyl transferase, were significantly decreased in animals gestationally exposed to B(a)P (Table 8.4). In another study, a single injection of 150 mg/kg of B(a)P to pregnant mice at mid-gestation significantly reduced the number of T cells present in the thymus of newborn and in the spleens of 1-week-old progeny (Rodriguez et al. 1999). The percentage of newborn CD4+CD8+, CD4+CD8+Vg2+TCR+, and of CD4+CD8+Vb8+TCR+ thymocytes was significantly reduced following *in utero* exposure to B(a)P. Normal expression of the cell surface TCR molecules Vγ2and Vβ8 is necessary for routine T cell function (Rod-

**Table 8.3 Effect of Gestational Exposure of Mice to B(a)P on GD 18 Thymocyte, Populations Defined by CD4 and CD8 Antigens**

| Population | Control | 50 mg/kg | 100 mg/kg |
|---|---|---|---|
| CD4+8- | 2.8 ± 0.1 | 3.2 ± 0.1 | 2.8 ± 0.2 |
| CD4+8+ | 78.0 ± 4.9 | 57.6 ± 1.0[a] | 33.4 ± 5.3[a] |
| CD4-8- | 18.1 ± 2.3 | 37.2 ± 1.2[a] | 62.0 ± 5.3[a] |
| CD4-8+ | 1.1 ± 0.1 | 2.0 ± 0.2[a] | 3.6 ± 0.2[a] |

*Note:* Numbers are percentages (means ± SE) of cells within each phenotype, N = 5.

[a] Significantly different from control, $p < 0.05$.

**Table 8.4 Prolymphoid Cell Populations Present in GD 18 Fetal Liver from Vehicle- and B(a)P-Exposed Mice**

| | Marker Expression (% positive) | | Cell Number ($\times 10^{-6}$) | |
|---|---|---|---|---|
| Treatment | CD45R | TdT | CD45R | TdT |
| Vehicle | 14.5 + 2.2 | 8.4 + 1.8 | 1.7 + 0.3 | 1.0 + 0.2 |
| 100 mg/kg B(a)P | 19.7 + 1.6 | ND | 0.8 + 0.3 | ND |
| 150 mg/kg B(a)P | 20.8 ± 4.9 | 10.5 ± 2.0 | 0.8 ± 0.2[a] | 0.4 ± 0.2[a] |

*Note:* Fetal liver antigen expression in GD 18 fetal mice after maternal exposure to 100 or 150 mg/kg/day B(a)P from GD 13 to 17. Numbers are means + SE, N = 3. ND = not determined.

[a] Significantly different from control, $p < 0.05$.

*Source:* Modified from Holladay and Smith 1994.

riguez et al. 1999). In a third study, lymphocytes of B(a)P-exposed GD 19 fetuses showed decreased subpopulation frequencies in fetal liver of total T cells (from 56% to 16%), Ly1+ expressing cells (from 33% to 9%), and Ly2+ expressing cells (from 56% to 1%) compared with untreated controls (Lummus and Henningsen 1995). These data clearly show that B(a)P, in addition to producing fetal thymic hypocellularity, inhibits normal fetal thymocyte maturation.

An additional possible mechanism for the PAH-induced immunosuppression is increased immune cell apoptosis. B(a)P is metabolized to BPDE; once formed, BPDE can bind to the DNA of developing T cells. BPDE-DNA adducts have been detected in mouse fetal and maternal tissues at high levels (Lu and Wang 1990). Rodriguez et al. (2002) found BPDE-DNA adducts after a single injection of B(a)P in mouse maternal and placental tissues and in fetal splenocytes and CD4+ and CD8+ thymocytes. DNA adducts were present at day 19 of gestation, at birth, and 1 week after birth (Rodriguez et al. 2002). B(a)P and BPDE induce apoptosis of T lymphocytes *in vitro* and may be responsible for inducing apoptosis in the immune system, affecting positive and negative selection of T cells in the developing fetus (Rodriguez et al. 2002). The developing thymus is comprised of approximately 95% CD4+CD8+ T cells, which may present a large susceptible population for BPDE-DNA adduct formation. Apoptosis of T cells can be blocked by a-naphthoflavone, a metabolic inhibition of the conversion of B(a)P to BPDE, indicating that conversion to BPDE may be necessary to induce T cell apoptosis (Davila et al. 1996; Rodriguez et al. 1999). It is possible that BPDE-DNA adduct formation induces apoptosis in the fetal thymus to selectively reduce CD4+CD8+, CD4+CD8+ Vγ2+ TCR+, and CD4+CD8+ Vβ8 TCR+ cells with the net result of secondary lymphoid tissue reduction in the spleen.

Mechanisms involving DNA adduct formation cannot alone fully explain PAH immunosuppression. PAHs also appear to affect T cell function via binding to the *Ah* receptor, similar to TCCD, which can directly affect T cell function through altered regulation of gene expression (Davila et al. 1996). An additional interesting finding demonstrating the complexity of B(a)P induced immunosuppression is that maternal immune capability influenced the immunotoxicity of B(a)P in offspring (Wolisi et al. 2001). In this study, adult female mice were thymectomized at 6 weeks, mated and injected with 150 μg B(a)P)/g body weight at GD 12. Maternal thymectomy and B(a)P exposure reduced average litter size by 40%. Progeny from thymectomized mothers exposed to B(a)P showed enhanced cell mediated immunity (CMI) compared to progeny of both the control mothers and nonthymectomized B(a)P-treated mothers. CMI in progeny from nonthymectomized B(a)P treated dams was significantly reduced compared to progeny form controls. Humoral immunity as measured by PFC assay was suppressed equally in offspring from B(a)P-treated thymectomized and nonthymectomized mothers. Thus, thymectomy of the mother prevents CMI immunosuppression by B(a)P, while humoral immune suppression is unaffected by maternal thymectomy. These results indicate that the maternal thymus is somehow necessary for incurring the effect of B(a)P on progeny CMI (Wolisi et al. 2001).

The above collective data taken together indicate that postnatal immune suppression following *in utero* exposure to B(a)P may result from several intracellular mechanisms targeting immune cells at different hematopoietic levels. Possible mechanisms for immune suppression include increased apoptosis in developing hematopoietic tissues, production of various toxic metabolites of B(a)P, alterations to cell signaling and activation pathways, as well as *Ah* receptor involvement. Regardless of the mechanism involved, the end result of *in utero* B(a)P exposure is a reduction in developing T cells, which results in immunosuppression.

Surprisingly, along with causing immunosuppression, *in utero* exposure to PAHs may also induce an increased hypersensitivity response in the offspring. Recent work has correlated the mother's exposure to PAHs via cigarette smoke and vehicle exhaust with increased inhalant allergies and asthma in children. The bulk of the work has examined the effects of pre- and postnatal exposure to cigarette smoke on pulmonary function and bronchial hyperreactivity. The results of these studies are inconclusive regarding inhalant allergies and asthma with numerous data demonstrating either positive or no association of cigarette smoke and allergies or asthma in children (Oliveti et al. 1996; Schäfer et al. 1997; Kulig et al. 1999). The conflicting results could easily be due to examining immune status in different ages of children, or to the inability to delineate the contributions of pre- and postnatal exposure to altered immune function. Further work will be necessary to resolve this particular question.

In addition to possibly increasing inhalant allergies and asthma in children, evidence indicates that exposure to cigarette smoke (and thus PAHs) may cause other immune hypersensitivity reactions as well. In a study by Kulig et al. (1999), children three years of age who were pre- and postnatally exposed to tobacco smoke had a statistically higher risk of sensitization to food allergens than unexposed children. Children who were only exposed postnatally also had a higher risk of sensitization than unexposed children. The allergens investigated were cow's milk, hen's egg, soybean, and wheat. No significant effect of tobacco smoke was seen on the inhalant allergens, birch, grass pollen, mite or cat. Exposure to environmental tobacco smoke reaches its peak at 1 year of age (due to increased physical contact of a small infant with its parent), which is the same time a child is first exposed to food allergens. The authors suggest it may be the simultaneous exposure to the tobacco smoke and food allergens at 1 year of age that interferes with development of normal tolerance, thus facilitating sensitization to food (Kulig et al. 1999). Schäfer et al. (1997) reported that neonates of mothers who smoked during pregnancy showed elevated levels of IgE, IgA, and IgG3, and that maternal smoking in pregnancy and lactation was found to be associated with an increased risk for atopic eczema, but not respiratory atopy later in life.

## SUMMARY

PAH immunotoxicity is the result of a complex series of events with many interacting factors affecting the final immune status of the individual. The effect of PAHs on immune function has been studied extensively, yet many questions remain. Further work is needed to determine the intracellular mechanism of action causing both immune suppression and hyperreactivity in offspring following maternal exposure. Of particular interest would be further investigations into the effects of maternal immune status on PAH immunotoxicity in the fetus similar to the study by Wolisi et al. (2001), as little work has been done in this area. A better understanding of PAH immunotoxicity could potentially result in intervention methods to improve postnatal immune health status and decrease the incidence of immunological associated disease.

## REFERENCES

Aggarwal AL, Raiyani CV, Patel PD, Shah PG, and Chatterjee SK. 1982. Assessment of exposure to benzo(a)pyrene in air for various population groups in Ahmedabad. *Atmos Environ* 16:867–877.

Autrup H, Vestergaard AB, and Okkels H. 1995. Transplacental transfer of environmental genotoxins: polycyclic aromatic hydrocarbon-albumin in nonsmoking women and the effect of maternal GSTM1 genotype. *Carcinogenesis* 16:1305–1309.

Ball J and Dawson D. 1969. Biological effects of neonatal injection of 7,12-dimethylbenz(a)anthracene. *J Natl Cancer Inst* 42:579–591.

Ball J, Sinclair N, and McCarter J. 1966. Prolonged immunosuppression and tumor induction by a chemical carcinogen injected at birth. *Science* 152:650–651.

Blaylock BL, Holladay SD, Comment CE, Heindel JJ, and Luster MI. 1992. Exposure to tetrachlorodibenzo-p-dioxin (TCDD) alters fetal thymocyte maturation. *Toxicol Appl Pharmacol* 112: 207–213.

Chen H-S and Perdew G. 1994. Subunit composition of the heteromeric cytosolic aryl hydrocarbon receptor complex. *J Biol Chem* 269: 27554–27558.

Comment CE, Blaylock BL, Germolec DR, Pollock PL, Kouchi Y, Brown HW, Rosenthal GJ, and Luster MI. 1992. Thymocyte injury after *in vitro* chemical exposure: potential mechanisms for thymic atrophy. *J Pharmacol Exp Ther* 262:1267–1273.

Davila DR, Romero DL, and Burchiel SW. 1996. Human T cells are highly sensitive to suppression of mitogenesis by polycyclic aromatic hydrocarbons and this effect is differentially reversed by a-naphthoflavone. *Toxicol Appl Pharmacol* 139:333–341.

Davis D and Safe S. 1988. Immunosuppressive activities of polychlorinated dibenzofurans congeners: quantitative structure-activity relationships and interactive effects. *Toxicol Appl Pharmacol* 94: 141–149.

Dean JH, Luster MI, Boorman GA, Lauer LD, Leubke RW, and Lawson L. 1983. Selective immunosuppression resulting from exposure to the carcinogenic congener of benzopyrene in B6C3F1 mice. *Clin Exp Immunol* 52:199–206.

Dencker L, Hassoun E, d'Argy R, and Alm G. 1985. Fetal thymus organ culture as an *in vitro* model for the toxicity of 2,3,7,8-tetrachlorodibenzo-*p*-dioxin and its congeners. *Mol Pharmacol* 27: 133–140.

Faith RE and Moore JA. 1977. Impairment of thymus-dependent immune functions by exposure of the developing immune system to 2,3,7,8-tetrachlorodibenzo-p-dioxin. *J Toxicol Environ Health* 3: 451–464.

Fine JS, Gasiewicz TA, and Silverstone AE. 1988. Lymphocyte stem cell alterations following perinatal exposure to 2,3,7,8-tetrachlorodibenzo-p-dioxin. *Mol Pharmacol* 35: 18–25.

Gehrs BC, Riddle MM, Williams WC, and Smialowicz RJ. 1997. Alterations in the developing immune system of the F344 rat after perinatal exposure to 2,3,7,8-tetrachlorodibenzo-p-dioxin:II. Effects on the pup and the adult. *Toxicol* 122: 229–240.

Gehrs BC and Smialowicz RJ. 1999. Persistent suppression of delayed-type hypersensitivity in adult F344 rats after perinatal exposure to 2,3,7,8-tetrachlorodibenzo-p-dioxin. *Toxicol* 134: 70–88.

Greenlee WF, Dold KM, Irons RD, and Osborne R. 1985. Evidence for direct action of 2,3,7,8-tetrachlorodibenzo-p-dioxin (TCDD) on thymic epithelium. *Toxicol Appl Pharmacol* 79: 112–120.

Griciute L. 1979. Carcinogenicity of polycyclic aromatic hydrocarbons. In: Egan H. (Ed.), *Environmental Carcinogens. Selected Methods of Analysis*, Vol. 3. IARC; 3–15.

Grimmer G and Pott E. 1983. Occurrence of PAH. In: Grimmer G (Ed.), *Environmental Carcinogens: Polycyclic Aromatic Hydrocarbons*. Boca Raton, FL: CRC Press, pp. 61–128.

Hankinson O. 1995. The aryl hydrocarbon receptor complex. Annu. Rev. Pharmacol. Toxicol. 35:307–340.

Holbrook DJ. 1980. Chemical carcinogens. In: Hodgson E and Guthrie FE (Eds.), *Introduction to Biochemical Toxicology*. New York: Elsevier, pp. 310–329.

Holladay SD, Lindstrom P, Blaylock BL, Comment CE, Germolec DR, Heindell JJ, and Luster MI. 1991. Perinatal thymocyte antigen expression and postnatal immune development altered by gestational exposure to tetrachlorodibenzo-*p*-dioxin (TCDD). *Teratol* 44:385–393.

Holladay SD and Smith B.J. 1994. Fetal hematopoietic alterations after maternal exposure to benzo[a]Pyrene: a cytometric evaluation. *J Toxicol Environ Health* 42:259–273.

Holladay SD and Luster MI. 1994. Developmental immunotoxicity. In: *Developmental Toxicity.* 2nd ed. Kimmel C. and Buelke-Sam J (Eds.), Raven Press, New York, pp. 93–118.

Huhti E. 1981. Smoking and lung cancer now. *Eur J Resp Dis* 62:147–158.

Kerkvliet NI, Brauner JA, and Matlock JP. 1985. Humoral immunotoxicity of polychlorinated diphenyl ethers, phenoxyphenols, dioxins and furans present as contaminants of technical grade pentachlorophenol. *Toxicol* 36: 307–324.

Kerkvliet NI, Baecher-Steppan L, Smith BB, Youngberg JA, Henderson MC, and Buhler DR. 1990a. Role of the Ah locus in suppression of cytotoxic T lymphocyte activity by halogenated aromatic hydrocarbons (PCBs and TCDD): structure-activity relationships and effects in C57Bl/6 mice congenic at the Ah locus. *Fundam Appl Toxicol* 14: 532–541.

Kerkvliet NI, Steppan LB, Brauner JA, Deyo JA, Henderson MC, Tomar RS, and Buhler DA. 1990b. Influence of the Ah locus on the humoral immunotoxicity of 2,3,7,8-tetrachlorodibenzo-p-dioxin: evidence for Ah-receptor-dependent and Ah-receptor-independent mechanisms of immunosuppression. *Toxicol Appl Pharmacol* 105: 26–36.

Kulig M, Luck W, Lau S, Niggemann B, Bergmann R, Klettke U, Guggenmoos-Holzmann I, and Wahn U. 1999. Effect of pre- and postnatal tobacco smoke exposure on specific sensitization to food and inhalant allergies during the first three yeas of life. *Allergy* 54:220–228.

Lai ZW, Pineau T, and Esser C. 1996. Identification of dioxin-responsive elements (DREs) in the 5' regions of putative dioxin-inducible genes. *Chem-Biol Interact* 100:97–112.

Lai ZW, Fiore NC, Hahn PJ, Gasiewicz TA, and Silverstone AE. 2000. Differential Effects of diethylstilbestrol and 2,3,7,8-tetrachlorodibenzo-p-dioxin on thymocyte differentiation, proliferation and apoptosis in bcl-2 transgenic mouse fetal thymus organ culture. *Toxicol Appl Pharmacol* 168: 15–24.

Lu LJW and Wang MY. 1990. Modulation of benzo[a]pyrene-induced covalent DNA modifications in adult and fetal mouse tissues by gestational stage. *Carcinogenesis* 11:1367–1372.

Lummus ZL and Henningsen G 1995. Modulation of T-cell ontogeny by transplacental benzo(a)pyrene. *Int J Immunopharmacol* 17:339–50.

Luster MI and Blank JA. 1987. Molecular and cellular basis of chemically induced immunotoxicity. *Ann Rev Pharmacol Toxicol* 27:23–49.

Ma Q and Whitlock JP, Jr. 1997. A novel cytoplasmic protein that interacts with the Ah receptor, contains tetratricopeptide repeat motifs, and augments the transcriptional response to 2,3,7,8-tetrachlorodibenzo-p-dioxin. *J Biol Chem*: 8878–8884.

McConkey DJ, Hartzell O, Duddy SK, Hakansson H, and Orrenius S. 1988. 2,3,7,8-tetrachlorodibenzo-p-dioxin kills immature thymocytes by $Ca^{2+}$ -mediated endonuclease activation. *Science* 242: 256–260.

Okey AB, Riddick DS, and Harper PA. 1994. The Ah receptor: mediator of the toxicity of 2,3,7,8-tetrachlorodibenzo-p-dioxin (TCDD) and related compounds. *Toxicol Lett* 70: 1–9.

Oliveti JF, Kercsmer CM, and Redline S. 1996. Pre and perinatal risk factors for asthma in inner city African-American children. *Amer J Epidemiol* 143:570–577.

Rodriguez JW, Kirlin WG, Wirsiy YG, Matheravidathu S, Hodge TW, and Urso P. 1999. Maternal exposure to benzo[a]pyrene alters development of lymphocytes in offspring. *Immunopharmacol Immunotoxicol* 21:379–396.

Rodriguez JW, Kohan MJ, King LC, and Kirlin WG. 2002. Detection of DNA adducts in developing CD4+ CD8+ thymocytes and splenocytes following *in utero* exposure to benzo[a]pyrene. *Immunopharmacol Immunotoxicol* 24:365–381.

Rowlands JC and Gustafsson JA. 1997. Aryl hydrocarbon receptor-mediated signal transduction. *Crit Rev Toxicol* 27: 109–134.

Salhab AS, James MO, Wang SL, and Shiverick KT. 1987. Formation of benzo(a)pyrene-DNA adducts by microsomal enzymes: Comparison of maternal and fetal liver, fetal hematopoietic cells and placenta. *Chem Biol Interactions* 61:203–214.

Schäfer T, Dirchedl P, Kunz B, Ring J, and Überla B. 1997. Maternal smoking during pregnancy and lactation increases the risk for atopic eczema in the offspring. *J Amer Acad Dermatol* 36:550–556.

Scollay R, Bartlett P, and Shortman K. 1984. T cell development in Ly2, L3T4 and B2A2 during development from early precursor cells to emigrants. *Immunol Rev* 82: 79–98.

Silkworth JB and Grabstein EM. 1982. Polychlorinated biphenyl immunotoxicity: dependence on isomer planarity and the Ah gene complex. *Toxicol Appl Pharmacol* 65: 109–115.

Silkworth JB, Antrim L, and Kaminsky LS. 1984. Correlations between polychlorinated biphenyl immunotoxicity, the aromatic hydrocarbon locus and liver microsomal enzyme induction in C57Bl/6 and DBA/2 mice. *Toxicol Appl Pharmacol* 75: 156–165.

Silkworth JB, Antrim L, and Sack G. 1986. Ah receptor mediated suppression of the antibody response in mice is primarily dependent on the Ah phenotype of lymphoid tissue. *Toxicol Appl Pharmacol* 86: 380–390.

Silverstone AE, Frazier DE, Fiore NC, Soults JA, and Gasiewicz TA. 1994. Dexamethasone, b-estradiol and 2,3,7,8-tetrachlorodibenzo-p-dioxin elicit thymic atrophy through different cellular targets. *Toxicol Appl Pharmacol* 126: 248–259.

Staples JE, Murante FG, Fiore NC, Gasiewicz TA, and Silverstone AE. 1998a. Thymic alterations induced by 2,3,7,8-tetrachlorodibenzo-p-dioxinare strictly dependent on aryl hydrocarbon receptor activation in hemopoietic cells. *J Immunol* 160: 3844–3854.

Staples JE, Fiore NC, Frazier DE, Gasiewicz TA, and Silverstone AE. 1998b. Overexpression of the anti-apoptotic oncogene, bc;-2, in the thymus does not prevent thymic atrophy induced by estradiol or 2,3,7,8-tetrachlorodibenzo-p-dioxin. *Toxicol Appl Pharmacol* 151:200–210.

Szczeklik A, Szczeklik J, Galuszka Z, Musial J, Kolarzyk E, and Targosz D. 1994. Humoral immune suppression in men exposed to polycyclic aromatic hydrocarbons and related carcinogens in polluted environments. *Environ Health Perspect* 102:302–304.

United States Department of Health and Human Services (U.S. DHHS). Agency for Toxic Substances and Disease Registry. 1995. Toxicological profile of polycyclic aromatic hydrocarbons (PAHS) (Update).

Urso P and Gengozian N. 1980. Depressed humoral immunity and increased tumor incidence in mice following *in utero* exposure to benzo[a]pyrene. *J Toxicol Environ Health* 6:569–567.

Urso P and Gengozian N. 1982. Depressed humoral immunity and increased tumor incidence in mice exposed to benzo[a]pyrene and X-rays before or after birth. *J Toxicol Environ Health* 10:817–835.

Urso P and Gengozian N. 1984. Subnormal expression of cell-mediated and humoral immune responses in progeny disposed toward a high incidence of tumors after *in utero* exposure to benzo(a)pyrene. *J Toxicol Environ Health* 14:569–584.

Urso P and Johnson RA. 1987. Early changes in T lymphocytes and subsets of mouse progeny defective as adults in controlling growth of a syngeneic tumor after *in utero* insult with benzo(a)pyrene. *Immunopharmacol* 14:1–10.

Vecchi A, Sironi M, Canegrati MA, Recchia M, and Garattini S. 1983. Immunosuppressive effects of 2,3,7,8-tetrachlorodibenzo-p-dioxin in strains of mice with different susceptibility to induction of aryl hydrocarbon hydrolase. *Toxicol Appl Pharmacol* 68: 434–441.

Vorderstrasse BA, Steppan LB, Silverstone AE, and Kerkvliet NI. 2001. Aryl hydrocarbon receptor-deficient mice generate normal immune responses to model antigens and are resistant to TCDD-induced immune suppression. *Toxicol Appl Pharmacol* 171: 157–164.

Vos JG and Luster MI. 1989. Immune alterations. In: Kimbrough RD and Jenson J (Eds.), *Halogenated Biphenyls, Terphenyls, Naphthalenes, Dibenzodioxins and Related Products*. Elsevier Science Publishers. New York, NY, pp. 295–322.

White KL Jr, Lysy HH, and Holsapple MP. 1985. Immunosuppression by polycyclic aromatic hydrocarbons: a structure-activity relationship in B6C3F1 and DBA/2 mice. *Immunopharmacol* 9:155–164.

Wolisi GO, Majekodunmi J, Bailey GB, and Urso P. 2001. Immunomodulation in progeny from thymectomized primiparous mice exposed to benzo(a)pyrene during mid-pregnancy. *Immunopharmacol Immunotoxicol* 23:267–80.

CHAPTER 9

# Developmental Immunotoxicity of Chlordane

**John B. Barnett, Laura F. Gibson, and Kenneth S. Landreth**

## CONTENTS



## INTRODUCTION

Chlordane is a cyclodiene organochlorine insecticide introduced in the early 1950s. Commercial or technical preparations of chlordane are a mixture of at least 26 different components, but the primary ingredients are the *cis* and *trans* isomers of chlordane, heptachlor, and *trans*-nonachlor (Adamis et al. 1984; Saha et al. 2003). Throughout this review, the term "chlordane" will be used to denote technical-grade chlordane, and the individual purified isomers will be clearly identified as such.

Chlordane is a highly effective insecticide that saw both heavy acceptance and use from the time of its introduction until at least the late 1970s to late 1980s. Early uses were quite diverse, ranging from food crop, ornamental lawn, and shrubbery

0-415-28457-0/05/$0.00+$1.50

protection, to protection of buildings from termites. For termite control, its environmental stability (> 15 years) meant that it could be applied infrequently and still provide a substantial level of protection for an ample period of time. Within the U.S., it was often reapplied annually without regard to residual levels, resulting in substantial accumulation over time. Environmental accumulation thus became a major problem. As the damage that organochlorine compounds inflicted on the environment appreciated, the approved uses for chlordane began to be restricted. The first restriction on its use in the U.S. was to limit its use to architectural protection from termites; however, its widespread availability in garden shops and other similar places made this restriction virtually unenforceable. Later, its distribution was limited to professional pesticide applicators, which made this restriction more realistic. Finally, beginning in 1988, its use was phased out in the U.S. The history of chlordane's use, regulation, and eventual mandated discontinuance was similar in other developed countries, e.g., Japan and the European nations. At that point, however, millions of buildings worldwide had been treated with chlordane.

Even with the enforceable restriction on its use and distribution, the potential for human exposure was very high due to chlordane's environmental stability and widespread use. A testament to its widespread use and environmental stability are the recent publications that report measurable levels of chlordane and contaminants in human populations, as well as in wildlife worldwide. For example, Lewis et al. (1999) reported that house dust contained both α-chlordane and γ-chlordane isomers as well as heptachlor, a common major contaminant of technical chlordane. Similarly, in 2003, Whyatt et al. reported chlordane, *trans*-nonachlor, and heptachlor levels in the ambient air of homes in New York City. Thus, although the use of these chemicals have been banned in the U.S. for more than a decade, they are still present in home environments. In addition to the nonimmunological effects of chlordane on individuals exposed as adults, our work has indicated that prenatal exposure of BALB/c mice to chlordane results in a number of immunotoxic effects in the offspring (Barnett et al. 1990a, 1985, 1990b; Blaylock et al. 1990; Blyler et al. 1994; Spyker-Cranmer et al. 1982; Theus et al. 1992a, 1991, 1992b). In our hands (unpublished data) and as reported by others (Johnson et al. 1986, 1987), these effects could not be duplicated in adult-treated mice; however, a recent report by Tryphonas et al. (2003) indicates an adverse effect of chlordane on the immune response of rats after a 28-day oral treatment. The major concern is that nearly everyone in the U.S. from the age of approximately 50 years of age and younger was exposed to some level of chlordane during gestation due to the treatment of their home. The possible subtle health effects as a result of this exposure may be difficult to determine using conventional epidemiological techniques because of the difficulty of finding an unexposed cohort.

## CHLORDANE BIODEGRADATION AND ELIMINATION

Chlordane is a cyclodiene organochlorine insecticide. Figure 9.1 illustrates the structure of chlordane and its common major contaminants and metabolites. The pathway for metabolic degradation of chlordane in adult research animals is char-

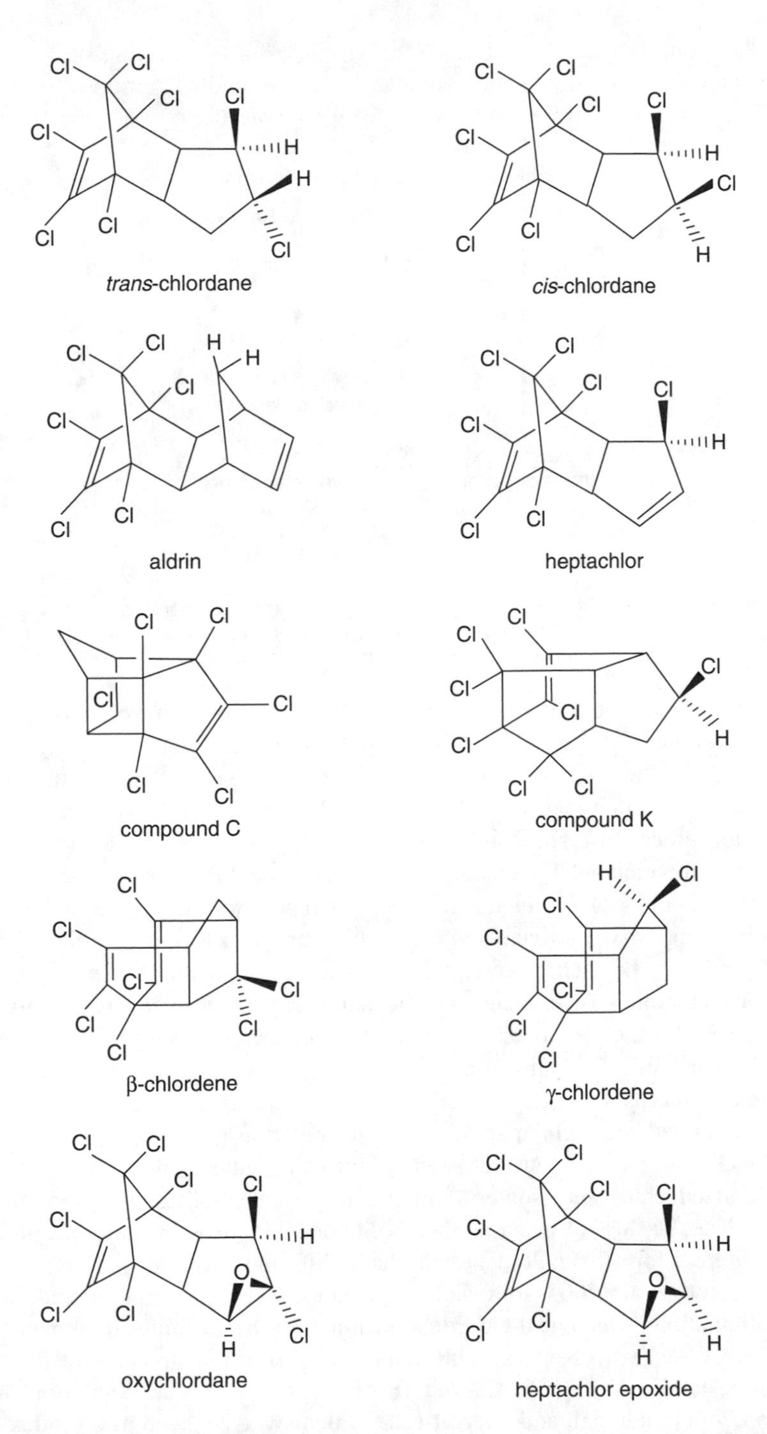


**Figure 9.1** Structure of the isomers, high-abundance contaminants, and metabolites of technical-grade chlordane.

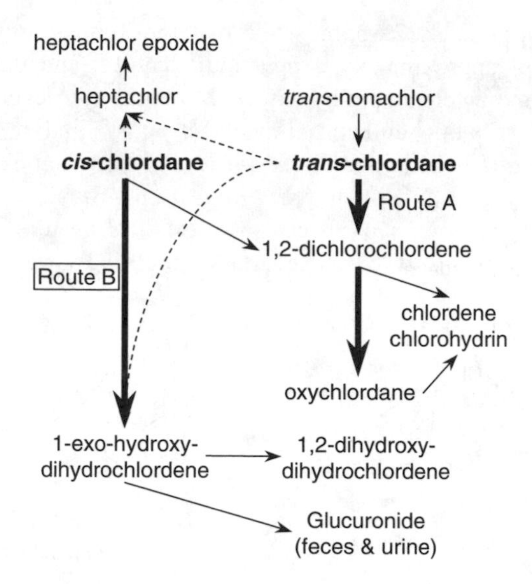


**Figure 9.2** Schematic of the metabolic breakdown of the chlordane isomers. Heavy bold arrows denote the major metabolic pathway, thin arrows denote a minor pathway, and dashed arrows denote a very minor pathway.

acterized (Brimfield et al. 1979; Tashiro et al. 1977) and an outline of this degradation pathway is shown in Figure 9.2. As noted by Tashiro and Matsumura (1977), as well as Brimfeld and Street (1979), while oxychlordane is the major metabolite for *trans*-chlordane, it is a minor metabolite for *cis*-chlordane with a majority of *cis*-chlordane degradation proceeding via Route B (Figure 9.2). Tashiro and Matsumura (1978) also studied the metabolic degradation of *trans*-nonachlor. *Trans*-nonachlor is converted to *trans*-chlordane, which is then degraded primarily by Route A (Figure 9.2) to oxychlordane. *Cis*-chlordane and *trans*-chlordane also form heptachlor, which is then degraded to heptachlor epoxide; however, this is a minority pathway. Thus, although metabolites other than oxychlordane may be formed after exposure to chlordane, reports on the bioaccumulation of chlordane usually report only oxychlordane as a metabolite—a rational approach since *trans*-chlordane is always present in technical chlordane.

It is assumed that similar or identical degradation pathways function in fetal metabolism; however, this has not been definitively studied. The fetus is exposed to unmetabolized chlordane isomers and the major metabolites of chlordane, e.g., oxychlordane, because of maternal metabolite degradation (Matsumoto et al. 2003). These different forms of chlordane may have differing toxicities.

Matsumoto et al. (2003) followed the gestational bioaccumulation and the postnatal elimination of technical chlordane in mice. In these studies, pregnant female mice were dosed with 8 mg/kg chlordane daily from the first day of pregnancy through gestation. At days 5, 12, and 18 of pregnancy, the accumulation of oxychlordane in fat, adrenal, and liver of treated dams was assayed as an index of the accumulation of the various components of chlordane during pregnancy. Maternal fat shows an essential linear accumulation of oxychlordane during gestation to a maximum of approximately 22 ppm, while maternal tissues showed a static level of oxychlordane of approximately 2 ppm until day 12, and then increased to a maximum of approximately 5 ppm by day 18. Maternal liver levels were essentially static throughout gestation at approximately 1 ppm. Fetal tissues show linear increases that essentially mimic maternal fat levels. Postnatal elimination of oxychlordane and *trans*-nonachlor show precipitous drops from 2 days of age until the age that the animals were weaned (3 weeks of age). After weaning, these compounds show a slow relatively steady decrease over time until 16 weeks of age, the last measurement point.

## RELATIONSHIP BETWEEN DOSING REGIMENS AND IMMUNE SYSTEM DEVELOPMENT

The time required for full development of the immune system varies from species to species. For the human, the various elements of the immune system are fully formed early in the second trimester (around 13 to 20 weeks of gestation) (reviewed in: Barnett 1995; Holladay and Smialowicz 2000, chap. 1). It is generally accepted that exposure to xenobiotics during the formative stages of immune development, i.e., prior to 20 weeks gestation in humans, could lead to either a more detrimental or permanent result. Although there is firm evidence to support this conclusion in experimental animal studies, it is difficult to obtain definitive data to support this conclusion for humans because of the large number of uncontrollable variables inherent in human studies. All of the reported developmental immunotoxicological studies on chlordane were performed using the mouse model. As the obvious reason for conducting experimental immunotoxicological studies is to determine an approximate risk assessment for humans, it is important to note the numerous differences between humans and mice in the timing of immune system development. Mice do not complete their immune system development until shortly after birth and do not transfer hematopoiesis from the fetal liver to the bone marrow until much later in gestation, i.e., approximately 1 to 3 days before birth. In contrast, humans transfer hematopoiesis from the fetal liver to the bone marrow at about 15 to 20 weeks of gestation. Thus, to be certain that all stages of immune system development were exposed to chlordane, for the studies reported herein, all animals were treated with chlordane from post-coital day (pcd) 1 through pcd 18.

## EFFECT OF PRENATAL CHLORDANE EXPOSURE ON THE IMMUNE SYSTEM OF THE OFFSPRING

Initial studies on the acute toxic effects of chlordane dealt primarily with its effects on the nervous system, liver, and kidneys. However, the focus of more recent toxicological studies on chlordane have centered on its effects on the immune system. One of the earliest indicators of chlordane's immunotoxicity is a 1976 report of significant hematological changes and reoccurring fevers in a human after documented exposure to chlordane (Furie et al. 1976). However, the first definitive experimental immunotoxicological study is by Spyker-Cranmer et al. in 1982.

The experimental paradigm used for our work was loosely predicated on the scenario of a human (mother) exposed to chlordane before and during pregnancy, who would nurse her own child. To duplicate this scenario within the realm of practicality, breeding female mice were checked twice daily for vaginal plugs, and once the plug was detected, the animal was assumed to be pregnant. The pregnant females were provided a daily ration of peanut butter spiked with the desired dose of chlordane. The females seldom refused this tasty ration and if they did so were eliminated from the study. The resulting offspring were allowed to nurse from their natural mother and were thus exposed to chlordane via the lactations.

Spyker-Cranmer et al. (1982) report that the female offspring of dams treated with chlordane throughout gestation show a profound deficit in contact hypersensitivity response (CHR) to a contact allergen, oxazolone, at both 0.16 and 8.0 mg/kg doses, while male offspring show a significant decrease at the 8.0 mg/kg dose at 100 days of age. Thus, females appear to be sensitive to lower doses of chlordane. Only after recent reanalysis of the CHR data (Spyker-Cranmer et al. 1982), in light of the gender effects noted in more recent studies (Blyler et al. 1994), were these dose-gender differences in the response revealed. Additional CHR analysis, using either an intermediate 4.0 mg/kg dose or a higher 16.0 mg/kg dose, showed significant differences with either dose at 100 days of age in both genders (Barnett et al. 1985). Neither gender demonstrated any significant defect in their humoral immune response (anti-sheep erythrocyte [SRBC] plaque-forming cells [PFC]) at either the 0.16 or 8.0 mg/kg doses (Spyker-Cranmer et al. 1982). The initially observed decrease in CHR was interpreted as an indication of a possible T cell defect; however, extensive investigations by Barnett et al. on T cell efferent functions such as mitogen responses to concanavalin A (Con A) and phytohemagglutinin (PHA) (Barnett et al. 1985), antigen-specific T cell blastogenesis (Barnett et al. 1985), and cytotoxic T lymphocyte (CTL) function (Blaylock et al. 1990), show none of these immune functions are decreased at doses as high as 16 mg/kg. Another cellular response, natural killer cell (NK) activity, is also not affected at a dose of 8.0 mg/kg (Blaylock et al. 1990). Thus, prenatal chlordane exposure does not appear to affect the T-effector cells as judged by these functional assays.

Further evidence for a lack of T-effector cell deficit was provided by host resistance studies. Offspring of dams treated with 0.16, 2, 4, or 8 mg/kg chlordane and then infected with virulent influenza virus [A/PR8/34] were able to withstand a higher infectious dose of virus than controls (Menna et al. 1985). Influenza-specific delayed-type hypersensitivity (DTH) responses in animals exposed prenatally to either 8 or 16 mg/kg of chlordane were also severely depressed (Barnett et al. 1985). In mice, the DTH response contributes to the pathology of an influenza infection, and the ultimate outcome of the infection largely depends on a balance between the immunoprotective effects of CTL cells and the degree of pulmonary damage mediated by $T_{DTH}$ cells (Wyde et al. 1978; Leung et al. 1980). Therefore, it follows that reduced DTH-induced pathology damage coupled with normal CTL responses in the chlordane-exposed offspring will increase the resistance to influenza virus.

**Table 9.1 Summary of the Changes in Macrophage Phenotype Associated with Prenatal Chlordane Exposure**

| Changed "Ground State" | Delayed Peak Activity[a] | Activity Unchanged |
|---|---|---|
| 5'-nucleotidase[b] | Tumoricidal activity | Antigen presentation |
| Transferin receptor expression[b] | Inositol trisphosphate | MHC II expression levels |
| Proteomic expression[c] | IL-1 secretion | |
| | TNF-α secretion | |
| | cAMP induction | |
| | $H_2O_2$ production | |
| | $NO_2$ production | |
| | C3b-RBC phagocytosis | |

[a] Peak activity was delayed in macrophages isolated from the offspring of animals treated with chlordane as compared to similarly isolated macrophages from control (vehicle-treated) animals.
[b] Resident macrophages were isolated from the offspring of animals treated with chlordane or vehicle during gestation. The phenotypic parameter indicated was at a level that was more characteristic of inflammatory macrophages than resident macrophages.
[c] Greater detail provided in Table 9.2.

Because of the lack of evidence for a direct effect on T cells, continuing studies on the mechanism of the decreased DTH function focused on the macrophage, a myeloid cell that functions both in the afferent and efferent portions of the DTH reaction. As we were interested in the mechanism of chlordane's action, in order to reduce the number of animals bred each time, we used a high frankly immunotoxic dose, i.e., 8.0 mg/kg, for most of these studies. As macrophages proceed from a quiescent stage to an activated state, several cytokines are produced (Tabor et al. 1988; Nathan 1987; Tabor et al. 1984), cell surface receptors are expressed and functional activity, such as lysis of tumor target cells, becomes readily measurable (Tabor et al. 1984; Vogel et al. 1983). Macrophages can be classified based on several phenotypic characteristics. Quiescent, nonthioglycolate-elicited macrophages are termed as "resident macrophages"; thioglycolate-elicited macrophages are termed "inflammatory"; inflammatory macrophages treated with interferon-γ (IFN-γ) are called "primed" macrophages; and primed macrophages that have been further stimulated with lipopolysaccharide (LPS) are called "activated." In adult mice exposed to chlordane *in utero*, the ability of macrophages to become fully activated shows a demonstrable lag in the peak time of several functions (summarized in Table 9.1). For example, inflammatory macrophages isolated from vehicle control animals and stimulated with IFN-γ and LPS demonstrate significant cytotoxicity at 2 and 24 hours of culture (Figure 9.3). Stimulated inflammatory macrophages isolated from animals prenatally exposed to chlordane require 48 hours to show equivalent tumoricidal activity (Theus et al. 1992b). Additional studies on enzymatic activity and receptor-expression events that characterize the activation state of macrophages indicate that the resident macrophages from chlordane-exposed animals appear to reside in a more elevated state of activation than macrophages from control animals (Theus et al. 1992a). Other changes included a significant down-regulation of 5'-nucleotidase activity and, in contrast, a significantly higher level of transferrin-receptor binding in resident macrophages from chlordane-exposed offspring. These

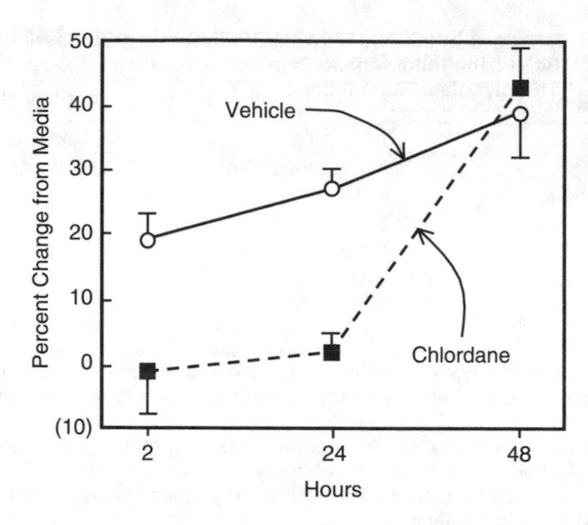


**Figure 9.3** Macrophage tumoricidal activity. Peritoneal exudates cells isolated after thioglycolate-stimulation from the offspring of chlordane or vehicle-treated mothers at 100 to 125 days of age were enriched for macrophages by adherence to plastic. These macrophages were activated by the addition of interferon-γ and lipopolysaccharide and the cytolytic activity against P815 cells was assessed using standardized techniques at the times noted.

phenotypes are more characteristic of inflammatory macrophages than resident macrophages. Macrophages from control offspring provided levels of these two molecules consistent with literature values. We also have preliminary data indicating that activation of some intracellular signaling molecules (e.g., inositol trisphosphate (IP3)) in macrophages from chlordane-exposed animals is temporally retarded as well. The ability of macrophages from chlordane-exposed animals to reach a similar level of cytolytic activity as macrophages from control animals with sufficient time, however, may explain why no overt adverse effects on the health of laboratory mice has been noted. We still have no explanation for the noted delay in intracellular IP3 formation and changes in the levels of 5'nucleotidase activity and transferin-receptor expression.

Further analysis of the macrophages from chlordane-exposed offspring indicated numerous changes in the protein complement (Theus et al. 1992a). Resident and inflammatory macrophages from the chlordane- and vehicle-exposed offspring were cultured overnight in the presence of $^{35}$S-methionine to label newly synthesized proteins and assess metabolic activity. Total cell protein extracts were prepared and subjected to two-dimensional polyacrylamide electrophoresis (2D-PAGE) and the resolved proteins detected by autoradiography. Table 9.2 details the changes noted in the proteins from the various macrophage populations. In the offspring of vehicle-exposed animals a total of 238 proteins were different between the quiescent and inflammatory macrophages, indicating at least some differences in protein content resulted from inducing the macrophages to become inflammatory via thioglycolate induction. A majority of these can be attributed to the induction of new proteins (149 proteins) or increased expression (38 proteins). The expression of a minority of proteins was either down-regulated by a statistically significant amount (16

**Table 9.2 Changes in 2D-PAGE Protein Patterns in Macrophages from Chlordane- and Vehicle-Exposed Animals as a Result of Thioglycolate Stimulation**

| Protein Status | Vehicle[a] Resident vs. Inflammatory[b] | Chlordane[a] Resident vs. Inflammatory[b] |
|---|---|---|
| Missing | 35 | 2 |
| Down-regulated | 16 | 21 |
| Up-regulated | 38 | 14 |
| New proteins | 149 | 49 |
| Total differences | 238 | 80 |

[a] Macrophage isolated from the offspring of vehicle- or chlordane-exposed animals.
[b] Resident macrophages versus inflammatory macrophages.
[c] The protein pattern of resident macrophages from animals exposed prenatally to either chlordane or vehicle was the reference point for these data.

proteins) or abrogated (35 proteins). A similar comparison of resident and inflammatory macrophages from animals exposed to chlordane prenatally showed far fewer changes *in toto* (80 proteins) and the number of proteins in each category showed substantial differences (Table 9.2). Spot patterns showed numerous differences; however, the identity of the changed proteins is unknown because the instrumentation needed to identify the protein(s) in each spot was not available. These data are interesting from two perspectives; first, as noted above, the changes in protein expression between macrophages from *in utero* exposed offspring with those from control offspring, and secondly, as an example of the application of proteomics to study both toxicological questions and basic biology correlated with phenotypic changes associated with activation.

## EFFECTS OF PRENATAL CHLORDANE EXPOSURE ON HEMATOPOIESIS

The demonstrable effects of chlordane on *in utero* development of macrophages prompted questions about the hematopoietic development of other cell types. The details of the hematopoietic hierarchy of differentiating progenitor cells and their relationship to the pluripotential stem cells have been detailed elsewhere in this book (Chapter 1). Pluripotent hematopoietic stem cells are first found intraembryonically (Le Douarin et al. 1977) and then in the yolk sac circulation (Moore et al. 1967) of developing mice. By day 8 of gestation, hematopoietic stem cells are resident in the developing fetal liver, and by day 15 the fetal liver appears to be the sole site of hematopoietic stem cell function (Moore et al. 1985). Between gestational days 17 and 20, hematopoiesis progressively increases in the developing bone marrow and postnatal hematopoietic function is primarily within the bone marrow. Hematopoietic stem cells proliferate, renew, and differentiate in the presence of a hierarchy of regulatory cytokines produced by stromal cells, macrophages and T lymphocytes.

The earliest detectable stem cells proliferate in the presence of IL-3 (Suda et al. 1985) and the extent of stem cell proliferation appears to be further regulated by IL-1 (Moore et al. 1987), stem cell factor (SCF; c-*kit*-ligand) (Zsebo et al. 1990), and IL-6 (Moore et al. 1985). Progeny of hematopoietic stem cells become committed to either myeloid (i.e., monocyte/macrophage, granulocyte, erythroid, and megakaryocytic) development or to lymphoid cell development. Proliferation and differentiation of cells destined to become monocytes/macrophages and granulocytes are regulated by granulocyte/macrophage colony stimulating factor (GM-CSF), macrophage colony stimulating factor (M-CSF), and granulocyte colony stimulating factor (G-CSF) (Metcalf 1989). Similarly, the progression of cells that enter the erythroid lineage is regulated by IL-3, erythropoietin (EPO), SCF and IL-11 (Cronkite et al. 1979; Quesniaux et al. 1992).

The proliferation of committed B lymphoid progenitor cells is primarily regulated by IL-7 (Namen et al. 1988) and this proliferative response is potentiated by two additional stromal cell products; SCF and insulin-like growth factor-1 (IGF-1) (Billips et al. 1992; Gibson et al. 1993; King et al. 1988; Landreth et al. 1992). IGF-1 (in the absence of IL-7) is required for B cell progenitors (pro-B cells) to express μ-heavy chain protein (Landreth et al. 1992), and IL-4 is required for efficient expression of light chain protein (King et al. 1988). Developing T lymphocytes also respond to the proliferative signal provided by IL-7 (Morrissey et al. 1989) as well as IL-2 and IL-1.

Our initial published work focused on myeloid stem cells (CFU-S) and mouse lung conditioned media (MLCM)-induced multipotent stem cell populations in adult animals and 18 pcd fetal liver (Barnett et al. 1990a; Barnett et al. 1990b). The CFU-S cell populations were assayed using the method of Till and McCulloch (Till and McCulloch 1961) by injecting bone marrow cells into an irradiated syngeneic host to produce splenic colony-forming units. GM-progenitor cells, which differentiate into either myeloid or erythroid cells, can be quantified by determining the number of colony-forming units (CFU)-in-culture with the addition of a source of granulocyte/macrophage-colony stimulating factor (GM-CSF) to the culture. In these studies, a persistent reduction in the number of both CFU-S and CFU-MLCM was noted. This was apparent at day 18 pcd (earliest time point assayed) as well as a number of time points during adulthood. The oldest age at which these stem/progenitor cell populations were measured was 300 days of age and deficits of similar magnitude were still detectable.

More recent data (Blyler et al. 1994) collected, using recombinant sources of growth factors, are summarized in Table 9.3. Two findings readily apparent from Table 9.3 are 1) that prenatal chlordane exposure results in a decrease in number of hematopoietic stem cells in adult offspring, and 2) that in the myeloid lineage, these decreases are apparent only in the female offspring.

**Table 9.3 The Effect of Prenatal Chlordane on Adult Stem and Progenitor Cells**

| Progenitor Type | Male | Female |
|---|---|---|
| CFU-IL-3 (total) | No change | Decreased |
| CFU-IL-3 (compact) | No change | Decreased |
| CFU-M | No change | Decreased |
| CFU-GM | No change | Decreased |
| CFU-B | Decreased | Decreased |

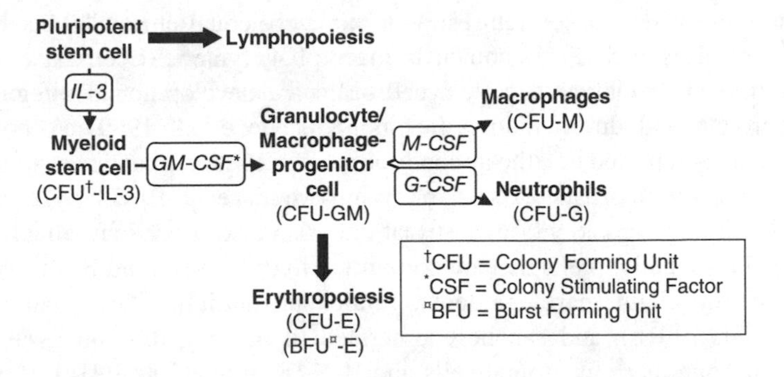


**Figure 9.4** Hematopoietic cell maturation scheme.

## EFFECT OF PRENATAL CHLORDANE EXPOSURE ON FETAL IMMUNE SYSTEM DEVELOPMENT IN MICE

Because the offspring are allowed to nurse from their natural mother, the question remained of whether defects measured were solely due to prenatal exposure or could have been caused, or at least contributed to, by the exposure to chlordane via the lactations. In part to answer this question, the effect of chlordane on the number of hematopoietic progenitor cells (Figure 9.4) was assessed by assaying the number of CFU using fetal liver cells from pcd 14, 16, and 18 fetuses. As shown in Figure 9.5, the CFU-interleukin (IL)-3, a very early progenitor cell, showed a significant elevation at pcd 16 but then dropped to approximately 80% of control by pcd 18. Within the myeloid lineage, the number of CFU-granulocyte/macrophage (CFU-GM) in the fetal liver was essentially equal in the chlordane-treated fetuses and control fetuses at pcd 14 and 16; however, by pcd 18, it had decreased to approximately 70% of control. The CFU-macrophage, a more mature myeloid progenitor, was elevated at

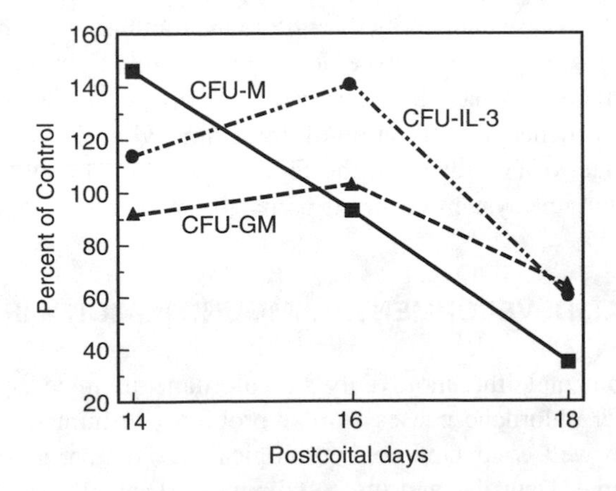


**Figure 9.5** The effect of chlordane on fetal liver hematopoietic progenitor cells. Liver cells were harvested from the fetuses on the stated post-coital day. These cells were processed and assayed using standard techniques to measure their ability to form colonies in semi-solid agar with the addition of the noted CSF.

pcd 14 in chlordane-exposed fetuses, approximately equal at pcd 16 but less than 40% of control by pcd 18. As shown in Figure 9.4, lymphoid cells also originate from multipotent hematopoietic stem cells, and are developmentally regulated by specific growth and differentiation factors (Kincade et al. 1989). Although our previous studies reported that the antibody responses to protein antigens were apparently normal in the chlordane-exposed offspring (Menna et al. 1985; Spyker-Cranmer et al. 1982), two observations suggest that chlordane exposure may affect humoral immunity. Pro-B cells—cells that are committed to the developing B lineage—have D-JH Ig-heavy chain rearrangements, retain substantial self-renewal potential (Rolink et al. 1991b), and can be enumerated in limiting dilution assays in the presence of bone marrow stromal cells and IL-7 (Rolink et al. 1991a). Pro-B cells are enumerated at cloning dilution as colonies of lymphoid cells initiated by single cells. Our experiments demonstrate that the number of pro-B cells is significantly reduced at 14 and 18 days pcd in chlordane treated animals. This suggests that prenatal chlordane exposure has substantial toxicity to the developing B lymphocyte lineage detectable at least 4 days prior to reductions of myeloid progenitors in the same tissue. Second, B cells detectable in the CFU-B assay were dramatically reduced in postnatal animals that had been exposed *in utero* to chlordane. This observation is intriguing because it closely approximates a known immunodeficiency. CBA/N mice have normal numbers of B cells and respond to protein antigen, but are immunodeficient and fail to mount an immune response to polysaccharide antigens (Kincade 1977). CBA/N mice also completely lack cells that will form colonies in the CFU-B assay (Kincade 1977). The congenital immunodeficiency of the CBA/N mouse is now known to be due to a point mutation in the Bruton's tyrosine kinase (BTK) gene, which results in failure of one intracellular signaling pathway necessary for appropriate B cell development and development of immune response to polysaccharide antigens (Rawlings et al. 1993; Kinnon et al. 1993). It is also known that interruption of BTK function in humans results in a more dramatic failure of B cell development, resulting in complete absence of B cells and humoral immune responses. These data imply that chlordane immunotoxicity to B cells found in mice may in fact suggest even more dramatic hematotoxicity in the human fetus.

Thus, while there may have been some contribution to the developmental immunotoxicity caused by neonatal lactation of contaminated milk, there is undoubtedly a measurable effect of chlordane on the fetus at pcd 18. This indicates that major damage to the immune system occurred prenatally.

## POTENTIAL DEVELOPMENTAL IMMUNOTOXICITY IN HUMANS

The developmental immunotoxicity of chlordane in mice is well described; however, whether chlordane causes similar problems in humans is far less clear. There have been well conducted epidemiological studies that touch on this possibility. For example, Dewailly and his collaborators (Dewailly et al. 1993; Muckle et al. 2002; Mulvad et al. 1996) have documented both the presence of chlordane and immunotoxic effects on Inuit infants exposed to organochlorine compounds during gestation and nursing. However, chlordane was not the only organochlorine compound in the food chain of these individuals. Thus, it is more accurate to state that organochlorine compounds, which contain some level of chlordane, cause a developmental immunotoxic effect; however, it is unlikely that this could be attributed to chlordane alone.

## SUMMARY

Chlordane is very clearly immunotoxic to the developing immune system of mice, while its immunotoxicity after an acute adult-only exposure is less well defined. The recent findings of Tryphonas et al. (2003) of an immunotoxic effect of subchronic exposure in adult rats calls into question whether this compound is solely a developmental immunotoxicant. The developmental immunotoxic effects in mice are, interestingly, specific for myeloid cell function, although there were also unexplained decrements in CFU-B numbers. Collectively, these studies suggest the immunotoxic effects of chlordane may have broader impact than previously appreciated.

## ACKNOWLEDGMENTS

This work was supported in part by the National Institutes of Health (NIH) grant ES06641.

## REFERENCES

Adamis Z, Akintonwa DA, Goulding R, Kashyap SK, Villeneuve DC, and Wassermann D. 1984. Chlordane. In: *Environmental Health Criteria World Health Organization*, Geneva.

Barnett JB. 1995. Developmental immunotoxicology. In: Smialowicz RJ and Holsapple MP (Eds.), *Experimental Immunotoxicology.* CRC Press, Boca Raton, FL, pp. 47–62.

Barnett JB, Blaylock BL, Gandy J, Menna JH, Denton R, and Soderberg, LS. 1990a. Alteration of fetal liver colony formation by prenatal chlordane exposure. *Fundam Appl Toxicol* 15, 820–822.

Barnett JB, Blaylock BL, Gandy J, Menna, JH, Denton R, and Soderberg LS. 1990b. Long-term alteration of adult bone marrow colony formation by prenatal chlordane exposure. *Fundam Appl Toxicol* 14, 688–695.

Barnett JB, Soderberg L, and Menna JH. 1985. The effect of prenatal chlordane exposure on the delayed hypersensitivity response of BALB/c mice. *Toxicol Lett* 25, 173–183.

Billips LG, Petitte D, Dorshkind K, Narayanan R, Chiu C-P, and Landreth KL. 1992. Differential roles of stromal cells, interleukin 7, and *kit*-ligand in the regulation of B lymphopoiesis. *Blood* 79, 1185–1192.

Blaylock BL, Soderberg LSF, Gandy J, Menna JH, Denton R, and Barnett JB. 1990. Cytotoxic T-lymphocyte and NK responses in mice treated prenatally with chlordane. *Toxicol Lett* 51, 41–49.

Blyler G, Landreth KS, and Barnett JB. 1994. Gender-specific effects of prenatal chlordane exposure on myeloid cell development. *Fund Appl Toxicol* 23, 188–193.

Brimfield AA and Street JC. 1979. Mammalian biotransformation of chlordane: *in vivo* and primary hepatic comparisons. *Annu NY Acad Sci* 320, 247–256.

Cronkite EP, Carsten AL, Cohen R, Miller ME, and Moccia G. 1979. The effects of humoral factors on amplification of nonrecognizable erythrocytic and granulocytic precursors. *Blood Cells* 5, 331–350.

Dewailly E, Ayotte P, Bruneau S, Laliberte C, Muir DC, and Norstrom RJ. 1993. Inuit exposure to organochlorines through the aquatic food chain in arctic Quebec. *Environ Health Perspect* 101, 618–620.

Furie B and Trubowitz S. 1976. Insecticides and blood dyscrasis. Chlordane exposure and self-limited refractory megaloblastic anemia. *J AMA* 235, 1720–1722.

Gibson LF, Piktel D, and Landreth KL. 1993. Pro-B cell proliferation is regulated by insulin-like growth factor-1. *Blood* 82, 3005–3011.

Holladay SD and Smialowicz RJ. 2000. Development of the murine and human immune system: differential effects of immunotoxicants depend on time of exposure. *Environ Health Perspect* 108 Suppl 3:463–73., 463–473.

Johnson KW, Holsapple MP, and Munson AE. 1986. An immunotoxicological evaluation of gamma-chlordane. *Fundam Appl Toxicol* 6, 317–326.

Johnson KW, Kaminski NE, and Munson AE. 1987. Direct suppression of cultured spleen cell responses by chlordane and the basis for differential effects on *in vivo* and *in vitro* immunocompetence. *J Toxicol Environ Health* 22, 497–515.

Kincade PW. 1977. Defective colony formation by B-lymphocytes from CBA/N and C3H/HEJ mice. *J Exper Med* 145, 249–263.

Kincade PW, Lee G, Pietrangeli CE, Hayashi S-I, and Gimble JM. 1989. Cells and molecules that regulate b lymphopoiesis in bone marrow. *Ann Rev Immunol* 7, 111–143.

King AG, Wierda D, and Landreth KS. 1988. Bone marrow stromal cell regulation of B-lymphopoiesis. I. The role of macrophages, IL-1, and IL-4 in pre-B cell maturation. *J Immunol* 141, 2016–2026.

Kinnon C, Hinshelwood S, Levinsky RJ, and Lovering RC. 1993. X-linked agammaglobulinemia—gene cloning and future prospects. *Immunol Today* 14, 554–558.

Landreth KS, Narayanan R, and Dorshkind K. 1992. Insulin-like growth factor-I regulates pro-B cell differentiation. *Blood* 80, 1207–1212.

Le Douarin NM, Houssaint E, and Jotere F. 1977. Differentiation of the primary lymphoid organs in avian embryos. In: *Avian Immunology*, pp. 29–37.

Leung KN and Ada GL. 1980. Cells mediating delayed-type hypersensitivity in the lungs of mice infected with an influenza A virus. S*cand J Immunol* 12, 393–400.

Lewis RG, Fortune CR, Willis RD, Camann DE, and Antley JT. 1999. Distribution of pesticides and polycyclic aromatic hydrocarbons in house dust as a function of particle size. *Environ Health Persp* 107, 721–726.

Matsumoto H, Schafer R, and Barnett JB. Prenatal and postnatal tissue distribution of technical grade chlordane in gestationally exposed BALB/c mice. 2003.

Menna JH, Barnett JB, and Soderberg L. 1985. Influenza type A virus infection of mice exposed *in utero* to chlordane; survival and antibody studies. *Toxicol Lett* 24, 45–52.

Metcalf D. 1989. The Molecular control of cell division, differentiation, commitment, and maturation in haemopoietic cells. *Nature* 339, 27–30.

Moore MAS, Gabrilove J, Lu L, and Yung YP. 1985. Myeloid and erythroid stem cells: regulation in normal and neoplastic states. In: Ford RJ and Maizel AL (Eds.), *Mediators In Cell Growth And Differentiation*. Raven Press, New York, pp. 147–157.

Moore MAS and Owen JJT. 1967. Stem-cell migration in developing myeloid and lymphoid systems. *Lancet* 658–659.

Moore MAS and Warren DJ. 1987. Synergy of interleukin 1 and granulocyte colony-stimulating factor. *Proc Natl Acad Sci USA* 84, 7134–7138.

Morrissey PJ, Goodwin RG, Nordan RP, Anderson D, Grabstein KH, Cosman D, Sims J, Lupton S, Acres B, Reed SGD, Mochizuki D, Eisenman J, Conlon PJ, and Namen AE. 1989. Recombinant interleukin 7, pre-B cell growth factor, has costimulatory activity on purified mature T cells. *J Exp Med* 169, 707–716.

Muckle G, Ayotte P, Dewailly E, Jacobson SW, and Jacobson JL. 2002. Prenatal exposure of the northern Québec Inuit infants to environmental contaminants. *Environ Health Perspect* 109, 1291–1299.

Mulvad G, Pedersen HS, Hansen JC, Dewailly E, Jul E, Pederson MB, Bjerregaard P, Malcom GT, Deguchi Y, and Middaugh JP. 1996. Exposure of Greenlandic Inuit to organochlorines and heavy metals through the marine food-chain: an international study. *Sci Total Environ* 186: 137–139.

Namen AE, Schmi AE, March CJ, Overell RW, Park LS, Urdal DL, and Mochizuki DY. 1988. B cell precursor growth-promoting activity. Purification and characterization of a growth factor active on lymphocyte precursors. *J Exp Med* 167: 988–1002.

Nathan CF. 1987. Secretory products of macrophages. *J Clinical Investigation* 79: 319.

Quesniaux VFJ, Clark SC, Turner K, and Fagg B. 1992. Interleukin 11 stimulates multiple phases of erythropoiesis *in vitro*. *Blood* 80: 1218–1223.

Rawlings DJ, Saffran DC, Tsukada S, Largaespada DA, Grimaldi JC, Cohen L, Mohr RN, Bazan JF, Howard M, Copeland NG, Jenkins NA, and Witte ON. 1993. Mutation of unique region of Bruton's tyrosine kinase in immunodeficient xid mice. *Science* 261: 359–361.

Rolink A, Kudo A, Karasuyama H, Kikuchi Y, and Melchers F. 1991a. Long-term proliferating early pre B cell lines and clones with the potential to develop to surface IG-positive, mitogen reactive B cells *in vitro* and *in vivo*. *Embo* 10: 327–336.

Rolink A, and Melchers F. 1991b. Molecular and cellular origins of B lymphocyte diversity. *Cell* 66: 1081–1094.

Saha JG, and Lee YW. 2003. Isolation and identification of the components of a commercial chlordane formulation. *Bull Environ Contam Toxicol* 4: 285–296.

Spyker-Cranmer JM, Barnett JB, Avery DL, and Cranmer MF. 1982. Immunoteratology of chlordane: cell-mediated and humoral immune responses in adult mice exposed *in utero*. *Toxicol Appl Pharmacol* 62: 402–408.

Suda T, Suda J, Ogawa M, and Ihle JN. 1985. Permissive role of interleukin 3 (IL-3) in proliferation and differentiation of multipotential hemopoietic progenitors in culture. *J Cell.Phys* 124: 182–190.

Tabor DR, Azadegan AA, Schell RF, and Lefrock JL. 1984. Inhibition of macrophage C3b-mediated ingestion by symphilitic hamster T cell-enriched fractions. *J Immunol* 135: 2698–2699.

Tabor DR, Burchett SK, and Jacobs RF. 1988. Enhanced production of monokines by canine alveolar macrophages in response to endotoxin-induced shock. *Proc Soc Exper Biol and Med* 187: 408.

Tashiro S and Matsumura F. 1977. Metabolic routes of cis- and trans-chlordane in rats. *J Agric Food Chem* 25: 872–880.

Tashiro S and Matsumura F. 1978. Metabolism of trans-nonachlor and related chlordane components in rat and man. *Arch Environ Contam Toxicol* 7: 113–127.

Theus SA, Lau KA, Tabor DR, Soderberg L, and Barnett JB. 1992a. *In vivo* prenatal chlordane exposure induces development of endogenous inflammatory macrophages. *J Leukocyte Biol* 51: 366–372.

Theus SA, Tabor DR, and Barnett JB. 1991. Alteration of macrophage TNF production by prenatal chlordane exposure. *FASEB J* 5: A1347.

Theus SA, Tabor DR, Soderberg LS, and Barnett JB. 1992b. Macrophage tumoricidal mechanisms are selectively altered by prenatal chlordane exposure. *Agents Actions* 37: 140–146.

Till JE and McCulloch EA. 1961. A direct measurement of the radiation sensitivity of normal bone marrow cells. *Radiat Res* 14: 213–222.

Tryphonas H, Bondy G, Hodgen M, Coady L, Parenteau M, Armstrong C, Hayward S, and Liston V. 2003. Effects of cis-nonachlor, trans-nonachlor and chlordane on the immune system of Sprague-Dawley rats following a 28-day oral (gavage) treatment. *Food Chem Toxicol* 41: 107–118.

Vogel SN., Finbloom DS, English KE, Rosenstreich DL, and Langreth SG. 1983. Interferon induced enhancement macrophage fc receptor expression: B-interferon treatment of C3H/HEJ macrophage results in increased numbers and density of Fc receptors. *J Immunol* 130: 1210.

Whyatt RM, Camann DE, Kinney PL, Reyes A, Ramirez J, Dietrich J, Diaz D, Holmes D, and Perera FP. 2003. Residential pesticide use during pregnancy among a cohort of urban minority women. *Environ Health Perspect* 110: 507–514.

Wyde P, Peavy D, and Cate T. 1978. Morphological and cytochemical characterization of cells infiltrating mouse lungs after influenza infection. *Infect and Immun* 21: 140–146.

Zsebo KM, Wypych M, McNiece IK, Lu HS, Smith KA, Karkare SB, Sachdev RK, Yuschenkoff VN, Birkett NC, Williams LR, Satyagal VN, Tung W, Bosselman RA, Mendiaz EA, and Langley KE. 1990. Identification, purification, and biological characterization of hematopoietic stem cell factor from buffalo rat liver-conditioned medium. *Cell* 63: 195–201.

CHAPTER 10

# Toxicity of Lead to the Developing Immune System

**Rodney R. Dietert and Ji-Eun Lee**

## CONTENTS



## INTRODUCTION

Lead is a pervasive environmental contaminant known to be toxic to numerous organs (Nolan and Shaikh 1992; Loghman-Adman 1997; Foster et al. 1996; He et al. 2000) and physiological systems. Sensitive targets include the neurological (Araki et al. 2000) and immune systems (Koller 1973; Zelikoff et al. 1994; Lawrence and McCabe Jr. 2002). Concern over the ramifications of early exposure to lead has existed for some time (Goyer 1993). During the past decade, it has been recognized that exposure of the fetus or infant to even low levels of lead presents a potential hazard. Levels of lead as measured in the blood of infants in the range of 10 to 15 μg/ml have been associated with deficits in cognitive and behavioral performance (Bellinger and Needleman 1983; Bellinger 1995; Dieterich 1991; Garavan et al. 2000).

0-415-28457-0/05/$0.00+$1.50

## ADULT IMMUNOTOXICITY

Seminal research has been performed by Dr. David Lawrence and colleagues characterizing the immunotoxicity to lead following adult exposure. Exposure to lead causes a shift in immune balance, with T helper 1 dependent functions depressed and T helper 2 functions elevated (McCabe and Lawrence 1991; Heo et al. 1997; 1998.). This is reflected in the cytokine production spectrum among stimulated splenic lymphocytes as well (Heo et al. 1996). Since the delayed-type hypersensitivity (DTH) reaction is linked to T helper 1 function, depressed DTH responses appear to represent a hallmark of lead-induced immunotoxicity (Muller et al. 1977; McCabe et al. 1999). In addition to lymphoid-dependent changes, lead appears to target macrophage function (Koller and Roan 1977; Zelikoff et al. 1993). A reduced capacity for nitric oxide production has been reported following exposure to lead *in vivo* (Kowolenko et al. 1988) and *in vitro* (Tian and Lawrence 1995; Chen et al. 1997), and changes in macrophage cytokine production (Cohen et al. 1994; Guo et al. 1996) have been noted following exposure. In some studies, changes in antibody titers have also been reported (Koller and Kovacic 1974).

Not surprisingly, reduced resistance to certain infectious diseases (Hemphill et al. 1971; Cook et al. 1975; Exon et al. 1979; Lawrence 1981; Kowolenko et al. 1991; Youssef et al. 1996; Gupta et al. 2002) has been reported following exposure to lead. Changes in host tumor resistance also occur with the impact seen on the incidence and growth of cancers (Kobayashi and Okamoto 1974; Kerkvliet and Baecher-Steppan 1982).

## LEAD TOXICITY AND HYPERSUSCEPTIBLE SUBPOPULATIONS

The likelihood that susceptibility to lead toxicity varies among individuals within a species has several lines of supporting evidence. First, it has been shown that the toxicokinetics of lead differs in the human population based on at least two genes. These genes can influence not only the amount of lead absorbed, but also the amount of lead which is chelatable (using dimercaptosuccinic acid [DMSA]) (Schwartz et al. 2000). Additionally, lead toxicity differences have been reported among mice differing in certain behavioral parameters (e.g., circling preferences). This led the authors to suggest that brain laterality and neuroimmune circuits may be important factors linked to individual susceptibility to environmentally induced immunomodulation (Kim and Lawrence 2000). Gender appears to influence the susceptibility and nature of the toxicity response to certain lead exposures. Vahter et al. (2002) reviewed the hypotheses and available literature that metal toxicity, including lead exposure, may pose a greater health risk to women than men; and Ronis et al. (1996) reported major gender differences in reproductive outcome following early developmental exposures to lead. Immunotoxicity differences among genders have been noted by investigators (Bunn et al. 2000; 2001a, b, c). Because these findings hold not only in mammals where the male is the heterogametic sex, but also in birds where the female is the heterogametic sex, the differences appear to be linked to sex and endocrine-based differences rather than chromosome heterogeneity. In addi-

**Table 10.1 Characteristics of Lead-Induced Immunotoxicity Following Early Life-Stage Exposures**

- Early exposure to lead produces persistent immunotoxic changes (mouse, rat, chicken).
- Lead induced immunotoxicity following fetal/neonatal exposure occurs at doses that are below the toxicity range for the adult immune system (mouse and rat).
- Gender differences exist in susceptibility and/or the pattern of immunotoxicity following early exposure to lead (rat and chicken).
- The pattern of immune alterations resulting from early lead exposure differs depending upon the specific life-stage of exposure (rat and chicken).
- Chelation of lead during gestational exposure can at least partially protect against lead-induced immunotoxicity (rat).
- For lead exposure to disrupt T helper balance, exposure may need to occur after the first half of embryonic development (apparently requiring some developmental maturation event).

tion to genotype and potential endocrine-related differences in susceptibility, life stage of exposure, as discussed in the following sections, is a critical factor relative to lead-associated immunotoxic risk and outcome.

## DIFFERENTIAL SUSCEPTIBILITY BASED ON LIFE STAGE

This chapter and book are concerned with the relative immunological risk of adults vs. nonadults following toxicant exposure. Several bases for probable differential risk are associated with the exposure of differential life stages to toxicants. First, different dose response relationships can exist where the no-observed effect levels (NOELs) may differ dramatically among life stages. Such differences could arise based on different pharmacokinetics across life stages or could be associated with differential susceptibility of the target organs (in this case the immune system) across life stages. Furthermore, it is possible that effects that are only transient following adult exposure might be permanent when exposure occurs to embryos, fetuses, or juveniles. Finally, because physiological systems such as the immune system are dynamically changing targets during early development, it is feasible that the nature of the changes to the immune system could differ based on the timing of exposure. In the specific case of lead, it appears that most of these distinctions are potentially applicable (Table 10.1). Not surprisingly, the immune system is not the only physiological system to display developmental differences for lead-induced toxicity (Ronis et al. 1996).

### Developmental Exposures and Lead-Induced Immunotoxicity

Unlike the situation for most chemicals where toxicological data may be limited to adult exposure hazard-identification data, immunotoxicity data for exposure to lead during early development have been obtained in at least three species and in at least two strains within a single species. The overall results support the fact that for some life stages of exposure, lead has the capacity to alter T helper 1 vs. T helper 2 balance; this is perhaps the single most significant hallmark of lead-induced immunotoxicity. Associated with the increase in Th2 function are increases in the production of Interleukin-4 and IgE.

Changes in Th1 vs. Th2 balance appear to be influenced by both environmental and genetic factors and have both biological and clinical ramifications. Th2 activity and IgE levels have been recognized as an important consideration in airway responsiveness and asthma (Sears et al. 1991, Umetsu and DeKruyff 1997; Lutz et al. 1999). Likewise, genetic-related risk factors have been identified for asthma and hyperreactive airways. At least some of the identified genetic polymorphisms seem to influence T helper cell balance as well (Lee et al. 2002a).

The interaction between genotype and environmental exposure is coming under increased examination relative to the risk of asthma (Sengler et al. 2002). There is a strong link between the lead-induced increase in Th2 activity and the associated increases in cytokine and IgE levels. Obviously, segments of the population can vary genetically relative to Th1 vs. Th2 bias. Therefore, for a subpopulation already genetically predisposed more toward Th2 responses than Th1, even modest lead-induced elevations in Th2 cytokines could shift that population into a symptomatology range. This has led several authors to suggest that lead is a potential risk factor contributing to the rise in both childhood asthma and allergic disease (Rabinowitz et al. 1990; Dietert and Hedge 1998; Miller et al. 1998; Chen et al. 1999; Snyder et al. 2000). However, the developmental data also indicate that the risk to the immune system posed by lead exposure is likely to vary widely depending upon the specific life stage of exposure (even within the nonadult).

In experiments using exposure of rats to lead acetate throughout gestation, lactation, and juvenile development, Luster and colleagues reported alterations in both humoral (Luster et al. 1978) and cell-mediated immunity (Faith et al. 1979). Both serum IgG titers and plaque-forming cells were decreased in Sprague–Dawley rats following exposure pre- and postnatally to 25 or 50 ppm lead acetate in the drinking water (Luster et al. 1978). Similarly, the DTH response and T lymphocyte mitogen responses were depressed by exposure to lead (Faith et al. 1979). These studies demonstrated that early exposure to lead at relatively low doses could alter immune function. In these studies, lymphoid organ changes were modest compared with the functional changes. This appears to be another hallmark of lead-induced immunotoxicity. For low to moderate levels of exposure, cell population/organ changes are modest in comparison to the lead-induced T helper-associated functional changes. One exception to this generalization may occur when postnatal environmental stressors are placed on prenatal lead-exposed animals (Lee et al. 2002b). In other early studies, Talcott and Koller (1983) exposed mice to combinations of lead and polychlorinated biphenyls (PCBs) throughout gestation and lactation. They found that lead exposure depressed antibody titers and that combined lead plus PCBs depressed the DTH response. While they did not detect a DTH response change with lead alone, the N numbers were quite modest and the data were pooled from both genders.

Figure 10.1 summarizes the relationship of early exposure to lead, immune functional development, and subsequent immune alterations in both the rat and the chicken. Landmarks of immune development for the rat and the chicken are shown for comparison with the timing of lead exposures and the subsequent immune alterations.

It should be noted that timing does matter relative to early lead exposure. Early gestational (or in the case of the chicken, *in ovo*) exposure of females to lead does

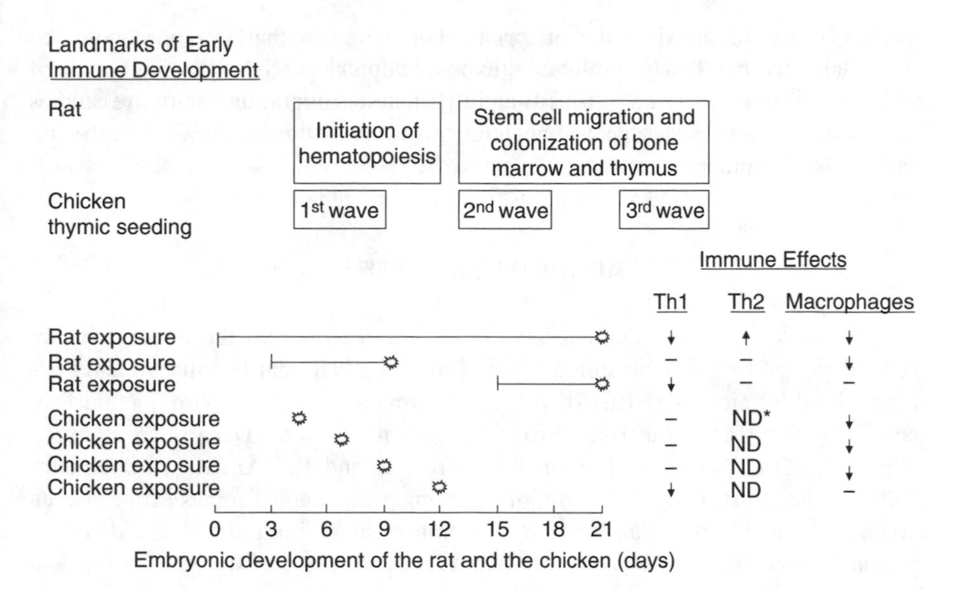


**Figure 10.1** The timeline illustrates functional-based windows of early immune development overlaying various embryonic lead-exposure regimes for the rat and the chicken. Ovals indicate the final day of embryonic exposure for the rat or the single day of *in ovo* lead administration for the chicken. Because cross-fostering was not performed, some low-level lactational transfer of lead was possible in the rat. Arrows indicate the direction of subsequent immune alterations detected in juveniles and adults. ND = not done. Distinctions are apparent among the functional changes induced by early embryonic vs. late embryonic exposure to lead. Information used for the developmental immune landmarks was derived from Dietert et al. (2000) and Gobel (1996).

not pose the same risk for depressing Th1 function as does late gestational exposure (Bunn et al. 2001b; Lee et al. 2001). As shown in Figure 10.1, the second half of embryonic development in the rat and the chicken appears to be a critical window in which the capacity to alter Th1/Th2 balance emerges. This has led to the proposal that even within embryonic development, windows of relative susceptibility and resistance are likely to exist (Bunn et al. 2001b; Dietert et al. 2002). With early exposure to lead, altered macrophage function and some changes in antibody (including autoantibody) production are the primary outcomes. In contrast, late embryonic/gestational exposure results in severely depressed Th1 function in the offspring. Exposure across the entirety of gestation seems to produce the combined alterations seen from early and late gestational exposures including depressed Th1 function (Miller et al. 1998; Chen et al. 1999; Bunn et al. 2001a, b, c; Lee et al. 2001; Dietert et al. 2002).

What is remarkable about these observations is that lead concentration at birth (in blood) does not appear to be a sole predictor of immunotoxic outcome. Instead, it is critical to know the window of active exposure and likely embryonic uptake of lead. With very early exposures, during the first half of embryonic development, lead is still retained in the late embryonic system at what are apparently sufficient concentrations (based on blood lead levels) to alter T helper function. Yet, T helper balance is not disrupted in the offspring. This suggests that lead might not be bioavailable to alter T helper balance if exposure/uptake precedes the emergence of some critical developmentally timed target. It may be important for future studies to consider active exposure/peak concentrations vs. body burden during specific life stages of development.

## ACKNOWLEDGMENTS

The results from the authors' laboratory described within the review chapter were supported by a combination of grants from the National Institute of Environmental Health Sciences (NIEHS) with funds provided via the Environmental Protection Agency (EPA) Superfund Program, the United States Department of Agriculture (USDA) Northeast Regional 60 Project, and the American Chemistry Council. The efforts of Terry Bunn-Gomez, Suping Chen and Thomas Miller, Karen Golemboski, and Forrest Sanders in the laboratory, in addition to the valued collaboration of Drs. Michael Holsapple, Greg Ladics, James Marsh, Syed Naqi, and Patrick Parsons, are greatly appreciated.

## REFERENCES

Araki S, Sato H, Yokoyama K, and Murata K. 2000. Subclinical neurophysiological effects of lead: a review on peripheral, central, and autonomic nervous system effects in lead workers. *Am J Ind Med* 37: 193–204.

Bellinger DC. 1995. Interpreting the literature on lead and child development: the neglected role of the "experimental system." *Neurotoxicol Teratol* 17(3): 201–212.

Bellinger DC and Needleman HL. 1983. Lead and the relationship between maternal and child intelligence. *J Pediatr* 102: 523–527.

Bunn TL, Marsh JA, and Dietert RR. 2000. Gender differences in developmental immunotoxicity to lead in the chicken: analysis following a single early low-level exposure *in ovo*. *J Toxicol Environ Health* Part A 61: 677–693.

Bunn TL, Ladics GS, Holsapple MP, and Dietert RR. 2001a. Developmental immunotoxicology assessment in the rat: age, gender and strain comparisons after exposure to Pb. *Toxicol Methods* 11: 41–58.

Bunn TL, Parsons PJ, Kao E, and Dietert RR. 2001b. Exposure to lead during critical windows of embryonic development: differential immunotoxic outcome based on stage of exposure and gender. *Toxicol Sci* 64: 57–66.

Bunn TL, Parsons PJ, Kao E, and Dietert RR. 2001c. Gender-based profiles of developmental immunotoxicity to lead in the rat: assessment in juveniles and adults. *J Toxicol Environ Health* Part A 64: 101–118.

Chen S, Miller TE, Golemboski KA, and Dietert RR. 1997. Suppression of macrophage metabolite production by lead glutamate *in vitro* is reversed by meso-2,3-dimercaptosuccinic acid (DMSA). *In Vitro Toxicol* 10(3): 351–357.

Chen S, Golemboski KA, Sanders FS, and Dietert RR. 1999. Persistent effect of *in utero* meso-2,3-dimercaptosuccinic acid (DMSA) on immune function and lead-induced immunotoxicity. *Toxicol* 132: 67–79.

Cohen MD, Yang Z, and Zelikoff JT. 1994. Immunotoxicity of particulate lead: *in vitro* exposure alters pulmonary macrophage tumor necrosis factor production and activity. *J Toxicol Environ Health* 42: 377–392.

Cook JA, Hoffman EO, and DiLuzio NR. 1975. Influence of lead and cadmium on the susceptibility of rats to bacterial challenge. *Proc Soc Exp Biol Med* 150: 741–747.

Dieterich KN. 1991. Human fetal lead exposure: intrauterine growth, maturation and postnatal neurobehavioral development. *Fundam Appl Toxicol* 16: 17–19.

Dietert RR and Hedge A. 1998. Chemical sensitivity and the immune system: a paradigm to approach potential immune involvement. *Neurotoxicol* 19: 253–258.

Dietert RR, Etzel RA, Chen D, Halonen M, Holladay SD, Jarabek AM, Landreth K, Peden DB, Pinkerton K, Smialowicz RJ, and Zoetis T. 2000. Workshop to identify critical windows of exposure for children's health: immune and respiratory systems work group summary. *Environ Health Perspect* 108 (Suppl. 3): 483–490.

Dietert RR, Lee J-E, and Bunn TL. 2002. Developmental immunotoxicology: emerging issues. *Human Exp Toxicol* 21(9–10): 479–485.

Exon JH, Koller LD, and Kerkvliet NI. 1979. Lead-cadmium interactions: effects on viral-induced mortality and tissue residues in mice. *Arch Environ Health* 34: 469–475.

Faith RE, Luster MI, and Kimmel CA. 1979. Effect of chronic developmental lead exposure on cell-mediated immune functions. *Clin Exp Immunol* 35: 413–420.

Foster WG, McMahon A, and Rice DC. 1996. Sperm chromatin structure is altered in cynomolgus monkeys with environmentally relevant blood lead levels. *Toxicol Ind Health* 12(5): 723–735.

Garavan H, Morgan RE, Levitshy DA, Herman-Vaquez L, and Strupp BJ. 2000. Enduring effects of early lead exposure: evidence for a specific deficit in associative ability. *Neurotoxicol Teratol* 22: 151–164.

Gobel TWF. 1996. The T-dependent immune system. In: Davidson TF, Morris TR, and Payne LN (Eds.), *Poultry Immunology.* Carfax Publishers, Abingdon, VA, pp. 31–45.

Goyer RA. 1993. Lead toxicity: current concerns. *Environ Health Perspect* 100: 177–187.

Guo TL, Mudzinski SP, and Lawrence DA. 1996. The heavy metal lead modulates the expression of both TNF-alpha and TNF-alpha receptors in lipopolysaccharide-activated human peripheral blood mononuclear cells. *J Leukoc Biol* 59: 932–939.

Gupta P, Husain MN, Shankar R, and Maheshwari RK. 2002. Lead exposure enhances virus multiplication and pathogenesis in mice. *Vet Hum Toxicol* 44: 205–210.

He L, Poblenz AT, Medrano CJ, and Fox DA. 2000. Lead and calcium produce rod photoreceptor cell apoptosis by opening the mitochondrial permeability transition pore. *J Biol Chem* 275, 16: 12175–12184.

Hemphill FE, Kaeberle ML, and Buck WB. 1971. Lead suppression of mouse resistance to Salmonella typhimurium. *Science* 172: 1031–2.

Heo Y, Parsons PJ, and Lawrence DA. 1996. Lead differentially modifies cytokine production *in vitro* and *in vivo*. *Toxicol Appl Pharm* 138: 149–157.

Heo Y, Lee WT, and Lawrence DA. 1997. *In vivo* the environmental pollutants lead and mercury induce oligoclonal T cell responses skewed toward type-2 reactivities. *Cell Immunol* 179: 185–195.

Heo Y, Lee WT, and Lawrence DA. 1998. Differential effects of lead and cAMP on development and activities of Th1-and Th2-lymphocytes. *Toxicol Sci* 43: 172–185.

Kerkvliet NI and Baecher-Steppan L. 1982. Immunotoxicology studies on lead: effects of exposure on tumor growth and cell-mediated tumor immunity after syngeneic or allogeneic stimulation. *Immunopharmacol* 4: 213–224.

Kim D and Lawrence DA. 2000. Immunotoxic effects of inorganic lead on host resistance of mice with different circling behavior preferences. *Brain Behav Immun* 14: 305–317.

Kobayashi N and Okamoto T. 1974. Effects of lead oxide on the induction of lung tumors in Syrian hamsters. *J. Natl Cancer Inst* 52: 1605–1610.

Koller LD. 1973. Immunosuppression produced by lead, cadmium, and mercury. *Am J Vet Res* 34: 1457–1458.

Koller LD and Kovacic S. 1974. Decreased antibody formation in mice exposure to lead. *Nature* 250: 148–150.

Koller LD and Roan JG. 1977. Effects of lead and cadmium on mouse peritoneal macrophages. *J Reticuloendoth Soc* 21(1): 7–12.

Kowolenko M, Tracy L, Mudzinski S, and Lawrence DA. 1988. Effect of lead on macrophage function. *J Leukoc Biol* 43: 357–364.

Kowolenko M, Tracy L, and Lawrence DA. 1991. Early effects of lead on bone marrow cell responsiveness in mice challenged with Listeria monocytogenes. *Fundam Appl Toxicol* 17: 75–82.

Lawrence DA. 1981. *In vivo* and *in vitro* effect of lead on humoral and cell-mediated immunity. *Infect Immun* 31(1): 136–143.

Lawrence DA and McCabe MJ Jr. 2002. Immunomodulation by metals. *Int Immunopharmac* 2: 293–302.

Lee J-E, Chen S, Golemboski KA, Parsons PJ, and Dietert RR. 2001. Developmental windows of differential lead-induced immunotoxicity in chickens. *Toxicol* 156: 161–170.

Lee SY, Lee YH, Shin C, Shim JJ, Kang KH, Yoo SH, and In KH. 2002a. Association of asthma severity and bronchial hyperreponsiveness with a polymorphism in the cytotoxic T-lymphocyte antigen-4 gene. *Chest* 122: 171–176.

Lee J-E, Naqi S, Kao E, and Dietert RR. 2002b. Embryonic exposure to lead: comparison of immune and cellular responses in unchallenged and virally stressed chickens. *Arch Toxicol* 75: 717–724.

Loghman-Adman M. 1997. Renal effects of environmental and occupational lead exposure. *Environ Health Perspect* 105: 928–939.

Luster MI, Faith RE, and Kimmel CA. 1978. Depression of humoral immunity in rats following chronic developmental lead exposure. *J Environ Pathol Toxicol* 1: 397–402.

Lutz PM, Wilson TJ, Ireland J, Jones AL, Gorman JS, Gale NL, Johnson JC, and Hewett JE. 1999. Elevated immunoglobulin E (IgE) levels in children with exposure to environmental lead. *Toxicol* 134: 63–78.

McCabe MJ Jr and Lawrence DA. 1991. Lead, a major environmental pollutant, is immunomodulatory by its differential effects on CD4+ T cell subsets. *Toxicol Appl Pharm* 111: 13–23.

McCabe MJ Jr, Singh KP, and Reiner JJ Jr. 1999. Lead intoxication impairs the generation of a delayed type hypersensitivity response. *Toxicol* 139: 255–264.

Miller TE, Golemboski KA, Ha RS, Bunn TL, Sanders FS, and Dietert RR. 1998. Developmental exposure to lead causes persistent immunotoxicity in Fischer 344 rats. *Toxicol Sci* 42: 129–135.

Muller S, Gillert KE, Krause C, Gross U, Age-Stehr JL, and Diamanstein T. 1977. Suppression of delayed type hypersensitivity of mice by lead. *Experientia* 33(5): 667–8.

Nolan CV and Shaikh ZA. 1992. Lead nephrotoxicity and associated disorders: biochemical mechanisms. *Toxicol* 73: 27–146.

Rabinowitz MB, Allred EN, Bellinger DC, Levitan A, and Needleman HL. 1990. Lead and childhood propensity to infectious and allergic disorders: is there an association? *Bull Environ Contam Toxicol* 44: 657–660.

Ronis MJ, Badger TM, Shema ST, Roberson RK, and Shaikh F. 1996. Reproductive toxicity and growth effects in rats exposed to lead at different periods during development. *Toxicol Appl Pharm* 136: 361–371.

Schwartz BS, Lee BK, Lee GS, Stewart WF, Simon D, Kelsey K, and Todd AC. 2000. Associations of blood lead, dimercaptosuccinic acid-chelatable lead, and tibia lead with polymorphisms in the vitamin D receptor and D-aminoleuvulinic acid dehydratase genes. *Environ Health Perspect* 108: 949–954.

Sears MR, Burrow B, Flannery EM, Hebison GP, Hewitt CJ, and Holdaway MD. 1991. *New Engl J Med* 325: 1067–1071.

Sengler C, Lau S, Wahn U, and Nickle R. 2002. Interaction between genes and environmental factors in asthma and atopy: new developments. *Respir Res* 3: 7.

Snyder JE, Filipov NM, Parsons PJ, and Lawrence DA. 2000. The efficiency of maternal transfer of lead and its influence on plasma IgE and splenic cellularity of mice. *Toxicol Sci* 57: 87–94.

Talcott PA and Koller LD. 1983. The effect of inorganic lead and/or a polychlorinated biphenyl on the developing immune system of mice. *J Toxicol Environ Health* 12: 337–352.

Tian L and Lawrence DA. 1995 Lead inhibits nitric oxide production *in vitro* by murine splenic macrophages. *Toxicol Appl Pharm* 132: 156–163.

Umetsu DT and DeKruyff RH. 1997. Th1 and Th2 CD4+ cells in the pathogenesis of allergic diseases. *Proc Soc Exp Biol Med* 215: 11–20.

Vahter M, Berglund M, Akesson A, and Liden C. 2002. Metals and women's health. *Environ Res* 88: 145–155.

Youssef SAH, El-Sanousi AA, Afifi NA, and El Brawy AM. 1996. Effect of subclinical lead toxicity on the immune response of chickens to Newcastle disease virus vaccine. *Res Vet Sci* 60: 13–16.

Zelikoff JT, Parsons E, and Schlesinger RB. 1993. Inhalation of particulate lead oxide disrupts pulmonary macrophage-mediated functions important for host defense and tumor surveillance in the lung. *Environ Res* 62: 207–222.

Zelikoff JT, Smialowicz R, Bigazzi PE, Goyer RA, Lawrence DA, Maibach HL, and Gardiner D. 1994. Immunomodulation by metals. *Fundam Appl Toxicol* 22: 1–7.

# CHAPTER 11

# Developmental Immunotoxicity of Therapeutic Immunosuppressive Drugs

**Robert M. Gogal, Jr. and Steven D. Holladay**

## CONTENTS



## INTRODUCTION: THERAPEUTIC IMMUNOSUPPRESSANT USE IN WOMEN OF CHILDBEARING AGE

Women receiving organ transplants or suffering from cancer or autoimmune disease are the most common candidates for therapeutic immunosuppressive drug regimens during pregnancy (Forastiere et al. 2003; Tournigand et al. 2003; Hurtova et al. 2001; Millan et al. 2003). The chemotherapeutic drugs used to treat cancer are not administered with the intent of causing immune suppression. However, many chemotherapeutic agents (e.g., cyclophosphamide, chlorambucil, methotrexate,

0-415-28457-0/05/$0.00+$1.50

melaphalan) have potent immunosuppressant properties (Furst et al. 1989; Folb et al. 1990; Connell and Miller 1999), which has led to use of some of these agents in humans for therapeutic immune suppression. The widely employed antineoplastic agent cyclophosphamide is perhaps best recognized in this regard, having also been used as an immunosuppressant for maintaining organ transplants and regulating autoimmunity (Deweys et al. 1970; Calabresi and Parks, 1985). Additional classes of powerful immunosuppressive drugs are seeing more frequent use to control immune-mediated disease, e.g., autoimmune disease, hypersensitivities, or inflammation, or for the prevention of organ transplant rejection. In many clinical cases involving childbearing women, the use of these drugs must be continued into and through pregnancy and is primarily directed toward improving or maintaining the health status of the mother. Fetal outcome from such maternal drug therapy then depends on a number of factors: the type of drug(s) employed, the stage of gestational/perinatal exposure, the frequency of exposure, the duration of exposure, and the ease in which the drug crosses the placenta (Yaffe 1978).

## MATERNAL-FETAL JUNCTION: THE PLACENTA

It is accepted that immune-modulating drugs can and do cross the hemochorial placenta, the type found in humans and rodents. Indeed, all presently used therapeutic immunosuppressive drugs cross the placental barrier and enter into fetal circulation, and may pose a risk for fetal development (Little 1997; Tenfron et al. 2002; Padgett and Seelig 2002). Passive diffusion appears to be the primary mode of such placental drug transfer. Active transport and pinocytosis also may contribute to drug transfer, although to a lesser degree (Rurak et al. 1991; Heutis and Choo 2002). The ease in which these compounds move across the placenta is greatly dependent on their physical properties. The most important of these properties are the lipophilic nature of the drug, molecular weight, and $pK_a$ (an acid's ionization constant [$K\cdot$], equal to the pH value at which equal concentrations of the acid and conjugate base forms of a substance [i.e., a buffer] are present). Compounds that are highly lipophilic readily diffuse across the cell membranes, whereas highly charged compounds are unable to diffuse across the cell membranes and pose less of a threat to the fetus. Thus, azathioprine and its metabolites, all highly charged molecules, are largely blocked at the placenta with minimal transport into the fetus; while glucocorticoids like prednisone readily transport across the placenta.

## MATERNAL HEALTH STATUS AND PLACENTAL DRUG TRANSPORT

Several important physiologic factors accompany pregnancy and may alter the ability of a drug to transfer across the placenta. For example, pregnancy can induce a doubling of blood volume, causing increased cardiac output by up to one-third, increased renal function, and decreased GI clearance (Koren 2001). This increased maternal blood volume is further accompanied by increased blood proteins that can bind a drug, limiting the amount of free drug available to cross the placenta. Also often overlooked, maternal health and nutritional status can directly influence drug availability, metabolism, and excretion (Heutis and Choo 2002), again influencing the amount of a given drug that transfers through the placenta.

## THERAPEUTIC IMMUNOSUPPRESSANT LEVELS IN MAMMARY GLAND SECRETIONS

Many of the immunosuppressive drugs used during pregnancy are also found in breast milk. Currently, the American Academy of Pediatrics (AAP) provides guidelines for breastfeeding based on the excretion rate and potential teratogenic effects of each drug (Tendron et al. 2002). For example, limited concentrations (<10%) of prednisone and prednisolone have been measured in breast milk, levels perceived by the AAP as nonthreatening to the breastfed infant (Coulam et al. 1982; Ost et al. 1985). Thus, mothers on prednisone therapy are allowed to breastfeed since the maternal passive immunity outweighs the minimal perceived risks of prednisone. Azathioprine and its metabolites are also detectable at low concentrations in breast milk. Interestingly, due to a paucity of data regarding azathioprine, no guidelines by the AAP for breastfeeding have yet been established. Likewise, tacrolimus (FK506), which has also been detected in breast milk at concentrations reported to be comparable to maternal blood concentrations, is lacking AAP guidelines (Tendron et al. 2002). This is not the case for other immunosuppressants.

Ablactation has been recommended in women treated with cyclosporine A, cyclophosphamide, doxorubicin, and methotrexate, because these compounds have been shown to have high rates of transfer into breast milk (Wiernik and Duncan 1971; Johns et al. 1972; Amato and Niblett 1977; Flechner et al. 1985; Nyberg et al. 1998). Some drugs, including doxorubicin, concentrate beyond maternal blood levels in human milk (Egan et al. 1985). As much as 5% of an immunosuppressive dose of cyclosporine A can transfer from serum into the breast milk, with largely unknown side effects on the child's immune system being a significant concern (Behrens et al. 1989). The blood of newborns, prior to initiation of breastfeeding, can contain cyclosporine A concentrations up to 65 to 85% of maternal levels (Gunter et al. 1989). Notwithstanding, a recent case study described a 35-year-old woman who, while on cyclosporine A, breastfed her child for 10.5 months. The child showed no clinical developmental abnormalities nor were detectable levels of cyclosporine A found in the infant (Munoz-Flores-Thiagarajan et al. 2001). This may be in contrast to laboratory rodent data. For instance, newborn rat pups, suckling from the dams who were dosed with cyclosporine A at either 15 or 25 mg/kg for 20 days, displayed thymic hypocellularity, inhibition of T-cell maturation, and depressed lymphocyte proliferation (Padgett and Seelig 2002). It is important to note that cyclosporine A blood levels in the rats were higher than what is normally reported in humans. Collectively, these studies raise questions regarding the ability of transplacental cyclosporine A exposure to impair neonatal development of the human immune system. Further, the possibility of chemotherapeutics (other than cyclosporine A) transmitting via mammary secretions and negatively impacting the developing immune system needs evaluation.

## ANTINEOPLASTIC DRUGS

Busulfan, an early antileukemic alkylating agent, was reported to cause severe thymic hypoplasia in mice receiving a single dose during midgestation (Pinto-Machado 1970). This observation resulted in concern over possible similar effects with other therapeutic alkylating agents, for instance, cyclophosphamide or chlorambucil. Cyclophosphamide treatment in adult rodents is particularly toxic to B lymphocytes (Poulter and Turk 1972; Misra and Bloom 1991), and results in impaired antibody production to T-dependent and T-independent antigens (Berenbaum 1979; Dean et al. 1979), decreased lymphoproliferative responses (Dean et al. 1979), impaired NK cell activity (Djeu et al. 1979), and impaired host resistance to infectious agents and syngeneic tumor cells (Luster et al. 1984). Chickens exposed to cyclophosphamide during embryonic/neonatal development displayed almost total loss of the lymphocyte population within the bursa of Fabricius, and complete suppression of humoral immunity (Liakopoulou et al. 1989; Lerman and Weidanz 1970). Interestingly, exposure of pregnant mice to cyclophosphamide resulted in no differences in offspring spleen or thymus organ weight or histology, hematological profiles, antibody responses, or DTH responses, even at doses of cyclophosphamide that significantly decreased fetal body weight (Luebke et al. 1986). The latter authors note that developmental immunotoxicity associated with cyclophosphamide in mice may be limited because this drug must be metabolized to an active alkylating agent, and fetal metabolism of cyclophosphamide does not appear to occur. However, an earlier report suggests that both parent compound and active metabolite cross the placenta (Glick 1971). It may be that fetotoxic doses of this drug are required before immunotoxic levels of metabolite are accumulated in the fetus (Luebke et al. 1986).

Thalidomide has recently been shown to be efficacious in treating a wide range of tumors (e.g., multiple myeloma, melanoma, prostatic cancer, Kaposi's sarcoma), as well as autoimmune disease (Crohn's disease) and inflammatory diseases (erythema nodosum leprosum). This drug is unique in that it is defined as a low-level toxic compound with the exception of being a well-known human teratogen. Thalidomide has profound immunomodulatory effects, including inhibition of monocyte/macrophage function and decreased TNF$\alpha$ and IFN$\gamma$ production (Tavares et al. 1997; Rowland et al. 1998). This drug is also very effective in blocking the enhanced angiogenesis that is important in solid tumor development (D'Amato et al. 1994). Although designated as an "orphan" drug by the FDA, a limited license is available for use in life-threatening cases (e.g., cancer in AIDS patients) (Diggle 2001; Cattelan et al. 2002). Numerous preventive measures are recommended to prevent inadvertent thalidomide use by pregnant women, including weekly pregnancy tests and contraceptive counseling. It is expected that these measures will reduce the number of pregnant women who take thalidomide; however, experience with other teratogenic drugs, e.g., Accutane®, suggests the likelihood that some women will use thalidomide while pregnant (Teratology Society 2000). Should thalidomide exposure occur in pregnant women either before or after the window of major morphologic teratogenicity (i.e., fetal limb, heart, eye, kidney, and GI development), developmental immunotoxicity associated with the drug may be an additional important issue.

## ORGAN TRANSPLANT IMMUNOSUPPRESSANTS

During the early years of organ transplant therapies, glucocorticoids and cytotoxic drugs were widely used separately or together to inhibit tissue allograft rejection. Many adverse side effects were associated with use of these compounds (Winkelstein 1991). Considerable research effort was therefore directed toward the development of new and more effective immunosuppressive drugs. Cyclosporine A was introduced into clinical trials in 1978 (Ptachcinski et al. 1985), and rapidly became the most widely used agent for prevention of human organ transplant rejection (Kosugi et al. 1989; Ryffel et al. 1983; Jenkins et al. 1988). A second potent immunosuppressant, tacrolimus (FK-506), has more recently joined cyclosporine A as a major driving force behind the current success of organ transplant surgery (Gaston 2001; Tendron et al. 2002; Bar et al. 2003). It is important to note that effective therapeutic immunosuppression with cyclosporine A or tacrolimus typically requires simultaneous administration of additional immunosuppressive drugs (e.g., azathioprine, 15-deoxyspergualin, mizoribine, didemnin B), as well as corticosteroids, including prednisolone (Montgomery et al. 1985; Tyden et al. 1989; Farley et al. 1991; Markowitz et al. 1993; Gaston 2001; Tendron et al. 2002). Thus, the pursuit of more powerful and less toxic antirejection drugs has continued to be a major goal in organ transplant medicine.

## USE OF THERAPEUTIC IMMUNOSUPPRESSANTS DURING PREGNANCY

Advances in available therapeutic immunosuppressive drug regimens, coupled with advances in organ transplant technology, have not only increased the life expectancy of recipients but also greatly improved quality of life. As a result, a population of women previously unable to conceive or maintain a conceptus, is now able to maintain a pregnancy following an organ or tissue (allograft) transplant procedure. Indeed, pregnancy as a rule is well tolerated after graft function has become stable and adequate, and many "normal" births have been documented in women undergoing immunosuppressive therapy throughout pregnancy (Lewis et al. 1983; Klintmalm et al. 1984; Pickrell et al. 1988; Ville et al. 1992; Tincani et al. 1992; Buchanan et al. 1992; Many et al. 1992; Markowitz et al. 1993; Farber and Nall, 1993). This is not to say that pregnancy after organ transplantation is no longer regarded as high-risk; clear associations with increased risk of hypertension, preeclampsia, intrauterine growth retardation, and prematurity exist (Riely 2001). For example, a study of 238 women receiving immunosuppressive drug regimens during pregnancy found that rates for both prematurity and small-for-gestational-age (SGA) infants were high (49% and 29%, respectively) (Pabelick et al. 1991). Regimens that included cyclosporine A resulted in 66% prematurity (43% with azathioprine), and 56% SGA-babies (19% with azathioprine). Nonetheless, successful pregnancies have been achieved following all kinds of solid organ transplantations, with thousands of pregnancies now reported (Armenti et al. 2000).

Pregnant women who have previously received allograph transplants require therapeutic immunosuppressants throughout pregnancy to maintain the allograft and to prevent rejection. Past and presently used drugs for immune suppression after organ transplant all cross the placenta (Little 1997; Scott et al. 2002; Prevot et al. 2002), a fact that has raised many concerns about possible harmful effects of the maternal immunosuppressant therapy on fetal development. Effectively evaluating such effects, or tracing adverse outcomes to individual drugs, is complicated by the previously mentioned general need for multiple drug therapy in transplant recipients. At the time of writing, immunosuppressive drug protocols to prevent allograph rejection typically combine use of one of the two most effective primary immunosuppressants (cyclosporine A or tacrolimus) with one or more adjunctive agents (azathioprine, mycophenolate mofetil, sirolimus, corticosteroids) (Gaston 2001; Tendron et al. 2002). Different individuals then require uniquely tailored drug combinations to prevent rejection responses, a part of immunosuppressant management that can prove daunting and continues to complicate tracing of adverse effects to individual drugs or drug combinations.

For the above reasons, case reports and prospective studies of pregnancy outcome with different drug therapies, and following different transplant procedures, continue to be of considerable interest. The first study estimating maternal and fetal risk from tacrolimus therapy after organ transplantation was published by Jain et al. in 1997. These authors studied 27 pregnancies in 21 female liver recipients who received tacrolimus before and throughout pregnancy. On the day of delivery, mean tacrolimus concentrations were 4.3 ng/ml in placenta versus 1.5, 0.7 and 0.5 ng/ml in maternal, cord, and child plasma, respectively, and 0.6 ng/ml in first breast milk. Two premature infants of 23 and 24 weeks gestation died shortly after birth, and one live infant displayed unilateral renal polycystic disease; however, all 25 of the surviving infants showed satisfactory postnatal growth and development.

Wu et al. (1998) similarly examined pregnancy outcome in 22 liver transplant patients who received tacrolimus, cyclosporine A, azathioprine, and/or a low-dose steroid therapy. Twenty-three children were born (one set of twins) with three preterm births; no congenital malformations or unusual infections were observed in the children; and postnatal growth and development were appropriate during the course of the study. Miniero et al. (2002) recently reported pregnancy outcome for 56 renal transplant recipients who were maintained on cyclosporine A, azathioprine, tacrolimus, or corticosteroids before and during pregnancy. There were four transplant rejections, two of which were irreversible; 36 infants were born, and 20 abortions reported. Complications occurred in 16 of 36 nonaborted pregnancies; 16 of these 36 infants were born at term, and 20 were pre-term. The children were followed up for periods ranging from 2 months to 13 years, and their development was described as normal.

The successful use of tacrolimus and prednisone in a pregnant lung transplant recipient was recently reported, with a healthy 2,208-g female infant born at 34 weeks gestation (Kruszka and Gherman 2002). The three-drug combination of mycophenolate mofetil, tacrolimus, and prednisone, was given to a kidney transplant patient throughout pregnancy; a female child was born prematurely at week 353/7, was reported to have possible teratogenic effects in the form of hypoplastic nails and short fifth fingers, but again was reported as growing and developing normally (Pergola et al. 2001).

Two of the studies mentioned previously reported a teratogenic outcome in pregnant women who received immunosuppressive therapy. Increased risk of such morphologic defects was an early concern of physicians monitoring pregnancies of transplant recipients. Early reports suggested that cyclosporine A administration during pregnancy may increase risk of some birth defects, including dysmorphic facial appearance (Reznik et al. 1987), cataracts (Tyden et al. 1989; Dieperink et al. 1987), and cleft palate (Bung and Dietmar 1991). However, recent retrospective investigations have examined considerably more births to immune-suppressed mothers, and concluded that structural birth defects are not increased by use of therapeutic immunosuppressants during pregnancy (Riely 2001; Armenti et al. 2002; Tendron et al. 2002).

Many collective reports and considerable data are available, therefore, showing increased pregnancy complications in organ transplant recipients, although apparently healthy children appear to be the most common result of such pregnancies. Relatively few studies have monitored immune development in these children, however, and later-life immune dysfunctions, including deficits or exacerbated responses (e.g., increased autoimmune diseases), remain a concern (Holladay 1999; Prevot et al. 2002). Long-term follow-up data from infants exposed to therapeutic immunosuppressive drug regimens throughout pregnancy also remain limited due to the relatively recent advent of this population, and will be required to determine if postpubertal or adult immune dysfunctions are more prevalent in humans exposed to immune-suppressive drugs during development.

## DEVELOPMENTAL IMMUNOTOXICITY IN RODENTS EXPOSED TO THERAPEUTIC IMMUNOSUPPRESSANTS DURING PREGNANCY

Studies employing rodent fetal thymic organ cultures have showed that this type of *ex vivo* cyclosporine A exposure alters early development of thymocytes within the fetal thymus, completely blocking the generation of mature T cells (Kosugi et al. 1989). Laboratory mice exposed perinatally to cyclosporine A were found to display hypoplastic peripheral lymphatic organs, impaired intrathymic thymocyte differentiation, absence of mature T cells in lymph nodes and spleens, and lack of functional T-cell reactivity (Heeg et al. 1989). In these studies, mice were exposed to cyclosporine A during the third trimester of pregnancy, with continued postnatal exposure for as long as 28 days. Thus, the consequences of earlier prenatal exposure, as well as the contribution of prenatal exposure alone, to these immune alterations could not be delineated. However, rat pups exposed to cyclosporine A by lactation only displayed significant reduction in thymus cellularity with almost complete loss of medullary cells (Padgett and Seelig 2002). The thymocyte population from these rats showed increased immature CD4+8+ cells and decreased mature CD4+CD3hi and TCRhi subsets.

Cyclosporine A has also been found to interfere with tolerization of developing T lymphocytes in irradiated rodents following syngeneic bone-marrow reconstitution, resulting in increased autoimmunity (i.e., syngeneic graft-vs.-host disease) in the host animals (Glazier et al. 1983). Such results raise clear questions about the ability of cyclosporine A to induce or exacerbate autoimmune disease in gestationally exposed individuals. In support of this possibility, newborn mice dosed daily with cyclosporine A for the first week of postnatal life developed organ-specific autoimmune disease (Sakaguchi and Sakaguchi 1988, 1989). Such disease elicited in rodents appeared to be related to interference by cyclosporine A with the production or expansion of self-reactive T cells in the thymus (Sakaguchi and Sakaguchi 1992). Similarly, Classen (1998) found that cyclosporine A exposure during pregnancy in mice greatly increased prevalence of autoimmune disease in the offspring. These collective observations in irradiated and perinatal rodent exposure models, where new immune systems are being established in the presence of cyclosporine A, may suggest a heightened sensitivity to this and other agents that disrupt selection for autoreactive cells. In this regard, Zacharchuk et al. (1991) found that susceptibility of thymocytes to clonal deletion changes during ontogeny. Studies by these authors indicate that there is a relatively synchronous wave of maturing thymocytes that are susceptible to deletion signals during fetal life and shortly after birth, but not seven days after birth. This observation is in agreement with current understanding of neonatal tolerance, and further suggests that failure to induce tolerization in glucocorticoid-exposed 1-week-old mice reflects an alteration in susceptibility to normal clonal deletion during that time (Zacharchuk et al. 1991).

## IMMUNOLOGIC OBSERVATIONS IN CHILDREN OF ORGAN TRANSPLANT RECIPIENTS

Transient neonatal leucopenia and hypoplasia of the lymphatic system were among the early observations in children exposed to cyclosporine A (Pickrell et al. 1988) or to azathioprine plus prednisolone (Lower et al. 1971; Cote et al. 1981) during gestation. Because of such reports, Takahashi et al. (1994) examined lymphocyte subpopulations in cord and peripheral blood of six infants born to mothers who had received renal transplants, compared to five control infants. All of the renal transplant mothers received cyclosporine A, azathioprine, and methylprednisolone throughout pregnancy. There were no differences between numbers of CD2+, CD4+, or CD8+ T cells in the immune-suppressant exposed offspring as compared to controls; however, a significant reduction in B cells was present and sometimes severe at 1 and 3 months of age in the children from renal transplant mothers. The authors concluded that the B cell line may be more sensitive than the T cell line to the immunosuppressive therapy studied, and that these children should be followed for possible insufficiency of immune function. In contrast, Ersay et al. (1995) found normal T, B, and NK cell levels at birth in an infant of a kidney transplant mother who was treated with cyclosporine A, azathioprine, and prednisolone during pregnancy. Follow-up of this child revealed higher than control numbers of B cells at 3 and 6 months of age; T cell populations were normal with the exception of a high CD4+/CD8+ ratio due to increased CD4+ cells. These authors also reported below normal numbers of NK cells in this child at 3 and 6 months of age.

Pilarski et al. (1994) analyzed peripheral blood T lymphocytes from seven children exposed to cyclosporine A and four children exposed to azathioprine during development. A slight delay in T cell development was found in children exposed to cyclosporine A, a slight acceleration of such development in children exposed to azathioprine, and no T cell development affect in children exposed to both cyclosporine A and azathioprine. From 1 to 6 years of age the children exposed to cyclosporine A showed a decreased proportion of T cells expressing a high density of CD29, the b1-integrin linked to ability to respond to recall antigens and to home sites of infection. However, the study children showed no outward signs of immune deficiency, and the authors concluded that the developmental immunosuppressant exposure was not likely to cause immune deficiency or autoimmune disease. Recently, however, Scott et al. (2002) reported autoimmune complications in the pregnancy of a 23-year-old daughter of a renal transplant recipient. The woman's mother had been maintained on 75 mg/day azathioprine and 5 mg/day prednisone throughout pregnancy. During the daughter's first pregnancy, she developed multiple autoantibodies and Raynaud's phenomenon; her second pregnancy was complicated by systemic lupus erythematosus. The authors concluded that no link could be verified between the woman's prenatal immunosuppressive drug exposure and her autoimmune manifestations; however, that these observations warranted further studies of offspring of organ transplant recipients. It should be noted that a growing body of laboratory animal data suggest developmental exposure to immunosuppressive drugs may increase risk of autoimmune disease; this subject is covered in detail in Chapter 13 of this textbook.

## SUMMARY

The use of potent therapeutic immunosuppressants during pregnancy has been increasing over the past several decades. The demonstrated heightened sensitivity of the immune system to immunotoxic exposures during its ontogeny, and the ability of immunosuppressive drugs to cross the placenta, suggest immunotoxicity will occur in developmentally exposed children. Currently available data suggest that immune alterations in these children are limited and transient. However, it will be important to follow these children to identify possible long-term adverse effects from *in utero* exposure to these immunotoxic agents. Increased autoimmune disease has been suggested, especially after perinatal cyclosporine A exposure, and is an endpoint especially worthy of consideration.

## REFERENCES

Amato D and Niblett JS. 1977. Neutropenia from cyclophosphamide in breast milk. *Med J Aust* 1:383–384.

Armenti VT, Moritz MJ, Cardonick EH, and Davidson JM. 2002. Immunosuppression in pregnancy: choices for infant and maternal health. *Drugs* 62:2361–2375.

Armenti VT, Herrine SK, Radomski JS, and Moritz MJ. 2000. Pregnancy and liver transplantation. *Liver Transpl* 6:671–685.

Bar J, Stahl B, Hod M, Wittenberg C, Pardo J, and Merlob P. 2003. Is immunosuppression therapy in renal allograft recipients teratogenic? A single-center experience. *Am J Med Genet* 116A:31–36.

Behrens O, Kohlhaw K, Gunter H, Wonigeit K, Neisert S. 1989. Detection of cyclosporine A in breast milk: is breast feeding contraindicated? [German] *Geburtshilfe Frauenheilkd* 49:207–209.

Berenbaum MC. 1979. Time dependence and selectivity of immunosuppressive agents. *Immunol* 36:355–365.

Buchanan NM, Khamashta MA, Morton KE, Kerslake S, Baguley EA, and Hughes GR. 1992. A study of 100 high risk lupus pregnancies. *Am J Repro Immunol* 28:192–194.

Bung P and Dietmar M. 1991. Pregnancy and postpartum after kidney transplantation and cyclosporin therapy: Review of the literature adding a new case. *J Perinat Med* 19:397–401.

Calabresi P and Parks RE. 1985. Chemotherapy of neoplastic diseases. In: Gilman AG and Goodman LS (Eds.), *The Pharmacological Basis of Therapeutics*, 7th ed. New York: MacMillan Publishing Co., p. 1240.

Cattelan AM, Trevenzoli M, and Aversa SM. 2002. Recent advances in the treatment of AIDS-related Kaposi's sarcoma. *Am J Clin Dermatol* 3:451–462.

Classen JB. 1998. Cyclosporine induced autoimmunity in newborns prevented by early immunization. *Autoimmunity* 27:135–139.

Connell W and Miller A. 1999. Treating inflammatory bowel disease during pregnancy: risks and safety of drug therapy *Drug Saf* 2: 311–323.

Cote CJ, Meuwissen HJ, and Pickering RJ. 1981. Effects on the neonate of prednisolone and azathioprine administration to the mother. *J Pediatr* 85:324–329.

Coulam CB, Myer TP, Jiang NS, and Zincke H. 1982. Breast-feeding after renal transplantation. *Transplant Proc* 14:605–609.

D'Amato RJ, Loughnan MS, Flynn E, Folkman J, and Hamel E. 1994. Thalidomide is an inhibitor of angiogenesis. *Proc Natl Acad Sci* 91:4082–4085.

Dean JH, Padarathsingh ML, and Jerrells TR. 1979. Assessment of immunobiological effects induced by chemicals, drugs, or food additives. II. Studies with cyclophosphamide. *Drug Chem Toxicol* 2:133–154.

Dewys WD, Gouldin A, and Mantel N. 1970. Hematopoietic recovery after large doses of cyclophosphamide: Correlation of proliferative state with sensitivity. *Cancer Res* 30:1692–1697.

Dieperink H, Steinbruchel D, Kemp E, Svendsen P, and Sarklint H. 1987. Cataractogenic effect of cyclosporin A: A new adverse effect observed in the rat. *Nephrol Dial Transpl* 1:251–256.

Diggle GE. 2001. Thalidomide: 40 years on. *Int J Clin Pract* 55:627–631.

Djeu JY, Heinbaugh JA, Viera WD, Holden HT, and Herberman RB. 1979. The effect of immunopharmacological agents on mouse natural cell-mediated cytotoxicity and its augmentation by poly I:C. *Immunopharmacol* 1:231–244.

Egan PC, Costanza ME, Dodion P, Egorin MJ, and Bachur NR. 1985. Doxorubicin and cisplatin excretion into human milk. *Cancer Treat Rep* 69:1387–1389.

Ersay A, Oygur N, Coskun M, Suleymanlar G, Trak B, and Yegin O. 1995. Immunologic evaluation of a neonate born to an immunosuppressed kidney transplant recipient. *Am J Perinatol* 12:413–415.

Farber EM and Nall L. 1993. Pustular psoriasis. *Cutis* 51:29–32.

Farley DE, Shelby J, Alexander A, and Scott JR. 1991. The effect of two new immunosuppressive agents, FK-506 and didemnin B, in murine pregnancy. *Transplantation* 52:106–110.

Flechner SM, Katz AR, Rogers AJ, Van Buren C, and Kahan BD. 1985. The presence of cyclosporine in body tissue and fluids during pregnancy. *Am J Kidney Dis* 5: 60–63.

Folb PI and Graham Dukes MN. 1990. Cytotoxic and immunosuppressive agents. *Drug Safety in Pregnancy* 1990;327–348 [DART].

Forastiere AA, Goepfert H, Maor M, Pajak TF, Wever R, Morrison W, Glisson B, Trotti A, Ridge JA, Chao C, Peters G, Lee DJ, Leaf A, Ensley J, and Cooper J. 2003. Concurrent chemotherapy and radiotherapy for organ preservation in advanced laryngeal cancer. *N Engl J Med* 349:2091–2098.

Furst DE, Clements PJ, Hillis S, Lachenbruch PA, and Paulus HE. 1989. Immunosuppression with chlorambucil, versus placebo, for scleroderma. *Arthritis Rheum* 32:584–593.

Gaston RS. 2001. Maintenance immunosuppression in the renal transplant recipient: an overview. *Am J Kidney Dis* 38:S25–35.

Glazier A, Tutschka A, Farmer ER, and Santos GW. 1983. Graft versus host disease in cyclosporin A-treated rats after syngeneic and autologous bone marrow reconstitution. *J Exp Med* 158:1–8.

Glick B. 1971. Morphological changes and humoral immunity in cyclophosphamide-treated chicks. *Transplantation* 11:433–439.

Gunter H, Frei U, Niesert S. 1989. Pregnancies following kidney transplantation and in immunosuppression with cyclosporine A. [German] *Geburtshilfe Frauenheilkd* 49:155–159.

Heeg K, Bendigs S, and Wagner H. 1989. Cyclosporin A prevents the generation of single positive (Lyt2+L3T4-, Lyt2-L3T4+) mature T cells but not single positive (Lyt2+T3-) immature thymocytes in newborn mice. *Scand J Immunol* 30:703–710.

Holladay SD. 1999. Perinatal immunotoxicant exposure: Increased risk of postnatal autoimmune disease? *Environ Health Perspect* 107 Suppl. 5:687–691.

Huestis MA and Choo RE. 2002. Drug abuse's smallest victims: *in utero* drug exposure. *Forensic Science Internal* 128; 20–30.

Hurtova M, Duclos-Vallee JC, Johanet C, Emile JF, Roque-Afonso AM, Feray C, Bismuth H, and Samuel D. 2001. Successful tacrolimus therapy for a severe recurrence of type 1 autoimmune hepatitis in a liver graft recipient. *Liver Transpl* 7:556–558.

Jain A, Venkataramanan R, Fung JJ, Gartner JC, Lever J, Balan V, Warty V, and Starzl TE. 1997. Pregnancy after liver transplantation under tacrolimus. *Transplantation* 64:559–565.

Jenkins MK, Schwartz RH, and Pardoll DM. 1988. Effects of cyclosporin A on T cell development and clonal deletion. *Science* 241:1655–1658.

Johns DG, Rutherford LD, Leighton PC, and Vogel CL. 1972. Secretion of methotrexate into human milk. *Am J Obstet Gynecol* 112:978–980.

Klintmalm G, Althoff P, and Appleby G. 1984. Renal function in a newborn baby delivered of a renal transplant patient taking cyclosporin. *Transplantation* 38:198–199.

Koren G. 2001. Changes in drug disposition in pregnancy and their clinical implications. In: Koren G (Ed.), *Maternal-Fetal Toxicology: A Clinician's Guide*. New York: Marcel Dekker, pp. 3–13.

Kosugi A, Sharrow SW, and Shearer GM. 1989. Effect of cyclosporin A on lymphopoiesis. I. Absence of mature T cells in thymus and periphery of bone marrow transplanted mice treated with cyclosporin A. *J Immunol* 142:3026–3032.

Kruszka SJ and Ghermann RB. 2002. Successful pregnancy outcome in a lung transplant recipient with tacrolimus immunosuppression. A case report. *J Reprod Med* 47:60–62.

Lerman SP and Weidanz WP. 1970. The effect of cyclophosphamide on the ontogeny of the humoral immune response in chickens. *J Immunol* 105:614–619.

Lewis GJ, Lamont CAR, and Leel HA. 1983. Successful pregnancy in a renal transplant recipient taking cyclosporin A. *Br Med J* 286:603–607.

Liakopoulou A, Buttar HS, Nera EA, and Fernando L. 1989. Effects on *in utero* exposure to cyclophosphamide in mice II. Assessment of immunocompetence of offspring from 5-10 weeks of age. *Immunopharmacol Immunotoxicol* 11:193–209.

Little BB. 1997. Immunosuppressant therapy during gestation. *Semin Perinatol* 21:143–148.

Lower GD, Stevens LE, Najarian JS, and Reemtsma K. 1971. Problems from immunosuppressives during pregnancy. *Am J Obstet Gyencol* 111:1120.

Luebke RW, Riddle MM, Rogers RJ, Garner DG, Rowe DG, and Smialowicz RJ. 1986. Immune function of young adult mice following *in utero* exposure to cyclophosphamide. *J Toxicol Environ Health* 18:25–39.

Luster MI, Dean JH, Boorman GA, Archer DL, Lauer L, Lawson LD, Moore JA, and Wilson RE. 1984. The effects of orthophenylphenol, tris (2,3-dichloropropyl) phosphate, and cyclophosphamide on the immune system and host susceptibility of mice following subchronic exposure. *Toxicol Appl Pharmacol* 58:252–261.

Many A, Pauzner R, Carp H, Langevitz P, and Martinowitz U. 1992. Treatment of patients with antiphospholipid antibodies during pregnancy. *Am J Repro Immunol* 28:216–218.

Markowitz J, Grancher K, Mandel F, and Daum F. 1993. Immunosuppressive therapy in pediatric inflammatory bowel disease (IBD): results of a survey of the North American Society for Pediatric Gastroenterology and Nutrition. *Am J Gastroenterol* 88:44–48.

Millan MT, Berquist WE, So SK, Sarwal MM, Wayman KI, Cox KL, Filler G, Salvatierra O Jr, and Esquivel CO. 2003. One hundred percent patient and kidney allograft survival with simultaneous liver and kidney transplantation in infants with primary hyperoxaluria: a single-center experience. *Transplantation* 76:1458–1463.

Miniero R, Tardivo I, Curtoni ES, Segoloni GP, La Rocca E, Nino A, Todeschini P, Tregnaghi C, Rosati A, Zanelli P, and Dall'Omo AM. 2002. Pregnancy after renal transplantation in Italian patients: focus on fetal outcome. *J Nephrol* 15:626–632.

Misra R and Bloom E. 1991. Roles of dosage, pharmacokinetics, and cellular sensitivity to damage in the selective toxicity of cyclophosphamide towards B and T cell development. *Toxicol* 66:239–256.

Montgomery DW, Zukoski CF, and Didemnin B. 1985. A new immunosuppressive cyclic peptide with potent activity *in vitro* and *in vivo*. *Transplantation* 40:49–54.

Munoz-Flores-Thiagarajan KD, Easterling T, Davis C, and Bond EF. 2001. Breastfeeding by a cyclosporine-treated mother. *Obstet Gynecol* 97:816–818.

Nyberg G, Halijamae, Frisenette-Fich C, Wennergren M, and Kjellmer I. 1998. Breast-feeding during treatment with cyclosporine. *Transplantation* 65:253–255.

Ost L, Wettrell G, Bjorkhem I, Rane A. 1985. Prednisolone excretion in human milk. *J Pediatr* 106:1008–1011.

Pabelick C, Kemmer F, and Koletzko B. 1991. Clinical findings in newborn infants of mothers with kidney transplants. *Monatsschr Kinderheilkd* (German) 139:136–140.

Padgett EL and Seelig LL Jr. 2002. Effects on T-cell maturation and proliferation induced by lactational transfer of cyclosporine to nursing pups. *Transplantation* 73:867–874.

Pergola PE, Kancharla A, and Riley DJ. 2001. Kidney transplantation during the first trimester of pregnancy: immunosuppression with mycophenolate mofetil, tacrolimus, and prednisone. *Transplantation* 71:994–997.

Pickrell MD, Sawers R, and Michael J. 1988. Pregnancy after renal transplantation: severe intrauterine growth retardation during treatment with cyclosporin A. *Br Med J* 296:825–829.

Pilarski LM, Yacyshyn BR, and Lazarovits AI. 1994. Analysis of peripheral blood lymphocytes populations and immune function from children exposed to cyclosporine or to azathioprine *in utero*. *Transplantation* 57:133–144.

Pinto-Machado J. 1970. Influence of prenatal administration of busulfan on the postnatal development of mice: production of a syndrome including hypoplasia of the thymus. *Teratol* 3:363–370.

Poulter LW and Turk JL. 1972. Proportional increase in theta-carrying lymphocytes in peripheral lymphoid tissue following treatment with cyclophosphamide. *Nature New Biol* 238:17–18.

Prevot A, Martini S, and Guignard JP. 2002. *In utero* exposure to immunosuppressive drugs. Biol Neonate 81:73–81.

Ptachcinski RJ, Burckart GJ, and Venkataramanan R. 1985. New Drug Evaluations: Cyclosporine. *Drug Intell Clin Pharm* 19:90–92.

Reznik VM, Jones KL, Durham BL, Mendoza SA. 1987. Changes in facial appearance during cyclosporin treatment. *Lancet* 1:1405–1409.

Riely CA. 2001. Contraception and pregnancy after liver transplantation. *Liver Transpl* 11:S74–76.

Rowland TL, McHugh SM, Deighton J, Dearman RJ, Ewan PW, and Kimber I. 1998. Differential regulation by thalidomide and dexamethasone of cytokine expression in human peripheral blood mononuclear cells. *Immunopharmacol* 40:11–20.

Rurak DW, Wright MR, and Axelson JE. 1991. Drug disposition and effects in the fetus. *J Dev Physiol* 15:33–44.

Ryffel B, Donatsch P, and Madorin M. 1983. Toxicological evaluation of cyclosporin A. *Arch Toxicol* 53:107–141.

Sakaguchi S and Sakaguchi N. 1988. Thymus and autoimmunity. Transplantation of the thymus from cyclosporin A-treated mice causes organ-specific autoimmune disease in athymic nude mice. *J Exp Med* 167:1479–1485.

Sakaguchi N and Sakaguchi S. 1989. Organ-specific autoimmune disease induced in mice by elimination of T cell subsets. V. Neonatal administration of cyclosporin A causes autoimmune disease. *J Immunol* 142:471–480.

Sakaguchi N and Sakaguchi S. 1992. Causes and mechanism of autoimmune disease: cyclosporine A as a probe for the investigation. *J Invest Dermatol* 98:70S–76S.

Sampaio EP, Sarno EN, Galilly R, Cohn ZA, and Kaplan G. 1991. Thalidomide selectively inhibits tumor necrosis factor alpha production by stimulated human monocytes. *J Exp Med* 173: 699–703.

Scott JR, Branch DW, and Holman J. 2002. Autoimmune and pregnancy complications in the daughter of a kidney transplant patient. *Transplantation* 73:815–816.

Takahashi N, Nishida H, and Hoshi J. 1994. Severe B cell depletion in newborns from renal transplant mothers taking immunosuppressive drugs. *Transplantation* 57:1617–1621.

Tavares JL, Wangoo A, Dilworth P, Marshall B, Kotecha S, and Shaw RJ. 1997. Thalidomide reduces tumour necrosis factor-alpha production by human alveolar macrophages. *Respir Med* 91: 31–39.

Tendron A, Gouyon JB, and Decramer S. 2002. *In utero* exposure to immunosuppressive drugs: experimental and clinical studies. *Pediatr Nephrol* 17:121–130.

Teratology Society Public Affairs Committee Position Paper: Thalidomide. 2000. *Teratol* 62:172–173.

Tincani A, Faden D, Tarantini M, Lojacono A, Tanzi PK, Gastaldi A, Di Mario C, Spatola L, Cattaneo R, and Balestrieri G. 1992. Systemic lupus erythematosus and pregnancy: a prospective study. *Clin Exp Rheumatol* 10:439–446.

Tournigand C, Louvet C, Molitor JL, Fritel X, Dehni N, Sezeur A, Pigne A, Cady J, Milliez J, and de Gramont A. 2003. Long-term survival with consolidation intraperitoneal chemotherapy for patients with advanced ovarian cancer with pathological complete remission. *Gynecol Oncol* 91:341–345.

Tyden G, Brattstrom C, Bjorkman U, Landgraf R, Baltzer J, Hillebrand G, Land W, Calne R, Brons IG, and Squifflet JP. 1989. Pregnancy after combined pancreas-kidney transplantation. *Diabetes* 38:43–45.

Ville Y, Fernandez H. Samuel D, Bismuth H, and Frydman R. 1992. Pregnancy after hepatic transplantation. *J Gynecol Obstet Biol Repro Paris* (French) 21:691–696.

Wiernik PH and Duncan JH. 1971. Cyclophosphamide in human milk. *Lancet* 1: 912.

Winkelstein A. 1991. Immunosuppressive therapy. In: Stites DP and Terr AI (Eds.), *Basic and Clinical Immunology*, 7th edition. Appleton and Lange, Norwalk, CT, pp. 766–779.

Wu A, Nashan B, Messner U, Schmidt HH, Guenther HH, Niesert S, and Pichmayr R. 1998. Outcome of 22 successful pregnancies after liver transplantation. *Clin Transplant* 12:454–464.

Yaffee SJ. 1978. Drugs and pregnancy. *Clin Toxicol* 13:523–533.

Zacharchuk CM, Mercep M, and Ashwell JD. 1991. Thymocyte activation and death: a mechanism for molding the T cell repertoire. *Ann NY Acad Sci* 636:52–70.

CHAPTER 12

# Mycotoxins and Other Miscellaneous Developmental Immunotoxicants

**Terry C. Hrubec and Steven D. Holladay**

## CONTENTS



## INTRODUCTION

Immunotoxicology is the study of adverse effects on the immune system resulting from exposure to xenobiotics such as environmental agents, drugs, or biological materials. The widespread, often large-scale use of many industrial and agricultural chemicals and the combustion of fossil fuels over the past half-century has resulted

0-415-28457-0/05/$0.00+$1.50

in considerable contamination of the environment. A number of these pollutants are known to produce immunological changes or alter immune responsiveness in humans and other animals. Additionally, the exposure to some of these immunotoxic agents during prenatal development has more dramatic or persistent effects on the immune system than does exposure during adult life (Ford et al. 1983; Ways et al. 1987; Noller et al. 1988; Fine et al. 1989; Holladay et al. 1991). The developing organism is not necessarily more sensitive to physical or chemical agents than the fully mature individual, but the consequences of exposure are often more severe, lasting longer or resulting in permanent immunologic change than when exposure occurs during adult life (Vos and Moore 1974; Ford et al. 1983; Thomas and Faith 1985; Ways et al. 1987; Noller et al. 1988; Fine et al 1989; Barnett et al. 1990; Holladay et al. 1991). As an example, women exposed *in utero* to the nonsteroidal estrogen, diethylstilbestrol (DES), appear to have long-term alterations to T- and natural killer (NK) cell function and an increased incidence of autoimmune diseases later in life (Herbst et al. 1971; Ford et al. 1983; Ways et al. 1987; Noller et al. 1988).

The developing immune system requires a series of sequential events that begin early in embryonic life and follow a regimented pattern to produce normal ontogenesis and function. Such events include differentiation of embryonic ectodermal, endodermal and mesodermal cell masses, sequential waves of hematopoietic cell production and migrations through hematopoietic organs; and cell-cell interactions under controlled intracellular microenvironmental conditions. The result of this tightly controlled developmental sequence is the production of highly specific cytodifferentiation and maturation (Schmidt 1984; LeDouarin et al. 1984). The temporal sequence of immune system ontogenesis through these various stages has been studied in humans and other animals. The major difference between species appears to be the stage of immune system development relative to parturition, and thus the degree of immunocompetence at birth (LeDouarin et al. 1984; Owen 1972; Tavassoli 1991; Paul et al. 1969; Velarde and Cooper 1984; Roberts and Chapman 1981). Studies in this relatively new field of developmental toxicology have demonstrated two basically different outcomes: altered postnatal immunocompetence, including decreased host resistance to infectious disease or neoplasia, and exacerbation of immune-mediated diseases, including hypersensitivity disorders and autoimmune disease.

The immune system is a frequent target of toxic insult following subchronic or acute exposure to selected environmental chemicals, therapeutics, abused drugs, or radiation. The complexity of the immune system and the redundancy of immune defense mechanisms make it difficult to determine the most appropriate tests to adequately assess the effects of xenobiotic exposure on overall immunocompetence. Thus, many laboratories employ a variety of assays that qualitatively and quantitatively test immune function (Roberts and Chapman 1981; Luster et al. 1988). These assays include NK cell activity, cell-mediated immunity (CMI) and humoral-mediated immunity (such as plaque-forming cell assays), specific antibody production and immunoglobulin levels, histology and cellularity of lymphoid tissues, and mitogen responses, as well as determining the ability of animals to resist challenge with infectious agents or transplantable tumor cells.

A number of miscellaneous compounds have demonstrated developmental immunotoxicity in a particular species, or have given indication for being toxic to the developing embryo. For many of these compounds, only a few studies have been conducted; thus, the full scope of immunologic damage and the mechanism(s) of action for the compound is not known. In some cases, separate studies find seemingly contradictory immunotoxic effects resulting from exposure to the same compound. This is not unexpected, as there are a great number of variables determining the outcome of exposure to a particular compound and frequently these variables are difficult to standardize. Different results can be expected from variation in amount of exposure, time in fetal development of exposure, duration of exposure, route of exposure, the species exposed, and the specific immune function tested. For example, contradictory results can be observed with compounds that can cause suppression of some aspects of immune function and disregulated stimulation (as in autoimmune diseases) of other functions. This chapter will give an overview of miscellaneous developmental immunotoxic substances, and will also indicate some compounds that may emerge as immunotoxic to the fetus as more specific research is conducted.

## MYCOTOXINS

Mycotoxins are compounds produced by molds, which are toxic when consumed or inhaled. They differ from other environmental chemical agents in that they are naturally occurring fungal metabolites and are found primarily on moldy grains. Mycotoxins in general are relatively heat-stable and persist as a toxin in cooked and processed foods. They are well recognized as a human health hazard, causing both acute toxicosis and carcinogenesis. Allowable mycotoxin levels in human and animal food products are regulated in most countries. A number of mycotoxins are immunotoxic to adults and a few are immunotoxic to the developing immune system. Trichothecenes (T-2 and vomitoxin), aflatoxins, and ochratoxin are discussed here.

### Trichothecenes

Trichothecenes are produced by *Fusarium* spp. and can be present in grains such as wheat, corn, oats, and barley. They are also present in the spoors of molds that grow on cellulose-based building materials, providing a route of human exposure through inhalation. Exposure to high doses of trichothecenes, both experimentally and accidentally, causes rapid decreases in lymphoid organ size, lymphopenia, and death from a circulatory shock-like syndrome (Bondy and Pestka 2000). Chronic exposure also seems to target the immune system and can cause either immunosuppression or immunostimulation, depending on the dose and length exposure to the toxin. Trichothecene toxins are closely related and classified into groups based on chemical structure. Group A trichothecenes include T-2 toxin, which is generally considered to be the most toxic trichothecene. Group B tricothecenes include deoxynivalenol or vomitoxin (VT), 3-acetyl deoxynivalenol, and nivalenol, and are less toxic than T-2 toxin (Bergsjø et al. 1993).

## T-2 Toxin

T-2 toxin is a metabolite of *Fusarium* spp. that grows on cereal grains in temperate climatic zones of North America, Europe, and Asia (Niyo et al. 1980; Scott et al. 1980; Bondy and Pestka 2000). Exposure to T-2 toxin can cause fatal alimentary toxic aleukia syndrome in humans, which targets mucosal surfaces and the immune system. T-2 toxin increases the susceptibility to infectious diseases and also increases tumor incidence. In adult animals, T-2 toxin is profoundly immunosuppressive and even lethal at relatively low concentrations. B and T-lymphocyte numbers are suppressed, as is the corresponding humoral and cellular immunity. T-2 toxin causes necrosis and depletion of lymphoid tissues resulting in thymic atrophy (Hayes et al. 1984; Corrier and Ziprin 1986, 1987).

The lymphoid depletion is caused in part by increased apoptosis in the lymphoid organs (Islam et al. 1998; Nagata et al. 2001). In adult mice, oral inoculation with T-2 toxin (10 mg/kg) induced apoptosis first in Peyer's patches, then mesenteric lymph nodes, and finally in the thymus in relation to the course of enteric absorption of orally inoculated T-2 toxin (Nagata et al. 2001). Lymphocyte subsets differ in their sensitivities to T2 toxin. Thymic T cells that were CD4+ CD8+ were most sensitive to T-2 toxin induced apoptosis followed by CD4+ CD8- T cells, while CD4- CD8+ T cells were least sensitive (Islam et al. 1998; Nagata et al. 2001). T-2 toxin also affects B cells. Among IgM+, IgG+, and IgA+ B cells, the IgA+ cells, those important for mucosal immunity, were most severely affected (Nagata et al. 2001).

T-2 toxin readily crosses the placenta (Holladay et al. 1993; LaFarge-Frayssinet et al. 1990), but the full effects of *in utero* exposure on the development of immune function are not known. Lafarge-Frayssinet et al. (1990) observed atrophic thymuses in newborn rats after maternal feeding at levels equivalent to that found in naturally contaminated foods. Histological examination of the spleens in the offspring at 6 days of age revealed reduced periarteriolar sheaths (T-dependent areas) as well as follicles (B-dependent areas). Further, splenic B- and T-lymphocyte mitogen responses were impaired postnatally. Similarly, in mice exposed prenatally to T-2 toxin at similar dose levels, severe fetal thymic atrophy and cellular depletion in offspring were again found in the absence of overt fetal or maternal toxicities (Holladay et al. 1993).

Fetal cell marker expression in mice prenatally exposed to T-2 toxin has also been examined. The cell markers CD44 (present on prothymocytes) and CD45 (common leukocyte antigen) are expressed by all leukocyte progenitors in mice (Holmes and Morse 1988). Granulocyte and macrophage progenitors stain brightly ("hi") with antibodies against these surface antigens, while lymphocytes stain more dimly ("lo") (Holmes and Morse 1988; Coffman 1982; Kincade et al. 1981). Prenatal exposure to T-2 toxin selectively decreased numbers of cells within CD44lo and CD45lo fetal liver pro-lymphoid cell subpopulations (Holladay et al. 1993, 1995). Further analysis showed that specifically the B-cell lineage was depleted by exposure to T-2 toxin (Holladay et al. 1995). The mechanism of action for immune alterations is not clear but may involve increased apoptosis as seen in adults. Fetal mice exposed to T-2 toxin *in utero* also showed signs of increased apoptosis. Ishigami et al. (1999) exposed pregnant mice to 3 mg/kg T-2 toxin orally at day 11 of gestation. Twenty-four hours after dosing, changes were observed in embryos, including pyknosis and karyorrhexis. The nuclei stained strongly by the TUNEL method used to detect *in situ* apoptosis. Thus, while only limited information is available on the influence of T-2 toxin on immune development, these preliminary reports indicate that the developing immune system may represent a sensitive target for T-2 toxin exposure, and that T-2 toxin causes increased apoptosis in fetal tissues as it does in adults.

### Vomitoxin (Deoxynivalenol)

Group B trichothecenes, such as vomitoxin (VT or deoxynivalenol) and similar metabolites 3-acetyl deoxynivalenol and nivalenol, are less toxic than the group A trichothecene T-2 toxin (Bergsjø et al. 1993). Vomitoxin is produced by *Fusarium graminearum* and *F. culmorum* and is a common contaminate of grain used to make human and animal feed. Because it is not destroyed by milling and processing, vomitoxin could be present in human food products at ppm levels (Islam et al. 2003). Swine and other monogastrics are the most sensitive to VT while chickens and turkeys have the highest tolerance to VT (Rotter et al.1996). Ingestion of moderate to low levels can cause anorexia, decreased nutritional efficiency, and immunotoxicity manifest by reduction in natural immunity and poor production performance in food animals. Exposure to high concentrations of VT causes nausea, emesis, leukocyte apoptosis, and circulatory shock (Rotter et al. 1996).

The effects of vomitoxin on immune function have been investigated. In adult mice, levels greater than 0.25 to 0.5 mg/kg/day causes immunotoxicity. A one-time exposure of 10 mg/kg VT caused necrosis of a number of tissues including the bone marrow and lymphoid tissues (Rotter et al. 1996). The effects of VT on cells of the adult immune system may be immunosuppressive or immunostimulatory depending on the length of exposure and dosage.

VT is cytotoxic to a number of cells, including fibroblasts, splenic lymphocytes, and human peripheral blood lymphocytes, because it inhibits protein synthesis at the ribosomal level during the elongation termination step (Rotter et al. 1996). VT at high doses induces rapid apoptotic loss of immature thymocytes and cytotoxic T-lymphocytes in thymus, mature-B lymphocytes in Peyer's patch, and pro/pre B-lymphocytes and mature B-lymphocytes in bone marrow in mice (Bondy and Pestka 2000; Islam et al. 2003). Exposure in the diet to VT at 2 ppm for 5 weeks or 5 ppm for 1 week suppressed mitogenic response of lymphocytes in humans, mice, and chickens (Rotter et al. 1996; Bondy and Pestka 2000). Host resistance to infectious disease is decreased due to a reduction in several cellular functions such as cell-mediated immunity, neutrophil migration, and macrophage phagocytosis (Rotter et al. 1996). In contrast to suppression of cellular responses, chronic dietary exposure of VT-induced elevation of serum IgA and increased renal deposition of IgA in mice by cyclooxygenase-2 mediated upregulation of interleukin (IL)-6 (Moon and Pestka 2003). Superinduction of IL-6 is also likely to be responsible for the shock-like and cytotoxic responses with acute high levels of VT exposure (Moon and Pestka 2003). Thus, VT both stimulates and inhibits differing aspects of immune function.

At the present, there are no studies demonstrating VT immunotoxicity in the developing fetus; however, there is evidence that VT is passed on to the egg in chickens, affecting embryo development (Bergsjø et al. 1993). VT at 1 to 5 mg/kg is not a health hazard to poultry or humans, but in the egg, seriously affected development and hatchability of the chicks. Hens exposed to 0, 2.5, 3, and 5 mg/kg VT in the feed showed no clinical signs of toxicity such as differences in egg production, body weight gain, percentage fertility, or percentage mortality of fertile eggs during incubation (Bergsjø et al. 1993). However numerous developmental abnormalities were observed in the hatched chicks. Chicks exposed to VT had unwithdrawn yolk sacs, cloacal atresia, delayed ossification of vertebra and extremities, and cardiac malformations (Bergsjø et al. 1993). This study did not evaluate any immunological parameters in the chicks; however, the tissues targeted by VT exposure included the developing hematopoietic tissues, e.g., yolk sac, bone marrow, and possibly bursa. It is not known if VT is toxic to these developing immune tissues; however, a fair amount of evidence suggests that it may be. VT and T-2 toxin are closely related structurally. Structure activity studies in a number of lymphocyte models suggest that translational arrest is an underlying mechanism for impaired proliferation in both VT and T-2 toxin (Rotter et al. 1996). Both VT and T-2 target immune function and immune tissues, and both induce apoptosis (Bondy and Pestka 2000). These data, although far from indicating that VT is immunotoxic to the developing fetus, show enough similarities between T-2 toxin (a known developmental immunotoxin) and VT that further studies are warranted investigating the developmental immunotoxicity of VT.

### Aflatoxins

Aflatoxins are toxic metabolites of the fungi genus *Aspergillus*, particularly *A. flavus* and *A. parasiticus*, found predominantly on corn, cornmeal, and other plant products such as nuts. The biologically active aflatoxins are termed AFB1 and AFG1 and are found in grain products; AFB1 can be metabolized to another toxic derivative, AFM1, which can be found in animal products including milk (Silvotti et al. 1997). Aflatoxins are lipophilic and disseminate in the fat and other soft tissues. Aflatoxins can depress immune function in humans, pigs, guinea pigs, lambs, chicks, and rats, with altered macrophage function and depressed cell mediated immunity resulting in decreased resistance to disease (Silvotti et al. 1997; Fernández et al. 2000; Çeílk et al. 2000; Doi et al. 2002; Theumer et al. 2003).

There is also evidence that aflatoxins are transferred to the fetus and newborn during pregnancy and lactation in both humans (Denning et al. 1990; Abdulrazzaq et al. 2002; Doi et al. 2002) and animals (Silvotti et al. 1997; Çeílk et al. 2000). The full immunological effects of transplacental aflatoxin exposure are not known; however, initial results indicate that transplacental aflatoxins are immunotoxic. Silvotte et al. (1997) fed pregnant sows diets containing AFB1 and AFG1 throughout gestation and lactation until the piglets were weaned. Concentrations of AFB1, AFG1, and AFM1 were detected in the milk and increased over time as the piglets suckled. Piglets exposed to aflatoxins throughout gestation and lactation had lower lymphocytic mitogenic response, macrophage respiratory burst activity, and neutrophil motility and chemotaxis (Silvotti et al. 1997). The separate effects of exposure during gestation and during lactation were not determined in this study.

In chickens, aflatoxins and their metabolites are passed on to the eggs in concentrations not harmful to humans who consume poultry products, but are harmful to the developing embryo. In one study (Qureshi et al. 1998), hens were fed diets amended with 0, 0.2, 1, 5, or 10 mg/kg of aflatoxin. Fertile eggs collected after 14 days on the experimental diet had aflatoxin residues of 0.15 to 0.60 ng/g of AFB1. This level of aflatoxin exposure resulted in embryonic mortality, reduced hatchability, depressed post-hatch humoral and cell-mediated immunity, and reduced phagocytosis and respiratory burst activity in macrophages as compared with the control chicks (Qureshi et al. 1998). In another study, 10 ng mixed aflatoxin or purified AFB1 administered to the egg did not induce any changes in the bursa of Fabricius or thymus of chicks, where as 100 ng/egg retarded both bursa and thymic development (Çeílk et al. 2000).

### Ochratoxin A

Ochratoxin A is a mycotoxin produced by *Aspergillus ochraceus* and *Penicillium verrucosum* and is found worldwide as a food contaminant. It is known to cause porcine nephropathy and may be associated with a human kidney disease endemic to the Balkan region (Bondy and Pestka 2000). Mice and rats exposed to ochratoxin A *in utero* had altered immune responses compared to controls. A single dose of ochratoxin A (10 to 500 μg/kg) to pregnant mice decreased thymic CD4+ and increased thymic CD4+CD8+ lymphocyte populations while suppressing splenic and thymic lymphocyte responses to mitogens in the pups (Thuvander et al. 1996a). In rat pups, thymic weights and cellularity were not affected, but splenic cell mitogen responses were suppressed with maternal exposure to ochratoxin A (Thuvander et al. 1996b).

## ALCOHOL

A number of studies have shown that consumption of alcohol inhibits immune function in both adult humans and animals (Jerrells et al. 1986). In adults, ethanol is immunosuppressive, decreasing both the cellularity of the thymus and spleen and interfering directly with B and T-lymphocyte responses. Additionally, alcohol suppresses the phagocytic activity of macrophages and reduces NK cell activity (Seelig et al. 1996). Ethanol consumption alters cytokine production, including IL-2, IL-5, IL-6, IL-10, and reduces the capacity of T cells to use IL-2. TNF-α production was impaired as was the distribution of TNF-α receptor on alveolar macrophages (Seelig et al. 1996).

Alcohol readily crosses the placenta and has a well-documented effect on the developing fetus. Maternal alcohol consumption can cause altered development ranging from a mild reduction in birth weight to severe manifestations of fetal alcohol syndrome (FAS). FAS is a multifaceted disorder characterized by growth deficits, CNS dysfunction, mental retardation, and cranial facial deformities. In addition, these children have defects in host defense and consequently an increase in the number and severity of infections (Johnson et al. 1981; Steinhausen et al. 1982; Moscatello et al. 1999). Some children with FAS have been shown to have long-lasting deficits in both humoral and cell-mediated immunity including lymphopenia, eosinophilia, decreased lymphocyte mitogenesis, lower IgG levels, and can suffer increased illness such as pneumonia and meningitis during the first few years of life (Johnson et al. 1981; Giberson and Weinberg 1997).

Experimental animals exposed to prenatal alcohol exhibit suppressed immune function. Specifically, animals exposed during development to alcohol displayed decreased lymphocyte response to the mitogens con A and LPS, reduced serum IgM and IgG levels, reduced specific antibody production to T-dependent antigen, suppressed IL-2 and TNF production, and reduced immune transfer resulting in decreased passive immune protection for the neonate (Adkins et al. 1987; Monjan and Mandell 1980; Ewald and Frost 1987; Wolcott et al. 1995; Seelig et al. 1996, 1999). This indicates prenatal ethanol exposure affects specific B- and T cell function as well as cytokine production in ethanol-exposed pups (Seelig et al. 1996). Many of these immunological deficits are long-lasting, including a reduced number of thymocytes, a reduced proliferative response of B- and T-lymphocytes to mitogens, reduced splenic and thymic T-lymphoblast response to IL-2, and deficits in CMI including graft vs. host response and contact hypersensitivity (Wolcott et al. 1995; Giberson and Weinberg 1997; Moscatello et al. 1999).

The suppression of immune function with prenatal alcohol exposure can be exacerbated by additional lactational exposure. Seelig et al. (1996) exposed the following four groups of mice to ethanol: 1.) dams fed ethanol during pregnancy and lactation and pups fed ethanol after weaning, 2.) dams received ethanol through pregnancy and lactation, 3.) dams received ethanol through delivery, and 4.) dams received ethanol only through lactation. Ethanol was passed from dam to pup in all four groups, and all pups had some degree of suppressed neonatal immunity. The greatest immunologic suppression was seen in pups receiving both prenatal and postnatal exposure. In a later study, mice exposed to maternal ethanol during pregnancy and lactation and who also consumed alcohol into adulthood demonstrated additional reductions in total T- and B-cells numbers, as well as specific IgM and IgG antibodies. The sequential reduction in cell numbers and antibody levels from unexposed mice, to first-generation ethanol animals (dams), to offspring exposed to ethanol prenatally and also through adulthood, suggests that there is a cumulative effect of *in utero* ethanol exposure and subsequent alcohol consumption (Seelig et al. 1999).

Fetal mice exposed to alcohol *in utero* have reduced thymic cellularity (Wolcott et al. 1995) resulting in thymic atrophy (Ewald and Frost 1987). Reduction in thymocytes in mice exposed to ethanol during fetal life may be due to altered development of T-lymphocyte populations resulting in thymic hypocellularity. Recent work using fetal thymus organ culture has shown that alcohol-exposed fetuses have decreased numbers of double positive CD4+ CD8+ T cells and increased number of single positive cells as well as increased apoptosis compared to control fetuses (Giberson and Weinberg 1997). The decrease of double positive CD4+ CD8+ T cells has also been seen in mouse fetuses *in vivo*. Because both T-helper and T-cytotoxic cells arise from this population of thymocytes with a decrease in double positive cells, one would expect reduced numbers of these mature T-lymphocytes.

B cell development is also affected by prenatal alcohol exposure. Mice exposed to ethanol *in utero* had decreased numbers of pre-B cells in the bone marrow and decreased numbers of total B cells in the spleen and bone marrow (Moscatello et al. 1999). In this study, the rate at which both mature and immature B cells accumulated in the spleen was reduced in prenatally ethanol-exposed animals. Immature and mature B cells did not reach control levels until 4 weeks of age in prenatally exposed animals. The pool of precursor B cells did not reach control levels by 5 weeks, the last point measured. It was hypothesized that the pool of B-lineage progenitors is specifically affected by the prenatal alcohol exposure (Moscatello et al. 1999).

Male offspring tend to be affected more than female offspring by *in utero* alcohol exposure (Giberson and Weinberg 1995, 1997). The sex differences and specific immune defects become more pronounced with stress. This indicates that the impaired immune system has some ability to compensate for the deficits resulting from prenatal ethanol exposure and may not demonstrate immune suppression when tested under nonstressed conditions. When rat pups, particularly males, were challenged by a stressor, the deficits in immune function became more apparent (Giberson and Weinberg 1995, 1997).

## HEAVY METALS

Heavy metals are the higher molecular weight metallic elements. Most are needed in trace amounts by biological systems for homeostatic functioning, but are toxic at higher concentrations. Heavy metals are a common byproduct of industrial processes resulting in environmental contamination where the metals can persist for long periods of time. Heavy metals are a health concern because they accumulate to high levels in body tissues as detoxification and elimination rates are low. Once heavy metals accumulate in tissues, biomagnification can occur; resulting in increasing concentration of a substance as one travels up the food chain. As a result, humans can potentially be exposed to highly toxic concentrations of heavy metals through their diet.

### Mercury

Contamination of the environment by mercury occurs mainly through industrial waste and through its use as a fungicide on crops. This inorganic or elemental mercury is then methylated to organic mercury by microorganisms and incorporated into the food chain where it biomagnifies. The primary exposure to humans comes from consuming fish with high levels of biomagnified mercury in their tissues.

Elemental, inorganic, and organic mercury are toxic to living organisms. The toxic effects mainly manifest as delays in neurological developmental and mental retardation, but immunologic effects are also seen. The immune effects of mercury are both inhibitory and stimulatory. Mercury decreases antibody titers in rabbits, lowers serum IgG, IgG1, and IgE levels in mice, and reduces the lymphoproliferative response to T cell mitogens in rats (Wild et al. 1997). In humans, inorganic mercury causes an autoimmune response with IgG and complement C3 deposition in the kidney, the presence of antiglomerular basement membrane antibodies, and an increase in anti-DNA antibody titer (U.S. DHHS 1999a; Cardenas et al. 1993). Additionally, T cells are stimulated with increases in CD3+, CD4+ and CD8+ cells. In mice, mercury vapor and mercuric chloride both induced a syndrome similar to the autoimmune response in humans characterized by general stimulation of the immune system, hyperimmunoglobulinemia, anti-nucleolar autoantibodies, and glomerular disease accompanied by vascular immune complex deposits.

Mercury stimulated or suppressed immune responses depending on genetics and susceptibility of the mouse strain. Immune hyperreactivity and autoimmunity were seen in SJL/N mice, but immune responses were suppressed in B6C3F$_1$ mice (U.S. DHHS 1999a). B6C3F$_1$ mice exposed to 2.9 or 14.3 mg Hg/kg/day as mercuric chloride in the drinking water demonstrated a suppression of lymphocyte proliferation to T cell mitogens. When SJL/N mice were exposed to mercuric chloride, antinucleolar antibodies were observed and deposition of granular IgG in the renal mesangium and glomerular blood vessels occurred (U.S. DHHS 1999a).

All forms of mercury transfer across the placenta to the fetus, but methyl mercury transfers to a greater extent than other forms. Mercury actually accumulates in placental tissues, and concentrations 2.25 times higher than in maternal tissues have been reported (Wild et al. 1997). Mercury is also transferred in milk during lactation (Wild et al. 1997). The fetus is more susceptible to mercury exposure than adults with 1/10 the adult toxic dose inducing toxic responses in the fetus. Abnormalities from gestational mercury exposure are dose-dependant and are greatest when exposure occurs in the second trimester. Developmental defects following *in utero* exposure include fetal death, dystocia, decreased fetal weight, immunologic alterations, skeletal deformities, brain lesions, hydronephrosis, and renal hypoplasia (U.S. DHHS 1999a).

Low-dose placental and lactational exposure to mercury affects thymocyte development and stimulates certain mitogen- or antigen-induced lymphocyte activities. Rat pups indirectly exposed to methyl mercury during gestation and lactation through their dams' drinking water at a dose recognized as safe for adults (5 or 500 μg/L methyl mercury chloride) exhibited depressed NK cytolytic activity and enhanced PWM-stimulated lymphocyte proliferation (Wild et al. 1997). In another study, female mice were exposed to methyl mercury at 0, 0.5, or 5 mg Hg/kg for 10 weeks prior to mating and during gestation and through 10 days of lactation (Thuvander et al. 1996c). The numbers of splenocytes in the pups were increased at day 10 and day 22, and thymocytes were increased at day 22 in the 0.5 mg/Kg-exposed animals. There were increased numbers and altered proportions of the lymphocyte subpopulations within the thymus at both concentrations of mercury. The proliferative response of splenocytes to LPS was increased in the high exposure group as was the antibody response to viral antigen in pups from the 0.5 mg Hg/kg exposure (Thuvander et al. 1996c). The significance of the immune stimulation is not known but may be a sign of induced autoimmunity.

## Cadmium

High doses of cadmium are immunotoxic, consistently causing decreased levels of immunoglobulin; suppressing phagocytosis, humoral, and CMI responses; and decreasing host resistance (Blakley 1988; Ritz et al. 1998). Inhalation exposure to lower doses of cadmium caused lymphoid hyperplasia and increased monocyte counts, but did not affect serum IgG, IgM, IgA, lymphocyte, neutrophil, or eosinophil cell counts (Kutzman et al. 1986; Karakaya et al. 1994).

There is little information available on the developmental immunotoxic effects of cadmium. Many authors have demonstrated cadmium transfer across the placenta and alterations in development of human and animal fetuses following maternal exposure to cadmium (U.S. DHHS 1999b). Even low-dose exposure to cadmium during gestation and lactation results in fetotoxicity, characterized predominantly by decreased birth weight, neurophysiologic dysfunction, and skeletal deformities (Branski 1985; Kostial et al. 1993; Nagymajtenyi et al. 1997; Desi et al. 1998). Given that cadmium exposure affects adult immune responses, and that cadmium readily crosses the placenta affecting neurophysiologic responses of the offspring, the possibility exists for immunologic deficits in neonates resulting from *in utero* cadmium exposure. The close association and interrelatedness in function between neurological and immunologic processes makes this possibility even greater. Further studies are warranted to determine what if any changes in immune function are associated with exposure to this heavy metal.

## Chromium

Chromate exposure in humans occurs mainly from occupational exposure to welding, printing, glues, dyes, wood ash, foundry sand, match heads, machine oils, timber preservative, engine coolants, boiler linings, television screens, magnetic tapes, tire fitting, and chrome plating. Exposure to chromium from industrial waste can also occur (U.S. DHHS 2000). Exposure is mainly to hexavalent chromium, which is reduced to trivalent chromium in the skin. Trivalent chromium is found in the epithelium and is transported bound to erythrocytes and plasma proteins. Exposure to chromium can cause an allergic dermatitis and other immunologic changes, fatal nephritis, loss of taste and smell, and hepatic cell injury. It is also a pulmonary carcinogen (U.S. DHHS 2000).

In adults, exposure to chromium by inhalation, oral or dermal contact can cause skin sensitization and eruptions, urticaria, a chromium-induced asthma, and bronchospasms accompanied with tripling of plasma histamine levels (U.S. DHHS 2000). Cellular changes include leucocytosis or leucopenia, eosinophilia, monocytosis, and changes in the number and function of alveolar macrophages. Additionally, chromium exposure increases serum immunoglobulin levels and stimulates T cell mitogen responses (U.S. DHHS 2000). Overall, chromium causes a dysregulation of immune function resulting in immunostimulation.

Chromium crosses the placenta and concentrates in fetal tissues, and can be found in breast milk during lactation. This generally gives fetuses higher tissue concentrations than adults. Tissue concentrations in the fetus decline rapidly with age except in the lungs (U.S. DHHS 2000). Hexavalent chromium exposure during gestation results in fetal developmental abnormalities, including ossification changes and facial deformities, and morphological changes in the developing reproductive tracts (Junaid et al. 1996a, 1996b; Kanojia et al. 1998). No studies have yet investigated immunological changes in offspring resulting from *in utero* exposure to chromium, but as with cadmium, there is a possibility for some form of developmental immunotoxicity based on the current literature.

## DIETHYLSTILBESTROL AND STEROID HORMONES

The sensitivity of the thymus and T cell-mediated immunity to certain steroidal compounds, including glucocorticoids such as hydrocortisone, has been well established (Hendrickx et al. 1975; Kalland et al. 1978; Ways et al.1979). However, the role of other steroids such as sex hormones in the development of immunity remains poorly understood. A variety of reports have indicated that exogenously administered hormones, including testosterone (Warner et al. 1969), estrogen, and related compounds possessing estrogenic properties, can alter the normal development of the immune system (Luster et al. 1979; Kalland and Forsberg 1981; Ways et al. 1980; Blair 1981; Whitehead et al. 1981; Kalland 1982; Ford et al. 1983; Noller et al. 1988; Ways et al. 1987).

The use of diethylstilbestrol (DES) from the 1940s to the 1970s to prevent threatened miscarriage in pregnant women was found to cause development of genital tract anomalies and neoplasia in a number of adult women exposed *in utero* (Ways et al. 1987). Several lines of evidence also indicate a link between DES exposure and alterations in the development of the immune system. Perinatal DES exposure in mice results in a long-term impairment of both cell-mediated and humoral-mediated immunity (Ways et al. 1987). Postnatal immunosuppression following pre- and/or perinatal DES exposure includes altered T- and B-lymphocyte mitogen responsiveness, graft-vs.-host reaction, delayed hypersensitivity, NK cell activity, and antibody production (Ways et al. 1987; Luster et al. 1979; Kalland and Forsberg 1981; Ways et al. 1980). The consequences of immunosuppression in rodents resulting from DES exposure during development include increased susceptibility to virally induced mammary tumors and to transplanted and primary carcinogen-induced tumors (Blair 1981; Kalland 1982). In studies of women exposed *in utero* to DES, unconfirmed evidence exists for altered T-lymphocyte and NK cell activity (Ford et al. 1983; Ways et al. 1987), as well as an increased incidence of autoimmune diseases (Noller et al. 1988; Chapter __ in this book). Several investigators have further stressed the potential relationship between prenatal DES exposure and postnatal immunocompetence and neoplastic growth in humans, including clear-cell adenocarcinoma of the vagina (Kalland et al. 1978; Luster et al. 1978).

Analysis of a variety of fetal liver markers in mice exposed prenatally to DES has indicated that fetal thymic atrophy may similarly be the result of selective prothymocyte targeting by this compound (Holladay et al. 1993). DES caused a reduction in the number of double positive CD4+ CD8+ T cells and enrichment in CD4- CD8- double negative cells. DES alters the intrathymic development of thymocytes in the double negative compartment at a late stage of development and may block the expression of pre-T cell receptor and its ability to respond to signals for proliferation and differentiation. It is also possible that DES causes apoptosis in cells entering the CD44- CD25- compartment (Lai et al. 1998).

Triamcinolone acetonide is a potent corticosteroid used clinically to treat a variety of inflammatory disorders. It is also immunotoxic to developing fetuses, inducing thymic hypoplasia and lymphocyte deficiency in monkeys following exposure at doses of 3 to 28 mg/kg/day during early, mid-, and late gestation (Hendrickx et al. 1975). Each group was exposed for 4 consecutive days during the following time periods: early gestation (GD 37 to 48) during the time of early proliferation of thymic tissue and separation from pharyngeal pouches, mid-gestation (GD 50 to 73) during the time of differentiation of lymphoid elements within the thymus, and late gestation (GD 100 to 133) during the time for maturation of the fetal lymphoid system. Seventy-six percent of the fetuses demonstrated a marked depletion of the thymic lymphocytes and a marked reduction in the epithelial component. A depletion of the lymphocytes from the thymic dependent areas of the spleen and lymph nodes was also observed.

## X-IRRADIATION

During organogenesis of the immune system, as well as after birth, the hematopoietic tissues are highly susceptible to damage from irradiation. The effects of irradiation have been described as varying from abnormal immune responses to imbalances in blood cell populations and leukemias (Boniver and Janowski 1986; Miller and Benjamin 1985; Nold et al. 1987). Lymphocytes, and in particular lymphoid progenitor cells in the bone marrow, are known to be among the most sensitive cells to radiation injury (Miller and Benjamin 1985; Nold et al. 1987; Platteau et al. 1989). Dose-response studies on the effects of prenatal exposure to single or subchronic low-dose X-ray exposure on subsequent development of the immune system in experimental animals or humans is, as is the case with many of the preceding developmental immunotoxic agents, extremely limited. An early study reported impaired antibody production in adult mice irradiated with a single 0.5 Gy dose of x-rays at either day 5 or between days 12 through 15 of gestation (Hooghe et al. 1988). Platteau et al. (1989) looked for impaired immune function in rats exposed to 0-2 Gy X-rays during prenatal or early postnatal life. At 8 weeks of age, however, histology of the spleen was found to be normal in exposed animals, as was the distribution of B- and T-lymphocytes and serum immunoglobulin levels. Similarly, Urso and Gengozian (1982) observed no appreciable differences in immune competence, in terms of antibody production or tumor incidence, in adult mice exposed to 150 R X-irradiation during mid-(day 11 to 13) or late (day 16 to 18) gestation.

The results in experimental animals have not always been consistent with the effects of radiation exposure in the human neonatal population. For example, alterations in peripheral lymphocyte counts and lymphoproliferative responses were found to persist in adults irradiated in infancy for thymus enlargement (Reddy et al. 1976). In additional experiments, low-dose irradiation of adult mice has been shown to alter T cell mediated immunologic responses in what appears to be selective killing of sensitive T-lymphocyte subpopulations (Anderson et al. 1982). Such T cells are generated in large numbers near birth in rodents (Ptak and Skorwon-Cendrzak 1977) and during late gestation in humans (Hayward and Kurnick 1981), suggesting that T-lymphocytes may be particularly sensitive to radiation exposure during certain periods of maturation. In a study designed to examine the effects of ionizing radiation administered during this sensitive period of T cell development, Miller and Benjamin (1985) irradiated pregnant beagle dams with 200 R at 30, 40, or 45 days of gestation. All exposed groups had thymic lobules and lobular cortices that were significantly reduced at 5 and 10 days post-irradiation when compared with age-matched controls. At 10 days post-irradiation, partial repopulation of the outermost cortices by thymocytes was observed. Repopulation of the thymus also occurs following irradiation in adult mice (Takada and Takada 1973). However, significant radiation-induced damage to the thymic epithelium of treated animals occurred, an unanticipated result in light of the radioresistant nature of thymic epithelium in adult animals (Maisin et al. 1971). As the thymic epithelium and stroma microenvironment are required for the orderly differentiation of distinct subsets of T-lymphocytes and elimination of autoreactive T cells, damage to these components during development may result in additional immune system deficits including impaired immunological recognition mechanisms and autoimmune disease (Miller and Benjamin 1985; Egerton et al. 1990). While additional studies will be necessary to determine acceptable radiation exposure levels during and around critical periods of development of the immune system, reports such as these suggest the possibility for long-term immunological deficits in humans resulting from prenatal exposure to ionizing radiation.

## SUMMARY

Exposure to certain xenobiotics during immune system development can alter immune functions in humans and laboratory animals. In some cases, the alterations in immune function can be long-lived or possibly permanent. Evidence from experimental studies indicate that the developing immune system is not necessarily more sensitive to damage than the mature immune system, but rather the damage is dependent on which cellular maturation and differentiation steps are disrupted during immune system ontogeny. In adults, long-term secondary immunodeficiency requires continued exposure to the inducing agent or damage to primitive hematopoietic cells, the latter being a relatively rare event. In contrast, the developing immune system may experience long-lasting alterations in immune function from a single point exposure to an immunotoxic compound if the exposure occurs during a critical development step. The health risks associated with human exposure to most potentially immunotoxic chemicals during the prenatal period remain largely undetermined. Further clinical and epidemiological studies using large, well-defined cohorts need to be undertaken to establish whether acute or chronic low-level exposure during prenatal life contributes to human immunological dysfunction and subsequent clinical disease, including the development of hypersensitivity and autoimmune diseases.

## REFERENCES

Abdulrazzaq YM, Osman N, and Ibrahim A. 2002. Fetal exposure to aflatoxins in the United Arab Emirates. *Ann Trop Paediatr* 22:3–9.

Adkins B, Mueller C, Okada C, Reichert R, Weissman IL, and Spangrude GJ. 1987. Early events in T-cell maturation. *Ann Rev Immunol* 5:325–365.

Anderson RE, Tokuda S, Williams WL, and Warner NL. 1982. Radiation-induced augmentation of the response of A/J mice to SaI tumor cells. *Am J Pathol* 108:24–33.

Baranski B. 1985. Effect of maternal cadmium exposure on postnatal development and tissue cadmium, copper and zinc concentrations in rats. *Arch Toxicol* 58:255–60.

Barnett JB, Blaylock BL, Gandy J, Menna JH, Denton R, and Soderberg LSF. 1990. Long-term alteration of adult bone marrow colony formation by prenatal chlordane exposure. *Fundam Appl Toxicol* 4:688–695.

Bergsjø B, Herstad O, and Nafstad I. 1993. Effects of feeding deoxynivalenol-contaminated oats on reproduction performance in white leghorn hens. *British Poultry Science* 34:147–159.

Blair PB. 1981. Immunological consequences of early exposure of experimental rodents to diethylstilbestrol and steroid hormones. In: Herbst AL and Bern HA (Eds.), *Developmental Effects of Diethylstilbestrol (DES) in Pregnancy.* New York: Thieme-Stratton, p. 167.

Blakley BR. 1985. The effect of cadmium chloride on the immune response in mice. *Can J Comp Med* 49:104–108.

Boniver J and Janowski M. 1986. Symposium on radiation carcinogenesis: molecular and biological aspects. *Leukemia Res* 10:703–936.

Bondy GS and Pestka JJ. 2000. Immunomodulation by fungal toxins. *J Toxicol Environ Health* 3:109–143.

Cardenas A, Roels H, Bernard AM, Barbon R, Buchet JP, Lauwerys RR, Rosello J, Hotter G, Mutti A, and Franchini I. 1993. Markers of early renal changes induced by industrial pollutants. I. Application to workers exposed to mercury vapour. *Br J Ind Med* 50:17–27.

Çeílk Í, Ouz H, Demet Ö, Boydak M, Dönmez HH, Sur E, and Nízamlíolu F. 2000. Embryotoxicity assay of aflatoxin produced by aspergillus parasiticus NRRL 2999. *British Poultry Science* 41:401–409.

Coffman B. 1982. Surface antigen expression and immunoglobulin rearrangement during mouse pre-B cell development. *Immunol Rev* 69:5–12.

Corrier DE and Ziprin RL. 1986. Immunotoxic effects of T-2 toxin on cell-mediated immunity to listeriosis in mice: comparison with cyclophosphamide. *Am J Vet Res* 47:1956–1960.

Corrier DE and Ziprin RL. 1987. Immunotoxic effects of T-2 toxin on cell-mediated resistance to *Listeria monocytogenes*. *Vet Immunol Immunopathol* 14:11–21.

Desi I, Nagymajtenyi L, and Schulz H. 1998. Behavioural and neurotoxicological changes caused by cadmium treatment of rats during development. *J Appl Toxicol* 18:63–70.

Denning DW, Allen R, Wilkinson AP, and Morgan MR. 1990. Transplacental transfer of aflatoxin in humans. *Carcinogenesis* 11:1033–1035.

Doi AM, Patterson PE, and Gallagher EP. 2002. Variability in aflatoxin B1-macromolecular binding and relationship to biotransformation enzyme expression in human prenatal and adult liver. *Toxicol and Appl Pharm* 181: 48–59.

Egerton M, Scollay R, and Shortman K. 1990. Kinetics of mature T-cell development in the thymus. *Proc Natl Acad Sci* 87:2579–2582.

Ewald SJ and Frost W. 1987. Effect of prenatal exposure to ethanol and development of the thymus. *Thumus* 9:211–215.

Fernández A, Hernández M, Verde MT, and Sanz M. 2000. Effect of aflatoxin on performance hematology and clinical immunology in lambs. *Can J Vet Res* 64:53–58.

Fine JS, Gasiewicz TA, and Silverstone AE. 1989. Lymphocyte stem cell alterations following perinatal exposure to 2,3,7,8-tetrachlorodibenzo-*p*-dioxin. *Mol Pharmacol* 35:18–25.

Ford CD, Johnson GH, and Smith WG. 1983. Natural killer cells in *in utero* diethylstilbestrol-exposed patients. *Gynecol Oncol* 16:400–404.

Giberson PK and Weinberg J. 1995. Effect of prenatal ethanol exposure and stress in adulthood lymphocyte populations in rats. *Alcoholism: Clin and Exp Res* 19:1286–1294.

Giberson PK and Weinberg J. 1997. Effect of surrogate fostering on splenic lymphocytes in fetal exposed rats. *Alcoholism: Clin and Exp Res* 21:44–55.

Hayes MA, Bellamy JEC, and Schiefer HB. 1984. Subacute toxicity of dietary T-2 toxin in mice: morphological and hematological effects. *Can J Comp Med* 44:203–218.

Hayward AR and Kurnick J. 1981. Newborn T cell suppression: early appearance, maintenance in culture, and lack of growth factor suppression. *J Immunol* 126:50–59.

Hendrickx AG, Sawyer RH, Terrell TG, Osburn BI, Henrickson RV, and Steffek AJ. 1975. Teratogenic effects of triamcinolone on the skeletal and lymphoid systems in nonhuman primates. *Fed Proc* 34:1661–1667.

Herbst AL, Ulfelder H, and Poskanzer DC. 1971. Adenocarcinoma of the vagina. Association of maternal stilboestrol therapy with tumor appearance in young women. *N Engl J Med* 284:878–881.

Holladay SD, Lindstrom P, Blaylock BL, Comment CE, Germolec DR, Heindel JJ, and Luster MI. 1991. Perinatal thymocyte antigen expression and postnatal immune development altered by gestational exposure to tetrachlorodibenzo-*p*-dioxin (TCDD). *Teratology* 44:385–393.

Holladay SD, Blaylock BL, Comment CE, Heindel JJ, and Luster MI. 1993. Fetal thymic atrophy after exposure to T-2 toxin: Selectivity for lymphoid progenitor cells. *Toxicol Appl Pharmacol* 121:8–14.

Holladay SD, Smith BJ, and Luster MI. 1995. B lymphocyte precursor cells represent sensitive targets of T2 mycotoxin exposure. *Toxicol Appl Pharmacol* 131:309–315.

Holmes KL and Morse HC. 1988. Murine hematopoietic cell surface antigen expression. *Immunology Today* 11:344–350.

Hooghe RJ, Maisin JR, Vander Plaetse F, Urbain J, and Urbain-Vansanten G. 1988. The effect of prenatal or early postnatal irradiation on the production of anti-arsonate antibodies and cross-reactive idiotypes. *International J Radiat Biol* 53:153–157.

Ishigami N, Shinozuka J, Katayama K, Uetsuka K, Nakayama H, and Doi K. 1999. Apoptosis in developing mouse embryos from T-2 toxin inoculated dams. *Histol Histopathol* 14:729–733.

Islam Z, Nagase M, Yoshizawa T, Yamauchi K, and Sakato N. 1998. T-2 induces thymic apoptosis *in vivo* in mice. *Toxicol App Pharmacol* 148: 205–214.

Islam Z, King LE, Fraker PJ, and Pestka JJ. 2003. Differential induction of glucocorticoid-dependent apoptosis in murine lymphoid subpopulations *in vivo* following coexposure to lipopolysaccharide and vomitoxin (deoxynivalenol). *Toxicol Appl Pharmacol* 187:69–79.

Jerrells TR, Marietta CA, Eckardt MJ, Majchrowicz E, and Weight FF. 1986. Effects of ethanol administration on parameters of immunocompetency in rats. *J Leukocyte Biol* 39:499–510.

Johnson S, Knoght R, Marmer DJ, and Steele RW. 1981. Immune deficiency in fetal alcohol syndrome. *Pediatric Res* 15:908–911.

Junaid M, Murthy RC, and Saxena DK. 1996a. Embryo- and fetotoxicity of chromium in pregestationally exposed mice. *Bull Environ Contam Toxicol* 57:327–34.

Junaid M, Murthy RC, and Saxena DK. 1996b. Embryotoxicity of orally administered chromium in mice: exposure during the period of organogenesis. *Toxicol Lett* 84:143–8.

Kalland T, Forsberg TM, and Forsberg MG. 1978. Effect of estrogen and corticosterone on the lymphoid system in neonatal mice. *Exp Mol Pathol* 28:76–95.

Kalland T and Forsberg JG. 1981. Natural killer cell activity and tumor susceptibility in female mice treated neonatally with diethylstilbestrol. *Cancer Res* 41:5134–5141.

Kalland T. 1982. Long-term effects on the immune system of an early life exposure to diethylstilbestrol. In: Hunt VR, Smith MK, Worth D, eds. *Environmental Factors in Human Growth and Development*. Cold Spring Harbor Lab, p.217.

Kanojia RK, Junaid M, and Murthy RC. 1998. Embryo and fetotoxicity of hexavalent chromium: a long-term study. *Toxicol Lett* 95:165–72.

Karakaya A, Yucesoy B, and Sardas OS. 1994. An immunological study on workers occupationally exposed to cadmium. *Hum Exp Toxicol* 13:73–75.

Kincade PW, Lee GL, Watanabe T, Sun L, and Scheid M. 1981. Antigens displayed on murine B lymphocytic precursors. *J Immunol* 127:2262–2268.

Kostial K, Blanusa M, Schonwald N, Arezina R, Piasek M, Jones MM, and Singh PK. 1993. Organ cadmium deposits in orally exposed female rats and their pups and the depleting efficiency of sodium N-(4-methoxybenzyl)-D-glucamine-N-carbodithioate monohydrate (MeOBDCG). *J Appl Toxicol* 13:203–207.

Kutzman RS, Drew RT, Shiotsuka RN, and Cockrell BY. 1986. Pulmonary changes resulting from subchronic exposure to cadmium chloride aerosol. *J Toxicol Environ Health* 17:175–189.

Lafarge-Frayssinet C, Chakor K, and Frayssinet C. 1990. Transplacental transfer of T-2 toxin: Pathological effect. *J Environ Pathol Toxicol Oncol* 10:64–68.

Lai ZW, Fiore NC, Gasiewicz TA, and Silverstone AE. 1998. 2,3,7,8-Tetrachlorodibenzeno-p-dioxin and Diethylstilbestrol affects thymocytes at different stages of development in fetal thymus organ culture. *Toxicol Appl Pharmacol* 149:167–177.

Le Douarin NM, Dieterlen-Lievre F, and Oliver PD. 1984. Ontogeny of primary lymphoid organs and lymphoid stem cells. *Am J Anat* 170:261–299.

Luster MI, Faith RE, and McLachlan JA. 1978. Modulation of the antibody response following *in utero* exposure to diethylstilbestrol. *Bull Environ Contam Toxicol* 20:433–437.

Luster MI, Faith RE, McLachlan JA, and Clark GC. 1979. Effect of *in utero* exposure to diethylstilbestrol on the immune response in mice. *Toxicol Appl Pharmacol* 47:287–296.

Luster MI, Munson AE, Thomas PT, Holsapple MP, Fenters JD, White KL, Lauer LD, Germolec DR, Rosenthal GJ, and Dean JH. 1988. Development of a testing battery to assess chemical-induced immunotoxicity: National Toxicology Program's guidelines for immunotoxicity evaluation in mice. *Fundam Appl Toxicol* 10:2–19.

Maisin J, Dunjic A, and Maisin JR. 1971. Lymphatic system and thymus. In: Berdjis CC, ed. *Pathology and Irradiation*. Baltimore, MD: Williams and Wilkins Co., p. 496.

Miller GK and Benjamin SA. 1985. Radiation-induced quantitative alterations in prenatal thymic development in the beagle dog. *Lab Invest* 52:224–231.

Monjan AA and Mandell W. 1980. Fetal alcohol and immunity: Depression of mitogen-induced lymphocyte blastogenesis. *Neurobehav Toxicol* 2:213–215.

Moon Y and Pestka JJ. 2003. Cyclooxygenase-2 mediates interleukin-6 upregulation by vomitoxin (deoxynivalenol) *in vitro* and *in vivo*. *Toxicol Appl Pharmacol* 187:80–88.

Moscatello KM, Biber KL, Jennings SR, Chervenak R, and Wolcott RM. 1999. Effects of *in utero* alcohol exposure on B cell development in neonatal spleen and bone marrow. *Cellular Immunol* 191:124–130.

Nagata T, Suzuki H, Ishigami N, Shinozuka J, Uetsuka K, Nakayama H, and Doi K. 2001. Development of apoptosis and changes in lymphocyte subsets in thymus, mesenteric lymph nodes and Peyer's patches of mice orally inoculated with T-2 toxin. *Exp Toxicol Pathol* 53:309–15.

Nagymajtenyi L, Schulz H, and Desi I. 1997. Behavioural and functional neurotoxicological changes caused by cadmium in a three-generational study in rats. *Hum Exp Toxicol* 16:691–9.

Niyo KA, Richard JL, Niyo Y, and Tiffany LH. 1980. Pathologic, hematologic, and serologic changes in rabbits given T-2 mycotoxin orally and exposed to aerosols of *Aspergillus fumigatus conidia*. *Am J Vet Res* 49:2151–2160.

Nold JB, Miller GK, and Benjamin SA. 1987. Prenatal and neonatal irradiation in dogs: hematologic and hematopoietic responses. *Radiation Res* 112:490–499.

Noller KL, Blair PB, O'Brien PC, Melton LJ, Offord JR, Kaufman RH, and Colton T. 1988. Increased occurrence of autoimmune disease among women exposed *in utero* to diethylstilbestrol. *Fertil Steril* 49:1080–1087.

Owen JJT. 1972. The origins and development of lymphocyte populations. In: Porter R and Knight J (Eds.), *Ontogeny of Acquired Immunity, A Ciba Foundation Symposium*. New York: Elsevier, pp. 35–54.

Paul J, Conkie D, and Freshney RI. 1969. Erythropoietic cell population changes during the hepatic phase of erythropoiesis in the foetal mouse. *Cell Tissue Kinet* 2:283–294.

Platteau B, Bazin H, Janowski M, and Hooghe R. 1989. Failure to detect immune deficiency in rats after prenatal or early postnatal irradiation. *Int J Radiat Bio* 55:7–14.

Ptak W and Skowron-Cendrzak A. 1977. Fetal suppressor cells: their influence on the cell-mediated immune responses. *Transplantation* 24:45–53.

Qureshi MA, Brake J, Hamilton PB, Hagler WM Jr, and Nesheim S. 1998. Dietary exposure of broiler breeders to aflatoxin results in immune dysfunction in progeny chicks. *Poult Sci* 77:812–9.

Reddy MM, Goh K, and Hempelmann LH. 1976. B and T lymphocytes in man. I. Effect of infant thymic irradiation on the circulating B and T lymphocytes. In: *Radiation and the Lymphatic System*. Proceedings of the 14th Annual Hanford Symposium, Oak Ridge, Tennessee, ERDA Technical Information Center, pp. 192–196.

Ritz B, Heinrich J, Wjst M, Wichmann E, and Krause C. 1998. Effect of cadmium body burden on immune response of school children. *Arch Environ Health* 53:272–280.

Roberts DW and Chapman JR. 1981. Concepts essential to the assessment of toxicity to the developing immune system. In: Kimmel CA and Buelke-Sam J (Eds.), *Developmental Toxicology*. New York; Raven Press.

Rotter BA, Prelusky DB, and Pestka JJ. 1996. Toxicology of deoxynivalenol (vomitoxin). *J Toxicol Environ Health*. 48:1–34.

Schmidt RR. 1984. Altered development of immunocompetence following prenatal or combined prenatal-postnatal insult: A timely review. *J Am College Toxicol* 3:57–72.

Scott PM, Harwig J, and Blanchfield BJ. 1980. Screening Fusarium strains isolated from overwintered Canadian grains for trichothecene. *Mycopathologia* 72:175–180.

Seelig Jr, Steven WM, and Stewart GL. 1996. Effect of maternal ethanol consumption on the subsequent development of immunity to *Thichinella spiralis* in rat neonates. *Alcoholism: Clin and Exp Res* 20: 514–522.

Seelig LL Jr, Steven WM, and Stewart GL. 1999. Second generational effects of maternal ethanol consumption on immunity to *Trichinella spiralis* in female rats. *Alcohol and Alcoholism* 34:520–528.

Silvotti L, Petterino C, Bonomi A, and Cabassi E. 1997. Immunotoxicological effects on piglets of feeding sows diets containing aflatoxins. *The Vet Rec* 141:469–472.

Steinhausen HC, Nestler V, and Spohr HL. 1982. Development and psychopathology of children with fetal alcohol syndrome. *Develop Behav Pediatr* 3:49–54.

Takada A and Takada Y. 1973. Proliferation of donor marrow and thymus cells in the myeloid and lymphoid organs of irradiated syngeneic host mice. *J Exp Med* 137:543–552.

Tavassoli M. 1991. Embryonic and fetal hemopoiesis: An overview. *Blood Cells* 1:269–281.

Theumer MG, Lopez AG, Masih DT, Chulze SN, and Rubinstein HR. 2003. Immunobiological effects of AFB1 and AFB1-FB1 mixture in experimental subchronic mycotoxicoses in rats. *Toxicol* 186:159–170.

Thomas PT and Faith RE. 1985. Adult and perinatal immunotoxicity induced by halogenated aromatic hydrocarbons. In: Dean JH, Luster MI, Munson AE, and Amos H (Eds.), *Immunotoxicology and Immunopharmacology*. New York: Raven Press, pp. 305–314.

Thuvander A, Breitholtz-Emanuelsson A, Brabencova D, and Gadhasson I. 1996a. Prenatal exposure of Balb/c mice to ochratoxin A: Effects on the immune system in the offspring. *Food Chem Toxicol* 34:547–554.

Thuvander A, Breitholtz-Emanuelsson A, Hallen I, and Oskarrson A. 1996b. Effects of ochratoxin A on the rat immune system after perinatal exposure. *Nat Toxins* 4:141–147.

Thuvander A, Sundberg J, and Oskarrson A. 1996c. Immunomodulating effect after perinatal exposure to methylmercury in mice. *Toxicol* 114:163–175.

United States Department of Health and Human Services (U.S. DHHS). Agency for Toxic Substances and Disease Registry. 1999a. Toxicological profile for Mercury (Update).

United States Department of Health and Human Services (U.S. DHHS). Agency for Toxic Substances and Disease Registry. 1999b. Toxicological profile for Cadmium (Update).

United States Department of Health and Human Services (U.S. DHHS). Agency for Toxic Substances and Disease Registry. 2000. Toxicological profile for Chromium (Update).

Urso P and Gengozian N. 1982. Alterations in the humoral immune response and tumor frequencies in mice exposed to benzo(a)pyrene and x-rays before and after birth. *J Toxicol Environ Health* 10:817–835.

Velarde A and Cooper MD. 1984. An immunofluorescence analysis of the ontogeny of myeloid, T, and B lineage cells in mouse hemopoietic tissues. *J Immunol* 133:672–677.

Vos JG and Moore JA. 1974. Suppression of cellular immunity in rats and mice by maternal treatment with 2,3,7,8-tetrachlorodibenzo-*p*-dioxin. *Int Arch Allergy* 47:777–794.

Warner NL, Uhr JW, Thorbecke GJ, and Ovary Z. 1969. Immunoglobulins, antibodies, and the bursa of Fabricius: Induction of agammaglobulinemia and the loss of all antibody-forming capacity by hormonal bursectomy. *J Immunol* 103:1317–1330.

Ways SC and Bern HA. 1979. Long-term effects of neonatal treatment with cortisol and/or estrogen in the female BALB/c mouse. *Proc Soc Exp Biol Med* 160:94–98.

Ways SC, Blair PB, Bern HA, and Staskawicz MO. 1980. Immune responsiveness of adult mice exposed neonatally to DES, steroid hormones, or vitamin A. *J Environ Pathol Toxicol* 3:207–216.

Ways SC, Mortola JF, Zvaifler NJ, Weiss RJ, and Yen SSC. 1987. Alterations in immune responsiveness in women exposed to diethylstilbestrol *in utero*. *Fertil Steril* 48:193–197.

Whitehead ED and Leiter E. 1981. Genital abnormalities and abnormal semen analysis in male patients exposed to diethylstilbestrol *in utero*. *J Urol* 125:47–56.

Wild LG, Ortega HG, Lopez M, and Salvaggio JE. 1997. Immune system alteration in the rat after indirect exposed to methyl mercury chloride or methyl mercury sulfide. *Environ Res* 74:34–42.

Wolcott RM, Jennings SR, and Chervenak R, 1995. *In utero* exposure to ethanol affects postnatal development of T- and B-lymphocytes, but not natural killer cells. *Alcoholism: Clin and Exp Res* 19: 170–176.

# Part IV

# Developmental Immunotoxicant Exposure and Exacerbated Postnatal Immune Responses

CHAPTER 13

# Perinatal Immunotoxicant Exposure and Autoimmune Disease

**Benny L. Blaylock, S. Ansar Ahmed, and Steven D. Holladay**

## CONTENTS



## INTRODUCTION

Immunologists have long recognized that the immune system may react against autologous antigens and cause tissue injury. The term *horror autotoxicus* was coined by Paul Ehrlich in the early 1900s to describe the harmful, or "toxic," immune reactions against self. During maturation, lymphocytes expressing autoreactive receptors are produced. Normally, there are several mechanisms that either eliminate or inactivate these cells, producing self-tolerance. Abnormal selection processes or altered regulation of self-reactive lymphocytes may result in the loss of this self-tolerance, resulting in autoimmune disease. A number of mechanisms may be responsible for altering the selection or regulation processes, including exposure to toxic chemicals.

Much of the selection process for removal of autoreactive lymphocytes occurs during prenatal immune system ontogenesis, requiring a sequential series of carefully timed and coordinated developmental events beginning early in embryonic life.

0-415-28457-0/05/$0.00+$1.50

Using experimental animals, observations have been made that developing organisms are often more sensitive to toxic effects from exposure to physical or chemical agents than the fully mature individual, and the consequences are often more persistent (Ford et al. 1983; Ways et al. 1987; Noller et al. 1988; Fine et al. 1989; Schlumpf et al. 1989; Holladay et al. 1991). Although considerable research efforts have been devoted to identification and mechanistic investigation of altered postnatal immunocompetence, including decreased resistance to infectious disease or neoplasia after perinatal xenobiotic exposure and to the development and validation of sensitive tests and screening procedures for detecting developmental immunotoxicants, the possibility that developmental chemical exposure may be related to increased incidence or severity of autoimmune responses has received limited attention.

## CHEMICAL TARGETING OF IMMUNE DEVELOPMENT

A growing literature base indicates that laboratory animals exposed *in utero* (during immune system development) to immunotoxic chemicals may result in severe effects on the immune system that last into adult life. For example, mice exposed to the organochlorine insecticide chlordane *in utero* have been found to display reduced numbers of granulocyte-macrophage colony-forming units (GM-CFU) and colony-forming units/spleen (CFU-S) as early as gestational day (GD) 18 and the suppression extended to 200 days of age (Barnett et al. 1990). Previous studies with *in utero* chlordane exposure demonstrated long-term depression of the contact hypersensitivity response to oxazolone in mice at 101 days of age (Spyher-Cranmer et al. 1982). It is noteworthy that these immune effects are either reduced or not observed in adult mice exposed to chlordane at dose levels equal to those given to the pregnant mice (Johnson et al. 1986). Similar results have been reported in mice with *in utero* exposure to benzo[a]pyrene (B[a]P), a polycyclic aromatic hydrocarbon (PAH). Offspring of pregnant mice treated with B[a]P have been found to display depressed antibody, graft-vs.-host, and mixed lymphocyte responses at 18 months of age (Urso and Gengozian 1984). These mice further exhibited an 8-to10-fold higher tumor incidence than control mice who did not experience *in utero* B[a]P exposure.

Low-level prenatal exposure to some of the halogenated aromatic hydrocarbons (HAH), notably dioxins, also results in severe, long-lasting immunologic incompetence in rodents (Ball et al. 1969; Faith and Moore 1977; Poland et al. 1982; Dencker et al. 1985; Corrier and Ziprin 1987; Fine et al. 1989). Collectively, these reports demonstrate that prenatal exposure to certain immunotoxic compounds may alter fetal development of immunity in mice, causing severe and sustained postnatal immunosuppression in the absence of overt toxicity. Additional agents producing developmental immunotoxicity in rodents are diverse and include: PAH other than B[a]P (e.g., 7,12-dimethylbenzanthracene; 3-methylcholanthrene); pesticides other than chlordane (e.g., hexachlorocyclohexane; DDT); polycyclic halogenated hydrocarbons such as 2,3,7,8-tetrachlorodibenzo-*p*-dioxin (TCDD); heavy metals (cadmium; mercury); hormonal substances (diethylstilbestrol [DES], testosterone, cortisone); mycotoxins (most notably T2 toxin); and therapeutic agents (acyclovir, cyclophosphamide) (reviewed in: Holladay and Luster 1994). A target of several of these agents is the fetal thymus, inducing thymic atrophy as well as altered differentiation of fetal thymocytes. Exposure of fetal thymocytes to such chemicals during the maturation process could alter or modulate critical periods of "self-learning," resulting in potentially detrimental consequences on immune function in postnatal life, including possible expression of autoimmunity (Silverstone et al. 1998).

Most available data suggesting a link between induction of autoimmune reactivity and prenatal chemical exposure concerns estrogenic chemicals such as DES, which produce altered prenatal hormonal environment and which directly target developing immune cells, and HAH such as TCDD. Cyclosporin A (CsA), a widely used therapeutic immunosuppressant, has been shown to produce a T cell-mediated autoimmunity in rodents by effects on T cell development similar to those caused by TCDD, and will also be discussed.

## PERINATAL ESTROGEN EXPOSURE: DIETHYLSTILBESTROL

Estrogenic steroids exert regulatory actions on immune function in adult animals that are well documented but not well understood mechanistically. Autoimmune disease may be caused, at least in part, by endogenous estrogen, as has been observed with increased serum estradiol levels (e.g., during pregnancy) and infections resulting from depression of cell-mediated immunity (Hamilton and Hellstrom 1977; Mathur et al. 1979). Pharmacologic or suprapharmacologic levels of steroidal and nonsteroidal estrogenic compounds have been shown to further result in numerous alterations of immune function, particularly when administered perinatally during lymphoid organ organogenesis. Effects of such administration in rodents include myelotoxicity (Fried et al. 1974; Boorman et al. 1980), suppression of cell-mediated immunity (Kalland et al. 1978; Ways and Bern 1979), significant thymic atrophy (Greenman et al. 1977; Aboussaouira et al. 1991), suppression of natural killer (NK) cell activity (Seaman et al. 1979; Kalland 1980), and reticuloendothelial system stimulation (Kicol et al. 1964; Ford et al. 1983).

DES, a synthetic nonsteroidal compound possessing estrogenic activity, may have possible adverse effects on the postnatal human immune system after *in utero* exposure in women. As an example, altered NK and T lymphocyte function in women exposed to DES *in utero* has been reported (Kalland and Fossberg 1981; Ways et al. 1987). In addition, Noller et al. (1988) have demonstrated an increased incidence of autoimmune diseases in women exposed *in utero* to DES. While it appears that estrogens mediate certain of their immune effects at the thymic level by altering thymic epithelium-dependent mechanisms (Grossman and Roselli 1983), little is understood about mechanisms by which estrogenic chemicals may influence immune responses to foreign or self-antigens.

An increased risk of developing autoimmune disease in mice has been associated with altered prenatal hormonal environment (Walker et al. 1996). It has also been suggested that humans exposed *in utero* to DES may display a hyperreactive immune response (Anderson et al. 1982). A retrospective study of DES-exposed (1711 individuals) and unexposed (922 individuals) cohorts examined the possibility that prenatal DES may affect the prevalence of autoimmune disease and found a positive correlation when autoimmune diseases were grouped (Noller et al. 1988). Specifically, the overall frequency of any autoimmune disease among exposed women was 28.6 per 1000 compared to 16.3 per 1000 among the controls (significantly different at $p$=0.02). Autoimmune diseases evaluated included systemic lupus erythematosus, scleroderma, Grave's disease, Hashimoto's thyroiditis, pernicious anemia, myasthenia gravis, thrombocytopenic purpura, rheumatoid arthritis, regional enteritis, chronic ulcerative colitis, multiple sclerosis, chronic lymphocytic thyroiditis, Reiter's syndrome, and optic neuritis. However, only Hashimoto's thyroiditis occurred significantly more often in the exposed women ($p$=0.04) when these autoimmune diseases were considered individually. A similar evaluation of 1173 humans exposed to DES during development (1079 daughters and 94 sons) found increased rates of asthma, arthritis, and diabetes mellitus compared to the general population (Wingard and Turiel 1988). In a more recent study by Baird et al. (1996) evaluating rates of allergy, infection, and autoimmune disease in DES-exposed sons and daughters (253 men and 296 women) matched with similar unexposed individuals (241 men and 246 women), no differences in disease occurrence were detected. These authors concluded that a larger sample was needed to evaluate DES-associated risk of autoimmunity since autoimmune diseases are relatively rare in the human population.

Data from humans in the preliminary studies above suggest the possibility that exposure to DES *in utero* may result in postnatal immune alterations, including increased autoimmune disease. However, the difficulty with continued surveillance of DES-exposed sons and daughters required for more definitive statements will become more difficult as this cohort ages and members are lost. This makes laboratory rodent studies important to determine if prenatal exposure to chemicals such as DES may predispose an individual to postnatal autoimmunity via alteration of development of immune cells. Animal models will also be necessary to answer questions regarding specific immune cell targets and mechanisms of action resulting from such exposures.

Hybrid B6C3F1 (C57Bl/6N x C3H) mice exposed to 8 mg/kg/day DES from GD 10 to16 displayed significant thymic hypocellularity in late gestation, as well as limited but significant inhibition of thymocyte maturation (Table 13.1) (Holladay et al. 1993). In these studies, the observed thymic involution was related to a reduction of fetal liver prothymocytes by DES.These fetal liver prothymocytes are responsible for colonizing the fetal thymus. These authors also reported that fetal liver prothymocytes expressed estrogen receptors at about 290 fmol/100 mg DNA, a level approximately 50% of that found in the uterus and sufficient to suggest an estrogen-responsive cell. These and other reports indicate the developing mouse immune system is sensitive to estrogen exposure, and that such exposure may contribute to postnatal immunosuppression (Holladay et al. 1994). However, similar to most of the well-established rodent developmental immunotoxicants, very limited information is available addressing possible relationships between gestational DES exposure and altered expression of postnatal autoimmune disease.

Silverstone et al. reported that a single fetal exposure to DES in $SNF_1$ mice induced autoimmune lupus-like nephritis in male offspring between 5 and 10 months of age (1998). Female $SNF_1$ mice develop this autoimmune syndrome spontaneously

**Table 13.1 Fetal Thymocyte CD Surface Marker Expression and Cellularity after DES Treatment**

| CD Marker Expression (% Positive) | Vehicle | DES | |
|---|---|---|---|
| | | 3 µg/kg | 8 µg/kg |
| $CD4^{+}8^{-}$ | 4.1 ±0.2 | 3.2 ± 0.4 | 4.0 ± 0.4 |
| $CD4^{-}8^{+}$ | 2.7 ± 0.2 | 4.0 ± 0.3[a] | 3.3 ± 0.3 |
| $CD4^{+}8^{+}$ | 71.4 ± 0.7 | 64.0 ± 2.5[a] | 63.6 ± 1.6[a] |
| $CD4^{-}8^{-}$ | 22.0 ± 0.7 | 28.8 ± 2.5[a] | 29.2 ± 1.5[a] |
| Cellularity ($\times 10^6$) | 40.4 ± 2.6 | 18.9 ± 1.9[a] | 8.7 ± 1.9[a] |

*Note:* CD4 and CD8 surface antigen expression determined in GD 18 fetal mice after maternal exposure to 3 or 8 µg/kg/day DES from GD 10 to 16. Values represent mean ± SEM of 5 mice per treatment group.

[a] $p < 0.05$ vs. vehicle controls.

*Source:* Modified from Holladay et al. 1993.

in their first year of life. The male mice, however, typically do not display significant autoimmunity before one year of age. These data suggest that pharmacologic exposure to DES may contribute to early expression of autoimmunity in genetically predisposed mice, and that autoimmune rodent models such as the $SNF_1$ model may prove valuable for identification of biological markers for human risk assessment. Because the etiology of autoimmune diseases is generally considered to be multifactorial (genetic, environmental, hormonal, infectious) (Talal 1993), clearly, continued research will be required to determine potential relationships between prenatal estrogen exposure and postnatal development of autoimmune disease.

## ENDOCRINE-DISRUPTING COMPOUNDS

A current public health issue concerns the fact that endocrine-disrupting chemicals (EDCs) may adversely affect human health in both children and adults (Colborn et al. 1993; Kavlock 1999; Landrigan et al. 2003; Longnecker et al. 2003; Rice et al. 2003). Recent experimental data suggest that EDCs affect not only the reproductive system but also the immune system, thereby raising additional concerns (Barnett and Rodgers 1994; Blake et al. 1997; Ahmed 2000). Fetal exposure to EDCs is of particular concern, since these chemicals could potentially alter critical immune developmental pathways, which may have long-lasting consequences. The development of the fetal immune system is tightly regulated to favor the selection of competent immune cells and to eliminate dangerous autoreactive lymphocytes. Any alterations in the normal fetal immune development (e.g., exposure to EDCs) may have permanent or long-lasting adverse consequences.

The synthetic estrogen DES can cross placenta, is considered a model EDC, and serves as an excellent prototype to study the immunological consequences of prenatal exposure to EDC since both clinical and experimental data are available. It is estimated that 3 to 5 million Americans have been exposed to DES during fetal life (Giusti et al. 1995). As indicated in the preceding section, this cohort, after sexual maturity, may be at increased risk for developing a number of maladies, including immune abnormalities. Ahmed et al. (1999, 2000) have recently proposed that individuals prenatally exposed to EDCs, including DES, may have a deviant immunological response when exposed to estrogenic compounds during adult life. In support of this hypothesis, these authors found that activated splenocytes from prenatal DES-exposed mice secrete augmented levels of IFNγ when exposed to a second dose of DES during adult life (as late as 1 to 1.5 years of age). This suggests that prenatal DES exposure may preprogram the highly sensitive fetal immune system for augmented IFNγ secretion when exposed to a second dose of EDC later in life. This is termed "EDC-induced immunological imprinting" (Karpuzoglu-Sahin et al. 2001). Understanding this alteration of IFNγ secretion by splenocytes from mice that received both prenatal and adult-life DES exposure (DES/DES mice) is crucial, since IFNγ is a key T helper-1 (Th-1) cytokine that regulates the functions of all cells of the immune system, and plays a major role in autoimmune diseases and in infection and tumor immunity. These results imply that fetal exposure to other immunomodulatory compounds may preprogram these individuals to respond abnormally to subsequent immunomodulatory endocrine disruptors during adult life. These studies are therefore likely to be applicable to other EDCs and other immune-altering chemicals and thus have broad health implications.

## TCDD: THE PROTOTYPIC HALOGENATED AROMATIC HYDROCARBON

The halogenated aromatic hydrocarbon most studied for suppressive effects on the immune system is undoubtedly TCDD. However, other studies indicate that TCDD may also induce changes in immune cells and immune support cells that may potentiate development of autoimmune diseases. Greenlee et al. (1985) and Schuurman et al. (1992) described thymic epithelium as targets of TCDD. This suggests that TCDD may have the potential to alter critical epithelium-dependent selective events in the thymus that could lead to decreased or defective negative selection of developing thymocytes expressing autoreactive T cell receptors (TCR). DeWaal et al. (1992) further observed altered thymic epithelial distribution of major histocompatability complex (MHC) class II molecules in TCDD-treated mice. This effect was hypothesized as having potential to cause defective thymocyte-epithelial cell interactions. Specifically, MHC class I and class II molecules act as thymic self-antigen presenting molecules in a process whereby thymocytes expressing TCRs with high affinity to self-antigen are eliminated ("negative selection"). It has also been noted that similar patterns of inhibited thymic T cell differentiation occur spontaneously in autoimmune mice (Kakkanaiah et al. 1990), in TCDD-treated mice (Blaylock et al. 1992), and in mice treated *in vivo* with monoclonal antibodies to MHC class I and class II molecules (Kruisbeek et al. 1985), suggesting the importance of these MHC molecules in thymocyte differentiation. In this regard, TCDD was recently found to down-regulate expression of a MHC class I gene (*$Q1^b$*) in a mouse hepatoma cell line (Dong et al. 1997). Specifically, these authors observed that MHC *$Q1^b$* cDNA encoded for the α3 domain and transmembrane domain of the *$Q1^b$* class I protein, implying that the MHC gene product could interact with β2-

microglobulin. These observations led to the hypothesis that the MHC $Q1^b$ molecule down-regulated by TCDD may function in antigen presentation (Dong et al. 1997).

The above effects of TCDD on thymocytes, thymic epithelium, and MHC molecules associated with antigen presentation, taken together, raise the possibility that TCDD has the ability to alter normal T cell self-tolerance development in a way that may increase autoimmunity. Fan et al. observed that male rats injected with TCDD elicited a positive response in the popliteal lymph node (PLN) assay (Fan et al. 1995). The PLN has been proposed by some as a tool to predict autoimmune reactions induced by chemicals (Descotes 1992). Based on the observation of a positive PLN assay, these authors joined others suggesting that TCDD may have the potential to induce or exacerbate autoimmune-like reactions.

Based on the potential that TCDD can induce an altered intrathymic negative selection of potential autoreactive T cells and that this may result in increased release of such cells to the periphery, other investigators have looked for emergence of T cells carrying TCR-variable regions associated with self-reactivity in TCDD-treated animals. The TCR variable β (Vβ) chains are usually deleted in the thymus by reaction with self-MHC and minor lymphocyte stimulatory antigens (Okuyama et al. 1992; Hanawa et al. 1993) and have been associated with autoimmunity in some experimental mouse models (Rocha et al. 1992). Therefore, deHeer et al. (1995) examined the thymus, spleen, and mesenteric lymph nodes of adult mice dosed with TCDD for autoreactive mature $V\beta6^+$ T cells, and were not able to demonstrate emergence of such cells. However, in related studies, Silverstone et al. (1994) found that both TCDD and estradiol induce extrathymic T cell differentiation in the liver of young adult mice, and that such extrathymic cells expressed elevated levels of $V\beta^+$ T cell receptors. Such an increase in T cells associated with autoreactivity has been suggested as a mechanism by which estrogen may promote autoimmunity (Okuyama et al. 1992). Silverstone et al. (1994) have similarly suggested that these findings with TCDD (increased extrathymic autoreactive T cells) may relate to the ability of this HAH to induce autoimmunity.

### Common Effects of TCDD and CsA on T Cell Development

The developing immune system has been demonstrated to have a high sensitivity to TCDD. This, coupled with evidence that increased autoimmune disease may result following TCDD exposure in adult animals, has raised questions regarding possible relationships between prenatal exposure to TCDD and increased postnatal autoimmunity. Low-level maternal TCDD exposure results in atrophy of the developing thymus as well as inhibition of thymocyte differentiation (Table 13.2). T-progenitor cells from the fetal liver seed the developing thymus and are initially double negative (DN) with respect to CD4 and CD8 surface antigens. Subsequently, thymocytes develop sequentially through immature $CD8^{lo}$ and $CD4^+8^+$ double positive stages in the thymic cortex, to mature $CD4^+$ SP or $CD8^+$ SP thymocytes in the thymic medulla by GD 18 to 19 (Hussman et al. 1988; Penit and Vaddeur 1989). TCDD produces a significant maturational delay in fetal thymocyte development as evidenced by these CD4 and CD8 surface antigens, which has been described as similar to the maturational inhibition produced in fetal thymic organ culture by the therapeutic

**Table 13.2 Fetal Thymocyte CD Surface Marker Expression and Thymus Weight after TCDD Treatment**

| CD Marker Expression (% Positive) | Vehicle | TCDD 1.5 μg/kg | TCDD 3 μg/kg |
|---|---|---|---|
| $CD4^+8^-$ | 1.8 ± 0.2 | 1.5 ± 0.1 | 2.0 ± 0.2 |
| $CD4^-8^+$ | 8.1 ± 0.7 | 15.5 ± 0.9* | 17.5 ± 0.9[a] |
| $CD4^+8^+$ | 69.1 ± 1.2 | 52.6 ± 2.5* | 43.2 ± 4.5[a] |
| $CD4^-8^-$ | 21.1 ± 0.7 | 30.3 ± 1.8* | 37.3 ± 3.7[a] |
| Thymus wt. (%) Body wt. | 0.24 | 0.14* | 0.10[a] |

*Note:* CD4 and CD8 surface antigen expression determined in GD 18 fetal mice after maternal exposure to 1.5 or 3.0 μg/kg/day TCDD from GD 6 to 14. Values represent mean ± SEM of 5 mice per treatment group.

[a] $p < 0.05$ vs. vehicle controls.

*Source*: Modified from Holladay et al. 1991.

immunosuppressive drug cyclosporin A (CsA) (Blaylock et al. 1992). Fetal thymic organ culture was used rather than *in vivo* exposure to study the effect of CsA on developing T cells because CsA crosses the placenta poorly (Nandakumaran and Eldeen 1990). Fetal mouse exposure to TCDD or *ex vivo* exposure of fetal mouse thymi to CsA decreased the percentage of DP cells (the most mature phenotype present in significant numbers in the end-gestation mouse fetus) and increased the percentage of both DN cells and immature (i.e., $TCR^-$) $CD8^{lo}$ thymocytes. Positive and negative selection occur in the developing thymus and any intrinsic (hormonal) or extrinsic (chemical) insult on thymocyte maturation during critical periods of thymocyte selection for induction of self-tolerance may have significant and detrimental consequences on immune function in postnatal life (Billingham 1966); thus, this pattern of inhibition by TCDD or CsA raises questions about interference with neonatal development of tolerance by these chemicals.

MHC class II antigen expression in the thymus is reduced in rodents by CsA (Kosugi et al. 1989), an effect associated with development of autoimmune disease in Lewis rats exposed to CsA following lethal irradiation and syngeneic bone marrow reconstitution (Hess et al. 1985). Specifically, these studies demonstrated a T cell-mediated autoimmune disease in the rats treated with CsA after bone marrow transplant, manifested as a syngeneic graft-vs.-host response (SGVHR). Briefly, the development of a chronic GVH-like disease, typical of that seen in rodents or humans following allogeneic marrow transplantation, was observed in the CsA-exposed rodents who had received syngeneic bone marrow transplants. This observation of a SGVHR was in fact the demonstration of an immune system rejecting "genetic self" (i.e., autoimmunity), indicating that CsA produced a fundamental disruption in development of self-tolerance. These authors went on to demonstrate that the CsA-induced autoreactivity was transferred with the $CD8^+$ subpopulation of T cells, suggesting that CsA interfered with deletion of these cells during the establishment of a new immune system in the irradiated animals (Hess et al. 1990). More recently, human patients have been observed to develop SGVHRs after autologous hematopoietic stem cell transplantation and CsA treatment, indicating a similar altered pattern of T cell selection by CsA during reestablishment of the human immune system (Baron et al. 2000).

Relatively few papers have evaluated potential long-term immunologic alterations in offspring of human mothers who received therapeutic immune suppression with CsA during pregnancy. The papers that do exist are difficult to interpret in terms of effects of individual immunosuppressive drugs, since the pregnant women were most often exposed to drug combinations. It is clear that the developing human immune system can be at least transiently affected by maternal immunosuppressive therapies. For instance, numbers of B and mononuclear cells were significantly lower in neonates from transplant-recipient mothers who received CsA, azathioprine (AZA), and methylprednisolone during pregnancy (Takahashi et al. 1994). Ersay et al. (1995) similarly reported reduced B cell numbers and depressed serum IgG values at 3 and 6 months in infants whose mothers received CsA, AZA, and prednisolone during pregnancy. In one clinical report, the 23-year-old daughter of a renal transplant recipient developed multiple autoantibodies during her first pregnancy, while her second pregnancy was complicated by systemic lupus erythematosus (Scott et al. 2002). It was uncertain whether these autoimmune manifestations were related to fetal exposure to immunosuppressive drugs. Lytton et al. (2002) also recently observed autoantibodies against cytochrome P450 enzymes (e.g., CYP3A4, CYP3A5) in sera of children treated with immunosuppressive drug regimens that included CsA. For reasons such as these, other authors have suggested that long-term follow-up for autoimmune disease is warranted in individuals exposed *in utero* to immunosuppressive drugs (Prevot et al. 2002). This subject is covered in more detail in Chapter 11 of this textbook, "Developmental Immunotoxicity of Therapeutic Agents." The heavy metal lead is an additional developmental immunotoxicant that may increase risk of autoimmune responses following *in utero* exposure; this subject is covered in detail in Chapter 10 of this textbook, "Toxicity of Lead to the Developing Immune System."

## REFERENCES

Aboussaouira T, Marie C, Brugal G, Idelman S. 1991. Inhibitory effect of 17b-estradiol on thymocyte proliferation and metabolic activity in young rats. *Thymus* 17:167–173.

Ahmed SA, Hissong BD, Verthelyi D, Donner K, Becker K, and Karpuzoglu-Sahin E. 1999. Gender and risk of autoimmune diseases: possible role of estrogenic compounds. *Environ Health Perspect* 107 Suppl 5, 681–6.

Ahmed SA. 2000. The immune system as a potential target for environmental estrogens (endocrine disrupters): a new emerging field. *Toxicol* 150, 191–206.

Anderson RE, Tokuda S, Williams WL, and Warner NL. 1982. Radiation-induced augmentation of the response of A/J mice to SaI tumor cells. *Am J Pathol* 108:24–33.

Baird DD, Wilcox AJ, and Herbst AL. 1996. Self-reported allergy, infection, and autoimmune diseases among men and women exposed *in utero* to diethylstilbestrol. *J Clin Epidemiol* 49:263–266.

Ball J and Dawson D. 1969. Biological effects of neonatal injection of 7,12- dimethylbenz(a)anthracene. *J Natl Cancer Inst* 42:579–591.

Baron Fl, Gothot A, Salmon JP, Hermanne JP, Pierard GE, Fillet G, and Beguin Y. 2000. Clinical course and predictive factors for cyclosporin-induced autologous graft-versus-host disease after autologous haematopoietic stem cell transplantation. *Br J Haematol* 111:745–753.

Barnett JB, Blaylock BL, Gandy J, Menna JH, Denton R, and Soderberg LSF. 1990. Alteration of fetal liver colony formation by prenatal chlordane exposure. *Fundam Appl Toxicol* 15:820–822.

Barnett JB, Blaylock BL, Gandy J, Menna JH, Denton R, and Soderberg LSF. 1990. Long-term alteration of adult bone marrow colony formation by prenatal chlordane exposure. *Fundam Appl Toxicol* 14:688–695.

Barnett JB and Rodgers KE. 1994. In Pesticides. Dean JH, Luster MI, Munson AE, and Kimber I (Eds.), *Immunotoxicology and Immunopharmacology*. New York: Raven Press, pp. 191–212.

Billingham RE. 1966. The biology of graft-versus-host reactions. *Harvey Lect* 62:21–28.

Blake CA, Nair-Menon JU, and Campbell GT. 1997. Estrogen can protect splenocytes from the toxic effects of the environmental pollutant 4-tert-octylphenol. *Endocrine* 6, 243–9.

Blaylock BL, Holladay SD, Comment CE, Heindel JJ, and Luster MI. 1992. Modulation of perinatal thymocyte surface antigen expression and inhibition of thymocyte maturation by exposure to 2,3,7,8-tetrachlorodibenzo-*p*-dioxin (TCDD). *Toxicol Appl Pharmacol* 112:207–213.

Boorman GA, Luster MI, Dean JH, and Wilson RE. 1980. The effect of adult exposure to diethylstilbestrol in the mouse on macrophage function. *J Reticuloendothel Soc* 28: 547–555.

Colborn T, vom Saal FS, and Soto AM. 1993. Developmental effects of endocrine-disrupting chemicals in wildlife and humans. *Environ Health Perspect* 101, 378–84.

Corrier DE and Ziprin RL. 1987. Immunotoxic effects of T-2 toxin on cell-mediated resistance to *Listeria monocytogenes*. *Vet Immunol Immunopathol* 14:11–21.

Dencker L, Hassoun E, d'Argy R, and Alm G. 1985. Fetal thymus organ culture as an *in vitro* model for the toxicity of 2,3,7,8-tetrachlorodibenzo-*p*-dioxin and its congeners. *Mol Pharmacol* 27:133–140.

Descotes J. 1992. The popliteal lymph node assay: A tool for studying the mechanisms of drug-induced autoimmune disorders. *Toxicol Lett* 64/65:101–107.

DeHeer C, van Driesten G, Schuurman H-J, Rozing J, and van Loveren H. 1995. No evidence for emergence of autoreactive Vb6$^+$ T cells in Mls-1$^a$ mice following exposure to a thymotoxic dose of 2,3,7,8-tetrachlorodibenzo-*p*-dioxin. *Toxicol* 103:195–203.

DeWaal EJ, Schuurman H-J, Loeber JG, Van Loveren H, and Vos JG. 1992. Alterations in the cortical thymic epithelium of rats after *in vivo* exposure to 2,3,7,8-tetrachlorodibenzo-*p*-dioxin (TCDD): An (immuno)histological study. *Toxicol Appl Pharmacol* 115:80–88.

Dong L, Ma Q, and Whitlock JP. 1997. Down-regulation of major histocompatibility complex Q1b gene expression by 2,3,7,8-tetrachlorodibenzo-*p*-dioxin. *J Biol Chem* 272:29614–29619.

Ersay A, Oygur N, Coskun M, Suleymanlar G, Trak B, and Yegin O. 1995. Immunologic evaluation of a neonate born to an immunosuppressed kidney transplant recipient. *Am J Perinatol* 12:413–415.

Faith RE and Moore JA. 1977. Impairment of thymus-dependent immune functions by exposure of the developing immune system to 2,3,7,8-tetrachlorodibenzo-*p*-dioxin (TCDD). *J Toxic Environ Health* 3:451–464.

Fan F, Pinson DM, and Rozman KK. 1995. Immunoregulatory effect of 2,3,7,8-tetrachlorodibenzo-*p*-dioxin tested by the popliteal lymph node assay. *Toxicol Pathol* 23:513–517.

Fine JS, Gasiewicz TA, and Silverstone AE. 1989. Lymphocyte stem cell alterations following perinatal exposure to 2,3,7,8-tetrachlorodibenzo-*p*-dioxin. *Mol Pharmacol* 35:18–25.

Ford CD, Johnson GH, and Smith WG. 1983. Natural killer cells in *in utero* diethylstilbestrol-exposed patients. *Gynecol Oncol* 16:400–407.

Fried W, Tichler T, Dennenberg I, Barone J, and Wang F. 1974. Effects of estrogens on hematopoiesis of mice. *J Lab Clin Med* 83: 807–815.

Giusti RM, Iwamoto K, and Hatch EE. 1995. Diethylstilbestrol revisited: a review of the long-term health effects. *Ann Intern Med* 122, 778–88.

Greenman DL., Dooley K, Breeden CR. 1977. Strain differences in the response of the mouse to diethylstilbestrol. J Toxicol Environ Health 3: 589–597.

Greenlee WF, Dold KM, Irons RD, and Osborne R. 1985. Evidence for direct action of 2,3,7,8-tetrachlorodibenzo-*p*-dioxin (TCDD) on thymic epithelium. *Toxicol Appl Pharmacol* 79:112–120.

Grossman CJ and Roselli GA. 1983. The interrelationship of the HPG-thymic axis and immune system regulation. *J Steroid Biochem* 119:461–470.

Hamilton MS and Hellstrom I. 1977. Altered immune responses in pregnant mice. *Transplantation* 23: 423–431.

Hanawa H, Tsuchida N, Matsumoto Y, Watanabe H, Abo T, Sekikawa H, Kodama M, Zhang S, Izumi T, and Shibata A. 1993. Characterization of T cells infiltrating the heart in rats with experimental autoimmune myocarditis: Their similarity to extrathymic T cells in mice and the site of proliferation. *J Immunol* 150:5682–5695.

Hess AD, Horwitz L, Beschorner WE, and Santos GW. 1985. Development of a graft-versus-host disease-like syndrome in cyclosporine-treated rats after syngeneic bone marrow transplantation. I. Development of cytotoxic T lymphocytes with apparent polyclonal anti-Ia specificity, including autoreactivity. *J Exp Med* 161:718–724.

Hess AD, Fischer AC, and Beschorner WE. 1990. Effector mechanisms in cyclosporine A-induced syngeneic graft-versus-host disease. Role of $CD4^+$ and $CD8^+$ T lymphocyte subsets. *J Immunol* 145:526–533.

Holladay SD, Lindstrom P, Blaylock BL, Comment CE, Germolec DR, Heindel JJ, and Luster MI. 1991. Perinatal thymocyte antigen expression and postnatal immune development altered by gestational exposure to tetrachlorodibenzo-*p*-dioxin (TCDD). *Teratol* 44:385–393.

Holladay SD, Blaylock BL, Comment CE, Heindel JJ, Fox WM, Korach KS, and Luster MI. 1993. Selective prothymocyte targeting by prenatal diethylstilbestrol exposure. *Cell Immunol* 152:131–142.

Holladay SD and Luster MI. 1994. *Developmental Immunotoxicology.* In: Kimmel C and Buelke-Sam J (Eds.), *Developmental Toxicology,* 2$^{nd}$ ed. New York: Raven Press, pp. 93–118.

Hussman LA, Shimonkevitz RP, Crispe IN, and Bevan MJ. 1988. Thymocyte subpopulations during early fetal development in the BALB/c mouse. *J Immunol* 141:736–740.

Johnson KW, Holsapple MP, and Munson AE. 1986. An immunotoxicological evaluation of gamma-chlordane. *Fundam Appl Toxicol* 6:317–326.

Kalland T, Fossberg T, and Fossberg J. 1978. Effect of estrogen and corticosterone on the lymphoid system in neonatal mice. *Exp Mol Pathol* 28:76–95.

Kakkanaiah VN, Pyle RH, Nagarkatti M, and Nagarkatti PS. 1990. Evidence for major alterations in the thymocyte subpopulations in murine models of autoimmune diseases. *J Autoimmunity* 3:27–288.

Kalland T. 1980. Reduced natural killer activity in female mice after neonatal exposure to diethylstilbestrol. *J Immunol* 124:1297–1306.

Kalland T and Fossberg J. 1981. Natural killer cell activity and tumor susceptibility in female mice treated neonatally with diethylstilbestrol. *Cancer Res* 41:5134–5141.

Karpuzoglu-Sahin E, Hissong B D, and Ansar Ahmed S. 2001. Interferon-gamma levels are upregulated by 17-beta-estradiol and diethylstilbestrol. *J Reprod Immunol* 52, 113–27.

Kavlock RJ. 1999. Overview of endocrine disruptor research activity in the United States. *Chemosphere* 39, 1227–36.

Kosugi A, Zuniga-Pflunker JC, Sharrow SO, Kruisbeek AM, and Shearer GM. 1989. Effect of cyclosporin A on lymphopoiesis. II. Developmental effects on immature thymocytes in fetal thymus organ culture treated with cyclosporin A. *J Immunol* 143:3134–3140.

Kruisbeek AM, Bridges S, Carmen J, Longo DL, and Mond JJ. 1985. *In vivo* treatment of neonatal mice with anti-I-A antibodies interferes with the development of the class I, class II, and Mls-reactive proliferating T cell subset. *J Immunol* 134:3597–3604.

Landrigan P, Garg A, and Droller DB. 2003. Assessing the effects of endocrine disruptors in the National Children's Study. *Environ Health Perspect* 111, 1678–82.

Longnecker MP, Bellinger DC, Crews D, Eskenazi B, Silbergeld EK, Woodruff TJ, and Susser ES. 2003. An approach to assessment of endocrine disruption in the National Children's Study. *Environ Health Perspect* 111, 1691–1697.

Lytton SD, Berg U, Nemeth A, and Ingelman-Sundberg M. 2002. Autoantibodies against cytochrome P450s in sera of children treated with immunosuppressive drugs. *Clin Exp Immunol* 127:293–302.

Mathur S, Mathur RS, Goust JM, Williamson HO, and Fudenburg HH. 1979. Cyclic variation in white cell subpopulations in the human menstrual cycle: correlations with progesterone and estradiol. *Clin Immunol Immunopathol* 13:246–254.

Nandakumaran M and Eldeen AS. 1990. Transfer of cyclosporin in the perfused human placenta. *Dev Pharmacol Ther* 15:101–105.

Nicol T, Bilbey DLJ, Charles LM, Cordingley JL, and Vernon-Roberts B. 1964. Oestrogen: the natural stimulant of body defense. *J Endocrinol* 30:277–285.

Noller KL, Blair PB, O'Brien PC, Melton LJ, Offord JR, Kaufman RH, and Colton T. 1988. Increased occurrence of autoimmune disease among women exposed *in utero* to diethylstilbestrol. *Fertil Steril* 49:1080–1087.

Okuyama R, Abo T, Seki S, Ohteki T, Sugiura K, Kusumi A, and Kumagai K. 1992. Estrogen administration activates extrathymic T cell differentiation in the liver. *J Exp Med* 175:661–669.

Penit C and Vaddeur F. 1989. Cell proliferation and differentiation in fetal and early postnatal mouse thymus. *J Immunol* 142:3369–3377.

Poland A and Knutson JC. 1982. 2,3,7,8-Tetrachlorodibenzo-*p*-dioxin and related halogenated aromatic hydrocarbons: examination of the mechanism of toxicity. *Annu Rev Pharmacol Toxicol* 22:517–554.

Prevot A, Martini S, and Guignard JP. In utero exposure to immunosuppressive drugs. 2002. *Biol Neonate* 81:73–81.

Rice C, Birnbaum LS, Cogliano J, Mahaffey K, Needham L, Rogan WJ, and vom Saal FS. 2003. Exposure assessment for endocrine disruptors: some considerations in the design of studies. *Environ Health Perspect* 111, 1683–90.

Rocha B, Bassalli P, and Guy-Grand D. 1992. The extrathymic T-cell development pathway. *Immunol Today* 13:449–454.

Schlumpf M, Ramseier H, and Lichtensteiger W. 1989. Prenatal diazepam induced persisting depression of cellular immune responses. *Life Sci* 44:493–501.

Schuurman H-J, Van Loveren H, Rozing J, and Vos JG. 1992. Chemicals trophic for the thymus: Risk for immunodeficiency and autoimmunity. *Int J Immunopharmacol* 14:369–375.

Seaman WE, Merigan TC, and Talal N. 1979. Natural killing in estrogen-treated mice responds poorly to poly IC despite normal stimulation of circulating interferon. *J Immunol* 123:2903–2910.

Scott JR, Branch DW, and Holman J. 2002. Autoimmune and pregnancy complications in the daughter of a kidney transplant patient. *Transplantation* 73:677–678.

Silverstone AE, Frazier DE Jr, and Gasiewicz TA. 1994. Alternate immune system targets for TCDD: Lymphocyte stem cells and extrathymic T-cell development. *Exp Clin Immunogenet* 11:94–101.

Silverstone AE, Gavalchin J, and Gasiewicz TA. 1998. TCDD, DES, and estradiol potentiate a lupus-like autoimmune nephritis in NZB x SWR ($SNF_1$) mice. *The Toxicologist* 42:403.

Spyker-Cranmer JM, Barnett JB, Avery DL, and Cranmer MF. 1982. Immunoteratology of chlordane: Cell-mediated and humoral immune responses in adult mice exposed *in utero*. *Toxicol Appl Pharmacol* 62:402–408.

Takahashi N, Nishida H, and Hoshi J. 1994. Severe B cell depression in newborns from renal transplant mothers taking immunosuppressive agents. *Transplantation* 57:1617–1621.

Talal N. 1993. Lessons from autoimmunity. *Ann NY Acad Sci* 690:19–23.

Urso P and Gengozian N. 1994. Subnormal expression of cell-mediated and humoral immune responses in progeny disposed toward a high incidence of tumors after *in utero* exposure to benzo(a)pyrene. *J Toxicol Environ Health* 14:569–584.

Walker SE, Keisler LW, Caldwell CW, Kier AB, and vom Saal FA. 1996. Effects of altered prenatal hormonal environment on expression of autoimmune disease in NZB/NZW mice. *Environ Health Perspect* 104, Supp. 4:815–821.

Ways SC, Mortola JF, Zvaifler NJ, Weiss RJ, and Yen SSC. 1987. Alterations in immune responsiveness in women exposed to diethylstilbestrol *in utero*. *Fertil Steril* 48:193–201.

Ways SC and Bern HA. 1979. Long-term effects of neonatal treatment with cortisol and/or estrogen in the female BALB/c mouse. *Proc Soc Exp Biol Med* 160:94–98.

Wingard DL and Turiel J. 1988. Long-term effects of exposure to diethylstilbestrol. *West J Med* 149:551–554.

CHAPTER 14

# Developmental Immunotoxicant Exposure and Exacerbated Postnatal Immune Responses: Asthma

**John M. Armstrong, Deborah Loer-Martin, and Ramona Leibnitz**

## CONTENTS



0-415-28457-0/05/$0.00+$1.50

## INTRODUCTION

### Asthma: An Allergic Response

Like an autoimmune disease or the rejection of a transplanted organ, allergic asthma is the result of an unwanted immune response. Characterized by episodes of usually reversible obstruction of the airways, bronchial hyperresponsiveness, and chronic inflammation with lung infiltration by lymphocytes, eosinophils, and mast cells, asthma is a respiratory response to a variety of stimuli. This disease is manifested by thickening of the bronchial mucosa and narrowing of airways, and most commonly results in coughing, wheezing, chest tightness, and shortness of breath (Koren and O'Neill 1998; Saltini et al. 1998).

### Asthma: An Increasingly Common Childhood Disease with an Unknown Cause

More than one-third of the 15 million Americans with asthma are children, and the biggest increase in cases since 1980 has been in preschool-age children (Donovan and Finn 1999). Indeed, most asthmatics are diagnosed by the age of five, with symptoms first occurring during infancy and early childhood (Croner and Kjellman 1992; Yunginger et al. 1992). Although the reason for the increasing prevalence of asthma in young children is unclear, several factors play an exacerbating role in the incidence of disease. These include environmental triggers, such as house dust mites, cockroaches, mold, and animal dander, as well as genetic factors.

## Asthma: The Role of Atopy, Risk Factors, and Maternal Exposures

People with atopy are genetically predisposed to produce IgE antibodies in response to common household allergens and have at least one atopic disease (i.e., asthma, allergic rhinitis, or atopic eczema) (Kay 2001). Most patients with asthma are atopic, although a minority have intrinsic, nonatopic asthma that often has a later onset and a more protracted course than atopic asthma. Recent studies indicate that there are more similarities than differences in the airway abnormalities of atopic and nonatopic asthma (Humbert et al. 1999). A large body of evidence shows that allergen exposure influences the atopic phenotype, and many studies indicate that allergen or toxicant exposure during critical windows of immune system development (*in utero* and early infancy) may influence a sensitization process that can lead to childhood asthma (reviewed in: Landrigan 1998). Epidemiological data support the findings from these studies and highlight the role of maternal factors (Donovan and Finn 1999; Warner et al. 1998). Clearly, an interaction exists between inherited genetic characteristics and the *in utero* and postnatal environment, such that early toxicant exposure may predispose an individual to asthma, but postnatal exposure affects disease occurrence (Warner and Warner 2000).

The disease occurrence can also be seen along racial lines. Although asthma is a problem of all races, based on the results of the 1993-1996 National Health Interview Survey data, asthma-related morbidity is reportedly higher among black and poor children (Akinbami et al. 2002), and poverty, racism, and inequitable access to medical care have all been reported to contribute to the cause of asthma (Landrigan et al. 1998). Nonetheless, it is believed that environmental toxicants play a major role, and more than half of the more than 15,000 synthetic chemicals widely used in consumer products and dispersed in the environment remain untested for toxicity (Landrigan et al. 1998).

## Asthma: Toward a Molecular Understanding

The purpose of this chapter is to address the influence of maternal/fetal immunotoxicant exposure on the development of asthma. To understand how developmental exposure can lead to postnatal asthma onset or affect the sensitivity to triggers of asthmatic attacks, one must have an awareness of the timeline of immunologic development, as well as an understanding of the cellular and molecular characteristics of asthma. Because developmental timelines were discussed in detail earlier in this textbook, this chapter will examine the relationship between developmental exposures to common allergens and toxicants and the cellular and molecular factors involved in asthmagenesis. The data on common asthma triggers such as dust mites and molds will be explored, and those on less common allergens such as foods and vaccines will be reviewed. In addition, the knowledge base on maternal exposure to environmental immunotoxicants such as heavy metals, pesticides, and industrial chemicals will be discussed, and the evidence linking childhood asthma to maternal tobacco smoke exposure will be summarized.

A comprehensive review of the developmental immunotoxicants involved in asthma cannot be written due to a paucity of experimental evidence. Indeed, the mechanisms related to immunotoxicants can only rarely be detailed. Nonetheless, data are available to implicate several environmental toxicants in the deviation of the immune response towards the development of asthma so that the interplay between genetics, environmental toxicants, and asthma onset can be appreciated and, in time, manipulated for better treatment or cure of this disease. The immunoregulatory mechanisms vulnerable to alteration by the factors listed above will now be discussed, including the cells, cytokines, and signaling pathways involved in asthmatic inflammation.

## SENSITIZATION TO ALLERGENS AND THE DEVELOPMENT OF ASTHMA

### Sensitization vs. Triggering

Encountering any particular antigen (Ag) for the first time sensitizes, or primes, a developed immune system. This elicits a primary T cell response characterized by a relatively slow clonal expansion of antigen-specific T cells secreting a variety of cytokines of no particular biological focus. Sensitization of T cells also allows the immune system to react with a memory, or anamnestic, response upon a second exposure—and all subsequent exposures—to the same antigen. A memory T cell response is very different from a primary response. Thus, maturation of the immune response requires this initial exposure or sensitization.

Memory T cells proliferate faster, require fewer activation signals than their naïve counterparts, and are able to migrate specifically, or home, to tissues in the body where an antigen-specific response is required (Caret et al. 1998; Carter and Swain 1998; Dutton et al. 1999; Early and Reen 1999a; Hayward and Groothuis 1991). Besides these differences, memory T cells are characterized by differentiation to two major types of effector cells (Figure 14.1). $CD4^+$ T effector cells secrete cytokine profiles that distinguish them as T Helper 1 (TH1) cells, TH2 cells, or TGF-β-producing T regulatory (Treg) cells, a category that includes Tr1 (Groux et al. 1997) and TH3 (Chen et al. 1994) type cells. $CD8^+$ T effectors cells are characterized by their cytolytic function. Moreover, memory T cells express different cell surface molecules than naïve T cells, and the functionality of these distinguishing markers even differs between neonates and adults (Early and Reen 1999b). It is the memory response that is elicited when an immune system is said to be "triggered," as opposed to the primary response that is evoked when the immune system is "sensitized."

Asthma is a triggered response most often associated with differentiation to the TH2 T cell subtype, although the involvement of other subsets has also been reported (Punnonen et al. 1997). TH2 cytokines include interleukin-4 (IL-4), IL-5, IL-6, IL-9, IL-10, and IL-13. The roles played by these and other cytokines in fetal sensitization and the pathogenesis of asthma will be discussed under the "Mechanisms of Asthma" section of this chapter.

A primary B cell response requires help from T cells and is characterized by the production of high titers of inherently low-affinity and cross-reactive antibodies of the IgM isotype (reviewed in: Lane 1996). Like a primary T cell response, this response is also relatively slow to start and is short-lived. However, primed B cells then undergo isotype switching and secrete low levels of other antibody isotypes (Lane 1996), including IgE. Tissue mast cells bearing high-affinity $Fc_{\varepsilon}RI$ become armed with antigen-specific IgE antibodies. Triggering a memory B cell response upon subsequent antigen exposure results in high titers of high affinity antibodies of isotypes other than IgM, and these antibodies remain in the blood and tissues for longer periods than in a primary response. Among the Ag that induce switching to the IgE isotype are allergens, nonpathogenic Ags that evoke a nonproductive immune response. In cases where B cells have undergone isotype switching to IgE, secondary exposure to any particular allergen triggers an allergic, and sometimes an asthmatic, response. In these cases, the allergen-specific, $Fc_{\varepsilon}RI$-bound IgE antibodies on mast cells become cross-linked. Allergen-induced cross-linking causes mast cell degranulation leading to the release of a variety of allergy- and asthma-associated chemical mediators (reviewed in: Black 2002; Brightling et al. 2002; Carroll et al. 2002) (Figure 14.2). These mediators, their receptors, and other soluble factors playing roles in the pathogenesis of childhood asthma are also discussed later.

An asthmatic response is a memory response. Salek-Ardakani and coworkers recently provided strong evidence that asthma is caused by memory TH2 cells (Salek-Ardakani et al. 2003). Using sensitized animals, they demonstrated that the costimulatory molecule OX40 (CD134) is expressed on memory CD4 cells. Blocking OX40-OX40L interactions at the time of inhalation of aerosolized antigen was shown to inhibit the accumulation of memory effector cells in lung-draining lymph nodes and lung. Moreover, inhibition of the ligation of OX40 by OX40L was shown to prevent eosinophilia, airway hyperreactivity, mucus secretion, and TH2 cytokine production; this emphasizes that memory TH2 cells drive lung inflammation and that this aspect of the asthmatic response is, at least in part, under regulation by OX40.

Depending on the amount of the allergen, the asthmatic response can occur in two phases: an acute phase and a late phase. The acute-phase response is the result of immediate hypersensitivity and can include anaphylaxis, the cause of death when some asthmatic patients encounter an allergen (Fallon et al. 2001; Holt et al. 1999). Late-phase responses peak 6 to 9 hours after an acute response and can also be life-threatening. Although acute- and late-phase responses are brought on by different sets of mediators, both are linked to a single triggering exposure to allergen. In the lung, late-phase reactions are characterized by further wheezing (Kay 2001), eosinophil and neutrophil accumulation, and then $CD4^+$ T cell infiltration of the lower airways (MacFarlane et al. 2000; Robinson et al. 1993; Ying et al. 1999). In addition, late-phase allergic responses are responsible for the remodeling of airways in asthma patients. These changes in tissue architecture are characterized by epithelial hypertrophy, fibrosis, goblet cell hyperplasia, and airway occlusion associated with mucus, cellular infiltrate, and extracellular needle-like crystal structures known as Charcot-Leyden crystals that are deposited by activated eosinophils (Fallon et al. 2001; Holt et al. 1999). Interestingly, in a recent study of the cross-section of the apical bronchus of the right upper lobes of 45 stable asthmatic patients with (n = 22) and without (n = 23) inhaled steroid treatment, Niimi and colleagues demonstrated that, although the thickened airway walls of asthmatic patients correlated with airway reactivity, they did not correlate with airway sensitivity, suggesting that airway wall thickening may serve a protective effect against airway narrowing by attenuating airway hyperresponsiveness in asthmatic patients (Niimi et al. 2003). The remodeling of airway structure and function, as well as how allergic sensitization and triggering lead to the complex cellular interactions associated with asthmatic inflammation, will be discussed next.

## Asthmatic Inflammation

Asthma is a chronic inflammatory disease of the airways. Asthmatic inflammation involves complex interactions between the cytokines and signaling pathways of many cell types. In adults and children, T cells are known to be integrally involved in this process (Azzawi et al. 1990; Corrigan and Kay 1990; Wardlaw et al. 1988). The maintenance of airway inflammation is attributed, in adults, to chronically activated memory T cells sensitized against a variety of allergens (Corrigan and Kay 1990) and, in children, to elevated soluble IL-2 receptor levels (Warner et al. 1998). In general, asthma is associated with cytokines secreted by TH2 cells; for instance, IL-5 is involved with the recruitment and activation of eosinophils, while IL-4 plays a role in programming B cells to secrete IgE. It should be noted, however, that $CD8^+$ T cells have also been implicated in IgE class switching through the actions of IL-13 (rather than IL-4) (Punnonen et al. 1997). Moreover, TH1 cytokines such as IL-2, interferon γ (IFNγ), TNFα, and IL-15 have been reported to promote allergic airway inflammation (Donovan and Finn 1999; Hessel et al. 1997; Krug et al. 1996),

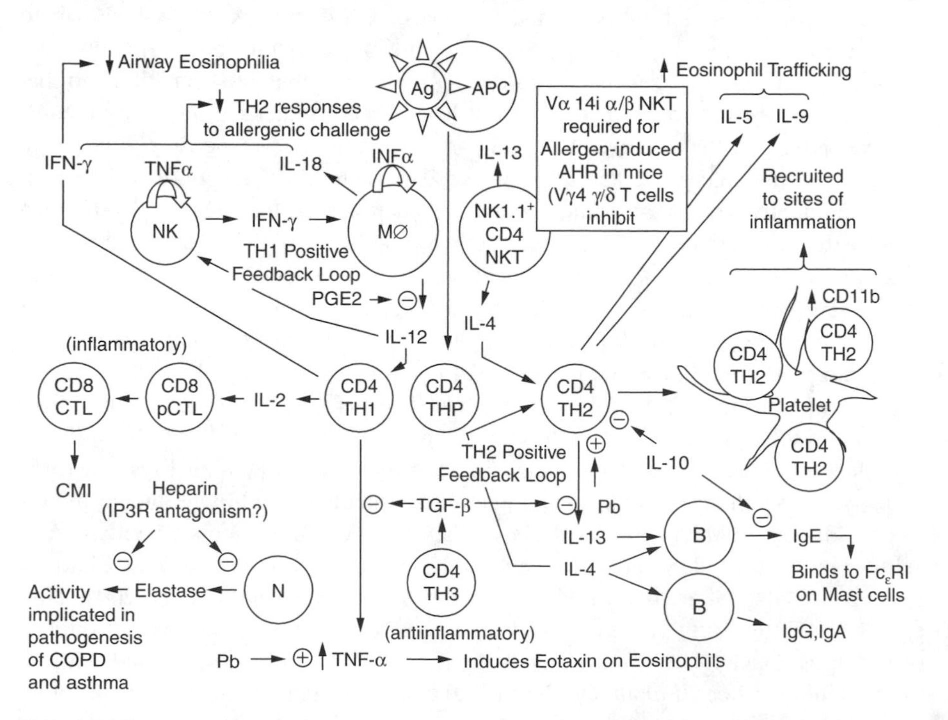


**Figure 14.1**

so ultimately, all T cell subsets probably contribute to the disease process of asthmatic inflammation. In a study to elucidate how circulating lymphocytes infiltrate into the airways of asthmatic patients, Tsumori and colleagues transplanted human bronchial xenografts into asthmatic huPBMC-SCID mice and found that the number of CD3-, CD4-, and CD8-positive cells in the xenografts of these mice was higher than those of dermatitis, rheumatic, and normal huPBMC-SCID mice (Tsumori et al. 2003). Moreover, expression of the mRNA of IL-4 and IL-5, but not of IL-2 or IFNγ, were significantly elevated in the xenografts of asthmatic huPBMC-SCID mice compared to those in the xenografts of normal huPBMC-SCID mice. Together, these results suggest that in asthmatics, T cells, especially TH2-type T cells, preferentially infiltrate into human bronchi.

It is pertinent at this point to discuss a relationship between leukocyte infiltration in asthmatics and circulating platelets. Platelet-leukocyte interactions have been demonstrated in cardiovascular disease (Ehlers et al. 2003), culminating in enhanced

**Figure 14.1 (See figure facing page.)** The cell-cytokine networks and immune deviation of asthmagenesis. Each T cell subset can contribute to disease, but the cytokines from the TH2 subset dominate. Naïve $CD4^+$ T cells (THP) differentiate upon Ag-specific recognition into TH1 or TH2 cells under the influence of interleukin(IL)-12 or IL-4, respectively. The originating source of IL-4 may be from $CD4^+$ NKT cells, which also secrete asthma-associated IL-13. Vα14i α/β NKT cells have been shown to be required for allergen-induced airway hyperresponsiveness (AHR) in mice, in a process that is inhibited by Vγ4 γ/δ T cells. Macrophage-derived IL-12 is inhibited by prostaglandin E2 (PGE2), which signals through cAMP-elevating Gs-protein coupled receptors. Macrophages and NK cells secrete and are activated by TNFα. Macrophage-derived IL-12 causes NK cells to secrete IFNγ, which restimulates macrophages in a TH1-amplifying positive feedback loop. TH1 cells secrete IL-2 to assist Ag-recognizing precursor cytotoxic T cells (pCTL) to become activated CTL, the effectors of cell-mediated immunity (CMI). TH1 cells also secrete IFNγ, which synergizes with macrophage-derived IL-18 to inhibit TH2 responses to allergenic challenge, including allergen-induced airway eosinophilia and airway epithelial cell-derived secretion of eotaxin. TH2 cells secrete IL-4, -5, -9, and -13, among other cytokines. IL-4 and IL-13 synergize to cause Ag-stimulated B cells to switch from producing immunoglobulin M (IgM) molecules to IgE, the Ig isotype associated with mast cell degranulation, asthma, and allergy. TH2 cytokine secretion is inhibited by IL-10, although the source of IL-10 may come from TH2 cells themselves as well as from regulatory T cells. IL-10 induces long-term hyporesponsiveness of allergen-specific $CD4^+$ T cells, decreases circulating numbers of mast cells and eosinophils, and in a rat model of asthma has been shown to inhibit the late-phase response as well as the influx of eosinophils and lymphocytes to airways. Other regulatory T cells include TH3 cells secreting Transforming Growth Factor-beta (TGF-beta), which inhibits cytokine secretion from both TH1 and TH2 cells. IL-4 stimulates TH2 cells to self-amplify in a positive feedback loop. TH2 cells bind to circulating platelets via CD11b, a component of Mac-1 (CD11b/CD18) that is upregulated under inflammatory conditions. TH2/platelet aggregates are recruited to sites of inflammation, including sites of allergen-induced degranulation of mast cells, basophils, and eosinophils. IL-5 and IL-9 synergize to mobilize and differentiate eosinophils from the bone marrow. TNFα from TH1, NK, cells, and macrophages primes neutrophils for activation by IL-6 or platelet activating factor (PAF). Neutrophil-derived elastase, as well as its activity in asthma and chronic obstructive pulmonary disease (COPD), are inhibited by heparin and other IP3 receptor antagonists. The immunotoxic effects of lead (Pb) may be related to its ability to support TH1-derived TNFα secretion, which is known to induce the CC chemokine eotaxin on eosinophils. Lead is also shown to induce TH2-derived secretion of the asthma-associated cytokines, IL-4 and IL-13.

leukocyte recruitment. Circulating platelet-leukocyte aggregates have been observed in the blood of allergic asthmatic patients during the allergen-induced late asthmatic response and in sensitized mice after allergen exposure (Pitchford et al. 2003a). Because of these observations, Pitchford and coworkers conducted a study that demonstrated that leukocyte infiltration was reduced in the airways of allergen-challenged mice that had been depleted of platelets (Pitchford et al. 2003b). Airway infiltration was restored following adoptive transfer of platelets from allergic animals. Moreover, using blood taken at various time points from asthmatic patients to document the degree of leukocyte activation and the presence of platelet-leukocyte aggregates before and after allergen exposure, these authors confirmed an essential role for platelets in leukocyte recruitment. These researchers also demonstrated an upregulation of CD11b, a subunit of the Mac-1 molecule, on leukocytes that were involved in aggregates with platelets and not on free leukocytes. The results suggest an essential role for platelets in the recruitment of leukocytes in allergic inflammation (Figure 14.1).

Interestingly, a biochemical explanation for the participation of platelets in the activity of neutrophils in asthmatic lungs may be available. Heparin is known to not only inhibit the generation of thrombin but the enzymatic activity of elastase as well. Neutrophil-derived elastase is an enzyme implicated in the pathogenesis of chronic obstructive pulmonary disease (COPD) (Ohbayashi 2002) and asthma (Monteseirin et al. 2003). However, a recent report demonstrated that heparin can inhibit the release of elastase from TNFα-primed human neutrophils (Brown et al. 2003). The authors of this study speculated that because IP(3) receptor antagonist 2-aminoethoxydiphenylborate (2-APB) mimicked the effects of heparin, itself an established IP(3) receptor antagonist, perhaps IP(3) antagonism inhibits the release of elastase from TNFα-primed human neutrophils. Because of the role platelets have been shown to play in the recruitment of leukocytes to asthmatic airways, it is interesting that heparin inhibits the generation of thrombin, which is known to activate platelets, and that heparin inhibits the receptor for IP(3), which is also thought to be involved in the upregulation of the CD11b molecule central to platelet-mediated leukocyte trafficking (Klos et al. 1992). Perhaps IP(3) pathway dysregulation will be found to be a marker in future asthma immunotoxicant studies.

The physiologic response of asthma, however, begins with the generation and activity of immunoglobulin E. Although IgE synthesis directed by TH2-derived IL-4 helps to initiate the asthmatic response, it is the allergen-mediated cross-linking of IgE/$Fc_{\varepsilon}R$ complexes on mast cells that leads to the physiologic response: Upon mast cell degranulation, release of inflammatory mediators causes bronchoconstriction, shortness of breath, lowered blood pressure, edema, and recruitment of inflammatory cells (Cookson 1999; Corry and Kheradmand 1999; Fallon et al. 2001). Some of the mediators released by mast cell degranulation include the three cysteinyl leukotrienes $C_4$, $D_4$, and $E_4$ (Drazen et al. 1999), although eosinophils, macrophages, and monocytes are also important sources of these mediators (Holgate, 1999) (Figure 14.2). These leukotrienes bind to specific receptors and are responsible for the contraction of smooth muscles, vasodilation, increased vascular permeability, and the hypersecretion of mucus associated with asthma (Drazen et al. 1999). Since many of the leukotrienes are known to generate cAMP via activation of G protein-

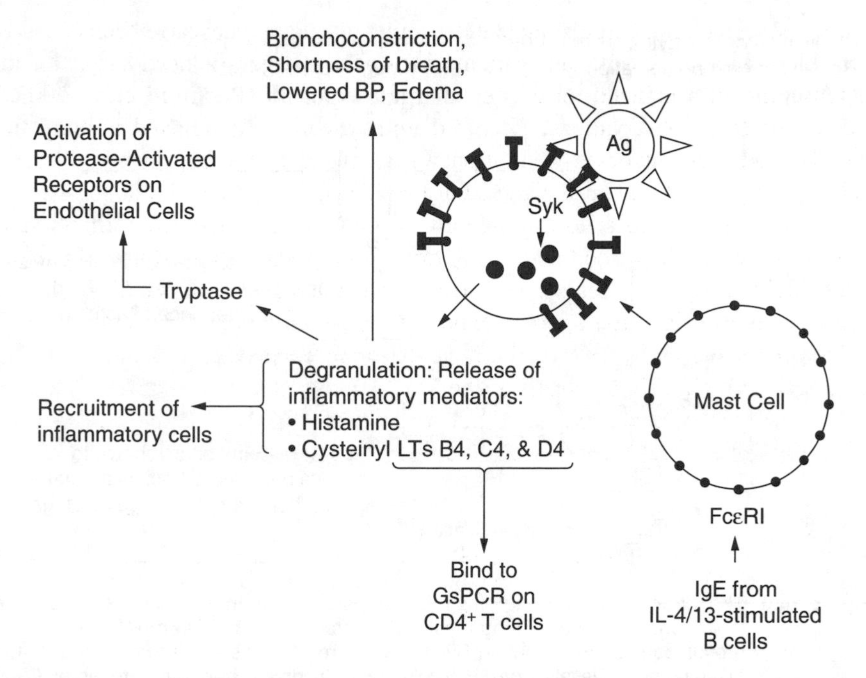


**Figure 14.2** Mast cell degranulation. Armed with B cell-derived allergen-specific IgE molecules bound to cell surface-associated, high-affinity $Fc_{\varepsilon}RI$ receptors, mast cells degranulate upon secondary and subsequent exposures to the same allergen in a process involving activation of the protein tyrosine kinase Syk. Upon degranulation, inflammatory mediators including histamine and cysteinyl leukotrienes B4, C4, and D4 (LTB4, LTC4, and LTD4) are released. These mediators recruit inflammatory cells, including circulating TH2/platelet aggregates, to sites of allergen-induced mast cell degranulation. Tryptase released during degranulation is responsible for activating endothelial receptors necessary for the recruitment of eosinophils to sites of inflammation. The entire process results in the bronchoconstriction, shortness of breath, lowered blood pressure, and edema associated with asthma.

coupled receptors (GPCR), and cAMP levels are regulated by the hydrolytic activity of phosphodiesterases (PDEs), it is interesting to note that PDE type IV inhibitors are showing promise as antiasthmatics. These agents have recently been shown to generate good efficacy against guinea pig respiratory tract inflammation and bronchoconstriction (Kim et al. 2003) (Figure 14.3).

Another mast cell-derived effector molecule released in response to IgE-crosslinking is tryptase, which is responsible for activation of the protease-activated receptors on endothelial and epithelial cells; this activation eventually leads to the upregulation of adhesion molecules that selectively attract basophils and eosinophils to the site of inflammation (Holgate 1999) (Figure 14.4). Importantly, eosinophilic inflammation is intimately involved in the inflammatory response of asthma (Robinson et al. 1992) and is driven by the TH2-derived cytokines IL-5 and IL-9, as well as by the chemotactic cytokine, or chemokine, eotaxin (Figure 14.5).

Mature and immature eosinophils are released from the bone marrow under the control of IL-5 (Palframan et al. 1998), which is also responsible for regulating the expression of the transmembrane isoform of its own receptor (Tavernier et al. 2000)

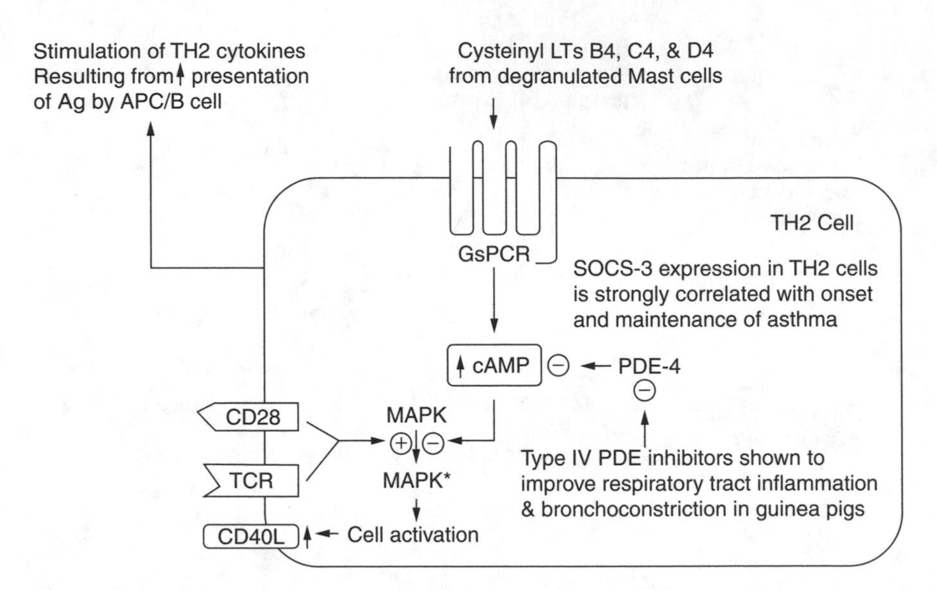


**Figure 14.3** The intracellular signaling of a TH2 cell participating in asthmagenesis. Cysteinyl leukotrenes released during mast cell degranulation bind to cAMP-elevating Gs-protein coupled receptors on TH2 cells and recruit those cells to sites of asthmatic inflammation. Elevated cAMP levels result in diminished T cell receptor (TCR)-induced mitogen-activated protein kinases (MAPKs) activation in a process countered by asthma-associated elevations in TH2 phosphodiesterase 4 (PDE4) levels. Type IV PDE inhibitors have been shown to be particularly promising in improving respiratory tract inflammation and bronchoconstriction in guinea pigs. Activation of the TH2 cells results from cognate interaction with an antigen presenting cell (APC) such as a B cell, macrophage, or dendritic cell, causing the release of asthma-associated TH2 cytokines.

and terminal differentiation of committed eosinophil precursors (Clutterbuck et al. 1989). After leaving the bone marrow and following a gradient of one or more chemokines, eosinophils migrate into the asthmatic airways via the bronchial post-capillary endothelium (Figure 14.4 and Figure 14.5). It is interesting to note that Katoh and coworkers have recently shown in a murine model of pulmonary eosinophilia that eosinophil and lymphocyte accumulation in the lung, as well as antigen-induced airway hyperresponsiveness, are completely prevented by treatment with anti-CD44 monoclonal antibodies (Katoh et al. 2003). The authors demonstrated that intraperitoneal administration of anti-CD44 antibodies inhibited the increased levels of hyaluronic acid (HA, hyaluronan) and leukotrienes in the bronchoalveolar lavage fluid (BALF) that typically result from antigen challenge. CD44 (also known as ECMR III; H-CAM; HUTCH-1; Hermes; Lu, In-related; Pgp-1; and gp85) serves as a recyclable receptor for hyaluronan and is involved in the homing of leukocytes to sites of inflammation (Liao and Patel 1999). Katoh's observation that CD44 blockade interferes with HA accumulation in BALF and inhibits eosinophil and lymphocyte accumulation in the lung confirms known biochemical and cellular activities of the CD44 hyaladherin molecule (Figure 14.5).

Following their migration into asthmatic airways, interactions with selective adhesion molecules ($\alpha_4\beta_1$ integrin and VCAM) (Denburg 1998) allow eosinophils

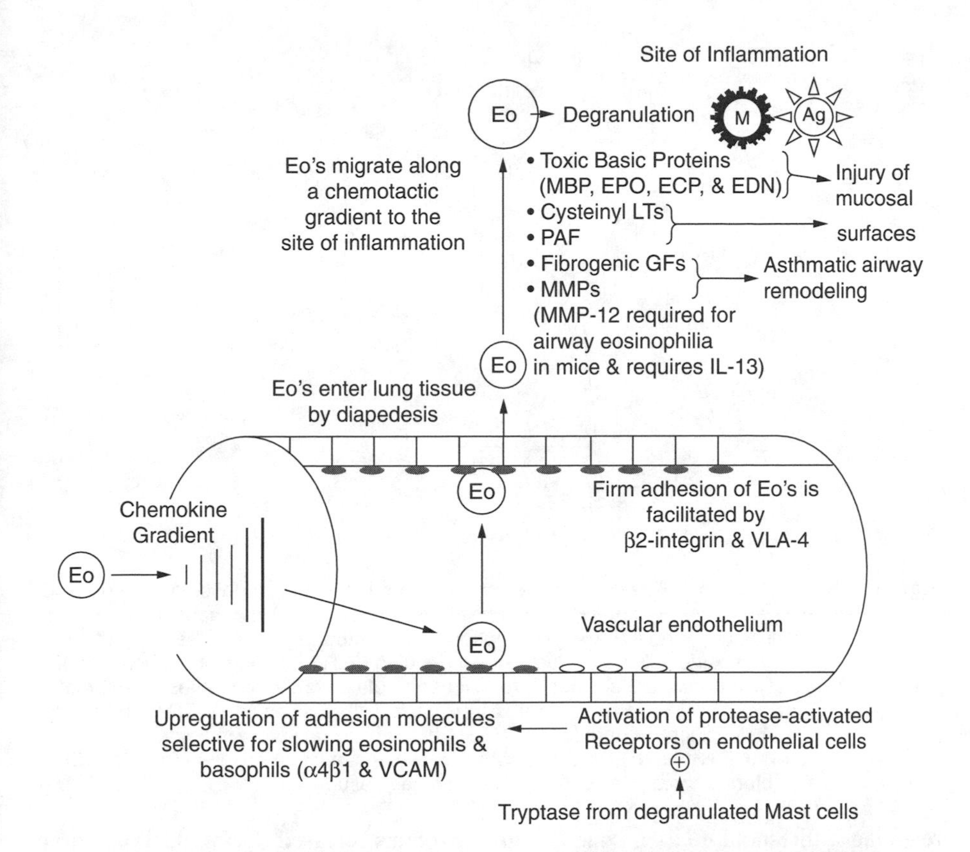


**Figure 14.4** Eosinophil (Eo) adhesion, diapedesis, and degranulation. Tryptase, released when mast cells degranulate, activates protease-activated receptors on vascular endothelium near sites of asthmatic inflammation. Activation of these receptors leads to an upregulation of the $\alpha 4\beta 1$ and VCAM adhesion molecules, which selectively slow the circulation of eosinophils and basophils. Eosinophils, mobilized from the bone marrow by IL-5, follow a CC chemokine gradient by virtue of their CCR3 receptors. Within the vasculature they are slowed by $\alpha 4\beta 1$ and VCAM but become firmly adhered to inflammatory site endothelium by the adhesion molecules b2-integrin and VLA-4. Eosinophils then enter lung tissue by diapedesis where they continue to migrate along a chemotactic gradient until they degranulate at the site of asthmatic inflammation. Mediators such as toxic basic proteins, cysteinyl leukotrienes, and platelet activating factor (PAF) are released and are known to cause injury of mucosal surfaces. Fibrogenic growth factors and matrix metalloproteinases (MMP), also released when eosinophils degranulate, lead to asthmatic airway remodeling.

to slow to a roll on blood vessel surfaces until they recognize eotaxin (or other chemokine) on the endothelial cell surface (Wardlaw 1999). Ligation of a class of chemokine receptors, called "CC" receptors, by eotaxin and other CC chemokines activates eosinophils, firmly adheres them to the vascular endothelium via adhesion molecules such as $\beta_2$-integrin and very late activation antigen-4 (VLA-4), and allows them to enter bronchial tissue in a process known as diapedesis. Once in the bronchial tissue, eosinophils migrate along a chemotactic gradient to the site of allergic inflammation where they degranulate (Wardlaw 1999), presumably in response to

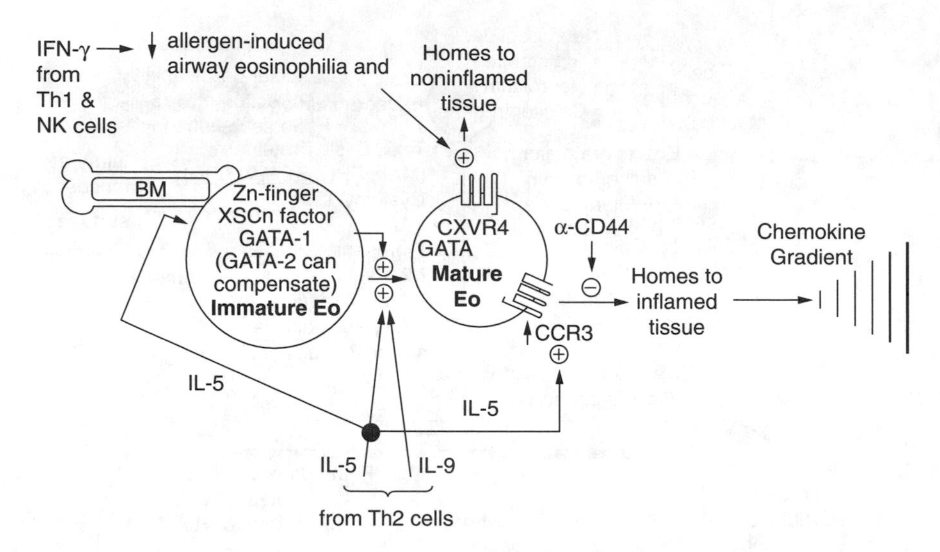


**Figure 14.5** Eosinophil mobilization from the bone marrow. Under the influence of TH2-derived IL-5 and IL-9, immature eosinophils are released from the bone marrow in a process requiring a GATA transcription (abbreviated as XSCn in the figure) factor. Mature eosinophils upregulate eotaxin-binding CCR3 receptors in response to IL-5. CCR3 allows the eosinophil to follow an eotaxin gradient to sites of asthmatic inflammation in a process that is inhibitable with antibodies to CD44. However, mature eosinophils upregulate CXCR4 chemokine receptors in response to TH1-derived IFNγ, and it is thought that these receptors compete with the CCR3 receptors to hold eosinophils in noninflamed tissue.

reaching a threshold eotaxin concentration. Proteins released during degranulation include major basic protein (MBP), eosinophil peroxidase (EPO), eosinophil cationic protein (ECP), and eosinophil-derived neurotoxin (EDN) (reviewed in: Menzies-Gow and Robinson 2001). The release of toxic basic proteins by eosinophils can injure mucosal surfaces, as can the release of cysteinyl leukotrienes and platelet activating factor (PAF). Interestingly, repair of these surfaces also comes from eosinophils, since they produce fibrogenic growth factors and matrix metalloproteinase (MMP), both of which play roles in asthmatic airway remodeling (Levi-Schaffer et al. 1999). MMPs, however, appear to play multiple roles in airway remodeling. A novel role for one MMP has recently been demonstrated by Pouladi and colleagues, where, using MMP-12-deficient and IL-13-deficient mice, MMP-12 was shown to be required for the development of airway eosinophilia in mice in a process that was dependent on IL-13 (Pouladi et al. 2003) (Figure 14.4).

Eotaxin and other chemokines have been implicated in eosinophil accumulation and activation within the bronchial mucosa of asthmatics, primarily due to the tissue-specific expression of their G-protein coupled receptors through which all chemokines signal (Zlotnik et al. 1999). Whereas the TH2 cytokines IL-4 and IL-13 have been shown to induce eotaxin expression by several cell types (including airway epithelial cells, endothelial cells, and fibroblasts (Gangur and Oppenheim 2000; Li et al. 1999; Menzies-Gow and Robinson 2001), the TH1 cytokine IFNγ has been demonstrated to inhibit eotaxin synthesis *in vitro* (Miyamasu et al. 1999). Within

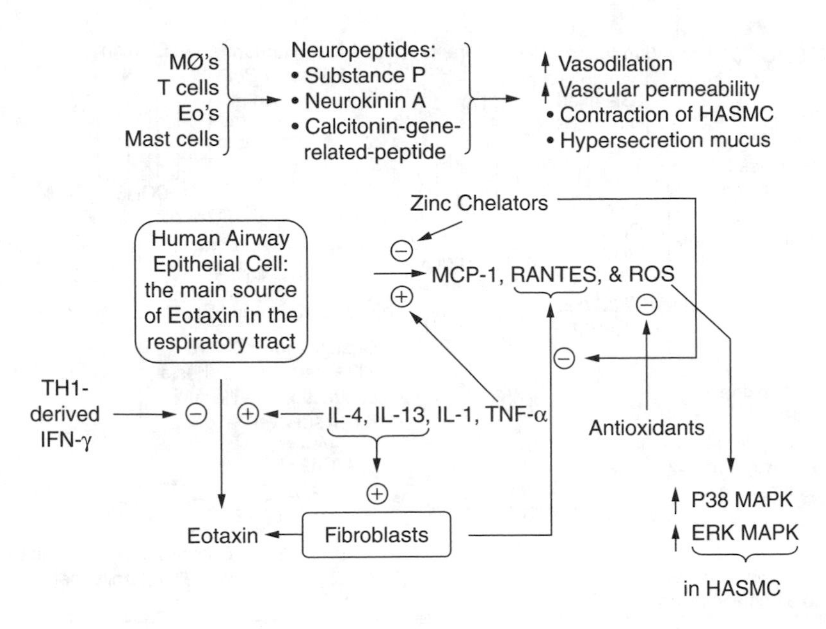


**Figure 14.6** The role of the human airway epithelial cell (HAEC) in asthmagenesis. The HAEC is the major source of eotaxin in the respiratory tract. Release of eotaxin from HAEC is inhibited by TH1-derived IFNγ, but it is induced by TH2-derived IL-4 and IL-13, which, like RANTES, also induce eotaxin release from airway fibroblasts. HAEC-derived eotaxin is also induced by the inflammatory cytokines IL-1 and TNFα that also induce the zinc-dependent release of the chemokines MCP-1 and RANTES, as well as reactive oxygen species (ROS) from HAEC. HAEC-derived ROS have been shown to induce activation of both p38 and ERK MAPK in human airway smooth muscle cells (HASMC). Within asthmatic bronchial airways, other cells also participate in asthmatic inflammation, including macrophages, T cells, eosinophils, and mast cells. Mediators released from these cells during asthmagenesis include the neuropeptides substance P, neurokinin A, and calcitonin-gene-related peptide. Release of these neuropeptides from the inflammatory cells of asthmatic bronchial airways results in an orchestration of increased vasodilation and vascular permeability, contraction of HASMC, and the hypersecretion of mucus.

the respiratory tract, the major source of eotaxin is the airway epithelial cell, where eotaxin mRNA and protein expression are increased by TNFα, IL-1, and IL-4 (reviewed in: Menzies-Gow and Robinson 2001) (Figure 14.6). The mechanism of action of IL-1 with respect to eotaxin expression has recently been investigated. In a report by Wuyts and colleagues (Wuyts et al. 2003), interleukin-1β (IL-1β)-induced chemokine release in human airway smooth muscle cells were shown to be inhibited by the antioxidative agent N-acetylcysteine (NAC). Both the expression and production of eotaxin and monocyte chemotactic protein 1 (MCP)-1 were inhibited by NAC. (As will be discussed further in the section on the transcription factor GATA-1, both eotaxin and MCP-1, as well as "regulated upon activation normal T cell expressed and secreted" [RANTES], have also been shown to be induced by TNFα.) Understanding that mitogen-activated protein kinases (MAPK) are often activated by reactive oxygen species (ROS), the authors investigated and determined that IL-1β-induced production of the ROS 8-isoprostane, as well as activation of p38 MAPK,

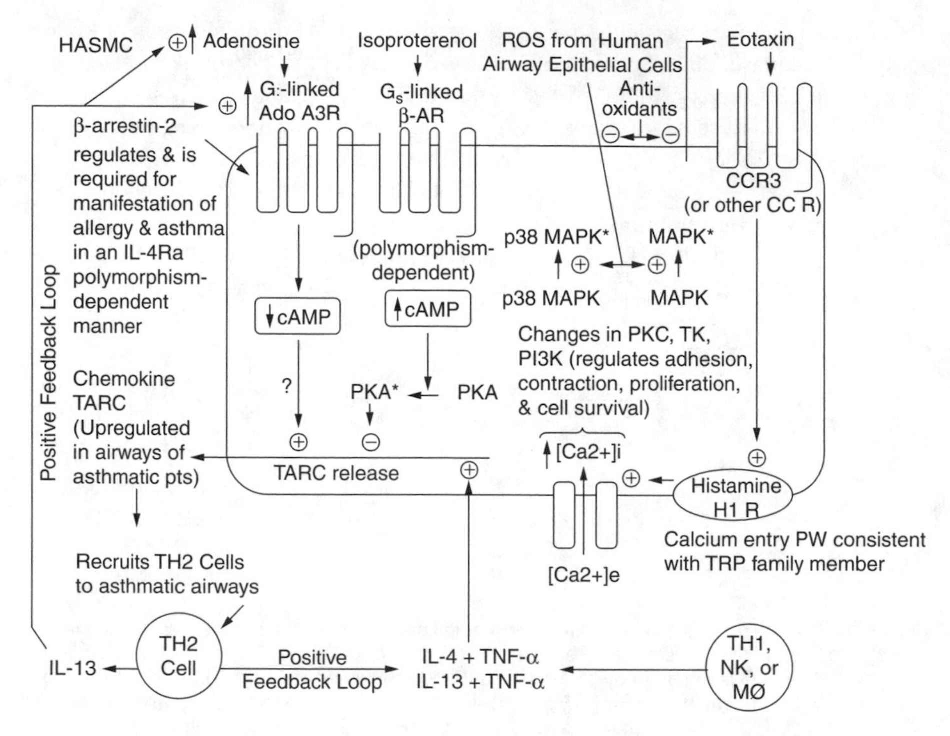


**Figure 14.7** The role of the human airway smooth muscle cell (HASMC) in asthmagenesis. Eotaxin derived from any number of sources stimulates the HASMC to produce more eotaxin, as well as to activate histamine H1 receptors leading to activation of a calcium entry pathway consistent with members of the transient receptor potential (TRP)-c family. Calcium then enters the HASMC, causing changes in PKC, TK, and PI3K. TH2-derived IL-4 or IL-13 synergize with TNFα presumably derived from macrophages, TH1 cells, or NK cells to induce the release of the chemokine "thymus- and activation-regulated chemokine" (TARC), which is upregulated in the airways of asthmatic patients. TARC, in turn, recruits more TH2 cells to asthmatic airways, which continue to secrete more IL-13 and IL-4 in a positive feedback loop. IL-13 also upregulates extracellular adenosine, which binds to Gi-protein coupled adenosine A3 receptors, themselves upregulated in response to IL-13 in another positive feedback loop. IL-13 receptors share the IL-4Rα subunit with IL-4 receptors, and IL-13-dependent upregulation of Adenosine A3 receptors is regulated by β-arrestin-2. β-arrestin-2 regulates, and is required for manifestation of allergy and asthma in an IL-4Rα polymorphism-dependent manner. IL-13-induced adenosine, in signaling through A3 receptors, decreases intracellular levels of the second messenger cAMP, which may serve to inhibit TARC release from HASMC. Conversely, cAMP levels are amplified in response to stimulation of Gs-coupled beta adrenergic receptors (β-AR) in a polymorphism-dependent manner. Elevated cAMP causes activation of cAMP-dependent protein kinase A (PKA), which has been shown to inhibit the release of TARC from HASMC.

were inhibited in a dose dependent manner by NAC (Figure 14.6 and Figure 14.7). The reduction of chemokine release by antioxidative agents via inhibition of p38 MAPK in human airway smooth muscle cells may have implications for the introduction of antioxidants during gestation.

Eosinophilic inflammation is directly tied to signaling through the CC chemokine receptors, and the biochemical events involved in this signaling are known to include intracellular calcium transients, pertussis toxin sensitive $G_i$-proteins, protein kinase C, tyrosine kinase, and phosphatidylinositol-3-kinase (PI3K) (Zlotnik et al. 1999). Indeed, the elevations of intracellular free calcium ([$Ca^{(2+)}$]i) induced by CC receptors are known to regulate many functional responses in airway smooth muscle, including contraction, proliferation, adhesion, and cell survival. Because entry from the extracellular environment, as opposed to release from intracellular stores, had been suggested, Corteling and coworkers (Corteling et al. 2003) sought to find the source of calcium. Using HASM and human bronchial epithelial cells (HBEC), the authors found a histamine H1 receptor-activated $Ca^{(2+)}$ entry pathway with characteristics typical of transient receptor potential (TRP)C family members. This led the researchers to demonstrate the expression of a range of TRP family homologues in the airway. As environmental toxicants are studied in the future for their effects on childhood asthma, it may be important to look at specific effects on chemokine receptor-associated pathways and second messengers, including calcium transients and TRP family members (Figure 14.7).

Although eotaxin was discovered in guinea pig studies (Jose et al. 1994), human eotaxin has since been cloned (Ponath et al. 1996). Eotaxin-2 (Forssmann et al. 1997) and eotaxin-3 (Kitaura et al. 1999) have been identified, shown to act on the same receptor (CCR3, expressed predominantly on eosinophils), and recognized as possessing the same cellular specificity. The importance of the CCR3 receptor cannot be overemphasized regarding its role in asthmagenesis. In support of the potential for inhibition of the eotaxin-CCR3 binding interaction as an anti-inflammatory treatment for asthma, Warrior and colleagues have recently reported identifying two compounds specifically inhibiting this interaction that also inhibit eosinophil chemotaxis in both *in vitro* and *in vivo* assays (Warrior et al. 2003). Besides eotaxin, other CC chemokines reported to recruit eosinophils *in vivo* are RANTES, monocyte chemoattractant protein-3 (MCP-3), MCP-4, and macrophage inflammatory protein-1a (MIP-1a) (Menzies-Gow and Robinson 2000; Menzies-Gow and Robinson 2001).

In addition to recruiting, activating, and degranulating eosinophils at sites of allergic inflammation, eotaxin is also known to recruit TH2 cells and basophils through the CCR3 receptor (Devouassoux et al. 1999). Moreover, eotaxin causes IgE-independent degranulation of basophils (Uguccioni et al. 1997a) and potentiates their production of IL-4 (Devouassoux et al. 1999). At the same time, IL-4 has been shown to induce eotaxin-3 expression in vascular endothelial cells (Shinkai et al. 1999). In this way, it appears that eotaxin may positively feedback to continue T cell-driven allergic inflammation *in vivo*. Sources of the chemokine eotaxin include not only T cells, alveolar macrophages, bronchial smooth muscle cells, cartilage chondrocytes, $CD68^+$ macrophages within the subendothelium and endothelial cells, but also eosinophils themselves (Rankin et al. 2000). Eotaxin expression can be upregulated by IL-3, IL-5, and TNFα (Rothenberg 1999) (Figure 14.8). Moreover, disparate chemokine receptors expressed on eosinophils may play functionally opposite roles regulating eosinophil trafficking in asthma. A model has been proposed that expression of the chemokine receptor CXCR4 holds eosinophils at noninflamed tissues, while CCR3 allows the cells to respond to eotaxin. The crux of the model is that TH1 cytokines such as TNFα and IFNγ induce CXCR4 expression, while TH2 cytokines such as IL-5 induce the expression of CCR3 (Nagase et al. 2000)

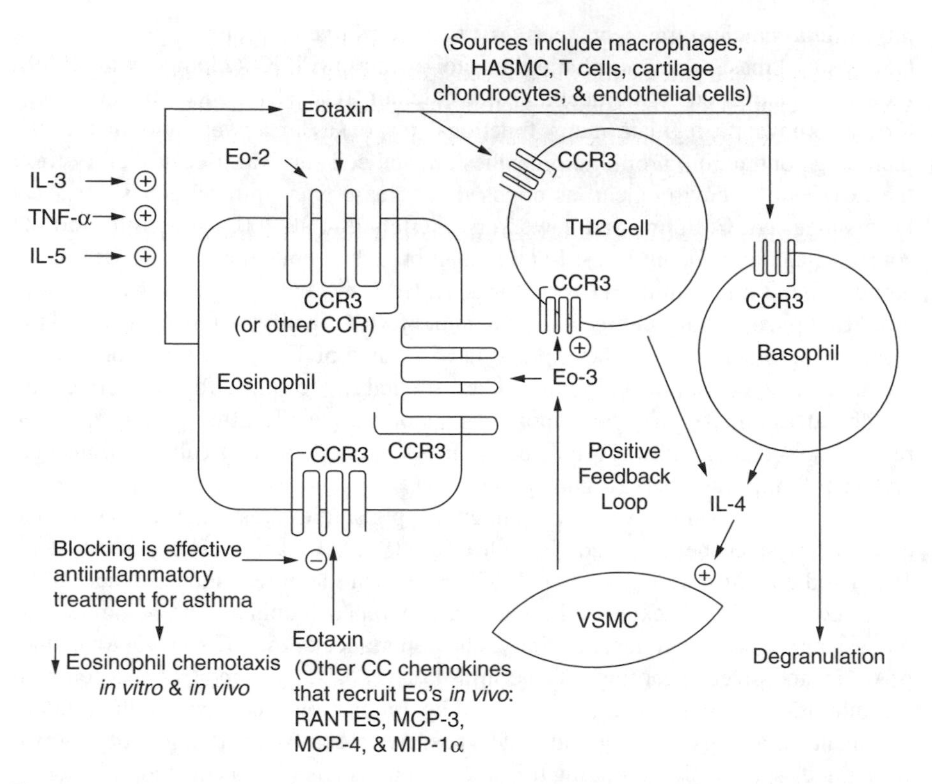


**Figure 14.8** The integral role of eotaxin in the positive feedback loops of asthmagenesis. Eotaxin as well as eotaxin-2 and eotaxin-3 stimulate the cells of asthmatic inflammation through CCR3 or other CC chemokine receptors. Eosinophils secrete eotaxin in response to IL-3, TNFα, and IL-5. Blocking the interaction of eotaxin with CCR3 receptors has been shown to be effective as an anti-inflammatory treatment for asthma, leading to reduced eosinophil chemotaxis both *in vitro* and *in vivo*. Eotaxin stimulates TH2 cells and basophils to secrete IL-4, in addition to causing basophils to degranulate. IL-4 stimulates vascular smooth muscle cells (VSMC) to secrete eotaxin-3, which in turn stimulates eosinophils and TH2 cells in a positive feedback loop. Other CC chemokines known to recruit eosinophils include RANTES, MCP-3, MCP-4, and MIP-1α. Other sources of eotaxin include macrophages, HASMC, T cells, cartilage chondrocytes, and endothelial cells.

(Figure 14.5). This model illustrates the complex nature of the regulation of eosinophil trafficking. Although complicated, the fact that eotaxin itself is upregulated by the TH1 cytokine TNFα (Figure 14.8) suggests that asthmagenesis may depend in part on the TH1 pathway as well as the TH2 pathway. Exactly how a balance between TH1- & TH2-derived cytokines affects eosinophil response remains to be seen. If such a balance is shown to be critical to the regulation of eosinophil trafficking, then the disruption of normal G-protein coupled receptor systems may be important endpoints to examine in the pathogenesis of asthma. Understanding the complexity of multiple redundant chemokines recruiting eosinophils, basophils, and TH2 lymphocytes to sites of allergic inflammation—i.e., when these cells secrete more TH2 cytokines that induce the synthesis of more eosinophil chemotactic factors—is just one of the challenges facing researchers trying to decipher or manipulate the mechanisms of asthmatic inflammation. In preparation for a discussion on how maternal-fetal interactions might sensitize a child for asthmatic responses to allergenic challenges, the soluble factors, receptors, and signaling pathways that mediate the cell-to-cell interactions and alterations in tissue architecture associated with asthma will be described next.

## Mechanisms of Asthmatic Inflammation: Inter- and Intracellular Mediators

### *Cytokines Involved in Promoting Allergic or Asthmatic Inflammation*

Inflammation is often described in the context of TH1 responses, particularly with regard to classical TH1-derived inflammatory cytokines such as TNFα and IFNγ. However, asthmatic inflammation involves cytokines whose actions typically oppose the actions of cytokines involved in the inflammation of other inflammatory diseases. Inflammation of the airways in asthma, although not exclusively so, is primarily a TH2-mediated phenomenon. Overproduction of GM-CSF in the airway mucosa of patients with asthma enhances MHC class II-restricted Ag presentation by dendritic cells and increases the accumulation of bronchoalveolar macrophages (Holt et al. 1999; Kay 2001), which further present allergen to $CD4^+$ T cells and stimulate the production of TH2-type cytokines (Larché et al. 1998). Reports of TH2 cytokines involved in asthmatic inflammation have implicated IL-4 and IL-13 in stimulating the production of IgE and vascular-cell adhesion molecule 1 (VCAM-1), IL-5 and IL-9 in the development of eosinophils, IL-4 and IL-9 in the development of mast cells, IL-9 and IL-13 in airway hyperresponsiveness, and IL-4, IL-9, and IL-13 in the overproduction of mucus (Wills-Karp et al. 1998). A role for IL-13 in asthmatic inflammation has been especially emphasized in recent literature. Using mice transgenic for and overexpressing IL-13, researchers have demonstrated phenotypic changes typical of asthma as a result of allergen exposure, including pulmonary fibrosis, goblet cell hyperplasia, elevated TH2 cytokines and IgE levels, eosinophilia, and airways occluded by mucus and Charcot-Leyden crystals (Fallon et al. 2001). Moreover, secondary exposure to Ag in these mice induces fatal anaphylaxis as a result of mast cell degranulation and histamine release, as well as marked pulmonary eosinophilia associated with hypertrophic bronchial epithelium.

The regulatory effects of cytokines alone or in combination in the development of atopic disease are even more complicated than presented above, however. Hahn and coworkers recently demonstrated some of this complexity using a mutated IL-4 variant as the basis for an IL-13/IL-4 inhibitor during allergic sensitization and in established disease in a murine model of asthma (Hahn et al. 2003a). This study indicates that while inhibition of the IL-4/IL-13 system efficiently prevents the development of the allergic phenotype, these cytokines play only a minor role in established allergy. Perhaps the disparity observed in the effects of an IL-4/IL-13 inhibitor on sensitization vs. established asthma lies in differences between signaling pathways not yet completely understood. In support of this idea, Zimmerman and colleagues used microarray analysis to conduct expression profiling of lung tissue from mice with experimental asthma (Zimmerman et al. 2003). The authors identified

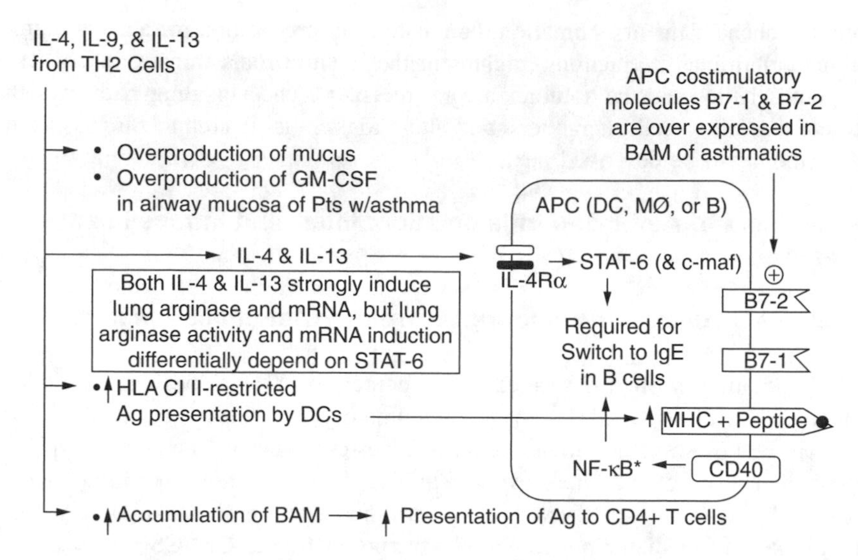


**Figure 14.9** The role of the antigen-presenting cell in asthmagenesis. Ag presentation to TH2 cells causes the release of the cytokines IL-4, IL-9, and IL-13, all of which are associated with the overproduction of mucus and GM-CSF found in the airway mucosa of patients with asthma. These cytokines increase HLA class II-restricted Ag presentation by dendritic cells (DCs) to CD4+ T cells and also increase the accumulation of bronchoalveolar macrophages (BAM) that, in turn, lead to increased Ag presentation to CD4+ T cells. In addition, the BAM from asthmatics have been shown to overexpress the APC co-stimulatory molecules B7-1 and B7-2. IL-4 and IL-13 stimulate the APC through receptors that share the IL-4R alpha chain in a process requiring activation of the STAT-6 transcription factor. However, while both IL-4 and IL-13 strongly induce lung arginase and mRNA, lung arginase activity and mRNA induction appear to differentially depend on STAT-6 activation in the APC. If the APC is a B cell, STAT-6 activation is required for isotype switching to allergy- and asthma-associated IgE production, but this process also requires NF-κB activation secondary to CD40 ligation by the interacting T cell.

6.5% of the tested genome whose expression was altered in asthmatic lung, and prominent among the asthma signature genes were those related to metabolism of basic amino acids, specifically the cationic amino acid transporter 2, arginase I, and arginase II. In particular, lung arginase activity and mRNA expression were strongly induced by IL-4 and IL-13 but were differentially dependent on signal transducer and activator of transcription 6 (STAT-6) (Figure 14.9).

Elucidating the differences between IL-4 and IL-13 signaling pathways as they relate to the pathogenesis of asthma may not be enough, however, to explain the differences in outcomes observed between phenotypically similar asthma patients treated under similar circumstances. Faffe and coworkers determined that human airway smooth muscle cells (HASMC) produce the chemokine thymus and activation regulated chemokine (TARC), known to induce selective migration of TH2, but not TH1 lymphocytes and known to be upregulated in the airways of asthmatic patients (Faffe et al. 2003) (Figure 14.7). These authors found that, alone, none of the cytokines IL-4, IL-13, IL-1β, IFNγ, or TNFα stimulated TARC release into the supernatant of cultured HASMC, whereas both IL-4 and IL-13 increased TARC protein and messenger RNA expression when administered in combination with

TNFα but not IL-1β or IFNγ. Of particular relevance to asthma patients is that the authors determined that polymorphisms of the IL-4Rα chain impact the ability of IL-4 or IL-13 to induce TARC release from HASMC. Genotype studies indicated that cells expressing the Val50/Pro478/Arg551 haplotype had significantly greater IL-4- or IL-13-induced TARC release than cells with other IL-4Rα genotypes. TH2 cytokines appear to enhance TARC expression in these cells, resulting in a positive feedback loop for the recruitment of TH2 cells to asthmatic airways in an IL-4Rα genotype-dependent fashion. Thus, research approaches that inhibit IL-13/IL-4 pathways with receptor inhibitors, as discussed earlier, are likely to have variable results depending on the receptor genotype of the patient (Figure 14.7).

### *IL-13 and G Protein-Coupled Receptor Systems*

Interestingly, Faffe and coworkers also determined that agents known to activate protein kinase A (PKA), including the beta-adrenergic receptor agonist isoproterenol, inhibited TARC release induced by TNFα plus IL-13, or TNFα plus IL-4 (Figure 14.7). In other words, biochemical signal transduction pathways involved in asthma pathogenesis could be inhibited by stimulation of cAMP-elevating G protein-coupled receptors. Particularly regarding IL-13 signaling, this concept is supported by another recent study that may go a long way toward helping us understand the potential effects of perinatal environmental toxicant exposure on asthma sensitization. Blackburn and colleagues recently demonstrated that adenosine and adenosine signaling contribute to and influence the severity of IL-13–induced tissue responses (Blackburn et al. 2003). Of particular relevance to this chapter is the work of Sitkovsky and colleagues at the National Institutes of Health, who have explored the causes and effects of the phenomenon that thymic T cells (thymocytes) are bathed in adenosine during the process of thymic education (reviewed in: Sitkovsky 1998). Although the effects of adenosine depend on the type of adenosine receptor expressed on different T cell subtypes, manipulation of this environment by altering the physiologic mechanisms for lowering extracellular adenosine levels can have dramatic results. In support of the findings of Sitkovsky's group, the findings of Faffe and colleagues indicate that IL-13 causes a progressive increase in adenosine accumulation, inhibits the activity and mRNA accumulation of adenosine deaminase (which is responsible in part for lowering extracellular adenosine levels), and augments the expression of the $A_1$, $A_{2B}$, and $A_3$ but not the anti-inflammatory $A_{2A}$ adenosine receptors in mice. While the murine adenosine $A_{2A}$ receptor is associated with cAMP elevation due to signaling through Gs protein-coupled receptors, the other murine adenosine receptors mentioned signal through cAMP-lowering Gi-protein coupled receptors. Since PKA is dependent upon cAMP for its activation, the findings presented by Blackburn and colleagues that adenosine mediates IL-13–induced inflammation and remodeling in the lung and interacts in an IL-13–adenosine amplification pathway are in agreement with Faffe's findings that TARC release is inhibited by PKA-activating pathways. Together, these reports suggest a novel asthmagenic pathway associated with IL-13 signaling. In this paradigm, asthma-promoting TARC release is indirectly facilitated by IL-13-induced, adenosine-mediated inhibition of cAMP and PKA. In support of these findings, it is interesting to note that our own experiments using developing murine thymocytes have shown that elevated cAMP levels resulting from $A_{2A}$ adenosine receptor-mediated signaling resulted in a specific downregulation of the TCR-triggered MAPK pathway (Armstrong et al. 1988). It may be particularly relevant that agents that inhibit cAMP levels in pathogenic T cells are known to facilitate and amplify the effects of IL-13 by recruiting TH2 cells to asthmatic airways, turning on these T cells in a self-amplifying proliferation cycle in response to antigen. As Blackburn and colleagues noted, the human $A_3$ adenosine receptor is known to be expressed in an exaggerated fashion in tissues from patients with asthma and COPD. This same group has recently demonstrated that lung mast cells are signaled to degranulate by activation through the murine adenosine A3 receptor (Zhong et al. 2003). Because the human adenosine A3 receptor is known to be Gi-linked, it is reasonable to speculate that the IL-13-mediated increase in expression of the Gi-linked murine adenosine receptors reported in this study may explain some of the adverse effects associated with IL-13 in human airway disorders such as asthma and COPD.

All chemokine receptors are GPCR and, as such, are known to be regulated by a protein called β-arrestin. A particularly interesting report by Walker and colleagues demonstrated a requirement for β-arrestin-2 in the manifestation of allergic asthma (Walker et al. 2003). In that study, allergen-sensitized mice with a targeted deletion of their β-arrestin-2 gene were incapable of accumulating T lymphocytes in their airways. In addition, these mice failed to demonstrate other physiological and inflammatory features characteristic of asthma, thus emphasizing a pivotal role for β-arrestin-2 in the regulation of the development of allergic inflammation at a proximal step in the inflammatory cascade (Figure 14.7).

The presentation by APC of cognate Ag directs the development of nascent T cells into either TH1 cells in the presence of IL-12, or IL-4/13-producing TH2 cells in the presence of IL-4 (Figure 14.1). The initial source of IL-4 is obscure but appears to be dependent on the presence of a cell type expressing an invariant TCR. NKT cells producing IL-4 and IL-13 have recently been shown to play an essential role in the development of allergen-induced AHR. Using NKT cell-deficient mice, Akbari and colleagues found that the inability to develop allergen-induced AHR in the absence of Vα14i NKT cells was not due to an inability to produce TH2 responses (Akbari et al. 2003). Instead, the authors demonstrated that the failure to develop AHR was reversible either by the adoptive transfer of tetramer-purified NKT cells producing IL-4 and IL-13 or by the administration of recombinant IL-13, emphasizing that regulation of the development of asthma and TH2-biased respiratory immunity against nominal exogenous antigens is dependent on pulmonary Vα14i NKT cells, and presumably their production of the cytokines IL-4 and IL-13.

Other cells known to express an invariant TCR have also been shown to regulate airway hyperreactivity in allergen-sensitized and allergen-challenged mice. In a recent report by Hahn and colleagues, the investigators used ovalbumin (OVA)-sensitized mice to demonstrate that $V\gamma4^+ \gamma\delta$ T cells act as negative regulators of AHR and that their regulatory effect allows much of the allergic inflammatory response coincident with AHR, including eosinophilic inflammation of the lung and airways, to be bypassed (Hahn et al. 2003b). Thus, invariant chain-expressing T cells, of both the $\alpha/\beta$ and $\gamma/\delta$ types, appear to be involved in regulating allergen-induced AHR in mice. It will be interesting to see if these results are confirmed in humans.

Noninterleukin cytokines have also been implicated in asthmatic inflammation, and these include neurotrophins secreted by macrophages, T cells, eosinophils, and mast cells (Braun et al. 2000). Aspects of allergic inflammation, including vasodilation, increased vascular permeability, contraction of the smooth muscles of the airway and hypersecretion of mucus in the lung, have been attributed to several neuropeptides, particularly including substance P, calcitonin-gene-related peptide, and neurokinin A (Braun et al. 2000) (Figure 14.6). Although these neuronal cytokines are all primarily of sensory neuron origin, they are also expressed in inflammatory cells (Belvisi and Fox 1997), and they have been reported to cause histamine release from mast cells in the lungs (Forsythe et al. 2000).

### *Cytokines Involved in Regulating or Inhibiting Allergic or Asthmatic Inflammation*

Asthma is an inflammatory disease of the peripheral airway, and several groups have been able to demonstrate an association of asthma severity with TH2 polarization in peripheral blood (Fingerle-Rowson et al. 1998; Gately et al. 1998; Humbert et al. 1997; Nurse et al. 1997). Involved in this polarization is the TH2 cytokine IL-10, but this anomalous interleukin should be considered a regulatory cytokine more than a promoting cytokine because of its inhibitory effects on allergic reactions (Figure 14.1). IL-10 has been shown to induce long-term hyporesponsiveness of allergen-specific $CD4^+$ T cells and to decrease mast cell numbers in addition to inhibiting the production of eosinophils (Akdis et al. 1998; Borish 1998; Mackay and Rosen 2001). Moreover, IL-10 has been shown to inhibit eosinophil survival and IL-4-induced IgE synthesis (Takanaski et al. 1994), and in an animal model, IL-10 was shown to inhibit the late-phase response and the influx of eosinophils and lymphocytes after allergen challenge in Brown-Norway rats (Humbert et al. 1997; Nurse et al. 1997; Woolley et al. 1994). Lastly, the phenotype of IL-10 knockout mice includes increased eosinophilic airway inflammation in sensitized and exposed mice and an increased synthesis of IL-5 (Humbert et al. 1997; Nurse et al. 1997; Woolley et al. 1994).

It is well known that IL-12 inhibits TH2 responses by selectively favoring the development of IFN$\gamma$–producing TH1 cells during immune deviation (Visser et al. 1998; Fingerle-Rowson et al. 1998; Gately et al. 1998; Manetti et al. 1993), so regulation of IL-12 is likely an important factor in immune sensitization and the development of asthma. In support of this idea, granulocytes, including basophils and eosinophils, have been shown to produce large amounts of prostaglandin $E_2$ ($PGE_2$), which is a potent inhibitor of human IL-12 production (Manetti et al. 1993) (Figure 14.1). In addition to its secretion from TH1 cells, IFN$\gamma$ has been reported to be secreted by $CD8^+$ T cells, and in this context has been shown to inhibit allergen-induced airway eosinophilia, although not the late-phase airway responses (Suzuki et al. 2002). The dissociation of eosinophilia from the late airway responses in asthma makes this report particularly interesting from a mechanistic point of view. Lastly, TH1-derived IFN-$\gamma$ has been shown to work together with

IL-18, produced by macrophages (Yoshimoto et al. 1998); these cytokines inhibit TH2 responses to antigenic challenge and suppress production of IgE antibodies (Robinson et al. 1997; Kay 2001; Yoshimoto et al. 1998) (Figure 14.1). Synergies of this sort appear to be the rule, rather than the exception, for cytokines involved in asthmatic inflammation.

## *Signaling Pathways and Transcription Factors Involved in Asthma Pathogenesis*

Soluble mediators may play important roles in the sensitization and triggering of asthmatic responses, but they only do so by acting through specific receptors. Binding to a specific receptor by any cognate ligand normally transduces a signal to the nucleus to transcribe, or block the transcription of, a gene or genes via one or more intracellular biochemical pathways. Steps along a pathway are referred to as pathway intermediates, and those intermediates directly responsible for binding to DNA elements in the nucleus and affecting mRNA expression are called transcription factors. When a toxicant is encountered during pregnancy, the potential exists for signal transduction pathways involved in allergic sensitization to be altered. Alterations could exist in many forms, including receptor blockade, receptor antagonism (sending a negative signal), inhibition or activation of pathway intermediates, or interference with the activity of transcription factors. Any of these events has the potential to change gene expression patterns and initiate pathogenesis, including the generation of asthma.

Allergic sensitization of T cells requires an allergen-specific signal transduced by the TCR and a nonspecific costimulatory signal (Mueller et al. 1989). TCR signals received from APC in the absence of costimulation result in allergenic tolerance or anergy (Schwartz 1990). At least two costimulatory pathways are involved in allergic sensitization. The first is the CD28/B7 pathway. CD28 is constitutively expressed on T cells (Figure 14.3), and its receptors are B7-1 (CD80) and B7-2 (CD86) on APC (Figure 14.9). The induction of airway eosinophilia and inflammation requires the ligation of both B7-1 and B7-2 by T cells (Harris et al. 1997; Mark et al. 1998; Tsuyuki et al. 1997). Overexpression of B7-1 and B7-2 in alveolar macrophages from asthmatics has been reported (Agea et al. 1998), and allergen-induced T cell activation involves B7-2, as does expression of IL-5 in human asthmatic airways (Larche et al. 1998). The second costimulatory system involved in allergic sensitization is the CD40/CD40L (CD154) system. When CD40L is expressed on activated T cells and binds to CD40 on B cells, B cell isotype switch machinery is turned on. This machinery includes activation of the nuclear transcription factor called nuclear factor κB (NFκB), which is required for IgE production (Corry and Kheradmand 1999; Kay, 2001). The cascade of events involved in the switch to IgE production also relies on the cytokines IL-4 and IL-13. Interestingly, each of these TH2 cytokines binds to the high-affinity $\alpha$ chain of the IL-4 receptor (reviewed in Wills-Karp et al. 1998). STAT-6 and c-maf are two transcription factors that are subsequently activated (Kay 2001; Mackay and Rosen 2001) (Figure 14.3 and Figure 14.9).

The biochemical signaling pathway of the IgE receptor $Fc_{\varepsilon}RI$, which includes activation of the protein-tyrosine kinase (PTK) Syk (Mackay and Rosen 2001), should also be scrutinized when assessing immunotoxicant effects on allergic asthma sensitization (Figure 14.2). Toxicants affecting Syk activity might increase $Fc_{\varepsilon}RI$-α signaling, a process central to the involvement of mast cells in the triggering of an asthmatic response. Other pathway intermediates intimately involved in allergic inflammation but yet to be investigated in the development of asthma include GATA transcription factors. Expression of GATA-1 has recently been shown to promote the development and terminal differentiation of eosinophils, with eosinophil progenitors failing to develop in the fetal livers of GATA-1 deficient mice (Hirasawa et al. 2002). In the same report, GATA-2 was demonstrated to be able to compensate for GATA-1 deficiency in terms of eosinophil development *in vivo* (Figure 14.5). Recognizing that GATA-1 is a zinc-finger transcription factor, Richter and coworkers found that zinc chelators are able to modulate the labile pool of zinc and regulate gene expression and protein synthesis of C-C chemokines (Richter et al. 2003). In particular, TNFα-stimulated human airway epithelial cells and fibroblasts decreased their production of eotaxin, RANTES and MCP-1 in response to zinc chelators (Figure 14.6). Because of the major roles played by eotaxin and eosinophils in the pathogenesis of asthma, dysregulation of the GATA-1 and GATA-2 transcription factors may also be important to investigate as endpoints in the effects of environmental toxicants on the developing immune system.

Other potential intracellular players in asthma pathogenesis include members of the suppressor of cytokine signaling (SOCS) family. Because this family of proteins is known to be involved in the pathogenesis of many inflammatory diseases, and SOCS-3 is predominantly expressed in TH2 cells, Seki and colleagues investigated the role of SOCS-3 in TH2-related allergic diseases. The authors showed a strong correlation between SOCS-3 expression, serum IgE levels in allergic human patients, and the pathology of asthma and atopic dermatitis (Seki et al. 2003). Moreover, SOCS-3 transgenic mice developed pathological features characteristic of asthma. These transgenic mice exhibited increased TH2 responses, whereas decreased TH2 responses were elicited in dominant-negative mutant SOCS-3 transgenic mice, as well as in mice with a heterozygous deletion of Socs3. Because these data suggest a critical role for SOCS-3 in the onset and maintenance of asthma, it would be prudent to monitor this marker in response to gestational exposures during critical windows of development (Figure 14.3).

We have shown that the multiple levels of complexity of asthmatic inflammation involve the bronchial airways, the circulation and lymphoid systems, cell-cytokine networks, and intracellular signaling. Understanding how maternal influences may affect fetal asthma sensitization will require an integrated view of the components of the asthma sensitization engine detailed earlier. For an integrated overview of how all of the parts of that engine fit together, Figure 14.1 through Figure 14.9 can be laid out into a grid, as shown in Figure 14.10. The contribution of the cells and molecules described above to fetal asthma sensitization will be discussed in the next section.

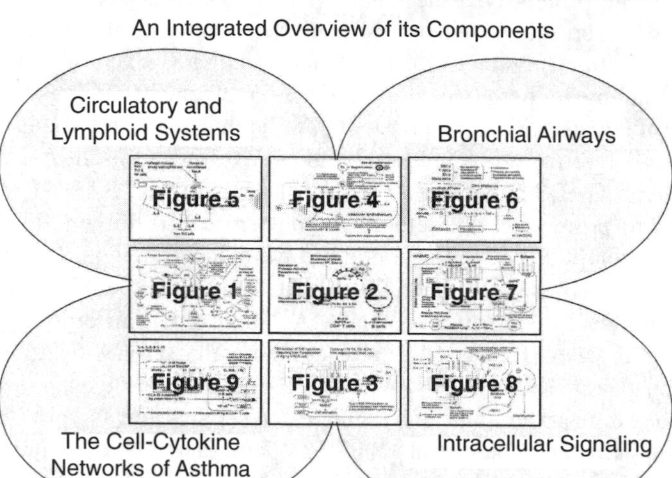


**Figure 14.10** The asthma sensitization engine. For an integrated, birdseye view of how the bronchial airways, circulatory and lymphoid systems, cell-cytokine networks, and intracellular signal transduction pathways involved in asthmagenesis interact, Figure 14.1 through Figure 14.9 can be laid out in the grid pattern depicted here.

## SECTION SUMMARY

Although the cellular and molecular processes involved in allergic sensitization differ from those initiated when an asthmatic response is triggered, allergic asthma is a pathologic secondary immune response resulting from re-exposure to an allergen encountered earlier. Chronic activation of allergen-specific T cells and associated airway eosinophilia cause the remodeling of airway tissue architecture characteristic of asthmatic inflammation. Mechanistically, these processes result from the activity of B cell-derived, allergen-specific IgE molecules and the degranulation of $Fc_{\varepsilon}RI$-bearing mast cells. The subsequent release of inflammatory mediators leads to contraction of smooth muscle cells and bronchoconstriction, vasodilation and lowered blood pressure, and increased vascular permeability and edema. An important feature of the pathogenesis of asthma is the recruitment by cytokines of eosinophils and other inflammatory cells. In general, the hypersecretion of mucus, shortness of breath, and wheezing associated with asthma are the products of a complex network of interactions involving cytokines, receptors, signaling pathways, and transcription factors that both promote and regulate allergic and asthmatic inflammation. It is hoped that these discussions have adequately set the stage for the following discussion of how maternal–fetal interactions might affect allergen sensitization in the fetus and neonate.

## THE INFLUENCE OF MATERNAL-FETAL INTERACTIONS ON ALLERGEN SENSITIZATION IN THE HUMAN FETUS AND NEONATE

Although much is known about sensitization and triggering of immune responses in human adults, a paucity of information exists regarding the normal course of these processes in fetuses. What is known about the interplay of antigens and allergens with the developing immune system in the intrauterine environment will now be presented, but much research is still required into the mechanisms of basic fetal sensitization and the neonatal triggering of normal and pathological immune responses.

Developmental studies have detailed the emergence of the human fetal immune system (reviewed in Chapter 2). Although stem cells are seen in the human yolk sac at 21 days of gestation (Hayward 1981), lymphocytes do not populate the fetal thymus until 8 to 9 weeks of gestation. They then migrate to the circulation by approximately 14 to 16 weeks of gestation (Donovan and Finn 1999). B lymphocytes appear in several organs, including the lungs and gut, from 14 weeks of gestation, and by fetal week 19-20, circulating B cells have detectable surface IgM (Hayward 1981). Circulating T cells at 22 weeks of gestation present a naïve phenotype with low expression of the cell surface antigen CD45RO and high concentrations of CD45RA (Byrne et al. 1994; Peakman et al. 1992); few third-trimester fetuses and neonates have circulating T cells expressing high levels of CD45RO, a classic "memory" cell marker (Frenkel and Bryson 1987; Hannet et al. 1992; Lewis 1998; Sanders et al. 1988).

In third-trimester fetuses and neonates, memory T cells and the production of cytokines such as IFN-gamma gradually increase. This has been attributed to an increase in antigenic exposure and T cell activation (Frenkel and Bryson 1987; Hannet et al. 1992; Lewis 1998; Sanders et al. 1988), yet the observation that both neonatal T cells and adult naïve T cells have a reduced capacity to proliferate and produce cytokines compared to memory or effector T cells may be due, in part, to their lack of antigenic experience (Donovan and Finn 1999). This review of the anatomy of—but absence of responses in—the developing immune system in fetuses provides only a general understanding of the nascent defense system. An overarching question is whether human fetuses, neonates, and newborns are innately compromised in the ability of their immune systems to respond to antigen because of defects in T cell activation, costimulation, or antigen processing (Burchett et al. 1992; Paryani and Arvin 1986; Starr et al. 1979)—i.e., developmental issues—or if their reduced capacity to respond to antigen is due to a lack of antigenic experience—i.e., an exposure issue (Lewis 1998). Obviously, these two lines of argument cannot be studied using a reductionist approach, as developing and gaining experience occurs concomitantly. Immunologic "maturation" happens around the time of birth until about 6 months of age when the T helper environment within the infant is established (Hanrahan and Halonen 1998; Prescott et al. 1998). The framework for this period of maturation may be best understood by studying maternal-fetal interactions.

A somewhat compelling argument is that the maternal-fetal environment in developed countries differs from that of developing countries in the level of pathogenic exposure. Epidemiologic studies indicate that "clean," developed countries have an increase in asthma rates, suggesting a direct relationship between pathogen exposure and normal (nonasthmatic) immune system maturation. However, lack of pathogen exposure does not necessarily mean lack of antigen exposure. The debate then focuses on whether the body recognizes pathogens vs. innocuous antigens, or — more specifically for the developing immune system of a fetus—whether the mother's immune system conditions the fetus to "interpret" the antigen it encounters as a threat to be reacted against or an agent towards which tolerance should be developed. In the first scenario, the "Danger Theory" of antigenic stimulation (Matzinger 1998) envisions that sensitization of the immune system occurs to dangerous, or necrotic, events. In the second scenario, the intrauterine environment influences later maturational pathways and provides a form of imprinting on the fetus, i.e., the state of the mother's immune system influences the fetus's developing immune system. Again, resolution of these theories is difficult, as the response of a fetus to a pathogen or antigen cannot be measured out of context of the mother's immune system. Moreover, coupling native vs. maternally influenced fetal immune responses to asthma development requires a time element that allows impact of later environmental exposures to cloud the issue. Understanding the influences on fetal immune system development must be obtained despite omnipresence of antigens (pathogenic or not) and the unique pregnancy environment. It is into this realm that we focus and provide a review of the scientific data.

### Fetal Sensitization and Subsequent Immune Responses in Childhood

Several groups have shown that responses set up during pregnancy can be predictive of subsequent allergic disease (Kondo et al. 1992; Prescott et al. 1998; Warner et al. 1994). Epidemiological data have been of particular importance in showing associations between various maternal factors and childhood asthma. Such maternal factors that are associated with an increased risk of childhood asthma include younger age (Anderson et al. 1986; Martinez et al. 1992; Schwartz et al. 1990), smoking (Martinez et al. 1992; Oliveti et al. 1996; Weitzman et al. 1990), lack of prenatal care, and weight gain of less than 20 pounds (9.1 Kg) during pregnancy (Oliveti et al. 1996; reviewed in: Donovan and Finn, 1999).

Experimental studies have shown that peripheral blood mononuclear cell sensitivity to allergens exists at birth (Kondo et al. 1992; Prescott et al. 1998; Warner et al. 1994; reviewed in: Warner and Warner 2000). In particular, specific allergen-induced responses can be measured in the peripheral blood mononuclear cells as early as 22 weeks into gestation (Jones et al. 1996). Moreover, events after birth are believed to modify the developing immune response in newborns; allergens, infections, diet, and gut microbial flora have all been implicated in the development, or not, of subsequent allergy (Warner and Warner 2000). The impact of diet on the development of allergies in newborns is now being recognized. The health benefits cited for breastfeeding include a reduction in childhood asthma (Oddy et al. 2002) and may be directly tied to gut microbial flora (Bjorksten et al. 1999; Holt et al.

1997; Kirjavainen and Gibson 1999). Moreover, it has been reported that allergic children are more likely to have a low colonization of *Lactobacilli* in their guts than non-allergic children (reviewed in: Warner and Warner 2000).

Finally, disease due to infection affects immune system maturation. The hygiene hypothesis proposed by Strachen (1989) is based on the inverse relationship between birth order in families and the prevalence of hay fever, as well as on an awareness that infections in early infancy brought home by older siblings might prevent sensitization. Further support comes from a recent report correlating reduced rates of asthma and wheezing among 812 rural European children with exposure to bacterial substances in dust from mattresses (Weiss 2002). The hygiene hypothesis and the gut microbial flora, diet, and allergen exposure issues likely all play roles in determining whether the events set up during pregnancy result in the development of allergic disease (Warner and Warner 2000). However, consideration must be taken of the unique intrauterine environment, which may actively dampen the neonate's immune system. One study demonstrates that birch and timothy grass pollen exposure via the mother sensitizes fetuses if it takes places in the first six months of pregnancy. Thus, exposure in later pregnancy appears to result in either immune suppression or tolerance (Van Duren-Schmidt et al. 1997). Obviously, much remains to be known about the myriad factors uniquely surrounding the development of the fetal and newborn immune system.

## Fetal IgE

Although T cells act as a driving force in the asthmatic response, the IgE-mast cell interaction is a central component, as it is for all immediate-type hypersensitivity responses. Therefore, understanding the production of IgE in the fetus provides a key for not only elucidating asthma development in children, but also in exploring the maternal influence on fetal and neonatal sensitization to allergens.

Several studies indicate that increased cord blood levels of IgE are associated with an increased risk of atopy or asthma (Croner and Kjellman 1990; Edenharter et al. 1998; Halonen et al. 1992; Hansen et al. 1992; Hide et al. 1991; Kjellman and Croner 1984; Michel et al. 1980; Ruiz et al. 1991). Other studies do not bear out this association (Hide et al. 1991; Martinez et al. 1995). However, a general belief is that an early and inappropriate immune response can develop toward allergens and that young children with asthma or allergic rhinitis have higher than normal levels of antibodies to inhaled allergens (Okahata et al. 1990). The evidence that fetal B cells are capable of isotype switching supports this idea (Punnonen et al. 1993; Punnonen and de Vries 1994). Furthermore, production of IgM, IgG subclasses, and IgE response to CD40 ligation and cytokines from T cells is similar in neonatal B cells and antigenically naïve adult B cells (Briere et al. 1994; Servet-Delprat et al. 1996), indicating equivalent maturational states of these cell types. However, T cell-dependent isotype switching and immunoglobulin production is limited in a fetus in that it has been documented only for certain pathogens such as in toxoplasmosis; congenital infections such as varicella zoster virus do not induce specific immunoglobulin by birth or early childhood (Paryani and Arvin 1986; Pinon et al. 1990).

Even if the capacity to respond to antigen exists in fetuses and newborns, the immune response to innocuous antigen could be due to developmental issues. One line of reasoning states that altered fetal and neonatal B cells, responses result from reduced expression of CD40L on T cells or reduced function of neonatal dendritic cells (Lewis 1998; von Hoegen et al. 1995) (Figure 14.3 and 14.9). Another theory proposes that the allergic response occurs when the amniotic fluid contains a high concentration of cytokines produced by decidual tissues (Jones et al. 2000). Maternal atopy has been shown to increase IgE levels in amniotic fluid (Jones et al. 1998) and, in fact, the allergens Der p1 of house dust mites and ovalbumin from hen's eggs have been detected in amniotic fluid (Bloomfield and Harding 1998). Fetuses swallow the amniotic fluid as well as aspirating it into their respiratory tract; furthermore, fetal skin is highly permeable and direct exposure may occur (Bloomfield and Harding 1998; Jones et al. 1998). Fetal gut has the most mature immune active tissues; several types of fully effective APC concentrate in rudimentary Peyer's patches early in the second trimester of pregnancy. The necessary costimulatory signals facilitate antigen presentation by these APC to T lymphocytes, and the APC have high- and low-affinity receptors for IgE as well as for IgG. Thus, the potential exists for sensitization to occur *in utero* (Warner and Jones 2000).

### The TH2-Bias of the Neonatal Immune System

Before the fetus has even a rudimentary immune system, the maternal immune response has had to adapt to permit development of a child whose paternal antigens are cause for immune rejection. Evidence exists in both murine and human studies that during pregnancy, TH2 immune responses dominate in the mother, reducing TH1 type reactions that would presumably mediate rejection of the fetus (reviewed in: Donovan and Finn 1999; Warner and Warner 2000). In support of this idea, Sills and coworkers demonstrated a relationship between the TH1-associated cytokine TNF and infertility. In that report, chronic therapy for infertility using biologic agents to block TNF was associated with successful induction of ovulation, conception, and normal delivery (Sills et al. 2001). Pregnant women have been shown to have compromised cell-mediated immunity (Weinberg 1984), impaired lymphocyte proliferation (Matthiesen et al. 1996), and a loss of peripheral blood mononuclear cell proliferative responses to recall antigens (Bermas and Hill 1997). However, the TH2 response, which promotes antibody production, functions effectively; production of maternal antibodies specific for paternal antigens occurs. A lack of these antibodies in women with spontaneous abortion is associated with a TH1 autoimmune response (Bermas and Hill 1997; Donovan and Finn 1999). Overall, data indicate that successful pregnancy requires a TH2-biased maternal immune response.

Fetal development also supports a TH2 environment during pregnancy. The fetal trophoblast only expresses low-affinity receptors for TH1 cytokines, such that TH2 responses predominate at the maternal-fetal interface (Lin et al. 1993; Raghupathy 1997). Moreover, the TH2 cytokine IL-4 is produced by human amnion epithelium (Jones et al. 1995), and IL-10 and IL-13 from the placenta are detectable during the second trimester of pregnancy (Cadet et al. 1995; Jones et al. 1995; Williams et al. 2000). It has also been thought that common environmental allergens that cross the placenta prime T cells of the fetus *in utero* (Kay 2001) and that this contributes to TH2-domination of the immune response in virtually all newborns (Prescott et al. 1998).

The observations that the *in utero* environment supports TH2-type immune responses has given rise to a hypothesis that atopy in children derives from a persistence of this bias beyond birth (Donovan and Finn 1999; Holt et al. 1992; Holt and Macaubas 1997; Prescott et al. 1998a; Prescott et al. 1998b). Normally, a shift occurs in an infant's immune system shortly after birth to favor the adult TH1-mediated responses. This developmental phenomenon is termed "immune deviation" (Holt et al. 1999). Atopic infants instead may increase the TH2 cells that were primed *in utero* (Kay 2001). Because macrophages that engulf microbes secrete IL-12, a cytokine that induces Natural Killer (NK) cells to produce IFNγ and in general promotes TH1-type cells, it is believed that microbes are the chief stimulus of protective TH1-mediated immunity (Kay 2001). This idea leads into the hygiene hypothesis discussed earlier proposing that lack of immune deviation is due to an overly clean environment, that is, one deprived of microbial antigens that stimulate TH1 cells (Rook and Stanford 1998). A potential contributing factor to this situation in Western countries is the overuse of antibiotics for minor illnesses in early life. Another indication that repeated immune stimulation may protect against atopic allergy is the observation that atopic allergic diseases are less common in younger children who have three or more older siblings, and among children who have had measles or hepatitis A (Openshaw and Hewitt 2000). This hygiene hypothesis is supported by evidence that exposure of young children to older children at home or to other children at day-care can protect against development of asthma in childhood (Ball et al. 2000). However, in contrast, an increased prevalence among poor blacks in the United States of atopic asthma is associated with sensitization to cockroaches and house dust mites (Levy et al. 1997).

While many other factors may also favor the persistence of the TH2 phenotype, Kovarik and colleagues have shown that microbial CgP oligodeoxynucleotides (ODN) can circumvent this polarization in neonatal responses to vaccines, although they may fail to fully redirect TH2 responses established by neonatal priming (Kovarik et al. 1999).

## Genetics, Epigenetics, and Asthma

Without a doubt, genetics plays a role in the development of asthma. However, family and population studies show that asthma is a complex genetic disorder (Saltini et al. 1998). The genetic studies of asthma generally take one of two approaches: either phenotypic markers are used to search for asthma genes, or Human Leukocyte Antigen (HLA) genes are associated with responsiveness to specific proteins or peptide allergens. As noted in a review by Saltini and colleagues, the phenotypic markers most commonly used in the search for asthma gene(s) include: 1.) total serum IgE levels; 2.) immediate reactivity in skin tests to aeroallergens and specific IgE levels; 3.) bronchial hyperresponsiveness to physical and pharmacological stimuli; and 4.) a history of wheezing in the clinical diagnosis of asthma (Saltini et al. 1998). In that review, it was noted that the genetic loci tentatively associated with asthma are on chromosome 11q13 in close proximity to the IgE receptor, on chromosome 5q31 near a cytokine cluster comprising the IL-4, IL-5, and IL-13 genes, and near the beta-adrenergic receptor gene. These authors also reported that other studies do not confirm these findings (Sandford et al. 1996) and, in particular, linkage studies identify over a dozen genomic regions linked to asthma. In July 2002, Van Eerdewegh and colleagues reported finding a gene that is closely linked to bronchial hyperresponsiveness and asthma (2002). In that report, investigators analyzed the genomes of 460 families to link the ADAM33 gene to increased asthma susceptibility. ADAM proteins are membrane-anchored metalloproteases with functions including the shedding of cell-surface cytokines and cytokine receptors.

The second approach to establishing a genetic association with asthma emphasizes HLA molecules. Support for this approach comes from studies claiming a weak association of asthma with HLA-A1, B8, and DR3 (Apostolakis et al. 1996; Turner et al. 1977), but more studies are needed to confirm these associations. Narrowing the scope to defined allergens or to a type of asthma, i.e., occupational asthma, has helped define HLA associations. Moreover, it may be as likely in occupational asthma as in contact dermatitis that a subpopulation of exposed individuals with "leaky" airways due to atopy or injured bronchial epithelium are more sensitive to environmental agents (Brooks 1992). Occupational asthma provides a model to test the hypothesis on susceptibility because of the diagnostic requirement for a causative agent to be identified. HLA genes appear to be directly linked to chemical susceptibility leading to occupational asthma; ultimately, HLA genes provide a genetic basis for this disease.

Lastly, if the forces influencing susceptibility to asthma are visualized as components of a Venn diagram, factors such as environment and molecular dysregulation should not be solely supplemented by genetics per se. In light of information from recent reports, balancing an equation that fully integrates factors determining disease risk should include epigenetic factors (Figure 14.11). Defined as changes having no effect on the sequence of molecules that make up a genome, the study of *epigenetics* refers to the pattern of gene silencing and activation. Although every cell contains two copies of almost every gene, many genes in sperm and ova are biochemically altered to be silenced, or imprinted, throughout life. The potential relevance of epigenetics to the influence of perinatal toxicant exposure on asthma sensitization becomes clear when one considers recent data from experiments conducted in mice. In a report by researchers at Duke University (Waterland and Jirtle 2003), mice with identical genes for fur color displayed two different colors. The difference was that, in some mice, the gene for fur color had been silenced by what their mother ate during pregnancy. Because the genetic sequence remains the same, the changes associated with imprinting are epigenetic. Epigenetic changes are suspected of being at the root of *fetal programming*, which is the term applied when nutrition, and presumably exposure to environmental toxicants, during gestation seems to affect the risk of disease later in life. One example illustrating fetal programming as a result of epigenetic changes in imprinting involves children conceived by *in vitro* fertilization (IVF) who develop Beckwith-Wiedemann syndrome (BWS). In that disease, where the incidence is roughly six times higher than in non-IVF babies, a normally silenced gene (KCN10T) loses

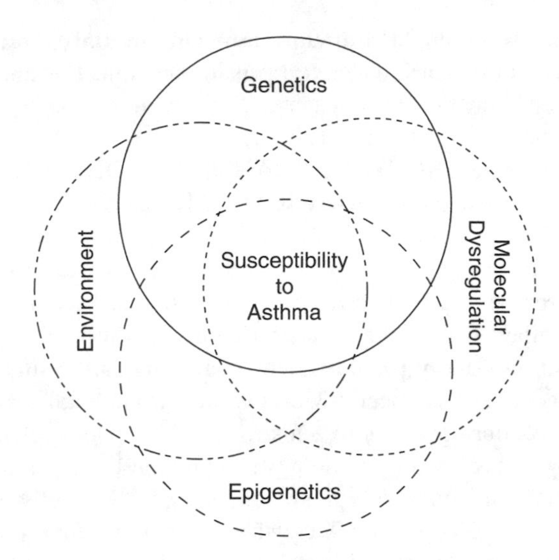


**Figure 14.11** The interplay of factors contributing to a susceptibility to asthma. An understanding of how environmental toxicants might affect components of the asthma sensitization engine *in utero* must derive from studies of more than genetics and the environment. In addition, elucidation of factors important in programming fetuses for susceptibility to asthmagenesis will likely involve studying dysregulation of many physiologic systems on a molecular scale, as well as epigenetic factors that affect genetic expression but do not change genomic sequences.

its imprinting. The result is an increased production of a growth factor and an associated increase in childhood and adult cancers (Gicquel et al. 2003). Other examples of adult diseases known to be influenced by fetal perturbations include type II diabetes, hypertension, and coronary artery disease (reviewed in: Sallout and Walker 2003). Whether epigenetic phenomena influence the development of asthma remains to be studied.

## SECTION SUMMARY

Antigen sensitization in the fetus and neonate is a normal part of immunologic maturation and can be affected by the timing and route of exposure during development. Genetics and environment both play roles in fetal sensitization, and the intrauterine environment may influence later maturational pathways and provide a form of imprinting on the fetus. A child is born with a TH2-biased immune system, and maturation to a TH1-biased system may be important in reducing the likelihood of childhood asthma. Sensitization *in utero* stems from maternal-fetal interactions, but the proportion of sensitization attributable to maternal vs. fetal exposures is uncertain. Clearly, a fetus makes its own IgE, but it can swallow and aspirate maternal IgE-rich amniotic fluid as well, the placental barrier to non-IgG isotypes notwithstanding. While the genetics of susceptibility can be concretely described in terms of family histories, the ability to identify susceptibility genes is only now becoming a reality, and much work remains to be done in this area. Just how these developmental sensitization issues might be affected by environmental exposures to allergens and toxicants encountered during development is the subject of the next section.

## FETAL EXPOSURE TO TOXICANTS AND DEVELOPMENT OF CHILDHOOD ASTHMA

As discussed in the previous section, when allergens are introduced to the intrauterine environment, a fetus can become sensitized. Maternal toxicant exposure has the potential to affect the sensitization process in various ways. Whereas *in utero* sensitization usually provides neonates with healthy immune responses, hyperresponsiveness can also occur. Moreover, environmental toxicants concomitantly encountered during *in utero* sensitization have the potential to deviate normal physiologic processes to pathological outcomes, including asthma. These processes potentially include: isotype switching by fetal B cells; functional maturation of fetal APCs, B cells, and T cells; expression of adhesion molecules on inflammatory cells, thus affecting migration to the airways; a change in the cytokine secretion profiles of T cell subsets; epigenetic changes affecting fetal programming; and modulation of biochemical signal transduction pathways and transcription factors in any of the cells involved in asthmatic inflammation and pathogenesis. The following section reviews what little is known regarding the development of asthma from maternal-fetal exposures to allergens, environmental toxicants, and tobacco smoke.

### Allergens and Maternal Exposure

#### *Relationship between Allergens and Development of Asthma*

In developed countries, such as the U.S., 80% of childhood asthmatics are allergic to indoor allergens (National Academy of Sciences 2000). A growing body of evidence suggests that increased exposure to these types of allergens may be relevant to the increased incidence and severity of asthma in children. Asthma manifests as a set of symptoms ranging from mild to severe. However, displays of asthma-like symptoms in infancy may or may not lead to persistent atopic childhood asthma. Furthermore, while the development of childhood asthma is agreed to result from an interplay of genetics and environment, that interplay is poorly understood. What is known is that most childhood asthma results from allergen sensitization in a genetically predisposed individual. Moreover, immunoregulatory mechanisms may be altered at mucosal surfaces in ways that promote a TH2-mediated allergic inflammatory response, and these alterations may be influenced by cofactors such as tobacco smoke and air pollutants (Holgate 1999; Kay 2001), as well as the frequency and type of respiratory infections, and the frequency (perennial or not) and intensity of allergen exposure (National Academy of Sciences 2000). While indoor allergens certainly exacerbate asthma in children, it must be understood that no specific allergens have been definitively linked to the development of asthma following exposure *in utero*. For the most part, these studies simply have not been done, and the effect of maternal or neonatal exposure to any allergen during the development of the immune system on enhanced susceptibility to asthma must be explored further. What is known of the influence of common indoor allergens associated with asthmagenesis (asthmagens) will now be discussed.

### Dust Mites and Cockroaches

The National Academy of Sciences (NAS) stated in 2000 that there is *sufficient evidence of causal relationship* between exposure to house dust mite allergen and the development of asthma in susceptible children (National Academy of Sciences 2000), and good correlation has been shown between early high level exposure to house dust mite and the subsequent increase in the prevalence and severity of asthma (Sporik et al. 1990). Moreover, in spite of many nonasthmatics testing positive in skin tests for dust mite sensitivity, early sensitization has been associated with a greater probability of the persistence of bronchial hyperresponsiveness and symptoms of asthma in late childhood and adolescence (Peat et al. 1990).

Similarly, cockroach antigen exposure is known to elicit a strong IgE response from B cells to induce sensitization, and the NAS has reported that 1.) *sufficient evidence of causal relationship* exists between cockroach (CR) allergen exposure and exacerbation of asthma in individuals specifically sensitized to CR, and 2.) there is *limited or suggestive evidence of an association* between CR exposure and the development of asthma in preschool-age children (National Academy of Sciences 2000).

Interestingly, the season of birth also seems to be relevant to CR antigen sensitization. The link made by authors Sarpong and Karrison is between CR sensitization and children born in winter, when certain viral infections are also common (Sarpong and Karrison 1998). While no studies have identified a link between infant wheezing with rhinovirus infections and development of asthma, since there is an increased incidence of respiratory syncytial virus (RSV) infections in winter, the authors have speculated that RSV (or other virus) infections concomitant with CR Ag exposure may promote IgE sensitization and the induction of asthma. Although this link is controversial (Roman et al. 1997), active research into these sorts of relationships holds the promise of unraveling some of the complexity involved when exploring the links between neonatal exposures, elevated IgE levels, infant wheezing, and the actual development of asthma.

### Pet Dander, Fungi/Mold, and Pollen

In general, evidence is weak or nonexistent supporting connections between pet dander or fungi and asthma development. Clearly, pet dander and fungi exacerbate asthma in individuals sensitized to a particular allergen, but evidence is lacking that pet dander contributes to asthma development, and only recently have preliminary studies started to make a connection towards fungi exposure and asthmagenesis (National Academy of Sciences 2000).

Perhaps the best evidence linking *in utero* exposures to allergen sensitization in infancy comes from a report showing that exposure to birch pollen during the last

2 to 3 months of pregnancy is associated with much less birch pollen reactivity of the offspring than occurs if the exposure was between 3 and 6 months' gestation (Van Duren-Schmidt et al. 1997). As the authors of that study pointed out, of particular relevance is that IgG transfer across the placenta is maximal during the latter few months of pregnancy, so maternal/fetal interactions may be playing a major role in this case. Moreover, other studies have shown that children are at increased risk for allergy to seasonal allergens if born shortly *before* the relevant pollen season, when protective maternal (and presumably fetal) IgG antibodies would be at their lowest levels (Jenmalm and Bjorksten 2000). It will be particularly interesting to learn from future studies if there is a concomitant increase in risk for the development of asthma, since many of the same allergens are involved.

## Environmental Toxicants and Maternal Exposure

This section reviews the available information on asthma and nonallergen environmental contaminants for which there is potential for exposure in both the general population and the developing fetus. Although there is great interest in the potential associations between maternal exposure to environmental contaminants, such as pesticides, polycyclic halogenated hydrocarbons, and heavy metals, and the development of asthma in offspring, there has been very little research in the area. One of the most prevalent environmental contaminants, environmental tobacco smoke, is reviewed in a subsequent section.

### *Pesticides*

Many pesticides are considered to be persistent chemicals, which means they are resistant to degradation and have become widespread contaminants throughout the environment. Because of their lipophilic qualities, they are known to accumulate in the fatty tissue of animals, including humans. The ability of these persistent compounds to bioaccumulate allows for transfer of the contaminant to the fetus through the placenta and to infants through breast milk (Tryphonas 1998; Vial et al. 1996). However, a recent study by Karmaus and colleagues indicates that contaminant pesticide in breast milk may negate the protective effect of breast milk on atopic manifestations (Karmaus et al. 2003).

Exposure to pesticides has been reported to cause a number of immunotoxic effects: hypersensitivity, autoimmunity, modulation of immune function, and clinical immunosuppression. Several authors have recently reviewed pesticide immunotoxicity (Banerjee 1999; Colosio et al. 1999; Thomas 1995; Vial et al. 1996; Voccia et al. 1999). The majority of studies investigating the effects of pesticides on the human immune system involve relatively high-level occupational exposures. Few studies have investigated pesticide immunotoxicity in the general population at relevant, environmental exposure levels (Thomas 1995).

More specifically, there has been very little research on the influence of exposure to persistent pesticides and either the development or exacerbation of the allergic asthma response or other atopic manifestation (Ernst 2002). A few epidemiological studies have found significant associations between asthma and other respiratory health problems and occupational (i.e., high-level) exposure to certain pesticides and various allergens in rural and agricultural settings (reviewed in: Vial et al. 1996; National Academy of Sciences 2000). The relevance of these studies to potential links between asthma exacerbation and lower-level pesticide exposures is not known. The mechanism of the reported asthma-like response in these studies is poorly understood, and there has only been one study indicating an immunologically mediated response directed against a pesticide molecule (see "Polycyclic Halogenated Hydrocarbons" section below).

Several studies in animals have shown that prenatal and neonatal exposure to organochlorine pesticides, such as chlordane, toxaphene, and hexachlorobenzene, does have significant effects on the developing immune system. These effects on the immune system are persistent and occur at dose levels that do not elicit effects in adult animals. However, the endpoints assessed in the studies are largely indicative of immunosuppressive effects rather than atopic or allergic responses (Thomas, 1995; Voccia et al. 1999). Whether the immunosuppression reported in these studies could be TH1-specific and thereby favor TH2-mediated diseases such as asthma will have to be borne out in further, more narrowly defined studies.

### *Polycyclic Halogenated Hydrocarbons*

Polycyclic halogenated hydrocarbons (PHH) are persistent synthetic or man-made chemicals that are or were heavily used in industry or are by-products of industrial processes. The PHHs of highest concern are halogenated aromatic hydrocarbons, including polychlorinated biphenyls (PCBs) and polychlorinated dibenzo-p-dioxins (dioxins). Both PCBs and dioxins have been shown to be highly toxic to many organ systems in animal studies and are potent fetotoxins (Carpenter 1998; Mukerjee 1998). PCB/dioxin immunotoxicity has been recently reviewed in two studies (Tryphonas 1998; Weisglas-Kuperus 1998). Occupational and transient high-level exposure from industrial accidents show some immunomodulations in humans, however, only *one* study has examined endpoints indicative of hypersensitivity, rather than immunosuppression. In 2001, Karmaus and Kruse (Karmaus et al. 2001) conducted an epidemiological study of 340 children that investigated associations between exposures to several persistent environmental contaminants and prevalence of otitis media, respiratory infections, asthma, and elevated IgE levels. Although the authors did not find any associations between PCBs or hexachlorobenzene and asthma symptoms or elevated IgE levels, exposure to the insecticide dichlorodiphenyldichloroethene (DDE) was found to be associated with both self-reported asthma symptoms and higher blood levels of IgE. In another study, Reichrtova et al. used placental contamination as a biological marker for exposure of the fetus via placental transfer (Reichrtova et al. 1999). The placental samples were assayed for 21 organochlorine compounds, including dioxin. Specimens of cord blood from neonates were analyzed for levels of total immunoglobulin E. Their results showed higher cord blood IgE levels following *in utero* dioxin/organochlorine exposure. Increased IgE levels may be associated with allergic sensitization, but it remains to be seen whether these increases can be used as a predictor of subsequent development of atopic asthma.

### *Other Indoor Chemical Contaminants*

The NAS 2000 report (National Academy of Sciences 2000) examined evidence for associations between various indoor chemical contaminants such as nitrogen dioxide, volatile organic compounds, and formaldehyde, and the development or exacerbation of asthma. The report concluded that there is *insufficient evidence to determine whether there is any link* between exposure to any of these environmental contaminants and the exacerbation or development of asthma.

### *Heavy Metals*

#### *Mechanisms of Immunotoxicity*

The immunotoxic effects of heavy metals, such as lead, arsenic, and mercury exposure are well recognized (Bernier et al. 1995; Schuppe et al. 1998). In comparison to most environmental toxicants, the mechanism of metal immunotoxicity is well studied (reviewed in: Schuppe et al. 1998). It is currently believed that metals initially elicit a cell-mediated immune response by binding to MHC class II molecules or MHC-bound self-peptides. The metal-MHC complex is then recognized by $CD4^+$ T cells. Additionally, metal-induced alteration of self-peptides may also cause activation of autoreactive T cells. In hypersensitivity responses, such as contact allergy, $CD8^+$ T cells may also be involved.

In addition, metals may alter the T cell repertoire and influence the TH1:TH2 ratio (see previous section "The TH2-bias of the Neonatal Immune System"). Results from recent *in vitro* studies are consistent with the hypothesis that metals, such as mercury and lead, inhibit the production of TH1 cells and enhance the production and activities of TH2 cells (Bernier et al. 1995; Selgrade, 1999). Selgrade and colleagues initiated a series of *in vivo* studies in mice to determine the mechanism by which metals influence the T cell balance (Selgrade 1999). Their results show that lead induces the expansion of *specific* $CD4^+$ T cells, which leads to a 10-fold higher TH2:TH1 ratio in lead-exposed mice than in controls (Figure 14.1). Results from subsequent experiments suggest that lead exposure can also enhance the expression of APC accessory molecules that preferentially activate TH2 cells. The ability of metals to alter the TH1:TH2 balance may have significant impact on the developing immune system, resulting in enhancement of the atopic response and potential susceptibility to development of childhood asthma.

#### *A Focus on Lead, and the Potential for Fetal/Neonatal Exposure*

The general population has the potential for chronic low-level exposure to heavy metals primarily through ingestion of contaminated food, and, to some extent, air pollution. Metals such as lead are very persistent once they enter the human body and can accumulate in bone deposits (Agency for Toxic Substances 1997). Mobilization of maternal bone lead deposits or lead circulating in maternal blood can result in exposure to the fetus through the placenta and to a breastfed infant through lactation (Snyder et al. 2000). Lead contamination in drinking water can also lead to exposure of the fetus in pregnant women, formula-fed infants, and children.

The extreme sensitivity of the fetus, infant, and young child to the effects of relatively low-level lead exposure is well documented (Agency for Toxic Substances 1997). In both developing humans and animals, even very low blood lead levels can cause the impairment of neurobehaviorial development and permanent brain damage. At higher levels, lead causes general growth and organ development impairment and abnormalities (Agency for Toxic Substances 1997).

### *Epidemiological Studies*

Although the effects of lead exposure on the developing central nervous system have been extensively studied, little work has been done on the effects of lead on other developing organ systems. In an important effort to fill the information gap, Lutz and colleagues have initiated a series of studies in a 5-year comprehensive evaluation of the effects of environmental lead exposure on the human developing immune system. Their most recently published study (Lutz et al. 1999) evaluated various immunological endpoints in a cohort of 279 children aged 9 months to 6 years who had been primarily exposed to lead-based paint and had blood lead levels ranging from 1 to 45 μg/dl. This range includes levels of lead that have been shown to cause neurotoxic effects in children. In contrast with previous studies of occupationally exposed adults, Lutz and colleagues found that higher levels of lead in children correlated with elevated levels of IgE. The authors note that although the correlation between IgE and blood lead levels is statistically robust, it is only a correlation and there is no direct evidence of any adverse immune system effects. The authors also note that the correlation between high blood lead levels and high IgE is not consistent with the findings of other investigators.

The Lutz study was designed to control for several important variables in sensitivity to lead toxicity, such as age, gender, race, nutrition, and socioeconomic status. Other variables could not be controlled, such as exposure to specific allergens that may induce elevated IgE levels. For example, cockroach allergens and cigarette smoke in the home are linked to aggravating atopic diseases, such as asthma. The authors note it is very likely that both cockroach allergens and lead paint are endemic to particular socioeconomic levels; therefore, the immunomodulations detected in this group may not be directly attributable to lead. Although the initial results reported in this study must be interpreted with caution, future studies should provide important data to fill information gaps in the effects of lead on the developing immune system of young children.

### *Animal Studies: Developing Immune System*

A few studies have investigated the effects of lead on the developing immune system using endpoints indicative of a hypersensitive or atopic response. A recent study by Miller and colleagues investigated whether the immune system of the fetus is more susceptible to lead exposure than that of the adult, and whether the immunotoxic effects seen in developing animals persist into adulthood (Miller et al. 1998).

The investigators designed the study to determine effects from prenatal exposure only. Female rats were fed low to moderate doses of lead acetate during pregnancy. Their offspring received no additional lead exposure after birth. Various immunotoxic endpoints, including IgE levels, were evaluated in the offspring at 13 weeks (a developmental stage analogous to that of human adulthood). Although dams in lead-exposed groups did not show immunotoxic effects compared to controls, a number of immunomodulations were detected in their adult offspring. The offspring that had been exposed to lower (100 ppm) doses displayed elevated IgE levels. Those that had been exposed to moderate and high dose levels (250 to 500 ppm) displayed altered cytokine and interferon levels and depression of cell-mediated immunity (CMI) function (250 ppm dose group) as manifested by a decrease in their delayed-type hypersensitivity (DTH) response.

Based on these results, the authors suggest that lead exposure *in utero* can cause persistent effects on the immune system in the adult animal. Dams were also assessed 13 weeks after the birth of their offspring, and no immunomodulations were noted. The study was designed to assess the effects of prenatal exposure vs. the effects of combined prenatal and lactation exposure. Although previous studies have found that maternal lead deposits can accumulate in breast milk, the authors did not find significant levels of lead in milk of dams used in this study. Also, while blood lead levels in the dams during pregnancy were relatively high, bone and blood lead levels in the offspring were relatively low, suggesting lead storage did not occur. Therefore, the authors concluded that the immunomodulations found in adult offspring are attributable to *in utero* lead exposure and were not induced by recirculation of stored lead deposits. The results of this study strongly suggest that even short-term exposure of the fetus to lead at a critical developmental window could permanently influence the immune system. In addition, the results indicate that dose levels that do not cause immune system effects in adults can cause effects in the fetus that persist into adulthood.

The impact of these findings on asthma is that they support the concept that lead influences the TH1:TH2 ratio in favor of a predominant TH2 environment. The DTH response (associated with a TH1 response) was depressed in the 250 ppm treatment group, while elevated IgE levels in the 100 ppm treatment group suggest an enhanced TH2 response. Miller and colleagues propose the possibility that lead induces TH2 activating/TH1 inhibiting cytokines to higher levels in the fetus and neonate than normal, which could permanently suppress the TH1 response. Because the baseline TH1 response is already far below adult functional capacity, a skewed TH1:TH2 that further favors TH2 may have significant implications for the neonate. Moreover, in this situation, it is possible that resistance to many infectious challenges could be compromised and the development of atopy and autoimmunity could be enhanced. The authors also found evidence of both increased TNF-α and altered macrophage function, as manifested by increased nitric oxide production. Because TNF-α has been shown to increase airway hypersensitivity, the authors propose that lead-induced elevated TH2 cytokine levels could synergistically interact with lead-induced elevated TNF levels to further increase airway hypersensitivity (Figure 14.1).

In further support of lead's influence on immune deviation, a study has been published in which plasma IgE levels of 2-week-old neonates exposed to lead before and after birth were measured as an index of atopy (Snyder et al. 2000). In this study, neonates exposed to lead either *in utero* or through lactation had significantly higher plasma IgE levels, as well as lower splenic white blood cell numbers, than age-matched controls. Carefully controlled studies aimed at separating the effects of lead and other heavy metals on disparate T cell subtypes and in tightly controlled exposure windows will be necessary if we are to begin to understand the immunologic dysregulation at the heart of the processes of asthma sensitization and exacerbation.

## Environmental Tobacco Smoke (ETS)

### *Effects of ETS on Respiratory Health*

There exists a significant body of research on the potential effects on respiratory health from exposure to environmental (passive) tobacco smoke (ETS) (reviewed in: National Academy of Sciences 2000). Many components of ETS are known lung irritants. There have been direct associations shown between exposure to tobacco smoke and the development of lung cancer, obstructive airway disease, chronic bronchitis, ear infections, and asthma. It is small wonder that ETS is associated with so many disparate disorders, considering that tobacco smoke components include the carcinogens benzene, toluene, and 1,3-butadiene (Mitacek et al. 2002), toxicants such as nickel (Tobacco Research Implementation Group 1998) and polycyclic aromatic hydrocarbons (PAHs) (Besaratinia et al. 2002), and common household chemicals including ammonia, formaldehyde, and acetone (Tobacco Research Implementation Group 1998). Moreover, the effects of tobacco smoke are not limited to the active smoker but to anyone exposed to passive or environmental tobacco smoke, including infants and children. Because ETS represents a large and diverse mixture of chemicals of many different types, it is difficult to ascribe ETS effects on the immune system to any one component. For this reason, most descriptions of relationships between developmental ETS exposure and asthma development require high levels of exposure. Individual molecular effects will remain elusive until complex "mixture" studies are conducted in a developmental setting. One study has shown direct effects of ETS and nicotine on lung function in the developing fetus. In this case, an interaction of tobacco-specific toxicants with the neuronal alpha (7) nicotinic acetylcholine receptor and its associated mitogenic signal transduction pathway was explored, particularly with regard to a potential role in lung carcinogenesis and pediatric lung disorders (Schuller et al. 2000). From that study's findings, the authors suggest that chronic nicotine and ETS exposure in pregnant women who smoke may upregulate the alpha (7) nicotinic receptor as well as components of its associated mitogenic signal transduction pathway. Upregulation of these receptors and downstream biochemical cascades are hypothesized to increase the susceptibilities of the infants for the development of pediatric lung disorders (Schuller et al. 2000). Whether those disorders include the development of asthma remains to be seen. Nonetheless, the adverse effects of maternal smoking on the health of the developing fetus and neonate are well recognized.

### *Effects of Maternal Smoking*

Active maternal smoking has been shown to be associated with many complications of fetal development, and several studies consistently link respiratory tract effects, the development of allergy, and impaired lung function in school-age children with maternal smoking (Cook and Strachan 1999; Lodrup Carlsen and Carlsen 2001; National Academy of Sciences 2000). Fetal exposure to ETS *in utero* differs both qualitatively and quantitatively from exposure of young children to airborne ETS. It must be emphasized that these types of studies are difficult to interpret because most mothers who smoke during pregnancy also continue to smoke after a child is born (National Academy of Sciences 2000). Having said that, however, at least one study has found a stronger influence of *in utero* exposure than exposure encountered postnatally (Hu et al. 1997). Although maternal smoking clearly negatively impacts the respiratory health of offspring, the effects of ETS on the development of allergy and asthma are less clear, a problem possibly attributable to the difficulty of accurately isolating fetal exposure. The NAS 2000 report (National Academy of Sciences 2000) found *sufficient evidence to conclude that there is an association* between ETS exposure and the development of asthma in younger children.

In the limited number of studies that have been able to separate the effects of maternal smoking during pregnancy from postnatal exposure, evidence suggests that smoking during pregnancy has a stronger adverse effect on subsequent development of asthma than exposure to ETS in early childhood. Finally, it should be noted that controversy still exists in this field, for although an association between ETS and asthma is clear, at least one study suggests that mechanisms other than those associated with atopy or allergy may be involved (Lodrup, Carlsen, and Carlsen 2001).

## SECTION SUMMARY

Mounting evidence suggests that increased exposure to allergens derived from dust mites, cockroaches, pet dander, molds, and pollens may be relevant to the increased incidence and severity of asthma in children. Seasonal exposures to perennial antigens as well as common viral infections may both play roles in the initiation or exacerbation of asthma. While there may be insufficient evidence to determine whether there is any link between exposure to pesticides, PHHs, or other indoor chemicals, and the exacerbation or development of asthma, the conclusion itself emphasizes the need for much more research and information in this area. Few epidemiological studies have investigated the associations between low-level exposures to environmental contaminants and the development or exacerbation of childhood asthma. Although a number of animal studies have investigated the effects of prenatal or neonatal exposures to persistent pesticides and contaminants on the developing immune system, most have only assessed endpoints indicative of immunosuppression rather than those indicative of a hypersensitive or atopic response. However, notable exceptions are the epidemiological and animal studies investigating the association between exposure to lead and atopic endpoints. The combined results of these studies suggest a role for environmental toxicants and non-allergen-specific immune responses in the prevalence of atopy and asthma in children. However, further investigations are needed to elucidate the relationship between endpoints indicative of allergic sensitization (elevated IgE levels) or hypersensitivity and the subsequent development of clinical symptoms of the asthmatic condition, especially in infants. The interaction of environmental exposures and an individual's genetic susceptibility in the development of asthma must also be further elucidated. Finally, while there exists a significant body of research on the potential effects on respiratory health from exposure to environmental tobacco smoke, the effects of ETS on the development of allergy and asthma still require clarification. What is clear is that there is an association between ETS exposure and the development of asthma in younger children. The goal of defining that association remains a challenge.

## CHAPTER SUMMARY

The purpose of this chapter was to address the influence of maternal/fetal immunotoxicant exposure on the development of asthma. Deciphering exactly how developmental exposures can lead to postnatal asthma onset or affect the sensitivity to triggers of asthmatic attacks remains an elusive goal due to insufficient evidence and a lack of mechanism studies. This chapter provided a molecular overview of asthma sensitization and triggering, asthmatic inflammation, and the inter- and intracellular mediators that may be affected by toxicants and allergens during asthma pathogenesis. Following this molecular review were discussions of the influences of maternal-fetal interactions during physiologic fetal sensitization and the development of normal immune responses. The processes involved included the generation of atopy-related fetal IgE, the bias towards a TH2 phenotype in the neonatal immune system, and the genetic components of asthma sensitization. The chapter concluded with discussions of how asthma as an exacerbated postnatal immune response may be acquired or influenced by developmental exposures to: 1.) aeroallergens such as dust mites, cockroaches, molds, pet dander, and pollens; 2.) toxicants such as pesticides, polycyclic halogenated hydrocarbons including PCBs and dioxins, and heavy metals, with particular emphasis on lead; and 3.) environmental tobacco smoke, with its sundry components including polycyclic aromatic hydrocarbons. It is our hope that this review will inspire further research into the effects of environmental toxicants on the development of asthma, particularly because of the paucity of studies currently able to provide insight into the mechanisms of interaction between molecular moieties of the environment and the various components of the immune system at discrete stages of development *in utero* and in the antenatal period.

## REFERENCES

Agea E, Forenza N, Piattoni S, et al. 1998. Expression of B7 co-stimulatory molecules and CD1a antigen by alveolar macrophages in allergic bronchial asthma. *Clin Exp Allergy* 28: 1359–1367.

Akbari O, Stock P, Meyer E, Kronenberg M, Sidobre S, Nakayama T, Taniguchi M, Grusby MJ, DeKruyff RH, and Umetsu DT. 2003. Essential role of NKT cells producing IL-4 and IL-13 in the development of allergen-induced airway hyperreactivity. *Nat Med* 9(5): 582–8.

Agency for Toxic Substances and Disease Registry PHS. 1997. Toxicological profile for lead. In: ATSDR's Toxicological profiles on CD-ROM. Boca Raton, FL: CRC Press, Inc.

Akdis CA, Blesken T, Akdis M, Wüthrich B, and Blaser K. 1998. Role of interleukin 10 in specific immunotherapy. *J Clin Invest* 102: 98–106.

Akinbami LJ, LaFleur BJ, and Schoendorf KC. 2002. Racial and income disparities in childhood asthma in the United States. *Ambul Pediatr* 2(5): 382–387.

Anderson HR, Bland JM, Patel S, et al. 1986. The natural history of asthma in childhood. *J Epidemiol Community Health* 40: 121–129.

Apostolakis J, Toumbis M, Konstantinopoulos K, et al. 1996. HLA and asthma in Greeks. *Respir Med* 90: 201–204.

Armstrong JM, Koshiba M, Apasov S, and Sitkovsky M. 1988. Studies of cross-talk between TCR-triggered and A2A extracellular adenosine-triggered signaling pathways in T-lymphocytes. Poster presentation, Exp Biol '98, San Francisco, CA.

Azzawi M, Bradley B, Jeffery PK, et al. 1990. Identification of activated T lymphocytes and eosinophils in bronchial biopsies in stable atopic asthma. *Am Rev Respir Dis* 142: 1407–1413.

Ball TM, Castro-Rodriguez JA, Griffith KA, Holberg CJ, Martinez FD, and Wright AL. 2000. Siblings, day-care attendance, and the risk of asthma and wheezing during childhood. *N Engl J Med* 343: 538–543.

Banerjee BD. 1999. The influence of various factors on immune toxicity assessment of pesticide chemicals. *Toxicol Lett* 107: 21–31.

Belvisi MG and Fox AJ. 1997. Neuropeptides. In: Kay AB (Ed.), *Allergy and Allergic Disease*. Vol. 1. Oxford, England: Blackwell Scientific. pp. 447–480.

Bermas BL and Hill JA. 1997. Proliferative responses to recall antigens are associated with pregnancy outcome in women with a history of recurrent spontaneous abortion. *J Clin Invest* 100: 1330–1334.

Bernier J, Brousseau P, Krzystyniak K, Tryphonas H, and Fournier M. 1995. Immunotoxicity of heavy metals in relation to Great Lakes. *Environ Health Perspect* 103(9): 23–34.

Besaratinia A, Maas LM, Brouwer EM, Moonen EJ, De Kok TM, Wesseling GJ, Loft S, Kleinjans JC, and Van Schooten FJ. 2002. A molecular dosimetry approach to assess human exposure to environmental tobacco smoke in pubs. *Carcinogenesis* 23(7): 1171–1176.

Bjorksten B, Naaber P, Sepp E, and Mikelsaar N. 1999. The intestinal microflora in allergic Estonian and Swedish 2-year-old children. *Clin Exp Allergy* 29(3): 342–346.

Black RE. 2002. The Role of Mast Cells in the Pathophysiology of Asthma. *N Engl J Med* 346(22): 1742–1743.

Blackburn MR, Lee CG, Young HWJ, Zhu Z, Chunn JL, Kang MJ, Banerjee SK, and Elias JA. 2003. Adenosine mediates IL-13–induced inflammation and remodeling in the lung and interacts in an IL-13–adenosine amplification pathway. *J Clin Invest* 112 (3): 332–344.

Bloomfield FH and Harding JE. 1998. Experimental aspects of nutrition and fetal growth. *Fetal Mat Med Rev* 10: 91–107.

Borish L. 1998. IL-10: evolving concepts. *J Allergy Clin Immunol* 101: 293–297.

Braun A, Lommatzsch M, and Renz H. 2000. The role of neurotrophins in allergic bronchial asthma. *Clin Exp Allergy* 30: 178–186.

Briere F, Servet-Delprat C, Bridon JM, et al. 1994. Human interleukin 10 induces naive surface immunoglobulin D+ (sIgD+) B cells to secrete IgG1 and IgG3. *J Exp Med* 179: 757–762.

Brightling CE, Bradding P, Symon FA, Holgate ST, Wardlaw AJ, and Pavord ID. 2002. Mast-cell Infiltration of Airway Smooth Muscle in Asthma. *N Engl J Med* 346(22): 1699–1705.

Brooks SM. 1992. Occupational and environmental asthma. In: Rom WN (Ed.), *Environmental and Occupational Medicine*. Boston, MA: Little Brown, pp. 481–524.

Brown RA, Lever R, Jones NA, and Page CP. 2003. Effects of heparin and related molecules upon neutrophil aggregation and elastase release *in vitro*. *Br J Pharmacol* 139(4):845–53.

Burchett SK, Corey L, Mohan KM, Westall J, Ashley R, and Wilson CB. 1992. Diminished interferon-gamma and lymphocyte proliferation in neonatal and postpartum primary herpes simplex virus infection. *J Infect Dis* 165: 813–818.

Byrne JA, Stankovic AK, and Cooper MD. 1994. A novel subpopulation of primed T cells in the human fetus. *J Immunol* 152: 3098–3106.

Cadet P, Rady PL, Tyring SK, et al. 1995. Interleukin-10 messenger ribonucleic acid in human placenta: implications of a role for interleukin-10 in fetal allograft protection. *Am J Obstet Gynecol* 173: 25–29.

Caret LL, Zhang X, Dubey C, Rogers P, Tsui L, and Swain AL. 1998. Regulation of T cell subsets from naive to memory. *J Immunotherapy* 21: 181–187.

Carpenter PO. 1998. Polychlorinated biphenyls and human health. *Int J Occup Med Environ Health* 11: 291–303.

Carroll NG, Mutavdzic S, and James AL. 2002. Distribution and degranulation of airway mast cells in normal and asthmatic subjects. *Eur Respir J* 19: 1–7.

Carter LL and Swain SL. 1998. From naive to memory. Development and regulation of CD4+ T cell responses. *Immunol Res* 18: 1–13.

Chen Y, Kuchroo VK, Inobe J, Hafler DA, and Weiner HL. 1994. Regulatory T cell clones induced by oral tolerance: suppression of autoimmune encephalomyelitis. *Science* 265(5176):1237–40.

Clutterbuck EJ, Hirst EM, and Sanderson CJ. 1989. Human interleukin-5 (IL-5) regulates the production of eosinophils in human bone marrow cultures: comparison and interaction with IL-1, IL-3, IL-6, and GMCSF. *Blood* 73: 1504–1512.

Colosio C, Corsini E, Barcellini W, and Maroni M. 1999. Immune parameters in biological monitoring of pesticide exposure: current knowledge and perspectives. *Toxicol Lett* 108(2–3): 285–295.

Cook DG and Strachan DP. 1999. Health effects of passive smoking -10: Summary of effects of parental smoking on the respiratory health of children and implications for research. *Thorax* 54(4): 357–366.

Cookson W. 1999. The alliance of genes and environment in asthma and allergy. *Nature* 402: B5–B11.

Corrigan CJ and Kay AB. 1990. CD4 T-lymphocyte activation in acute severe asthma. Relationship to disease severity and atopic status. *Am Rev Respir Dis* 141: 970–977.

Corry DB and Kheradmand F. 1999. Induction and regulation of the IgE response. *Nature* 402: B18–B23.

Corteling RL, Li S, Giddings J, Westwick J, Poll C, and Hall IP. 2003. Expression of TRPC6 and related TRP family members in human airway smooth muscle and lung tissue. *Am J Respir Cell Mol Biol* Jul 18 [July 18 Epub ahead of print].

Croner S and Kjellman N-IM. 1990. Development of atopic disease in relation to family history and cord blood IgE levels - 11 year follow-up in 1,654 children. *Pediatr Allergy Immunol* 1: 14–21.

Croner S and Kjellman NI. 1992. Natural history of bronchial asthma in childhood. A prospective study from birth up to 12-14 years of age. *Allergy* 47: 150–157.

Denburg JA. 1998. Hemopoietic progenitors and cytokines in allergic inflammation. *Allergy* 53:Suppl: 22–26.

Devouassoux G, Metcalfe D, and Prussin C. 1999. Eotaxin potentiates antigen-dependent basophil IL-4 production. *J Immunol* 163: 2877–2882.

Donovan CE and Finn PW. 1999. Immune mechanisms of childhood asthma. *Thorax* 54: 938–946.

Drazen JM, Israel E, and O'Byrne PM. 1999. Treatment of asthma with drugs modifying the leukotriene pathway. *N Engl J Med* 340: 197–206 [Errata, *N Eng J Med* 1999; 340:663, 341:1632.].

Dutton RW, Swain SL, and Bradley LM. 1999. The generation and maintenance of memory T and B cells. *Immunol Today* 20(7): 291–293.

Early E and Reen DJ. 1999b. Rapid conversion of naive to effector T cell function counteracts diminished primary human newborn T cell responses. *Clin Exp Immunol* 116(3): 527–533.

Edenharter G, Burgmann RL, Burgmann KE, et al. 1998. Cord blood IgE as risk factor and predictor of atopic diseases. *Clin Exp Allergy* 28: 671–678.

Ehlers R, Ustinov V, Chen Z, Zhang X, Rao R, Luscinskas FW, Lopez J, Plow E, and Simon DI. 2003. Targeting platelet-leukocyte interactions: identification of the integrin Mac-1 binding site for the platelet counter receptor glycoprotein Ibalpha. *J Exp Med* 98(7): 1077–88.

Ernst P. 2002. Pesticide Exposure and Asthma. *Am J Respir Crit Care Med* 165: 563–564.

Faffe DS, Whitehead T, Moore PE, Baraldo S, Flynt L, Bourgeois K, Panettieri RA, and Shore SA. 2003. IL-13 and IL-4 promote TARC release in human airway smooth muscle cells: Role of IL-4 receptor genotype. *Am J Physiol Lung Cell Mol Physiol* [Jul 18 Epub ahead of print].

Fallon PG, Emson CL, Smith P, and McKenzie NJ. 2001. IL-13 overexpression predisposes to anaphylaxis following antigen sensitization. *J Immunol* 166: 2712–2717.

Fingerle-Rowson G, Angstwurm M, Andreesen R, and Ziegler-Heitbrock HW. 1998. Selective depletion of CD14+CD16+ monocytes by glucocorticoid therapy. *Clin Exp Immunol* 112: 501–506.

Forssmann U, Uguccioni M, Loetscher P, Dahinden CA, Langen H, Thelen M, and Baggiolini M. 1997. Eotaxin-2, a Novel CC Chemokine That is Selective for the Chemokine Receptor CCR3, and Acts Like Eotaxin on Human Eosinophil and Basophil Leukocytes. *J Exp Med* 185: 2171–2176.

Forsythe P, McGarvey LPA, Heaney LG, MacMahon J, and Ennis M. 2000. Sensory neuropeptides induce histamine release from bronchoalveolar lavage cells in both non-asthmatic coughers and cough variant asthmatics. *Clin Exp Allergy* 30: 225–232.

Frenkel L and Bryson YJ. 1987. Ontogeny of phytohemagglutinin-induced gamma interferon by leukocytes of healthy infants and children: evidence for decreased production in infants younger than 2 months of age. *J Pediatr* 111: 97–100.

Gangur V and Oppenheim J. 2000. Are chemokines essential or secondary participants in allergic responses? *Ann Allergy Asthma Immunol* 84: 569–581.

Gately MR, Renzetti LM, Magram J, Stern AS, Adorini L, Gubler U, and Presky DH. 1998. The interleukin-12/interleukin-12-receptor system: Role in normal and pathologic immune responses. *Annu Rev Immunol* 16: 495–521.

Gicquel C, Gaston V, Mandelbaum J, Siffroi JP, Flahault A, and Le Bouc Y. 2003. In vitro fertilization may increase the risk of Beckwith-Wiedemann syndrome related to the abnormal imprinting of the KCN1OT gene. *Am J Hum Genet* 72(5):1338–41.

Groux H., O'Garra A, Bigler M, Rouleau M, Antonenko S, de Vries JE, and Roncarolo MG. 1997. A CD4+ T-cell subset inhibits antigen-specific T-cell responses and prevents colitis. *Nature* 389(6652): 737–42

Hahn C, Teufel M, Herz U, Renz H, Erb KJ, Wohlleben G, Brocker EB, Duschl A, Sebald W, and Grunewald SM. 2003a. Inhibition of the IL-4/IL-13 receptor system prevents allergic sensitization without affecting established allergy in a mouse model for allergic asthma. *J Allergy Clin Immunol* 111(6): 1361–1369.

Hahn Y-S, Taube C, Jin N, Takeda K, Park J-W, Wands JM, Aydintug MK, Roark CL, Lahn M, O'Brien RL, Gelfand EW, and Born WK. 2003b. $V\gamma4^+$ $\gamma\delta$T cells regulate airway hyperreactivity to methacholine in ovalbumin-sensitized and challenged mice. *J Immunol* 171: 3170–3178.

Halonen M, Stern D, Taussig LM, et al. 1992. The predictive relationship between serum IgE levels at birth and subsequent incidences of lower respiratory illnesses and eczema in infants. *Am Rev Respir Dis* 146: 866–870.

Hannet I, Erkeller-Yuksel F, Lydyard P, et al. 1992. Developmental and maturational changes in human blood lymphocyte subpopulations. *Immunol Today* 13:p. 215.

Hanrahan JP and Halonen M. 1998. Antenatal interventions in childhood asthma. *Eur Respir J* Suppl 27: 46s–51s.

Hansen LG, Host A, Halken S, et al. 1992. Cord blood IgE. II. Prediction of atopic disease. A follow-up at the age of 18 months. *Allergy* 47: 397–403.

Harris N, Peach R, Naemura J, et al. 1997. CD80 costimulation is essential for the induction of airway eosinophilia. *J Exp Med* 185: 177–182.

Hayward AR. 1981. Development of lymphocyte responses and interactions in the human fetus and newborn. *Immunol Rev* 57: 39–60.

Hayward AR and Groothuis J. 1991. Development of T cells with memory phenotype in infancy. In: Mestecky J, et al. (Eds.), *Immunology of Milk and the Neonate.* New York: Plenum Press, pp. 71–76.

Hessel EM, Van Oosterhout AJ, Van Ark I, et al. 1997. Development of airway hyperresponsiveness is dependent on interferon-gamma and independent of eosinophil infiltration. *Am J Respir Cell Mol Biol* 16: 325–334.

Hide DW, Arshad SH, Twiselton R, and Stevens M. 1991. Cord serum IgE: an insensitive method for prediction of atopy. *Clin Exp Allergy* 21: 739–743.

Hirasawa R, Shimizu R, Takahashi S, Osawa M, Takayanagi S, Kato Y, Onodera M, Minegishi N, Yamamoto M, Fukao K, Taniguchi H, Nakauchi H, and Iwama A. 2002. Essential and instructive roles of GATA factors in eosinophil development. *J Exp Med* 195(11): 1379–1386.

Holgate ST. 1999. The epidemic of allergy and asthma. Nature Suppl 402: B2–B4.

Holt PG, Clough JB, Holt BJ, Baron-Hay MJ, Rose AH, Robinson BW, and Thomas WR. 1992. Genetic 'risk' for atopy is associated with delayed postnatal maturation of T-cell competence. *Clin Exp Allergy* 22: 1093–1099.

Holt PG, Macaubas C. 1997. Development of long-term tolerance *versus* sensitisation to environmental allergens during the perinatal period. *Curr Opin Immunol* 9: 782–787.

Holt PG, Macaubas C, Stumbles PA, and Sly PD. 1999. The role of allergy in the development of asthma. *Nature* 402:Supp: B12–B17.

Holt PG, Sly PD, and Bjorksten B. 1997. Atopic *versus* infectious diseases in childhood: a question of balance? *Pediatr Allergy Immunol* 8: 53–58.

Hu FB, Persky V, Flay BR, Zelli A, Cooksey J, and Richardson J. 1997. Prevalence of asthma and wheezing in public schoolchildren: association with maternal smoking during pregnancy. *Ann Allergy Asthma Immunol* 79(1): 80–84.

Humbert M, Corrigan CJ, Kimmitt P, Till SJ, and Kay AB. 1997. Relationship between IL-4 and IL-5 mRNA expression and disease severity in atopic asthma. *Am J Respir Crit Care Med* 156: 704–708.

Humbert M, Menz G, Ying S, et al. 1999. The immunopathology of extrinsic (atopic) and intrinsic (nonatopic) asthma: more similarities than differences. *Immunol Today* 20: 528–533.

Jenmalm MC and Bjorksten B. 2000. Cord blood levels of immunoglobulin G subclass antibodies to food and inhalant allergens in relation to maternal atopy and the development of atopic diseases during the first 8 years of life. *Clin Exp Allergy* 30: 34.

Jones AC, Miles EA, Warner JO, et al. 1996. Fetal peripheral blood mononuclear cell proliferative responses to mitogenic and allergenic stimuli during gestation. *Pediatr Allergy Immunol* 7: 109–116.

Jones CA, Holloway JA, and Warner JO. 2000. Does atopic disease start in foetal life? *Allergy* 55: 2–10.

Jones CA, Warner JA, and Warner JO. 1998. Fetal swallowing of IgE. *Lancet* 351:p. 1859.

Jones CA, Williams KA, Finlay-Jones JF, and Hart PH. 1995. Interleukin 4 production by human amnion epithelial cells and regulation of its activity by glycosaminoglycan binding. *Biol Reprod* 52: 839–847.

Jose PJ, Griffiths-Johnson DA, Collins PD, Walsh DT, Moqbel R, Totty NF, Truong O, Hsuan JJ, and Williams TJ. 1994. Eotaxin: a potent eosinophil chemoattractant cytokine detected in a guinea pig model of allergic airways inflammation. *J Exp Med* 179: 881–887.

Karmaus W, Kuehr J, and Kruse H. 2001. Infections and atopic disorders in childhood and organochlorine exposure. *Arch Environ Health* 56(6): 485–492.

Karmaus W, Davis S, Chen Q, Kuehr J, and Kruse H. 2003. Atopic manifestations, breast-feeding protection and the adverse effect of DDE. *Paediatr Perinat Epidemiol* 17(2): 212–220.

Katoh S, Matsumoto N, Kawakita K, Tominaga A, Kincade PW, and Matsukura S. 2003. A role for CD44 in an antigen-induced murine model of pulmonary eosinophilia. *J Clin Invest* 111(10): 1563–70.

Kay AB. 2001. Allergy and allergic diseases. First of two parts. *N Engl J Med* 344(1): 30–37.

Kim E, Chun HO, Jung SH, Kim JH, Lee JM, Suh BC, Xiang MX, and Rhee CK. 2003. Improvement of therapeutic index of phosphodiesterase type IV inhibitors as anti-asthmatics. *Bioorg Med Chem Lett* 13(14):2355–8.

Kirjavainen PV and Gibson GR. 1999. Healthy gut microflora and allergy: factors influencing development of the microbiota. *Ann Med* 31(4): 288–292.

Kitaura M, Suzuki N, Imai T, Tagaki S, Suzuki R, Nakajima T, Hirai K, Nomiyama H, and Yoshie O. 1999. Molecular cloning of the novel human CC chemokine (eotaxin-3) that is a functional ligand of CC chemokine receptor 3. *J Biol Chem* 274: 27975–27980.

Kjellman NIM and Croner S. 1984. Cord blood IgE determination of allergy prediction: a follow-up to seven years of age in 1,651 children. *Ann Allergy* 53: 167–171.

Klos A, Bank S, Gietz C, Bautsch W, Kohl J, Burg M, and Kretzschmar T. 1992. C3a receptor on dibutyryl-cAMP-differentiated U937 cells and human neutrophils: the human C3a receptor characterized by functional responses and 125I-C3a binding. *Biochemistry* 31(46):11274–82.

Kondo N, Cubiyashi Y, Shinoda S, et al. 1992. Cord blood lymphocyte responses to food antigens for the prediction of allergic disorders. *Arch Dis Child* 67: 1003–1007.

Koren H and O'Neill M. 1998. Experimental assessment of the influence of atmospheric pollutants on respiratory disease. *Toxicol Lett* 102–103: 317–321.

Kovarik J, Bozzotti P, Love-Homan L, Pihlgren M, Davis HL, Lambert PH, Krieg AM, Siegrist CA. 1999. CgP oligodeoxynucleotides can circumvent the TH2 polarization of neonatal responses to vaccines but may fail to fully redirect TH2 responses established by neonatal priming. *J Immunol* 162(3): 1611–1617.

Krug N, Madden J, Redington AE, et al. 1996. T-cell cytokine profile evaluated at the single cell level in BAL and blood in allergic asthma. *Am J Respir Cell Mol Biol* 14: 319–326.

Landrigan PJ. 1998. Environmental hazards for children in USA. *Int J Occup Med Environ Health* 11(2): 189–194.

Landrigan PJ, Carlson JE, Bearer CF, Cranmer JS, Bullard RD, Etzel RA, Groopman J, McLachlan JA, Perera FP, Reigart JR, Robison L, Schell L, and Suk WA. 1998. Children's health and the environment: a new agenda for prevention research. *Environ Health Perspect* 106 (Suppl 3): 787–794.

Lane P. 1996. Development of B-cell memory and effector function. *Curr Opin Cell Biol* 8: 331–335.

Larche M, Till SJ, Haselden BM, et al. 1998. Costimulation through CD86 is involved in airway antigen-presenting cell and T cell responses to allergen in atopic asthmatics. *J Immunol* 161: 6375–6382.

Larché M, Till SJ, Haselden BM, et al. 1998. Costimulation through CD86 is involved in airway antigen-presenting cell and T cell responses to allergen in atopic asthmatics. *J Immunol* 161: 6375–6382.

Levi-Schaffer F, Garbuzenko E, Rubin A, et al. 1999. Human eosinophils regulate human lung- and skin-derived fibroblast properties *in vitro*: a role for transforming growth factor beta (TGF-β). *Proc Natl Acad Sci USA* 96: 9660–9665.

Levy F, Kristofic C, Heusser C, and Brinkmann V. 1997. Role of IL-13 in CD4 T cell-dependent IgE production in atopy. *Int Arch Allergy Immunol* 112: 49.

Lewis D. 1998. Cellular immunity of the human fetus and neonate. *Immunol Allergy Clin North Am* 18: 291–328.

Li L, Yiyang X, Nguyen A, Lai YH, Feng L, Mosmann TR, and Lo D. 1999. Effects of Th2 cytokines on chemokine expression in the lung: IL-13 potently induces eotaxin expression by airway epithelial cells. *J Immunol* 162: 2477–2487.

Liao H-X and Patel DD. 1999. PROW and IWHLDA present the GUIDE on CD44. In: Kincade PW (Ed.), Protein Reviews on the Web (PROW). prow@ncbi.nlm.nih.gov.

Lin H, Mosmann TR, Guilbert L, et al. 1993. Synthesis of T helper 2-type cytokines at the maternal-fetal interface. *J Immunol* 151: 4562–4573.

Lodrup Carlsen KC and Carlsen KH. 2001. Effects of maternal and early tobacco exposure on the development of asthma and airway hyperreactivity. *Curr Opin Allergy Clin Immunol* 1(2): 139–143.

Lutz PM, Wilson TJ, Ireland J, Jones AL, Gorman JS, Gale NL, Johnson JC, and Hewett JE. 1999. Elevated immunoglobulin E (IgE) levels in children with exposure to environmental lead. *Toxicol* 134: 63–78.

MacFarlane AJ, Kon OM, Smith SJ, et al. 2000. Basophils, eosinophils, and mast cells in atopic and nonatopic asthma and in late-phase allergic reactions in the lung and skin. *J Allergy Clin Immunol* 105: 99–107.

Mackay IR and Rosen FS. 2001. Allergic diseases and their treatment. *N Engl J Med* 344(2): 109–113.

Manetti R, Parronchi P, Giudizi MG, Piccinni MP, Maggi E, Trinchieri G, and Romagnani S. 1993. Natural killer cell stimulatory factor (interleukin 12 [IL-12]) induces T helper type 1 (Th1)-specific immune responses and inhibits the development of IL-4-producing Th cells. *J Exp Med* 117: 1199–1204.

Mark DA, Donovan CE, De Sanctis GT, et al. 1998. Both CD80 and CD86 co-stimulatory molecules regulate allergic pulmonary inflammation. *Int Immunol* 10: 1647–1655.

Martinez FD, Cline M, and Burrows B. 1992. Increased incidence of asthma in children of smoking mothers. *Pediatrics* 89: 21–26.

Martinez FD, Wright AL, Holberg CJ, et al. 1992. Maternal age as a risk factor for wheezing lower respiratory illnesses in the first year of life. *Am J Epidemiol* 136: 1258–1268.

Martinez FD, Wright AL, Taussig LM, et al. 1995. Asthma and wheezing in the first six years of life. The Group Health Medical Associates. *N Engl J Med* 332: 133–138.

Matthiesen L, Berg G, Ernerudh J, et al. 1996. Lymphocyte subsets and mitogen stimulation of blood lymphocytes in normal pregnancy. *Am J Reprod Immunol* 35: 70–79.

Matzinger P. 1998. An innate sense of danger. *Semin Immunol* 10: 399–415.

Menzies-Gow A, Robinson DS. 2000. Eosinophil chemokines and their receptors: an attractive target in asthma? *Lancet* 355: 1741–1743.

Menzies-Gow A and Robinson DS. 2001. Eosinophil Chemokines and Chemokine Receptors: Their Role in Eosinophil Accumulation and Activation in Asthma and Potential as Therapeutic Targets. *Journal of Asthma* 38(8): 605–613.

Michel FB, Bousquet J, Greillier P, et al. 1980. Comparison of cord blood immunoglobulin E and maternal allergy for the prediction of atopic diseases in infancy. *J Allergy Clin Immunol* 65: 422–430.

Miller TE, Golemboski KA, Ha RS, Bunn T, Sanders FS, and Dietert RR. 1998. Developmental exposure to lead causes persistent immunotoxicity in Fischer 344 rats. *Toxicol Sci* 42: 129–135.

Mitacek EJ, Brunnemann KD, Polednak AP, Limsila T, Bhothisuwan K, and Hummel CF. 2002. Rising leukemia rates in Thailand: The possible role of benzene and related compounds in cigarette smoke. *Oncol Rep* 9(6): 1399–1403.

Miyamasu M, Yamaguchi M, Nakajima T, Misaki Y, Morita Y, Matsushima K, Yamamoto K, and Hirai K. 1999. Th1-derived cytokine IFN-gamma is a potent inhibitor of eotaxin synthesis *in vitro*. *Int Immunol* 11: 1001–1004.

Monteseirin J, Bonilla I, Camacho MJ, Chacon P, Vega A, Chaparro A, Conde J, and Sobrino F. 2003. Specific allergens enhance Elastase release in stimulated neutrophils from asthmatic patients. *Int Arch Allergy Immunol* 131(3):174–81.

Mueller DL, Jenkins MK, and Schwartz RH. 1989. Clonal expansion *versus* functional clonal inactivation: a costimulatory signaling pathway determines the outcome of T cell antigen receptor occupancy. *Annu Rev Immunol* 7: 445–480.

Mukerjee D. 1998. Health impact of polychlorinated dibenzo-p-dioxins: a critical review. *J Air Waste Manage Assoc* 48: 157–165.

Nagase H, Miyamasu M, Yamaguchi M, Fujisawa T, Ohta K, Yamamoto K, Morita Y, and Hirai K. 2000. Expression of CXCR4 in eosinophils: Functional analyses and cytokine-mediated regulation. *J Immunol* 164: 5935–5943.

National Academy of Sciences IoM. 2000. Clearing the air: Asthma and indoor air exposure. Washington, D.C.: National Academy Press. pp. 1–262.

Niimi A, Matsumoto H, Takemura M, Ueda T, Chin K, and Mishima M. 2003. Relation of airway wall thickness to airway sensitivity and airway reactivity in asthma. *Am J Respir Crit Care Med*. [Epub ahead of print].

Nurse B, Haus M, Puterman AS, Weinberg e.g., Potter PC. 1997. Reduced interferon-gamma but normal IL-4 and IL-5 release by peripheral blood mononuclear cells from Xhosa children with atopic asthma. *J Allergy Clin Immunol* 100: 662–668.

Oddy WH, Peat JK, and deKlerk NH. 2002. Maternal asthma, infant feeding, and the risk of asthma in childhood. *J Allergy Clin Immunol* 110(1): 65–67.

Ohbayashi H. 2002. Novel neutrophil elastase inhibitors as a treatment for neutrophil-predominant inflammatory lung diseases. *Idrugs* 5(9): 910–23.

Okahata H, Nishi Y, Mizoguchi N, et al. 1990. Development of serum *Dermatophagoides farinae*-, ovalbumin- and lactalbumin-specific IgG, IgG1, IgG4, IgA and IgM in children with bronchial asthma/allergic rhinitis or atopic dermatitis. *Clin Exp Allergy* 20: 39–44.

Oliveti JF, Kercsmer CM, and Redline S. 1996. Pre- and perinatal risk factors for asthma in inner city African-American children. *Am J Epidemiol* 143: 570–577.

Openshaw PJM and Hewitt C. 2000. Protective and harmful effects of viral infections in childhood on wheezing disorders and asthma. *Am J Respir Crit Care Med* 162:Suppl: S40–S43.

Palframan RT, Collins PD, Severs NJ, Rothery S, Williams TJ, and Rankin SM. 1998. Mechanisms of acute eosinophil mobilization from the bone marrow stimulated by interleukin-5: the role of specific adhesion molecules and phosphatidylinositol 3-kinase. *J Exp Med* 188: 1621–1632.

Paryani SG and Arvin AM. 1986. Intrauterine infection with varicella-zoster virus after maternal varicella. *N Engl J Med* 314:1542–1546.

Peakman M, Buggins AG, Nicolaides KH, et al. 1992. Analysis of lymphocyte phenotypes in cord blood from early gestation fetuses. *Clin Exp Immunol* 90: 345–350.

Peat JK, Salome CM, and Woolcock AJ. 1990. Longitudinal changes in atopy during a 4-year period: Relation to bronchial hyperresponsiveness and respiratory symptoms in a population sample of Australian schoolchildren. *J Allergy Clin Immunol* 85: 65–74.

Pinon JM, Toubas D, Marx C, et al. 1990. Detection of specific Immunoglobulin E in patients with toxoplasmosis. *J Clin Microbiol* 28: 1739–1743.

Pitchford SC, Riffo-Vasquez Y, Sousa A, Momi S, Gresele P, Spina D, and Page CP. 2003a. Platelets are necessary for airway wall remodeling in a murine model of chronic allergic inflammation. *Blood* [Sep 22 Epub ahead of print.]

Pitchford SC, Yano H, Lever R, Riffo-Vasquez Y, Ciferri S, Rose MJ, Giannini S, Momi S, Spina D, O'Connor B, Gresele P, and Page CP. 2003b. Platelets are essential for leukocyte recruitment in allergic inflammation. *J Allergy Clin Immunol* 112(1): 109–118.

Ponath PD, Qin S, Ringler DJ, Clark-Lewis I, Wang J, Kassam N, Smith H, Shi X, Gonzalo J-A, Newman W, Gutierrez-Ramos JC, and Mackay CR. 1996. Cloning of the human eosinophil chemoattractant, eotaxin. *J Clin Invest* 97: 604–612.

Pouladi MA, Robbins CS, Swirski FK, Cundall M, McKenzie AN, Jordana M, Shapiro SD, and Stampfli MR. 2004. IL-13-dependent expression of matrix metalloproteinase-12 is required for the development of airway eosinophilia in mice. *Am J Respir Cell Mol Biol* 30(2): 202–211.

Prescott SL, Macaubas C, Holt BJ, Smallacombe TB, Loh R, Sly PD, and Holt PG. 1998a. Transplacental priming of the human immune system to environmental allergens: universal skewing of initial T cell responses toward the TH2 cytokine profile. *J Immunol* 160(10): 4730–4737.

Prescott SL, Macaubas C, Smallacombe T, Holt BJ, Sly PD, Loh R, and Holt PG. 1998b. Reciprocal age-related patterns of allergen-specific T-cell immunity in normal vs. atopic infants. *Clin Exp Allergy* 28: 39–49.

Punnonen J, Aversa G, and de Vries JE. 1993. Human pre-B cells differentiate into Ig-secreting plasma cells in the presence of interleukin-4 and activated CD4+ T cells or their membranes. *Blood* 82: 2781–2789.

Punnonen J and de Vries JE. 1994. IL-13 induces proliferation, Ig isotype switching, and Ig synthesis by immature human fetal B cells. *J Immunol* 152: 1094–1102.

Punnonen J, Yssel H, and de Vries JE. 1997. The relative contribution of IL-4 and IL-13 to human IgE synthesis induced by activated CD4+ or CD8+ T cells. *J Allergy Clin Immunol* 100: 792–801.

Raghupathy R. 1997. Th1-type immunity is incompatible with successful pregnancy. *Immunol Today* 18: 478–482.

Rankin SM, Conroy D, and Williams TJ. 2000. Eotaxin and eosinophil recruitment: Implications of human disease. *Mol Med Today* 6: 20–27.

Reichrtova E, Ciznar P, Prachar V, Palkovicova L, and Veningerova M. 1999. Cord serum immunoglobulin E related to the environmental contamination of human placentas with organochlorine compounds. *Environ Health Perspect* 107(11): 895–899.

Richter M, Cantin AM, Beaulieu C, Cloutier A, and Larivee P. 2003. Zinc chelators inhibit eotaxin, RANTES and MCP-1 production in stimulated human airway epithelium and fibroblasts. *Am J Physiol Lung Cell Mol Physiol* [May 23 Epub ahead of print].

Robinson D, Shibuya K, Mui A, Zonin F, Murphy E, Sana T, Hartley SB, Menon S, Kastelein R, Bazan F, and O'Garra A. 1997b. IGIF does not drive Th1 development but synergizes with IL-12 for Interferon-γ production and activates IRAK and NFκB. *Immunity* 7: 571–581.

Robinson DS, Hamid Q, Bentley A, Ying S, Kay AB, and Durham SR. 1993. Activation of CD4+ T cells, increased TH2-type cytokine mRNA expression, and eosinophil recruitment in bronchoalveolar lavage after allergen inhalation challenge in patients with atopic asthma. *J Allergy Clin Immunol* 92: 313–324.

Robinson DS, Hamid Q, Ying S, Tsicopoulous A, Barkans J, Bentley A, Corrigan C, Durham SR, and Kay AB. 1992. Predominant Th2-like bronchoalveolar T lymphocyte population in atopic asthma. *N Engl J Med* 326: 298–303.

Roman M, Calhoun WJ, Hinton KL, Avendano LF, Simon V, Escobar AM, Gaggero A, and Diaz PV. 1997. Respiratory syncytial virus infection in infants is associated with predominant Th-2-like response. *Am J Respir Crit Care Med* 156(1): 190–195.

Rook GAW and Stanford JL. 1998. Give us this day our daily germs. *Immunology Today* 19(3): 113–116.

Rothenberg ME. 1999. Eotaxin: An essential mediator of eosinophil trafficking into mucosal tissues. *Am J Respir Cell Mol Biol* 21: 291–295.

Ruiz RGG, Richards D, Kemeny DM, and Price JF. 1991. Neonatal IgE: a poor screen for atopic disease. *Clin Exp Allergy* 21: 467–472.

Salek-Ardakani S, Song J, Halteman BS, Jember AG, Akiba H, Yagita H, and Croft M. 2003. OX40 (CD134) Controls memory T helper 2 cells that drive lung inflammation. *J Exp Med* [Epub ahead of print].

Sallout B and Walker M. 2003. The fetal origin of adult diseases. *J Obstet Gynaecol* 23(5):555–60.

Saltini C, Amicosante A, Franchi A, Lombardi G, and Richeldi L. 1998. Immunogenetic basis of environmental lung disease: lessons from the berylliosis model. *Eur Respir J* 12: 1463–1475.

Sanders ME, Makgoba MW, and Shaw S. 1988. Human naive and memory T cells: reinterpretation of helper-inducer and suppressor-inducer subsets. *Immunol Today* 9: 195–199.

Sandford A, Weir T, and Paré P. 1996. The genetics of asthma. *Am J Respir Crit Care Med* 153: 1749–1765.

Sarpong SB and Karrison T. 1998. Season of birth and cockroach allergen sensitization in children with asthma. *J Allergy Clin Immunol* 101(4 Pt 1): 566–568 [Erratum in *J Allergy Clin Immunol* 1998 Sep; 10(3):448].

Schuller HM, Jull BA, Sheppard BJ, and Plummer III HK. 2000. Interaction of tobacco-specific toxicants with the neuronal α7 nicotinic acetylcholine receptor and its associated mitogenic signal transduction pathway: potential role in lung carcinogenesis and pediatric lung disorders. *Eur J Pharmacol* 393: 265–277.

Schuppe HC, Ronnau AC, von Schmiedeberg S, Ruzicka T, Gleichmann E, and Griem P. 1998. Immunomodulation by heavy metal compounds. *Clin Dermatol* 16: 149–157.

Schwartz J, Gold D, Dockery DW, et al. 1990. Predictors of asthma and persistent wheeze in a national sample of children in the United States. Association with social class, perinatal events, and race. *Am Rev Respir Dis* 142: 555–562.

Schwartz RH. 1990. A cell culture model for T lymphocyte clonal anergy. *Science* 248: 1349–1356.

Seki Y-I, Inoue H, Nagata N, Hayashi K, Fukuyama S, Matsumoto K, Komine O, Hamano S, Himeno K, Inagaki-Ohara K, Cacalano N, O'Garra A, Oshida T, Saito H, Johnston JA, Yoshimura A, and Kubo M. 2003. SOCS-3 regulates onset and maintenance of $T_H2$-mediated allergic responses. *Nat Med* 9(8): 1047–1054.

Selgrade MK. 1999. Use of immunotoxicity data in health risk assessments: uncertainties and research to improve the process. *Toxicol* 133(1): 59–72.

Servet-Delprat C, Bridon JM, Djossou O, Yahia SA, Banchereau J, and Briere F. 1996. Delayed IgG2 humoral response in infants is not due to intrinsic T or B cell defects. *Int Immunol* 8: 1495–1502.

Shinkai A, Yoshisue H, Koike M, Shoji E, Nakagawa S, Saito A, Takeda T, Imabeppu S, Kato Y, Hanai N, Anazawa H, Kuga T, and Nishi T. 1999. A novel human CC chemokine, Eotaxin-3 which is expressed in IL-4 stimulated vascular endothelial cells, exhibits potent activity toward eosinophils. *J Immunol* 163: 1602–1610.

Sills ES, Perloe M, Tucker MJ, Kaplan CD, and Palermo GD. 2001. Successful ovulation induction, conception, and normal delivery after chronic therapy with etanercept: a recombinant fusion anti-cytokine treatment for rheumatoid arthritis. *Am J Reprod Immunol* 46: 366–8.

Sitkovsky MV. 1998. Extracellular purines and their receptors in immunoregulation. Review of recent advances. *Nippon Ika Daigaku Zasshi*. 1998 65(5):351–7.

Snyder JE, Filipov NM, Parsons PJ, and Lawrence DA. 2000. The efficiency of maternal transfer of lead and its influence on plasma IgE and splenic cellularity of mice. *Toxicol Sciences* 57: 87–94.

Sporik R, Holgate ST, Plattes-Mills TAE, et al. 1990. Exposure to house dust mite allergen (Der p 1) and the development of asthma in childhood. *N Engl J Med* 323: 502–507.

Starr SE, Tolpin MD, Friedman HM, et al. 1979. Impaired cellular immunity to cytomegalovirus in congenitally infected children and their mothers. *J Infect Dis* 140: 500–505.

Strachan DP. 1989. Hay fever, hygiene, and household size. *BMJ* 299: 1259–1260.

Suzuki M, Maghni K, Molet S, Shimbara A, Hamid QA, and Martin JG. 2002. IFN-γ secretion by CD8+ T cells inhibits allergen-induced airway eosinophilia but not late airway responses. *J Allergy Clin Immunol* 109: 803–809.

Takanaski S, Nonaka R, Xing Z, O'Byrne P, Dolovich J, and Jordana M. 1994. Interleukin 10 inhibits lipopolysaccharide-induced survival and cytokine production by human peripheral blood eosinophils. *J Exp Med* 180: 711–715.

Tavernier J, Van der Heyden J, Verhee A, et al. 2000. Interleukin 5 regulates the isoform expression of its own receptor α-subunit. *Blood* 95: 1600–1607.

Thomas PT. 1995. Pesticide-induced immunotoxicity: are Great Lake residents at risk? *Environ Health Perspect* 103 Suppl 9: 55–61.

Tobacco Research Implementation Group. Tobacco Research Implementation Plan: Priorities for tobacco research beyond the year 2000. National Cancer Institute, NIH. 1998. Unpublished.

Tryphonas H. 1998. The impact of PCBs and dioxins on children's health: immunological considerations. *Canadian Journal of Public Health* 89 Suppl 1: S49–S52.

Tsumori K, Kohrogi H, Goto E, Hirata N, Hirosako S, Fujii K, Ando M, Kawano O, and Mizuta H. 2003. T cells of atopic asthmatics preferentially infiltrate into human bronchial xenografts in SCID mice. *J Immunol* 170(11):5712–8.

Tsuyuki S, Tsuyuki J, Einsle K, et al. 1997. Costimulation through B7-2 (CD86) is required for the induction of a lung mucosal T helper cell 2 (TH2) immune response and altered airway responsiveness. *J Exp Med* 185: 1671–1679.

Turner MW, Brostoff J, Wells RS, Stokes CR, and Soothill JF. 1977. HLA in eczema and hay fever. *Clin Exp Immunol* 27: 43–47.

Uguccioni M, Mackay CR, Ochensberger B, Loetscher P, Rhis S, LaRosa GJ, Rao P, Ponath PD, Baggiolini M, and Dahinden CA. 1997a. High expression of the chemokine receptor CCR3 in human blood basophils. Role in activation by Eotaxin, MCP-4, and other chemokines. *J Clin Invest* 100: 1137–1143.

Van Duren-Schmidt K, Pichler J, Ebner C, et al. 1997. Prenatal contact with inhalant allergens. *Pediatr Res* 41: 128–131.

Van Eerdewegh P, Little RD, Dupuis J, Del Mastro RG, Falls K, Simon J, Torrey D, Pandit S, McKenny J, Braunschweiger K, Walsh A, Liu Z, Hayward B, Folz C, Manning SP, Bawa A, Saracino L, Thackston M, Benchekroun Y, Capparell N, Wang M, Adair R, Feng Y, Dubois J, FitzGerald MG, Huang H, Gibson R, Allen KM, Pedan A, Danzig MR, Umland SP, Egan RW, Cuss FM, Rorke S, Clough JB, Holloway JW, Holgate ST, and Keith TP. 2002. Association of the ADAM33 gene with asthma and bronchial hyperresponsiveness. *Nature* 418 (6896): 426–430.

Vial T, Nicolas B, and Descotes J. 1996. Clinical immunotoxicity of pesticides. *J Toxicol Environ Health* 48: 215–229.

Visser J, van Boxel-Dezaire A, Methorst D, Brunt T, de Kloet ER, and Nagelkerken L. 1998. Differential regulation of interleukin-10 (IL-10) and IL-12 by glucocorticoids *in vitro*. *Blood* 91: 4255–4264.

Voccia I, Blakely B, Brousseau P, and Fournier M. 1999. Immunotoxicity of pesticides: a review. *Toxicol Ind Health* 15(1–2): 119–132.

von Hoegen P, Sarin S, and Krowka JF. 1995. Deficiency in T cell responses of human fetal lymph node cells: a lack of accessory cells. *Immunol Cell Biol* 73: 353–361.

Walker JKL, Fong AM, Lawson BL, Savov JD, Patel DD, Schwartz DA, and Lefkowitz RJ. 2003. β-Arrestin-2 regulates the development of allergic asthma. *J Clin Invest* 112: 566–574.

Wardlaw A. 1999. Molecular basis for selective eosinophil trafficking in asthma: A multistep paradigm. *J Allergy Clin Immunol* 104: 917–926.

Wardlaw AJ, Dunnette S, Gleich GJ, et al. 1988. Eosinophils and mast cells in bronchoalveolar lavage in subjects with mild asthma. Relationship to bronchial hyperreactivity. *Am Rev Respir Dis* 137: 62–69.

Warner JA, Jones CA, Williams TJ, et al. 1998. Maternal programming in asthma and allergy. *Clin Exp Allergy* 28: 1359–1367.

Warner JA, Miles EA, Jones AC, et al. 1994. Is deficiency of interferon gamma production by allergen triggered cord blood cells a predictor of atopic eczema? *Clin Exp Allergy* 24: 423–430.

Warner JA and Warner JO. 2000. Early life events in allergic sensitization. *British Medical Bulletin* 56(4): 883–893.

Warner JO and Jones CA. 2000. Fetal origins of lung disease. In: Barker DJP (Ed.), *Fetal Origins of Cardiovascular Lung Diseases. Monographs from Lung Biology and Health and Disease.* New York: National Heart Lung and Blood Institute. Marcel Dekker. pp. 297–321.

Warner JO, Marguet C, Rao R, et al. 1998. Inflammatory mechanisms in childhood asthma. *Clin Exp Allergy* 28: 71–75.

Warrior U, McKeegan EM, Rottinghaus SM, Garcia L, Traphagen L, Grayson G, Komater V, McNally T, Helfrich R, Harris RR, Bell RL, and Burns DJ. 2003. Identification and characterization of novel antagonists of the CCR3 receptor. *J Biomol Screen* 8(3): 324–31.

Waterland RA and Jirtle RL. 2003. Transposable elements: targets for early nutritional effects on epigenetic gene regulation. *Mol Cell Biol* 23(15): 5293–300.

Weinberg ED. 1984. Pregnancy-associated depression of cell-mediated immunity. *Rev Infect Dis* 6: 814–831.

Weisglas-Kuperus N. 1998. Neurodevelopmental, immunological, and endocrinological indices of perinatal human exposure to PCBs and dioxins. *Chemosphere* 37: 1845–1853.

Weiss ST. 2002. Eat dirt: the hygiene hypothesis and allergic diseases. *N Engl J Med* 347(12): 930–931.

Weitzman M, Gortmaker S, Walker DK, et al. 1990. Maternal smoking and childhood asthma. *Pediatrics* 85: 505–511.

Williams TJ, Jones CA, Miles EA, Warner JO, and Warner JA. 2000. Fetal and neonatal interleukin-13 production during pregnancy and at birth and subsequent development of atopic symptoms. *J Allergy Clin Immunol* 105: 951–959.

Wills-Karp M, Luyimbazi J, Xu X, Schofield B, Neben TY, Karp CL, and Donaldson DD. 1998. Interleukin-13: central mediator of allergic asthma. *Science* 282: 2258–2261.

Woolley MJ, Woolley KL, Otis J, Conlon PD, O'Byrne PD, and Jordana M. 1994. Inhibitory effects of IL-10 on allergen-induced airway inflammation and airway responses in Brown Norway rats. *Am J Respir Crit Care Med* 149, p. A760.

Wuyts WA, Vanaudenaerde BM, Dupont LJ, Demedts MG, and Verleden GM. 2003. N-acetylcysteine reduces chemokine release via inhibition of p38 MAPK in human airway smooth muscle cells. *Eur Respir J* 22(1): 43–9.

Ying S, Robinson DS, Meng Q, et al. 1999. C-C chemokines in allergen-induced late-phase cutaneous responses in atopic subjects: association of eotaxin with early 6-hour eosinophils, and of eotaxin-2 and monocyte chemoattractant protein-4 with the later 24-hour tissue eosinophilia, and relationship to basophils and other C-C chemokines (monocyte chemoattractant protein-3 and RANTES). *J Immunol* 163: 3976–3984.

Yoshimoto T, Nagai N, Ohkusu K, Ueda H, Okamura H, and Nakanishi K. 1998. LPS-stimulated SJL macrophages produce IL-12 and IL-18 that inhibit IgE production *in vitro* by induction of IFN-γ production from CD3intIL-2Rβ+ T cells. *J Immunol* 161: 1483–1492.

Yunginger JW, Reed CE, O'Connell EJ, et al. 1992. A community-based study of the epidemiology of asthma. Incidence rates, 1964-1983. *Am Rev Respir Dis* 146: 888–894.

Zhong H, Shlykov SG, Molina JG, Sanborn BM, Jacobson MA, Tilley SL, and Blackburn MR. 2003. Activation of murine lung mast cells by the adenosine A3 receptor. *J Immunol* 171(1): 338–45.

Zimmermann N, King NE, Laporte J, Yang M, Mishra A, Pope SM, Muntel EE, Witte DP, Pegg AA, Foster PS, Hamid Q, and Rothenberg ME. 2003. Dissection of experimental asthma with DNA microarray analysis identifies arginase in asthma pathogenesis. *J Clin Invest* 111 (12): 1863–1874.

Zlotnik A, Morales J, and Hedrick JA. 1999. Recent advances in chemokines and chemokine receptors. *Crit Rev Immunol* 19: 1–47.

# PART V

# Neuroimmune Interactions

# CHAPTER 15

# Developmental Neurotoxicity and Postnatal Immune Deficits

**Patricia V. Basta**

## CONTENTS



## INTRODUCTION

The nervous and immune systems and the neuroendocrine axis form an integrative system of checks and balances that maintain an organism's homeostasis. In the adult animal, the interactions of these systems are tightly connected and complex, and have been extensively studied (Besedovsky and Sorkin 1977; Besedovsky et al. 1979; Besedovsky et al. 1985; reviewed in: Elenkov et al. 2000; Salzet et al. 2000; Straub et al. 2001). Interactions occur via cytokines, hormones, neurotrans-

0-415-28457-0/05/$0.00+$1.50

mitters, and neuropeptides. During development, these systems are also highly integrated, although these interactions are much less understood. There are two major pathways by which the central nervous system (CNS) signals the immune system. The first pathway, the endocrine hypothalamic-pituitary-adrenal (HPA) axis, affects immune responses in both the adult and during ontogeny, and has been reviewed extensively elsewhere (Chrousos 1998; Chrousos 2000; Coe and Lubach 2000; Ng 2000; Morale et al. 2001; Shanks and Lightman 2001; Webster et al. 2002), and will not be further discussed. The second pathway, and the focus of this chapter, is via direct neural control. This can either occur by direct stimulation of neuroreceptors on lymphoid cells or by direct neural control of the autonomic nervous system (ANS).

The ANS is a two-neuron pathway with the preganglionic efferents originating in the CNS and terminating in peripheral autonomic ganglia. These neurons release acetylcholine and stimulate $\alpha3\beta2$ or $\alpha3\beta4$ nicotinic cholinergic receptors (nAChRs) on the postganglionic neurons. The postganglionic neurons innervate the peripheral organs and are composed of either noradrenergic sympathetic nervous system (SNS) or cholinergic parasympathetic nervous system (PNS) efferents. ANS innervation in immune organs occurs along the vasculature and also within the parenchyme of the lymphoid organs. Thus, the postganglionic terminals are in close contact with immune cells. Neurotransmitter is released in a nonsynaptic manner, i.e., it is released into the extraneuronal space and diffuses a relatively long distance to receptors on lymphoid cells. This type of transmission is considered "slow" and nonspecific in contrast to synaptic neurotransmission (Elenkov et al. 2000).

Primary and secondary immune organs receive ANS innervation prior to cell migration into these organs during gestation and the neonatal period (Ackerman et al. 1991; Felten et al. 1987; Stevens-Felten and Bellinger 1997). Immune organs primarily receive innervation from the SNS, although evidence exists that cholinergic innervation via the parasympathetic nervous system is also present (Niijima 1995; Elenkov et al. 2000). Peptadergic innervation also occurs in immune organs, although it is less well described. The most abundant neuropeptides observed are the tachykinins, calcitonin gene-related peptide, and vasoactive intestinal polypeptide (VIP). Although much is known about the innervation of the immune system it is still largely unknown how insults to the developing brain result in changes to the functioning of these neural pathways or how this ultimately induces long-lasting alterations to the immune system.

Immune function is programmed during fetal and early neonatal life and is closely coordinated by neural input (Ackerman et al. 1991; Felten et al. 1998; Madden et al. 1995). This assertion is based on the following two points. First, as discussed in detail above, nerve terminals directly innervate lymphoid organs and this innervation occurs prior to immune cell migration into these organs during gestation and the neonatal period (Felten et al. 1987; Felten et al. 1988; Stevens-Felten and Bellinger 1997). Second, the developing (and adult) CNS and immune systems share common receptors and ligands, and during development these ligand-receptor interactions perform essential roles in the growth and differentiation of both systems (Chambers et al. 1993; Haas and Schauenstein 1997; Kohm and Sanders 2000; Maestroni 2000; Sanders and Straub 2002).

Insults to interneuronal communications during development can be induced by psychological (maternal neglect, noise), chemical (drugs of abuse, pesticides), physically, or a combination of these factors, which can result in a complex set of neurotoxic-induced effects on the developing CNS. These toxic effects then result in alterations to peripheral systems and organs, including the peripheral nervous and immune systems (Slotkin et al. 1987; Lindstrom 1997; Song et al. 1999). Appropriate expression of neurotransmitters and neuromodulators are not only important for the neural trafficking of the embryo but also for the development of neuronal circuits and activation responses. Disruptions of these systems caused by inappropriate stimulation can result in long-term effects. In addition, unlike classical teratology, in which the first trimester of fetal development is the most sensitive target for adverse effects of drugs and chemicals, brain development is likely affected by exposures ranging from the early embryonic stage through adolescence. In fact, it has been suggested (Hofer 1983; Slotkin 1998), that insults during fetal, neonatal, and adolescent periods can modulate the physiological "setpoints" of the autonomic nervous system. Thus, it seems reasonable to assume that toxic insults during development that affect central control of ANS neurons, or the development of ANS neurons themselves, have the potential to disrupt the development of the immune system.

The complex nature of the interaction between the CNS, ANS, and immune system makes the study of how one system influences the other very difficult to perform. Studies designed to determine the influence of developmental neurotoxicity on offspring immunity have taken the following forms: 1.) analysis of the effects of *in vivo* stimulations or ablations of one system on one or more parameters of the other, and 2.) analysis of the effects of genetic manipulations in one system on one or more parameters of the other. The discussion that follows will describe studies in each of these categories in the most well understood systems to illustrate how developmental neurotoxic insults can result in long-term alterations to offspring immune function.

## CHOLINERGIC INSULTS

### Nicotine

Cholinergic input plays a critical role in brain development, and nicotine, a cholinergic agonist, can disrupt this normal development (Roy et al. 2002). Interference with this input (excessive or inappropriately timed) can have broad-ranging results both in the brain and the peripheral nervous system. Activation of cholinergic input by nicotine is suspected to cause many of the adverse effects of maternal smoking. This is particularly important since the fetus accumulates relatively high levels of nicotine even in situations where maternal exposure occurs via environmental tobacco smoke. Developmental disruptions to rat brains were seen using continuous nicotine infusion pumps in pregnant dams, a delivery route that avoids hypoxia-ischemia and delivers the same steady-state levels of nicotine as seen in smokers or nicotine patch users (reviewed in: Slotkin et al. 1992; Slotkin 1999).

Nicotine-induced disruptions are mediated by specific nicotinic receptors and include neuronal cell death, specific alterations of neuronal activity, and misprogramming of receptor signaling mechanisms. Developmental nicotinic receptor stimulation can lead to either an induction of apoptosis or a change from cell replication to cell differentiation. Contrast studies with adult rats indicate that nicotine exerts protective effects (Janson et al. 1988; Owman et al. 1989). In addition, since nicotinic receptors are located not only at postsynaptic sites but also at presynaptic terminals of a wide variety of neurotransmitter systems, particularly the catecholamines (norepinephrine and dopamine), developmental nicotine exposure can affect all brain regions and neurotransmitters that have cholinergic input resulting in both immediate and delayed defects.

Gestational nicotine exposure has been shown to induce brain disruptions in both noradrenergic tonic activity and noradrenergic responsiveness to cholinergic stimulation (Seidler et al. 1992; Slotkin 1998). Comparable alterations also occur in peripheral autonomic pathways. As expected these disruptions result in a number of neurobehavioral and other peripheral organ alterations and defects. Neuronal effects of nicotine exposure have been seen after developmental exposures from early gestation to adolescence. Gestational nicotine exposure disrupts synaptic function and produces behavioral anomalies in two phases (Roy et al. 2002). The first phase of disruptions (widespread apoptosis and disruption of mitotic organization) occur during the immediate neonatal period. Often these disruptions are repaired and resolved by weaning, but may recur during adolescence to cells located largely in the hippocampus and somatosensory cortex. These cells develop after the gestational nicotine exposure, indicating that these deficits were not due to direct effects on these cells but to early events occurring on progenitor cells or alterations to postmitotic cell migration or connectivity. Gestational nicotine exposure is thus not a general neuroteratogen but specifically affects certain subregions and cell types, including cells that mature after nicotine exposure. Thus, gestational nicotine exposure may cause misprogramming of neural development in regions that are still maturing during adolescence.

Nicotine infusion in adolescent rats also results in neurological disruptions. The disruptions seen after an adolescent exposure both have similarities and differences from those seen after either a gestational or adult exposure (reviewed in: Slotkin 2002). A 2-week infusion of nicotine in adolescent rats, at a dose that produced plasma nicotine levels resembling those of active human smokers, resulted (as with gestational exposure) in decreases in cell number in the hippocampus, cerebral cortex, and midbrain, with greater effects observed in the female hippocampus. Adolescent nicotine exposure also produced an up-regulation of nicotinic acetylcholine receptors (nAChR) that persisted for over a month, and the changes were greater in males than in females. These changes differ from both an adult exposure, where receptor changes are short-lived, and from a gestational exposure, where receptor changes were less pronounced and less prolonged. Adolescent nicotine exposure also resulted in significant reductions in choline acetyltransferase (ChAT) (the enzyme that synthesizes acetylcholine and is a marker for cholinergic innervation), in the midbrain, but not in the cerebral cortex or hippocampus. This was also accompanied by an increase in the high-affinity presynaptic choline transporter (the rate-limiting step in acetylcholine synthesis). This pattern is an indication of cholinergic hyperstimulation at the same time as neuronal damage. These midbrain effects were not seen immediately following gestational exposure, but were seen later at adolescence. Gestational nicotine exposure resulted in changes to cholinergic activity in the cerebral cortex, which was not seen following adolescent exposure. The one change consistent between gestational and adolescent exposure was the prolonged reduction in high-affinity presynaptic choline transporter in the hippocampus. As previously discussed, nAChR also control release of the catecholamines, norepinephrine, and dopamine. After adolescent nicotine exposure, it has been shown that there is persistent subsensitivity to acute cholinergic stimulation in both genders, particularly in the midbrain. Similar effects were seen following gestational exposure.

Immune deficits following developmental nicotine exposure from early gestation to adolescence have also been observed (Basta et al. 2000; Navarro et al. 2001a; Navarro et al.), and are suspected to be a result of a combined altered CNS/peripheral nervous system input. A 2-week gestational exposure to infused nicotine was shown to induce neonatal abnormalities of both T-lymphocyte and B-lymphocyte function (Basta et al. 2000). The pattern of immune deficits was similar to the pattern of neural effects produced by gestational exposure. In both the immune and nervous system, defects were observed during the early neonatal period in the treated animals, which resolved to some extent by weaning but which reemerged during adolescence and persisted into adulthood. Therefore, just as in the brain, immune deficits were maintained well after the cessation of nicotine exposure suggesting, as before, that nicotine-induced actions on the developing immune system permanently affect the "programming" of lymphocyte responses. A likely explanation for this finding is that excessive premature stimulation of cholinergic receptors causes disruptions to the normal development of neuroimmune homeostatic controls, potentially through direct alterations of the ANS in immune organs or secondary alterations to signaling pathways. Consistent with the idea that activation of nicotinic cholinergic receptors is a trigger for these events, we found similar effects from an early developmental exposure to chlorpyrifos (discussed later), a pesticide that enhances endogenous cholinergic activity through acetylcholinesterase inhibition (Navarro et al. 2001b). In addition, developmental nicotine exposure has been shown to result in other nonimmune long-lasting changes in neuronal controls, including deficits in peripheral organ sympathetic neuron activity, lasting alterations of adenylate cyclase responses (i.e., increased nonstimulated or basal activity), and minor transient changes in β-AR numbers (Navarro et al. 1990).

The exposure of adolescent rats to infused nicotine employing the 2.5-week adolescent nicotine exposure model described above for neural toxicity (6 mg/kg/day) also caused immune alterations (Navarro et al. 2001a). This study showed that, just as for gestational exposure, adolescent exposure elicits a long-term deficit in T-lymphocyte function, which persists into adulthood. The immune effects were less dramatic than seen with gestational exposure, since only T-lymphocyte function and not B-lymphocyte function was perturbed. Additional studies have examined the effects of adolescent nicotine exposure in terms of dose dependence, duration effect, and differing methods of nicotine administration (Navarro et al.). It was found that an exposure time, as short as one week, beginning on PN 30 (30 days postnatal) was sufficient to elicit the effects on T-lymphocyte function that occur with the longer infusion period. One month after exposure ended or in young adulthood, 6 mg/kg/day of nicotine resulted in a nearly 50% reduction in the mitogenic response of T-lymphocytes, a result virtually identical to that obtained with a 2.5-week infusion. These results are in contrast to those obtained with nicotine infusions given to adult rats, where adults required a much longer minimum exposure period (4 weeks) to produce any immediate effect on T-cell mitogenic responses despite higher plasma nicotine levels (Geng et al. 1995). Further, adults did not show a delayed reappearance of deficits (Geng et al. 1996). As with gestational exposure, this points to a developmental misprogramming.

This same study (Navarro et al. 2001a), also examined the effects of a lower dose (2 mg/kg/day) of infused nicotine over this short one-week period. This dose resulted in plasma nicotine levels well below those seen in active smokers or nicotine patch users, and is consistent with levels detected in occasional smokers or in individuals exposed to nicotine via environmental tobacco smoke (ETS). This low-level adolescent exposure resulted in significant reductions in T cell function. Interestingly, these T cell deficits were not observed when nicotine was delivered by a different mode (repeated injections) at either the 6 mg/kg or the 2 mg/kg dosage rates. This result supports the postulate that early inappropriate sustained action of nicotine at nicotinic cholinergic receptors can alter programming.

Earlier work with agents acting on other neurotransmitter receptors also indicates that a prolonged action is required to elicit long-term developmental deficits (Zagon and McLaughlin 1984), while episodic stimulation or inhibition of neuroreceptors allows for recovery and adaptation during the intervals between doses. This result indicates that hypoxia-ischemic insults associated with an injection model, which produced significant weight deficits into young adulthood, did not produce the immune deficits seen with infusion, which did cause hypoxia-ischemia. This result strongly suggests that nicotine itself is responsible for the adverse effects, not the secondary effects associated with injection. In summary, an adolescent exposure as short as one week, with doses that achieve plasma nicotine levels well below those in active smokers, was sufficient to elicit long-term deficits in mitogenic responses of T-lymphocytes.

## Chlorpyrifos

Chlorpyrifos (CPY), an organophosphorus pesticide, is a neurotoxicant that acts through acetylcholinesterase inhibition, preventing the breakdown of acetylcholine, prolonging the actions of acetylcholine in the synapse (Aldridge 1990; Ray 1991). CPY was commonly used in the United States until it was discovered that the developing animal was especially sensitive to CPY toxicity. As outlined below, the effects of chlorpyrifos on the developing brain have been shown to be mediated by both excessive cholinergic stimulation and by noncholinergic mechanisms (Whitney et al. 1995). For instance, direct cholinergic effects were seen when a single dose of chlorpyrifos administered to neonatal rats immediately induced a reduction in DNA synthesis, which was also seen with the cholinergic agonist, nicotine. At the end of the first postnatal week after this single exposure, brain cholinergic regional selectivity was observed that could be blocked with antagonists. Immediate effects, however (see below), showed no selectivity. More prolonged neonatal exposures (1mg/kg/day) over two separate time frames, PN 1 to 4 vs. PN 11 to 14, however, did elicit decreases (long-term) in cholinergic innervation and tone, similar to the effects produced by nicotine. The earlier exposure resulted in the most profound effects, which were greatest in the hippocampus. Again, these effects are similar to the effects seen with nicotine after either a gestational or adolescent exposure. This may indicate that CPY may preferentially affect the programming of cell development, axongenesis, or synaptogenesis for the neurons providing the input to the hippocampus; since the hippocampus develops later than the other regions, there is more opportunity for the wiring of this region after a neonatal exposure.

However, during the immediate postexposure period when a single dose of CPY was administered, unlike with nicotine, there was no brain region selectivity; even regions with low cholinergic stimulation were affected. Other studies demonstrating noncholinergic effects of CPY exposure include a neonatal CPY exposure that reduced cholinesterase activity by only 20%, but still caused neurotoxicity (Song et al. 1997). It has been postulated that the noncholinergic effects of CPY appear to involve effects on cell signaling intermediates common to multiple neuronal and hormonal inputs, especially the adenylyl cyclase transduction pathway and transcription factors involved in promoting proliferation and differentiation (Song et al. 1997; Crumpton et al. 2000). Therefore, CPY affects brain development by at least two mechanisms, an early noncholinergic effect on second messenger systems and a later cholinergic effect.

Immune deficits have been observed after developmental CPY exposure during the same neonatal period that caused nervous system anomalies (Navarro et al. 2001b). A delayed expression of immune deficits was seen after a PN 1 to PN 4 exposure period. Exposure of neonatal rats to 1mg/kg/day during this period had no immediate effects (PN 5) on T cell mitogenic responses. However, once the animals reached adulthood, T cell responses were significantly impaired. This same delayed appearance of the effect of CPY was seen in the brain and was determined to be a direct cholinergic effect. A higher dose of CPY exposure (5 mg/kg/day) produced similar effects on the mitogenic response in adults. Only the mitogenic response was impaired, not the basal rate of proliferation or cell viability. In addition, the mitogenic response was impaired at a maximally effective concentration, indicating that the intrinsic ability of T cells to respond was impaired and not merely that the dose response curve had shifted. In summary, these results indicate that developmental exposure to CPY, known to produce lasting changes in neural function, elicits corresponding, long-term deficits in immune competence.

## Summary

The results discussed above indicate that there may be common neural-induced immune effects after developmental exposure to cholinergic-based neurotoxicants. These defects are likely to be due to an excessive premature stimulation of cholinergic receptors that caused disruptions in the normal development of neuroimmune homeostatic controls, either through direct stimulation of lymphocyte nAChRs, or indirectly through stimulation of nAChRs on postganglionic efferents of the autonomic nervous system that terminate in immune organs. Either of these mechanisms could result in either specific receptor-related changes or secondary alterations to common signaling pathways.

## MULTIMODAL INSULTS

### Ethanol

Gestational exposure to ethanol in humans has been shown to cause numerous defects, and the brain is one of the most sensitive targets. These defects are widely diverse, extremely severe, and depend greatly on the dose and timing of ethanol exposure. In fact, recent reviews of the literature indicate that multiple mechanisms (all of which have not been fully elucidated) mediate the adverse outcomes of children prenatally exposed to ethanol. An extensive review of this literature is beyond the scope of this article, so we present highlights of the key neural effects of developmental ethanol exposure. The main postulated causative mechanisms for ethanol-induced brain alterations are as follows (reviewed in: Costa and Guizzetti 1999; Goodlett and Horn 2001; Guerri 1998):

1. Free radical formation resulting in cell damage
2. Alterations in GABA, NMDA, serotonin, muscarinic cholinergic neurotransmitters systems, and potentially other systems, particularly dopamine and neuropeptide Y
3. Interferes with growth factor functions, resulting in alterations to the proliferation of neuronal precursor, impairing their migration and inducing premature apoptosis
4. Effects on astrocyte formation resulting in misguidance of neuron growth
5. Altered glucose uptake
6. Alteration of cell adhesion molecule function, particularly L1, which results in interference with the cell clustering needed for appropriate brain development
7. Altered regulation of gene transcription

Neonatal treatment of rats during rapid brain growth (third trimester human equivalent), with ethanol, which acts as both a GABA agonist and a NMDA antagonist, causes extensive neurodegeneration. Prenatal exposure, even in modest amounts, has been shown to decrease the number and function of NMDA receptors (Olney et al. 2001). GABA- and NMDA-induced defects may play a role in the mental and behavioral abnormalities seen in alcohol-exposed human offspring, especially since GABA is a trophic factor in the development of other neurotransmitter systems (Lauder et al. 1998). Ethanol-induced serotonergic changes have also been seen in mice and rats treated with ethanol in liquid diets for the majority of the gestational period (Zhou et al. 2001). This form of exposure can have overwhelming effects, since serotonin neurons innervate and communicate with almost the entire population of neurons in the brain. Three types of changes are seen in serotonergic neurons in ethanol-exposed offspring, altered movement of their final location, changes in their nerve fiber growth, and neuron loss. Fewer signals from these serotonergic fibers can result in inappropriate formation of the cortical brain region. The brain muscarinic cholinergic neurotransmitter system can also be affected by exposure to ethanol during development. In fact, it has been shown that exposure of rats to ethanol from PN 4 to 10 (rapid brain growth period) causes the inhibition of muscarinic receptor-stimulated phosphoinositide (PI) metabolism (Balduni and Costa 1989). This effect was seen after a neonatal exposure but not after an adult exposure. It has been postulated that the molecular target for the action of ethanol on the muscarinic-receptor second messenger system is G-protein coupling. As outlined above, developmental exposure to ethanol can also alter the normal growth and migration of neuronal cells. This outcome can result from several ethanol-induced neural alterations. One mechanism by which this occurs was demonstrated by Miller and Robertson (1993), who observed that a prenatal exposure to ethanol caused premature differentiation of radial (guiding) glia cells into astrocytes, resulting in abnormal positioning of cerebral cortex neurons. In addition, ethanol exposure during development has been shown to alter neurotrophic factors and the adhesion molecule L1.

Immune deficits have also been seen after gestational exposure to ethanol. Human and animal studies have frequently demonstrated a decrease in T cell number (Ewald and Walden 1988; Ewald 1989) and function (Basham et al.1998; Chang et al. 1994; Gottesfeld and Ullrich 1995; Monjan and Mandell 1980; Norman et al. 1989; Redei et al. 1989; Seelig et al. 1996; Tewari et al. 1992; Weinberg and Jerrells 1991) with gestational or neonatal exposures of varying length. The majority of these studies delivered ethanol in a chronic dosage as a component of a liquid diet or in the drinking water. This chronic-exposure regimen produced more consistent immune deficits than have binge-drinking models (Basham et al. 1998). However, just how these immune alterations are related to the produced neural defects is unclear. One potential mechanism whereby the extensive CNS changes might affect offspring immune function is via the second messenger alterations to the muscarinic cholinergic system. Alternatively, it is also known that gestational and neonatal exposure to ethanol which cause immune deficits in offspring (review in: Chiappelli and Taylor 1995; Gottesfeld and Abel 1991; Jerrells 1991; Johnson et al. 1981) also cause changes in the sympathetic nervous system in immune organs (Basta et al. 2000; Gottesfeld et al. 1990). The immune changes might also be initiated by changes to the GABA, NMDA, or serotonin neurotransmitters in the brain due to additional downstream effects. Clearly, additional studies are necessary to determine which, if any, of the ethanol-induced nervous system deficits influence or induce the immune deficits seen after developmental exposure.

### Pyrethroid Insecticides: Cypermethrin

Cypermethrin is a widely used synthetic type II pyrethroid insecticide due to its high insect toxicity, low mammalian toxicity, and rapid metabolism (Crawford et al. 1981; Rhodes et al. 1984). Cypermethrin is primarily considered to be a sodium channel toxin. At high concentrations, pyrethroids induce a long-lasting depolarization and block nerve conduction. However, other secondary mechanisms of action are also involved in their toxicity, including antagonism of GABA inhibitory action, modulation of nicotinic cholinergic transmission, and enhanced noradrenaline and adrenaline release (reviewed in: Aldridge 1990; Ray 1991; Vijveberg and van den Bercken 1990). Gestational exposure to cypermethrin has been shown to cause severe and persistent alterations in the neural and physical development of rat offspring (Gomes et al. 1991; Malaviya et al. 1993; Biernacki et al. 1995).

Santoni et al. (1997) have shown that the prenatal exposure of rats to cypermethrin results in increased numbers of peripheral blood lymphocytes and bone marrow cells but decreases in the number of lymphoid cells in the spleen and thymus. This study also demonstrated similar compartmental effects on NK cell number and function. Santoni et al. (1998) demonstrated that T cell number (thymocyte depletion, altered distribution of thymocyte subsets) and function (decreased mitogen proliferation and IL-2 production) were affected by gestational cypermethrin exposure from GD 7 to 16. Finally, Santoni et al. (1999) suggested that these immune changes may be due to the increased plasma levels of adrenaline (neuroendocrine) and noradrenaline (SNS) observed in gestationally cypermethrin-exposed offspring.

### Other Neuroteratogens

A number of other neuroteratogens. including additional drugs of abuse (heroin and other opiates, cocaine, and so on), pesticides, and other toxicants, act diffusely in the brain, affecting many regions through neural pathways and neurotransmitter systems, resulting in many behavioral effects and additional peripheral organ defects. In addition, some of these agents, particularly the opioids, can act directly on immune cells. A great deal of information is available on the effects of these drugs on the immune system of adults; however, their effects on the developing immune system are less well described than the examples provided above.

## GENETIC MANIPULATIONS

### Knockout Mice

To help address the role that the extensive sympathetic innervation of primary and secondary lymphoid organs plays in modulating immune responses, Alaniz et al. (1999) analyzed the immune response in dopamine β-hydroxylase deficient mice. These mice cannot produce norepinephrine (NE) or epinephrine (E), but can produce the precursor dopamine. Compared to wild-type or heterozygous controls, these mice have been shown to have altered metabolism, thermoregulation, cardiovascular tone, and maternal behavior (Thomas et al. 1995; Thomas et al. 1997; Thomas et al. 1998). It was noticed that a third of these animals die during adolescence (Thomas et al. 1995); however, if housed under specific pathogen-free conditions (SPF), they were able to survive to adulthood. These results indicate that the development of the immune system was affected by the lack of norepinephrine or epinephrine.

Alaniz et al. (1999) performed additional studies to analyze in detail which aspects of the immune system were affected in these knockout mice. In these studies, mice kept under SPF conditions had no changes to the numbers of T- and B-lymphocytes, granulocytes, and monocytes present in blood, thymus, or spleen compared to wild-type or heterozygous controls. However, they observed severe thymic involution in animals that were not maintained under SPF conditions. These results suggest that these mice did not have the capacity to respond to an immune challenge. This was confirmed with both a primary and secondary (re-challenge) infection model. When -/- vs. +/+ mice were infected with *L. monocytogenes*, the numbers of bacteria in the liver were higher and deaths occurred more frequently in the -/- animals. A similar result was found after a rechallenge. Alaniz et al. (1999) determined that this down-modulated immune response was due to depressed cytokine production by T-lymphocytes, particularly Th1 cytokine production. The antibody response that depended on Th1 cells (IgG2a) was reduced in these animals, but the antibody response that depended on Th2 cells (IgG1) was not affected.

To determine which component of this reduced immune response might be due to effects on the HPA axis, as opposed to the SNS-induced effects, corticosterone and prolactin concentrations were measured in these mice. Corticosterone can inhibit an immune response and prolacton, which is induced by dopamine, can enhance T cell function. Alaniz et al. (1999) found that corticosterone was elevated during a primary infection but not during a secondary infection, indicating that corticosterone may be only partly responsible for the observed T cell suppression. In addition, prolactin concentrations were similar in both groups. These findings provide direct evidence that the SNS has a dramatic influence on the development of Th1 cell responsiveness. Interestingly, leptin-deficient knockout mice with reduced sympathetic outflow, show similar T cell deficiencies (Lord et al. 1998). These mice may also offer a potential explanation for why malnourished animals have suboptimal immune responses. In fact, morphometric studies have shown links between certain parameters of birth size and depressed immune function in offspring (Leadbitter et al. 1999; McDade et al. 2001).

Conversely, nerve growth factor transgenic mice, displaying a developmentally early hyperinnervation of immune organs, are also characterized by an immune system down-regulation (Carlson et al. 1995; Carlson et al. 1998). Similarly, a naturally occurring mutant, the Spontaneous Hypertensive Rat, which also has a hyperactivation of lymphoid organ adrenergic neurons, also manifests immune deficits (Purcell and Gattone 1992; Purcell et al. 1993). These rats have reduced numbers of immature T cells, decreased proliferative responses to T cell mitogens, decreased delayed-type hypersensitivity, allograft rejection, and altered antibody formation (Strausser 1983; Sauro and Hadden 1992; Purcell et al. 1993).

The studies discussed demonstrate that inappropriate (either too much or too little) sympathetic stimulation of immune organs during development can result in depressed immune responses. The exact mechanisms responsible for these alterations after either hyper- or hypoadrenergic stimulation remain to be determined.

## SUMMARY

This review has concentrated on the potential non-neuroendocrine routes by which the developing brain can modulate offspring immune responses. The information reviewed illustrates that this is a complex issue in that different insults affect different brain regions, and the specific effects can vary depending on the timing, duration, and dose of the insult. For these reasons, it is clear that additional studies are necessary to more exactly define how changes (injuries) to different parts of the developing brain may cause downstream effects on immune system development. The studies performed with nicotine have presented considerable evidence that hyperactivation of cholinergic input leads to long-term immune alterations. Studies are still needed, however, to determine the exact process that causes the immune misprogramming. Remaining questions include which brain regions are involved, what changes occur in these regions, what specific neural pathways (sympathetic, parasympathetic, or neuropeptides) are involved, and what signaling mechanisms in immune organs have been altered that mediate the adverse developmental outcome(s). Although the answers to these questions have yet to be defined, clues from the developmental exposures and genetic manipulation research described indicate that long-term changes to SNS and other neuropeptides' systems input into immune organs, as well as how immune cells respond to this input (i.e., changes to common second messenger systems), are likely involved. In fact, it is entirely possible that both common and specific alterations occur in the "setting of immune tone" after different developmental insults.

## ACKNOWLEDGMENTS

The author would like to thank my collaborator Dr. Hernan A. Navarro for providing invaluable suggestions and comments, and to Drs. Sherry Parker and Jeremy Taylor for their expert editing.

## REFERENCES

Ackerman KD, Madden KS, Linmat S, Felton SY, and Felton DL. 1991. Neonatal sympathetic denervation alters the development of *in vitro* spleen cell proliferation and differentiation. *Brain Behav Immun* 5: 235–261.

Alaniz RC, Thomas SA, Perez-Melgosa M, Mueller K, Farr AG, Palmiter RD, and Wilson CB. 1999. Dopamine β-hydroxylase deficiency impairs cellular immunity. *Proc Natl Acad Sc USA* 96: 2274–2278.

Aldridge WN. 1990. An assessment of the toxicological properties of pyrethroids and their neurotoxicity. *CRC Crit Rev Toxicol* 21:89–104.

Balduni W and Costa LG. 1989. Effect of ethanol on muscarinic receptor-stimulated phosphoinositide metabolism during brain development. *J Pharmacol Exp Ther* 250: 541–547.

Basham KB, Whitmore SP, Adcock AF, and Basta PV. 1998. Chronic and acute prenatal and postnatal ethanol exposure on lymphocyte subsets from offspring thymic, splenic and intestinal intraepithelial sources. *Alcohol Clin Exp Res* 22: 1501–1508.

Basta PV, Basham KB, Ross WP, Brust ME, and Navarro HN. 2000. Gestational nicotine exposure alone or in combination with ethanol down-modulates offspring immune function. *Int J Immunopharmacol* 22: 159–169.

Besedovsky H and Sorkin E. 1977. Network of immune-neuroendocrine interactions. *Clin Exp Immunol* 27(1): 1–12.

Besedovsky HO, del Rey A, Sorkin E, Da Prada M, and Keller HH. 1979. Immunoregulation mediated by the sympathetic nervous system. *Cell Immunol* 48(2): 346–55.

Besedovsky HO, del Rey AE, and Sorkin E. 1985. Immune-neuroendocrine interactions. *J Immunol* 135: 750s–754s.

Biernacki B, Wlodarczyk B, Minta M, and Juszkiewicz T. 1995. The influence of cypermethrin on the pregnant and prenatal development of rabbits. *Med Weter* 51:31–35.

Carlson SL, Albers KM, Beiting DJ, Parish M, Conner JM, and Davis BM. 1995. NGF modulates sympathetic innervation of lymphoid tissues. *J Neurosci* 15:5892–5899.

Carlson SL, Johnson S, Parrish ME, Cass WA. 1998. Development of immune hyperinnervation in NGF-transgenic mice. *Exp Neurol* 149: 209–220.

Chambers DA, Cohen RL, and Perlman RL. 1993. Neuroimmune modulation: signal transduction and catecholamines. *Neurochem Int* 22(2): 95–110.

Chang MP, Yamaguchi DT, Yeh M, Taylor AN, and Norman DC. 1994. Mechanism of the impaired T-cell proliferation in adult rats exposed to alcohol *in utero*. *Int J Immunopharmacol* 16(4): 345–357.

Chiapelli F and Taylor AN. 1995. The fetal alcohol syndrome and fetal alcohol effects on immune competence. *Alcohol Alcohol* 30: 259–263.

Chrousos GC. 1998. Stressors, stress, and neuroendocrine integration of the adaptive response: The 1997 Hans Selye memorial lecture. *Ann NY Acad Sci* 851: 311–335.

Chrousos GC. 2000. The stress response and immune function: Clinical implication: The 1999 Novera H. Spector Lecture. *Ann NY Acad Sci* 917: 38–67.

Coe CL and Lubach GR. 2000 Prenatal influences on Neuroimmune Set Points in Infancy. *Ann NY Acad Sci* 917: 468–477.

Costa LG and Guizzetti M. 1999. Muscarinic cholinergic receptor signal transduction as a potential target for the development of neurotoxicity of ethanol. *Biochem Pharm* 57: 721–726.

Crawford MJ, Croucher A, and Hutson DH. 1981. Metabolism of *cis*- and *trans*-cypermethrin in rats: Balance and tissue retention study. *J Agric Food Chem* 29: 130–135.

Crumpton TL, Seidler FJ, and Slotkin TA. 2000. Developmental neurotoxicity of chlorpyrifos *in vivo* and *in vitro*: effects on nuclear transcription factors involved in cell replication and differentiation. *Brain Research*, 857: 87–98.

Elenkov IJ, Wilder RL, Chrousos GP, and Vizi ES. 2000. The sympathetic nerve—an integrative interface between two supersystems: the brain and the immune system. *Pharmacol Rev* 52(4): 595–638.

Ewald SJ. 1989. T lymphocyte populations in fetal alcohol syndrome. *Alcohol Clin Exp Res* 13: 485–489.

Ewald SJ and Walden SM. 1988. Flow cyrometric and histological analysis of mouse thymus in fetal alcohol syndrome. *J Leukoc Bio.* 44: 434–440.

Felten DL, Felten SY, Bellinger DL, Carlson SI, Ackerman KD, Madden KS, Olschowska JA, Livnat S. 1987. Noradrenergic sympathetic neural interactions with the immune system: structure and function. *Immunol Rev* 100: 225–260.

Felten SY, Felten DL, Bellinger DL, Carlson SL, Ackerman KD, Madden KS, Olschowka JA, and Livnat S. 1988 Noradrenergic sympathetic innervation of lymphoid organs. *Prog Allergy* 43: 14–36.

Felten SY, Madden KS, Bellinger DL, Kruszewska B, Moynihan JA, and Felten DL. 1998. The role of the sympathetic nervous system in the modulation of immune responses. *Adv Pharmacol* 42: 583–587.

Geng Y, Savage SM, Johnson LJ, Seagrave JC, and Sopori ML. 1995. Effects of nicotine on the immune response. I. Chronic exposure to nicotine impairs antigen receptor-mediated signal transduction in lymphocytes. *Toxicol Appl Pharmacol* 135: 268–278.

Geng Y, Savage SM, Razani-Boroujerdi S, and Sopori ML. 1996. Effects of nicotine on the immune response. II. Chronic nicotine treatment induces T cell anergy. *J Immunol* 156: 2384–2390.

Goodlett CR and Horn KH. 2001. Mechanisms of alcohol-induced damage to the developing nervous system. *Alcohol Res Health* 25: 175–184.

Gomes M, Bernardi MM, and de Souza Spinosa H 1991. Pyrethorid insecticides and pregnancy: effect on physical and behavioral development of rats. *Vet Hum Toxicol* 33: 315–317.

Gottesfeld Z, Christie R, Felten DL, and Legrue SJ. 1990. Prenatal ethanol exposure alters immune capacity and noradrenergic synaptic transmission in lymphoid organs of the adult mouse. *Neuroscience* 35: 185–194.

Gottesfeld Z and Abel E. 1991. Maternal and paternal alcohol use: Effects on the immune system of the offspring. *Life Sci* 48:1–8.

Gottesfeld Z and Ullrich SE. 1995. Prenatal alcohol exposure selectively suppresses cell-mediated but not humoral immune responsiveness. *Int J Immunopharmacol* 17(3): 247–54.

Guerri C. 1998 Neuroanatomical and neurophysiological mechanisms involved in central nervous system dysfunctions induced by prenatal alcohol exposure. *Alcohol Clin Exp Res* 22: 304–312.

Haas HS and Schauenstein K. 1997. Neuroimmunomodulation via limbic structures—the neuroanatomy of psychoimmunology. *Prog Neurobiol* 51(2): 195–222.

Hofer MA. 1983. The mother-infant interaction as a regulator of infant physiology and behavior. In: Sosenblum LA and Moltz H (Eds.), *Symbiosis in Parent-Offspring Interactions,* pp. 61–76. Plenum; New York.

Janson AM, Fuxe K, Agnati LF, Kitayama L, Harfstrand A, Andersson K, and Goldstein M. 1988. Chronic nicotine treatment counteracts the disappearance of tyrosine-hydroxylase-immunoreactive nerve cell bodies, dendrites and terminals in the mesostriatal dopamine system of the male rat after hemitransection. *Brain Res* 455: 332–345.

Jerrells TR. 1991. Immunodeficiency associated with ethanol abuse. *Adv Exp Med Biol.* 288: 229–236.

Johnson S, Knight R, Marmer, DJ, and Steele R. 1981. Immune deficiency in fetal alcohol syndrome. *Pediatr Res* 15: 908–911.

Kohm AP and Sanders VM. 2000. Norepinephrine: a messenger from the brain to the immune system. *Trends Immunol Today* 21(11): 539–542.

Lauder JM, Liu J, Devaud L, and Morrow AL. 1998. GABA as a trophic factor for developing monoamine neurons. *Perspect Dev Neurobiol* 5: 247–259.

Leadbitter P, Pearce N, Cheng S, Sears MR, Holdaway MD, Flannery EM, Herbison GP, and Beasley R. 1999. Relationship between fetal growth and the development of asthma and atopy. *Thorax* 54: 905–910.

Lindstrom J. 1997. Nicotinic acetylcholine receptors in health and disease. *Mol Neurobiol.* 15: 193–222.

Lord GM, Matarese G, Howard JK, Baker RJ, Bloom SR, and Lechler RI. 1998. Leptin modulates the T-cell immune response and reverses starvation-induced immunosuppression. *Nature* (London) 394: 897–901.

Madden KS, Felten SY, Felten DL, and Bellinger DL. 1995. Sympathetic nervous system—immune system interactions in young and old Fischer 344 rats. *Ann N Y Acad Sci* 771: 523–534

Maestroni GJ. 2000. Neurohormones and catecholamines as functional components of the bone marrow microenvironment. *Ann NY Acad Sci* 917:29–37.

Malaviya M, Husain, Seth PK, and Husain R. 1993. Perinatal effects of two pyrethoid insecticides on brain neurotransmitter function in the neonatal rat. *Vet Hum Toxicol* 35: 119–122.

McDade TW, Beck MA, Kuzawa C, and Adair LS. 2001. Prenatal undernutrition, postnatal environments, and antibody response to vaccination in adolescence. *Am J Clin Nut* 74: 543–548.

Miller MW and Robertson S. 1993. Prenatal exposure to ethanol alters the postnatal development and transformation of radial glia to astrocytes in the cortex. *J Comp Neurol* 337: 253–266.

Monjan AA and Mandell W. 1980. Fetal alcohol and immunity: Depression of mitogen-induced lymphocyte blastogenesis. *Neurobehav Toxicol* 2: 213–215.

Morale MC, Gallo F, Tirolo C, Testa N, Caniglia S, Marletta N, Spina-Purrello V, Avola R, Caucci F, Tomasi P, Delitala G, Barden N, and Marchettie B. 2001. Neuroendocrine-immune (NEI) circuitry from neuron-glial interaction to function: Focus on gender and HPA-HPG interactions on early programming of the NEI system. *Immunol Cell Biol* 79(4): 400–417.

Navarro, H.A., Mills E, Seidler FJ, Baker FE, Lappi SE, Tayyeb MI, Spencer JR, and Slotkin TA. 1990. Prenatal nicotine exposure impairs beta-adrenergic function: persistent chronotropic subsensitivity despite recovery from deficits in receptor binding. *Brain Res Bull* 25: 233–237.

Navarro HN, Basta PV, Seidler FJ, and Slotkin TA. 2001a. Adolescent nicotine: deficits in immune function. *Dev Brain Res* 130: 253–256.

Navarro HN, Basta PV, Seidler FJ, and Slotkin TA. 2001b. Neonatal chlorpyrifos administration elicits deficits in immune function in adulthood: a neural effect? *Dev Brain Res* 130: 249–252.

Navarro HN, Basta PV, Seidler FJ, and Slotkin TA. 2003. Short-term adolescent nicotine exposure in rats elicits immediate and delayed deficits in T-lymphocyte Function: Critical periods, patterns of exposure, dose thresholds. Submitted to *Nicotine and Tobacco Research* 5: 859–868.

Ng PC. 2000. The fetal and neonatal hypothalamic-pituitary-adrenal axis. *Arch Dis Child* 82: F250–F254.

Niijima A. 1995. An electrophysiological study on the vagal innervation of the thymus in the rat. *Brain Res Bull* 38: 319–323.

Norman DC, Chang MP, Castle SC, Van Zuylen JE, and Taylor AN. 1989. Diminished proliferative response of con A-blast cells to interleukin 2 in adult rats exposed to ethanol *in utero*. *Alcohol Clin Exp Res* 13(1): 69–72.

Olney JW, Wozniak DF, Jevtovic-Todorovic V, and Ikonomidou C. 2001. Glutamate signaling and the fetal alcohol syndrome. *MRDD Res Reviews* 7: 267–275.

Owman C, Fuxe K, Janson AM, and Kahrstrom J. 1989. Chronic nicotine treatment eliminates asymmetry in striatal glucise utilization following unilateral transection of the mesostiatal dopamine pathway in rats. *Neurosci Lett* 102:279–283.

Purcell ES and Gattone VH. 2nd, 1992. Immune system of the spontaneously hypertensive rat. I. Sympathetic innervation. *Exp Neurol* 117: 44–50.

Purcell ES, Wood GW, and Gattone VH. 2nd 1993. Immune system of the spontaneously hypertensive rat: II. Morphology and function. *Anat Rec.* 237(2): 236–242.

Ray ED. 1991. Pesticides derived from plants and other organisms. In: Hayes WY Jr and Laws ER Jr (Eds.), *Handbook of Pesticide Toxicology.* Academic Press, San Diego, CA, pp. 585–636.

Redei E, Clark WR, and McGivern RF. 1989. Alcohol exposure *in utero* results in diminished T-cell function and alterations in brain corticotropin-releasing factor and ACTH content. *Alcohol Clin Exp Res* 13(3): 439–443.

Rhodes C, Jones BK, and Croucher A. 1984. The bioaccumulation and biotransformation of cis, trans-cypermethrin in the rat. *Pest Sci* 15: 471–480.

Roy TS, Seidler FJ, and Slotkin TA. 2002. Prenatal nicotine exposure evokes alteration of cell structure in the hippocampus and somatosensory cortex. *J Pharmacol Exp Ther* 300: 124–133.

Salzet M, Vieau D, and Day R. 2000. Crosstalk between nervous and immune systems through the animal kingdom: focus on opioids. *TINS* 23(11): 550–555.

Sanders VM and Straub RH. 2002. Norepinephrine, the beta-Adrenergic Receptor, and Immunity. *Brain Behav Immun* 16(4): 290–332.

Santoni G, Cantalamessa F, Mazzucca L, Romagnoli S, and Piccoli M. 1997. Prenatal exposure to cypermethrin modulates rat NK cell cytotoxic functions. *Toxicol* 120: 231–242.

Santoni G, Cantalamessa F, Cavagna R, Romagnoli S, Spreghini E, and Piccoli M. 1998. Cypermethrin-induced alteration of thymocyte distribution and functions in prenatally-exposed rats. *Toxicol* 125: 7–78.

Santoni G, Cantalamessa F, Spreghini E, Sagretti O, Staffolani M, and Piccoli M. 1999. Alteration of T cell distribution and functions in prenatally cypermethrin-exposed rats: possible involvement of catecholamines. *Toxicol* 138: 175–187.

Sauro MD and Hadden JW. 1992. Gamma-interferon corrects aberrant protein kinase C levels and immunosuppression in the spontaneously hypertensive rat. *Int J Immunopharmacol.* 14: 1421–1427.

Seelig LL Jr, Steven WM, and Stewart GL. 1996. Effects of maternal ethanol consumption on the subsequent development of immunity to *Trichinella spiralis* in rat neonates. *Alcohol Clin Exp Res* 20(3): 514–522.

Seidler FJ, Levin ED, Lappi SE, and Slotkin TA. 1992. Fetal nicotine exposure ablates the ability of postnatal nicotine challenge to release norepinephrine from rat brain regions. *Dev Brain Res* 69: 288–291.

Shanks N and Lightman SL. 2001. The maternal-neonatal neuro-immune interface: Are there long-term implications for inflammatory or stress-related disease? *J Clin Invest* 108 (11): 1567–1573.

Slotkin TA, Cho H, and Whitmore WL. 1987. Effects of prenatal nicotine exposure on neuronal development: selective actions on central and peripheral catecholaminergic pathways. *Brain Res Bull* 18: 601–611

Slotkin TA, McCook EC, Lappi SE, and Seidler FJ. 1992. Altered development of basal and forskolin-stimulated adenylate cyclase activity in brain regions of rats exposed to nicotine prenatally. *Brain Res Dev Brain Res* 68(2): 233–239.

Slotkin TA. 1998. Fetal nicotine or cocaine exposure: which one is worse? *J Pharmacol Exp Ther* 285: 931–945.

Slotkin TA. 1999. Developmental cholinotoxicants: nicotine and chlorpyrifos. *Environ Health Perspect* 107 Suppl 1: 71–80.

Slotkin TA. 2002. Nicotine and the adolescent brain. Insights from an animal model. *Neurotoxicol Teratol* 24(3): 369–384.

Song X, Seidler FJ, Saleh JL, Zhang J, Padilla S, and Slotkin TA. 1997. Cellular mechanisms for developmental toxicity of chlorpyrifos: targeting the adenylyl cyclase signaling cascade. *Toxicol Appl Pharmacol* 145(1): 158–174.

Song DK, Im YB, Jung JS, Suh HW, Huh SO, Song JH, and Kim YH. 1999. Central injection of nicotine increases hepatic and splenic interleukin 6 (IL-6) mRNA expression and plasma IL-6 levels in mice: involvement of the peripheral sympathetic nervous system. *FASEB J* 13:1259–1267.

Stevens-Felten SY and Bellinger DL. 1997. Noradrenergic and peptidergic innervation of lymphoid organs. *Chem Immunol* 69: 99–131.

Straub RH, Cutolo M, Zeitz B, and Scholmerich J. 2001. The process of aging changes the interplay of the immune, endocrine, and nervous systems. *Mech Ageing Dev* 122 (14): 1591–1611.

Strausser HR. 1983. Immune response modulation in the spontaneously hypertensive rat. *Thymus* 5: 19–33.

Tewari S, Diano M, Bera R, Nguyen Q, and Parekh H. 1992. Alterations in brain polyribosomal RNA translation and lymphocyte proliferation in prenatal ethanol-exposed rats. *Alcohol Clin Exp Res* 16(3): 436–442.

Thomas SA, Matsumoto AM, and Palmiter RD. 1995. Noradrenaline is essential for mouse fetal development. *Nature* (London) 374: 643–646.

Thomas SA and Palmiter RD. 1997. Thermoregulatory and metabolic phenotypes of mice lacking noradrenaline and adrenaline. *Nature* (London) 387: 94–97.

Thomas SA, Marck BT, Palmiter RD, and Matsumoto AM. 1998. Restoration of norepinephrine and reversal of phenotypes in mice lacking dopamine beta-hydroxylase. *J Neurochem* 70: 2468–76.

Vijveberg HP and van den Bercken J. 1990. Neurotoxicological effects and mode of action of pyrethroid insecticides. *Crit Rev Toxicol* 21: 105–126.

Webster JI, Tonelli L, and Sternberg EM. 2002. Neuroendocrine regulation of Immunity. *Annu Rev Immunol* 20: 125–163.

Weinberg J and Jerrells TR. 1991. Suppression of immune responsiveness: sex differences in prenatal ethanol effects. *Alcohol Clin Exp Res* 15(3): 525–31.

Whitney KD, Seidler FJ, and Slotkin TA. 1995. Developmental neurotoxicity of chlorpyrifos: cellular mechanisms. *Toxicol Appl Pharmacol* 134(1): 53–62.

Zagon IS and McLaughlin PJ. 1984. Duration of opiate receptor blockade determines tumorigenic response in mice with neuroblastoma: a role for endogenous opioid systems in cancer. *Life Sci* 35(4): 409–416.

Zhou FC, Sari Y, Zhang JK, Goodlett CR, Li T-K. 2001. Prenatal alcohol exposure retards the migration and development of serotonin neurons in fetal C57BL mice. *Dev Brain Res* 126: 147–155.

CHAPTER 16

# Neuroendocrine-Immune Interactions: Impact of Gender, Negative Energetic Balance Condition, and Developmental Stage on the Mechanisms of the Acute Phase Response of the Inflammatory Process in Mammals

**Eduardo Spinedi, Néstor B. Pérez, and Rolf C. Gaillard**

## CONTENTS



0-415-28457-0/05/$0.00+$1.50

## INTRODUCTION

The discovery of a negative feedback loop between the immune system and the brain (Besedovsky et al. 1986; Fauci 1979; Sapolsky et al. 1987) is one of the most exciting advances in the field of neuroendocrine-immunology. Immune cells can be stimulated by microorganism-derived toxins to secrete cytokines (Dinarello and Mier 1987). In turn, cytokines may induce many host responses associated with endotoxemia (Michalek et al. 1980), characterized by fever, stress hormone release, mineral redistribution, and increased acute phase protein synthesis (Dinarello 1988). The proinflammatory cytokines, interleukin (IL)-1 (Girardin et al. 1988), and tumor necrosis factor alpha (TNF) α (Michie et al. 1988) have been proposed to be the most important mediators for the development of the above-mentioned pathophysiological responses.

As stated above, reciprocal communication between the immune and endocrine systems is currently accepted. However, the relationships most deeply investigated up to now have been between the immune and the hypothalamo-pituitary-adrenal (HPA) and the HP-gonadal (HPG) axes (see Spangelo et al. 1995 for references). For instance, mitogen/antigen-activated immune cells secrete cytokines; in turn, these substances are able to stimulate the hypothalamus, thus inducing the activation of the CRH-ergic function (Spinedi et al. 1992a). Once activated, the CRH neuronal system is able to locally (centrally) inhibit the HPG axis function (Rivest and Rivier 1995) and, via the corticotrope, to stimulate HPA axis function (Bateman et al. 1989).

A sexual dimorphism in the neuroendocrine-immune response is strongly supported by evidence indicating that gonadal steroids modulate immunological function (Grossman 1984). Our own findings and the findings of other researchers suggest that while estrogens enhance the immune response (Erbach and Bahr 1991), androgens inhibit it (Spinedi et al. 1992b), and that gonadectomy alters this response (Graff et al. 1969; Roubinian et al. 1979). Skin allograft rejection time in mice is longer in males than in females, and orchidectomy significantly reduces the time for such rejection (Graff et al. 1969). Male F1 N2B/N2W mice are less susceptible to autoimmune lupus, but will die if gonadectomized (Roubinian et al. 1979). In addition, mitogen-driven plaque-forming cell response of B-lymphocytes *in vitro* is inhibited by androgens (Sthoeger et al. 1988). All this evidence is strongly supported by the presence of specific receptors for sex hormones in organs responsible for the immune response (Grossman 1984). However, gonadal steroids seem to influence not only immune activity but also HPA axis function, since they can either positively (estrogen) (Vamvakopoulos and Chrousos 1993) or negatively (androgen) (Bingaman et al. 1994) modulate hypothalamic CRH production, and as a consequence, close the interactive circuit between the HPA and HPG axes (Rivest and Rivier 1995). That a sex hormone basis is responsible for neuroendocrine-immunological sexual dimorphism finds support in previous studies (Grossman 1984; Spinedi et al. 1992b). The HPA axis function in rodents has been characterized as sexually dimorphic in both basal and poststress conditions (Spinedi et al. 1992b; Spinedi et al. 1995); moreover, sexual dimorphism in the HPA axis response holds true regardless of the type of stimulus since, for instance, both neuroendocrine (Spinedi et al. 1992a) and immune (Spinedi et al. 1992c) stresses induce a final release of glucocorticoid in plasma that is higher in female than in male adult animals. Glucocorticoid secretion is a hormone crucial for metabolic adaptations of the organism to stress (Munck et al. 1984) and a glucocorticoid hormone basis tallies with survival of individuals in sepsis (Munck et al. 1984; Evans and Zuckerman 1991).

Weight loss and anorexia frequently accompany infection. It is known that several cytokines are released during endotoxemia. They in turn induce anorexia and cachexia (Plata-Salaman and Borkoski 1994; Fantino and Wieteska 1993), and among them, the anorexigenic protein (Zhang et al. 1994), leptin (LEP), is also released during endotoxic shock (Grunfeld et al. 1996). Leptin physiology is in close relationship with, among other systems, the activity of HPA axis function. In fact, it is recognized that leptin administration in wild-type mice is able to blunt the fasting-induced increase in plasma adrenocorticotropic hormone (ACTH) and glucocorticoid levels (Ahima et al. 1996); reciprocally, adrenalectomy (ADX) induced a fall in circulating leptin and dexamethasone administration in ADX animals restored plasma leptin to levels found in intact animals (Spinedi and Gaillard 1998). Thus, communication between the HPA axis and adipose tissue functions could also play an important role in the physiopathological events occurring during the acute phase response after injury/infection.

Leptin is also directly involved in the regulation of the immune function. Interestingly, among the interrelationships between the organism metabolic status and the immune function, it is recognized that under unfavorable energetic conditions associated with hypoleptinemia, an augmented individual susceptibility to lipopolysaccharide (LPS)-induced lethality (Faggioni et al. 1999) exists. This observation, in models of leptin deficiency or starvation, has been an important clue for the understanding that leptin itself possesses an inherent activity that restores resistance to the inflammatory process (Lord et al. 1998). These characteristics find support in previous studies indicating that: a.) peripheral mononuclear cells (PMNC) and other components of the immune system express functional Ob-Rb receptors (Sanchez-Margalet and Martin-Romanero 2001), b.) a modulatory activity of leptin on T lymphocyte cell function could take place by the polarization of the Th cell response to the Th1 phenotype (Faggioni et al. 2001), c.) circulating leptin levels increase during inflammation and infection, which further suggests that this adipokine is involved in the response of both the innate and acquired immune functions (Faggioni et al. 2001), thus indicating that leptin plays a key role in the host defense mechanisms against infection/inflammation.

As previously stated, glucocorticoid production is the most important factor for the survival of the organism during infection/injury (Munck et al. 1984). A disturbed metabolic status of the individual clearly affects several endocrine system functions (Glass et al. 1986; Bronson 1986; Brady et al. 1990), and some mechanisms of neuroimmunomodulation are also probably altered, as a consequence of a condition characterized by low energetic balance, such as undernutrition or malnutrition. TNF$\alpha$ is a cachectic cytokine involved in the pathogenesis of septicemia (Tracey et al. 1987). Impaired cytokine production in malnourished children has been reported (Doherty et al. 1994; Bhaskaram and Sivakumar 1986). Malnutrition seems to induce an attenuation of the febrile (Doherty et al. 1989) and acute-phase protein (Doherty et al. 1993) responses. Since cytokines are the most important immune system-derived substances implicated in such responses, a reduction in proinflammatory cytokine production during inflammatory stress in undernourished/malnourished subjects could explain the attenuation of the above mentioned responses. In undeveloped countries, malnutrition and undernutrition, particularly in children under 4 years old, are associated with increased risk of mortality. In fact, significant synergy between wasting and infections, as predictors of mortality, was found in several field studies (Fauzi et al. 1997).

Cachexia, immune dysfunction, and other metabolic disturbances are linked to malnutrition of individuals with a poor prognosis after infection, injury, or many illnesses (Fauzi et al. 1997; Zaman et al. 1997). Both malnutrition (Jacobson et al. 1997a) and undernutrition (Giovambattista et al. 2000a) are accompanied by high levels of circulating glucocorticoid which can in turn induce immune (Doherty et al. 1993; Suttmann et al. 1994) and other metabolic disturbances (Giovambattista et al. 2000a) that contribute to the pathophysiological effects of different eating disorders. Calorie- and protein-malnutrition may arise from a variety of conditions and both are characterized by enhanced basal HPA axis activity (Jacobson et al. 1997b; Giovambattista et al. 2000a); however, several symptoms found in many cases of nutrition disturbances are connected with associated parasitic or bacterial infections, so that the relative contribution of calorie and protein malnutrition to enhanced glucocorticoid production has been controversial (Smith et al. 1975; Pimstone 1976). However, it has been demonstrated that neuroendocrine-metabolic-immune dysfunction induced by severe undernutrition, starting after weaning, can be rapidly reversed by placing individuals on an *ad libitum* food intake program (Bronson 1986; Giovambattista et al. 2000a; Gruaz et al. 1993), indicating that this serious allostatic state, characteristic of undeveloped countries, can be prevented by an improvement in socioeconomic conditions.

On the other hand, there is little information on the consequences of maternal undernutrition on the offspring's HPA axis and other metabolic functions. One study showed that in sheep, abnormal development of the fetal HPA axis occurred as a consequence of nutrient restriction of mothers in early gestation, resulting in HPA axis hyporesponsiveness to hypoxaemia (Hawkins et al. 2000). Whereas undernutrition during late gestation in rats has been reported to induce maternal adaptive mechanisms, which could in turn produce fetal endocrine pancreas dysfunction (Alvarez et al. 1997), poor nutrition of mothers during gestation and lactation was demonstrated to lead to impaired pancreatic β-cell activity during adulthood (Wilson and Hughes 1997). These findings clearly underscore the importance of maternal homeostasis for a normal adult life and for the development of appropriate metabolic-endocrine responses in acute/chronic allostatic states.

Individual allostasis (Sterling and Eyer 1988) as a consequence of undernutrition or malnutrition is characterized by immune-neuroendocrine disturbances (Glass et al. 1986; Bronson 1986; Brad et al. 1990), among other effects. It is already known that glucocorticoid is an important factor for the survival of the organism during allostatic states (Munck et al. 1984). A bidirectional relationship between the immune and neuroendocrine (HPA) functions (Bateman et al. 1989) is principally maintained by endogenous glucocorticoid (Besedovsky et al. 1986). TNFα is the pathognomonic cytokine of early stages of septicemia (Tracey et al. 1987). Impaired cytokine production in malnourished children has already been reported (Doherty et al. 1994; Bhaskaram and Sivakumar 1986), and malnutrition has been described as a factor responsible for attenuated febrile (Doherty et al. 1989) and acute-phase protein (Doherty et al. 1993) responses during infection. Malnutrition in children is associated with increased risk of mortality (Fauzi et al. 1997; Zaman et al. 1997); however, the impact of undernutrition of mothers on the neuroendocrine-immune and adipocyte functions in their offspring has not been fully investigated. TNFα,, the most relevant cachectic cytokine, is involved in the pathogenesis of septicemia (Tracey et al. 1987). Impaired cytokine production in malnourished children has been reported (Doherty et al. 1994; Bhaskaram and Sivakumar 1986). Malnutrition seems to induce an attenuation of the febrile (Doherty et al. 1989) and acute-phase protein (Doherty et al. 1993) responses. Since cytokines are the most important immune system-derived substances implicated in such responses, a reduction of cytokine production during inflammatory stress in malnourished subjects could explain the attenuation of the above-mentioned responses. Although the diminished production of cytokines by peripheral immune system-derived cells from malnourished individuals, incubated *in vitro* in the presence of different mitogens, has been widely studied and well documented (Doherty et al. 1994; Bhaskaram and Sivakumar 1986; Kauffman et al. 1986), the regulatory effect of peripheral cytokine levels of malnourished patients on the *in vitro* production of these substances is still lacking. Glucocorticoid production is the most important factor for the survival of the organism during infection/injury (Munck et al. 1984). It is known that there is a bidirectional relationship between the immune and endocrine systems (Bateman et al. 1989), in which the endogenous glucocorticoid environment forms the main link between the two systems (Besedovsky et al. 1986); in addition, a disturbed metabolic status of the individual clearly affects, among other systems, several endocrine system functions (Glass et al. 1986; Bronson 1986; Gruaz et al. 1993), and some mechanisms of neuroimmunomodulation are probably altered too, as a consequence of malnutrition.

## SEXUAL DIMORPHISM OF NEUROENDOCRINE-IMMUNE INTERACTIONS: EVIDENCES SUPPORTING A SEX HORMONE MOLECULAR BASIS

Adult (8- to 10-week-old) mice are the best laboratory animal model to study the influence of the genetic background on HPA response to immune challenge. A gender-dependent characteristic HPA axis response to endotoxin administration (i.e., 2 mg/Kg) is displayed in Figure 16.1. For instance, the time courses of plasma ACTH (Figure 16.1, upper left) and corticosterone (Figure 6.1, upper right) levels before and several times after LPS stimulus clearly indicate that the corticotrope function, while similar in both sexes in basal condition, was maximal 2 hours post-LPS, regardless of sex, but was greater in females than in males. Corticotrope resiliency (Kauffman et al. 1986) was completed 72 hours after challenge and was identical in both groups of mice. Adrenal function, in basal and post-endotoxin conditions, displayed a very clear gender-dependent characteristic, since female corticosterone plasma levels were significantly higher than male values, regardless of the time lapse. This pattern was also found in middle-aged (15-month-old) mice (Figure 16.1, lower left [ACTH] and lower right [glucocorticoid]), thus indicating the persistence of the gender-dependent difference in HPA axis function at more advanced ages. These characteristics were observed not only in circulating plasma hormone levels, but also in adrenal glucocorticoid content, regardless of age (Figure 16.2, upper panels).

However, some parameters indicate that enzymatic process could be involved in these patterns. For example, in studies performed in our laboratory on 8-week-old mice, females showed increased estrogen plasma concentrations 2 hours post-LPS ($90.4 \pm 5.3$ vs. basal of $35.4 \pm 3.5$ pg/ml), while males had a significant decrease (from $4.6 \pm 1.9$ ng/ml, in basal condition) in plasma testosterone levels (to $1.4 \pm 0.3$ ng/ml). This indicates that the gender-dependent differences in HPA axis function after immune challenge could have a sex-steroid hormone basis. An increased testicular aromatization of androgen to estrogen has been reported during endotoxicosis (Christeff et al. 1992). Therefore, these results may indicate that this increased aromatization in males, who are less immunoresponsive than females (Spinedi et al. 1992b), could result in enhanced immune response as an adaptation of the body's defense mechanisms shortly after injury. Interestingly, the gender-dependent characteristics of HPA axis function holds true not only in adult and middle-aged mice, but also pre- (30-day-old) and peri- (45-day-old) pubertal mice, who also displayed HPA axis response to immune stressor administration. In fact, LPS-stimulated corticosterone secretion, when studied over development, was higher in 30-day-old than in 45-day-old males, whereas in females, it was maximal in 45-day-old animals, indicating that puberty, with concomitant increases in gonadal activity, resulted in marked changes in immune-neuroendocrine interactions (Figure 16.2, lower panels).

To characterize the modulatory effects of sex steroids on HPA axis and immune system responses during endotoxemia, the influence of gonadectomy and sex hormone replacement therapy on LPS-activated neuroendocrine-immune function was explored. Experiments were performed in which adult BALB/c mice of both sexes were sham operated (m = male and f = female) or gonadectomized (Gnx) for 20

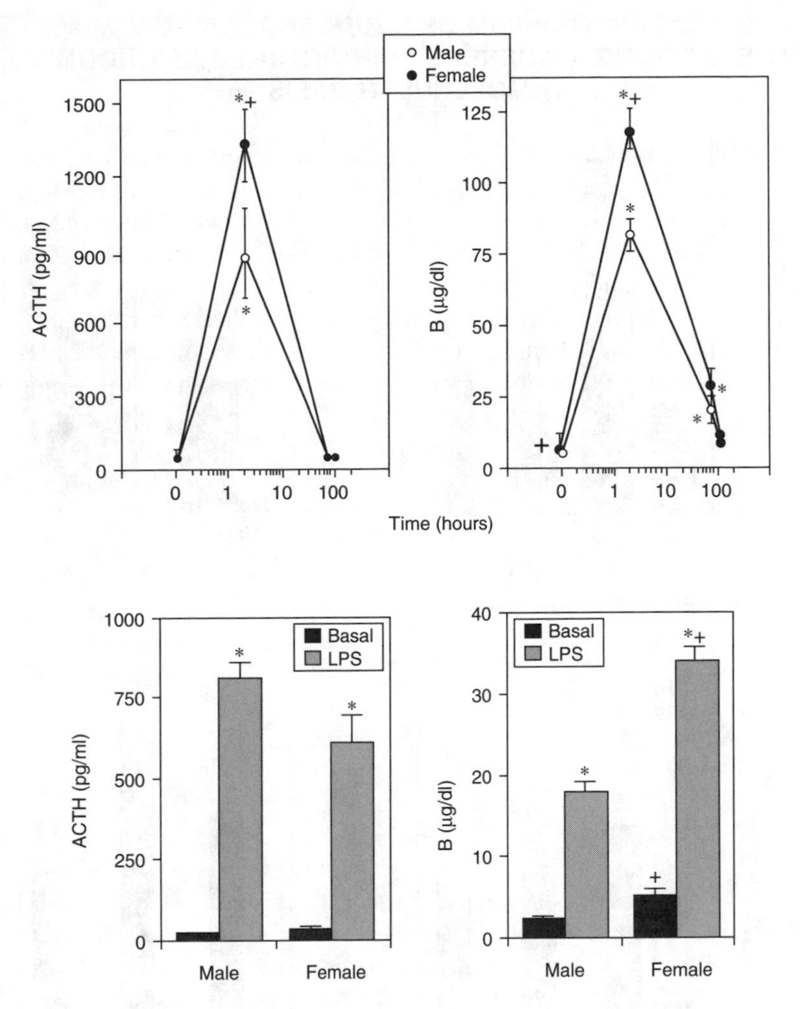


**Figure 16.1** Plasma ACTH (upper left) and corticosterone (upper right) concentrations before and several times after i.p. administration of endotoxin (2 mg/Kg) in 2-month-old mice of both sexes. Circulating ACTH (lower left) and glucocorticoid (lower right) 2 hours after i.p. injection of vehicle alone (basal) or containing LPS (2 mg/Kg) in 15-month-old mice of both sexes. Mean ± SEM (n = 6-9 animals per group). *$P < 0.05$ vs. the respective basal values; $^{+}P < 0.05$ vs. males in similar condition. (Spinedi E et al. *Front Horm Res* 29:91-107, 2002.)

days (Odx = orchidectomized and Ovx = ovariectomized). In order to determine whether the lack of endogenous sex steroid hormone or replacement therapy in Gnx mice with the homologous sex steroid (20 µg testosterone propionate/50 µl corn oil, Odx+T; or estradiol benzoate, 2 µg/50 µl corn oil, Ovx+E; on alternate days between days 1 and 19 after Gnx) influence HPA axis function during endotoxemia, on day 20 postsurgery, mice were injected intraperitoneally (i.p.) with vehicle (Veh) alone (sterile saline solution) or containing LPS (2 mg/Kg). Results indicate that: a.) Gnx enhanced LPS-stimulated glucocorticoid release regardless of sex group (Figure 16.3, upper panels), and b.) replacement therapy with the homologous sex steroid

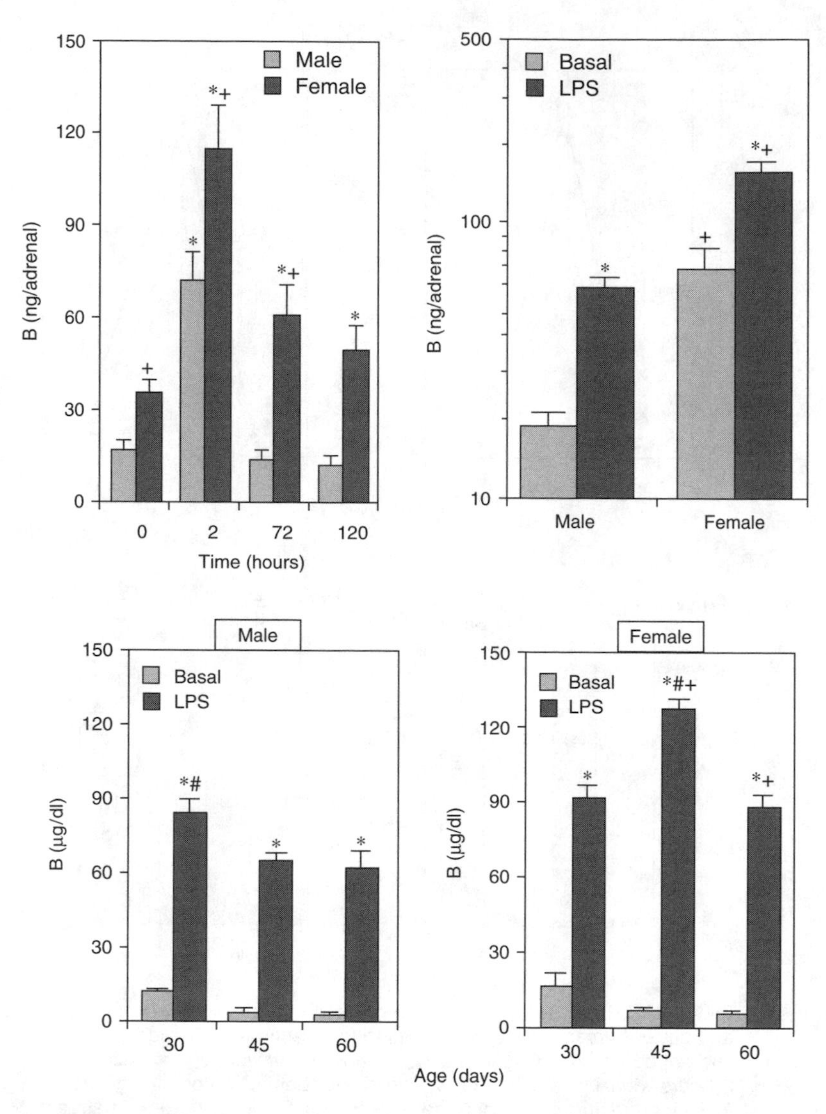


**Figure 16.2** Adrenal gland glucocorticoid content before and several times after i.p. administration of endotoxin (2 mg/Kg) in 2 month-old mice of both sexes (upper left). Basal and LPS-stimulated (at 2 hours) adrenal gland glucocorticoid content in middle-aged mice of both sexes (upper right). Circulating corticosterone in basal and post LPS (2 mg/Kg, i.p.) conditions in male (lower left) and female (lower right) mice of different ages. Mean ± SEM (n = 6-9 animals per group). *$P < 0.05$ vs. the respective basal values; +$P < 0.05$ vs. condition-matched males; #$P < 0.05$ vs. remaining LPS values in mice of same sex. (Spinedi E et al. *Front Horm Res* 29:91-107, 2002.)

fully reversed the effect of gonadectomy alone (Figure 16.3, upper panels) in animals of both sexes. Interestingly, when immune function was evaluated, TNFα release in plasma after LPS stimulus was also increased several fold (vs. intact animal values) post-Gnx, and replacement therapy with the respective sex steroid either fully or partially reversed this effect in males and females (Figure 16.3, lower panels).

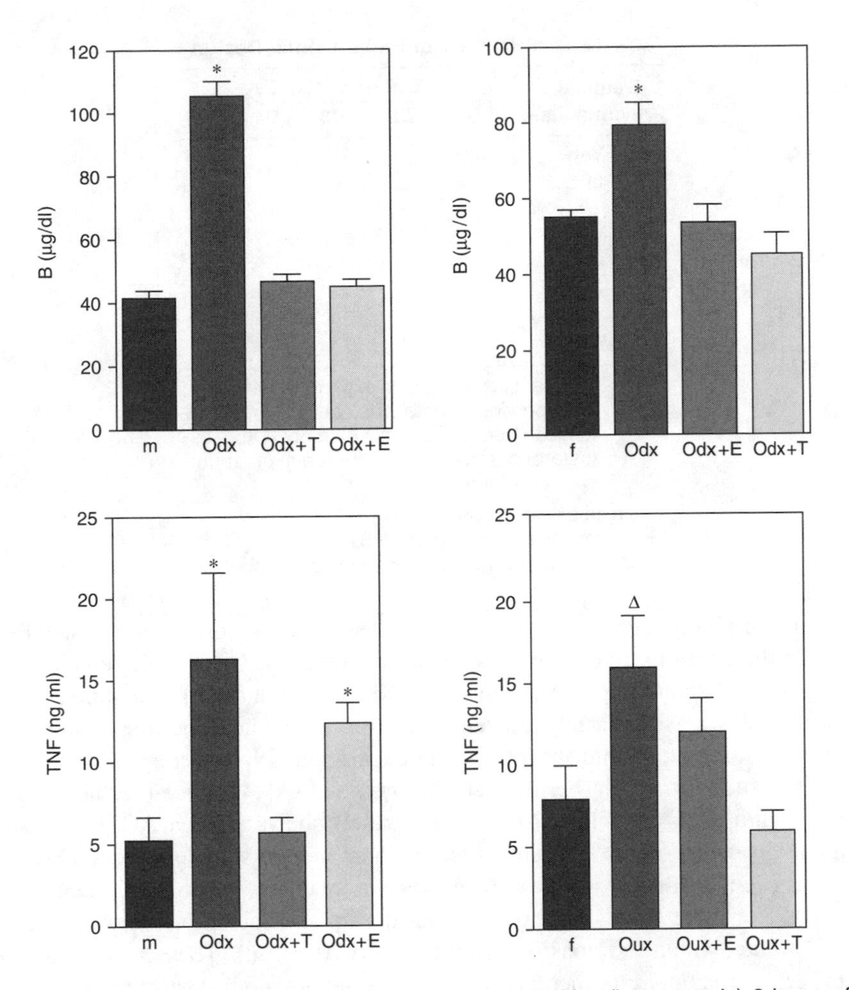


**Figure 16.3** Plasma glucocorticoid (upper panels) and cytokine (lower panels) 2 hours after i.p. administration of LPS (50 μg/mouse) in several groups of male (left panels) and female (right panels) 2 month-old mice. Mean ± SEM (n = 6–9 animals per group). All values poststressor are significantly ($P < 0.05$) higher than the respective basal (Veh-injected) values. *$P < 0.05$ vs. values in the remaining groups; $^{\Delta}P < 0.05$ vs. f and Ovx+T. (Spinedi E et al. *Front Horm Res* 29:91–107, 2002.)

We extended the evidence for a sex hormone basis of immunoneuroendocrine interaction when we developed a model of tolerance to repeated endotoxin administration. In these experiments, sham-operated (sham) and 20-day orchidectomized mice treated (Odx+T) or not (Odx) with testosterone propionate were tested. The dose of i.p. administered LPS was also 2 mg/Kg, and the schedule of the tolerance model, which started on day 20 postgonadectomy, was as follows: a) on day one (D1), mice were injected with Veh alone or containing LPS and animals were killed 2 hours post-treatment; b) mice were injected on D1 and D2 with LPS and on D3 were injected with Veh alone (LPS2-Veh) or containing LPS (LPS2-LPS); animals were then killed two hours after the last treatment; and c) mice were treated on D1, D2, D3, and D4 with LPS and on D5 were injected with either Veh alone (LPS4-

**Table 16.1 Summary of Experimental Design**

| Treatment Previous–Last | Experimental Day D1 | D2 | D3 | D4 | D5 |
|---|---|---|---|---|---|
| None–Veh | Δ, + | | | | |
| None–LPS | *,+ | | | | |
| LPS2–Veh | * | * | Δ, + | | |
| LPS2–LPS | * | * | *,+ | | |
| LPS4–Veh | * | * | * | * | Δ, + |
| LPS4–LPS | * | * | * | * | *,+ |

*Note:* Mice employed in this experimental design were previously (one week before) orchidectomized [receiving (Odx+T) or not (Odx) testosterone replacement therapy] or sham-operated (sham).

Δ= i.p. injection of NaCl 0.9% (Vehicle).
* = i.p. injection of 25 μg of LPS.
+= decapitation 2 hours after the injection.

Veh) or containing LPS (LPS4-LPS) and killed 2 hours postinjection (see Table 16.1). In the control group, sham mice were injected every day with vehicle only.

Figure 16.4 shows the decrease in mice body weight under LPS treatment. Mean body weight was significantly decreased on day 2 of the experiment; thereafter weights remained low until the end of the experiment. No differences were found in the loss of body weight between sham, Odx, and Odx+T mice (see also Figure 16.4). Figure 16.5 and Figure16.6 (upper panel) shows plasma ACTH levels in different groups of mice. Animals killed 2 hours after a vehicle injection without previous (Veh, D1) and with previous endotoxin treatments (LPS2-Veh, D3; LPS4-Veh, D5) (for more details see experimental design and Table 16.1) showed plasma ACTH values within the nonstress range (10 to 70 pg/ml). These results indicate that previous treatments (either 2 or 4) with LPS did not modify basal plasma ACTH levels measured 24 hours after the last endotoxin administration. Intact mice treated with LPS, in a single (LPS, D1) or repeated (LPS2-LPS, D3 and LPS4-LPS, D5) doses, had significantly higher plasma ACTH levels (2 hours after the last endotoxin administration) than those observed in control groups (killed 2 hours after Veh injection on D1, D3, and D5, respectively). However, the LPS-induced plasma ACTH secretion on D1 (2 hours after 1 injection) was significantly higher than values found 2 hours after the third (D3) and fifth (D5) LPS injections (Figure 16.5, upper panel). Figure 16.5 (middle panel) shows that plasma corticosterone (B) levels in control groups (those killed 2 hours after a last administration of Veh with or without previous LPS treatment) were also within the nonstress range (1 to 10 μg/dl). Two hours after the last LPS administration a significant increase over control values in adrenal B release has taken place, regardless of the day of treatment. However, LPS-elicited B secretion was significantly lower 2 hours after the fifth (D5) than after both the first (D1) and the third (D3) LPS administrations (Figure 16.5, middle panel). Figure 16.5 (lower panel) shows that 2 hours after vehicle injection, without (D1) or with previous LPS treatments (D3 and D5), plasma TNFα values were at

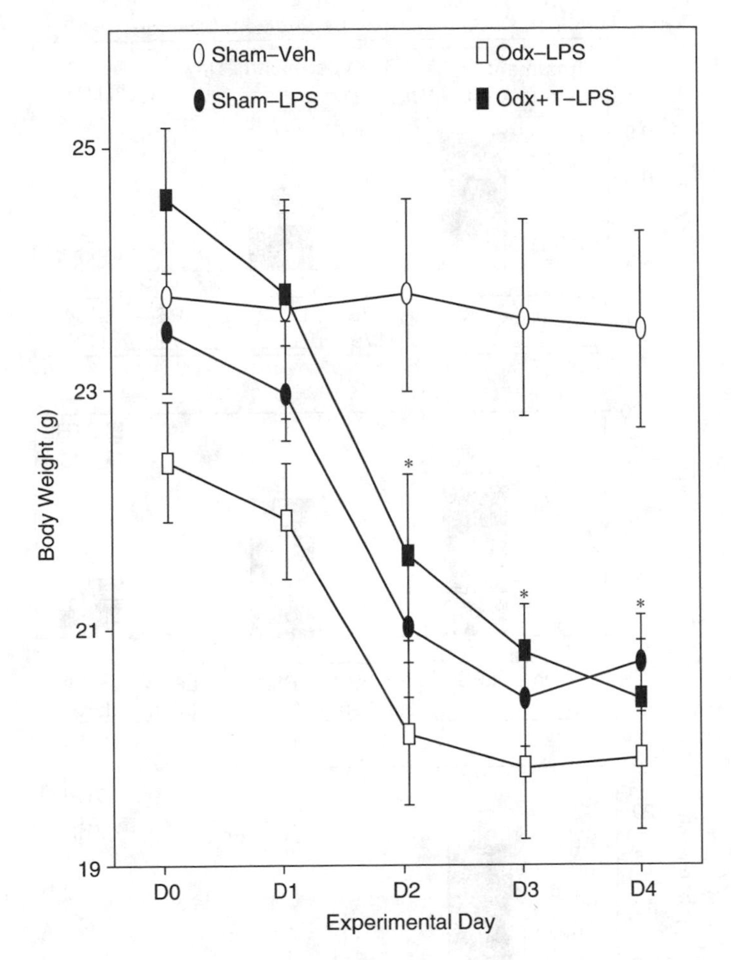


**Figure 16.4** Effects of i.p. injections of vehicle (Veh) or LPS (2mg/Kg) on the body weight of different groups of mice. Values registered 16 hours before killing. Sham-Veh mice were killed on D5. Mean ± SEM (n = 6–11 animals per group). * $P < 0.01$ vs. sham-Veh on same day. (Spinedi E et al. *Front Horm Res* 29:91–107, 2002.)

basal levels (0.05-0.8 ng/ml). A significant increase over baseline values in TNFα secretion was found 2 hours after the first, third ($P < 0.01$), and fifth LPS administrations (Veh injected on D1, D3, and D5, respectively). However, this increase was markedly higher 2 hours after the first (D1) than after the third (D3) and fifth (D5) LPS treatments (Figure 16.5, lower panel). Finally, while on D1 of treatment there is a high correlation between plasma ACTH ($r = 0.8$) or B ($r = 0.7$) and TNFα levels after LPS administration, this correlation is significantly lower on D3 and D5 of treatment ($r = 0.3$ for ACTH and $r = 0.4$ for B).

To determine whether orchidectomy alone or followed by testosterone therapy could influence the HPA axis and immune functions in our design, we examined these responses in 7-day-orchidectomized mice substituted with testosterone propionate (Odx+T) or not (Odx) (see Table 16.2). Figure 16.6 (upper panel) shows the results of plasma ACTH levels in our studies. As shown, Odx and Odx+T had no

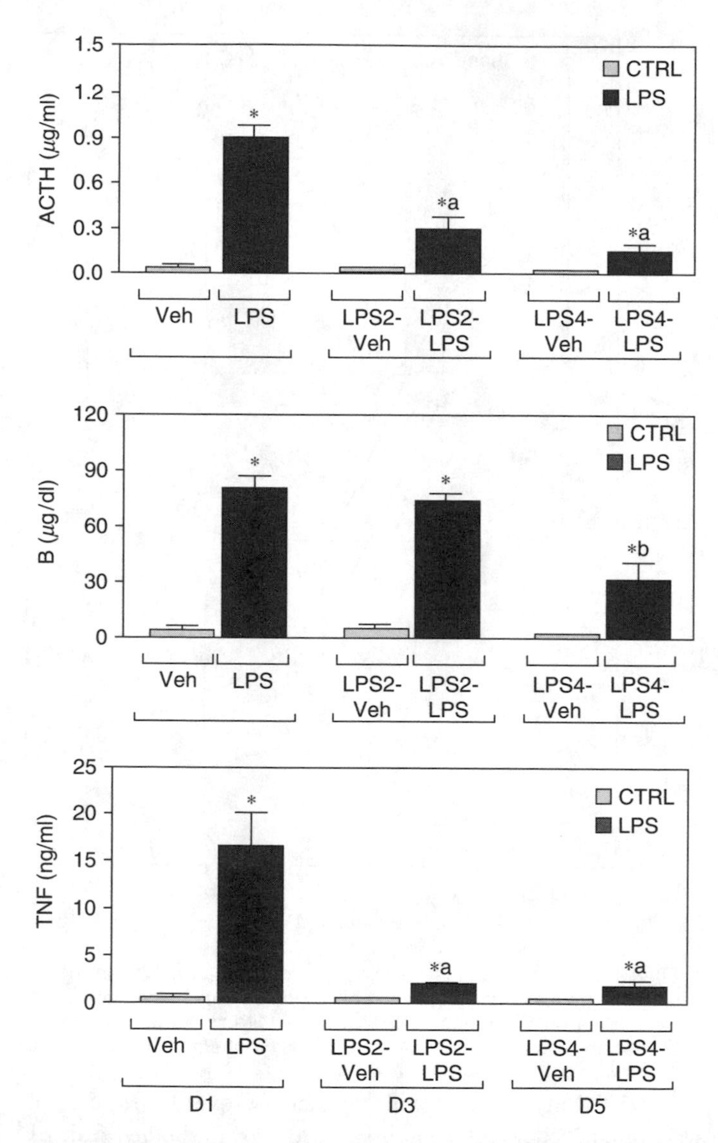


**Figure 16.5** Effects of single and repeated LPS (2mg/Kg) administration on plasma ACTH (upper), glucocorticoid (middle), and cytokine (lower) levels 2 hours after the last treatment with either Veh or LPS in sham male mice. Mean ± SEM (n = 6–11 animals per group). *P < 0.02 vs. the respective Veh; [a]P < 0.05 vs. LPS (D1); [b]P < 0.05 vs. LPS (D1) and LPS2-LPS (D3). (Spinedi E et al. *Front Horm Res* 29:91–107, 2002.)

significant effect on plasma ACTH levels in controls (animals killed 2 hours after Veh administration on D1, D3, and D5, respectively). However, Odx was able to significantly enhance the LPS-elicited ACTH secretion after the first LPS injection, and this effect was partially prevented by T therapy (Odx+T). Plasma B and TNF$\alpha$ levels (Table 16.2) in Veh-injected animals were not influenced by Odx and Odx+T. Similarly, orchidectomy or Odx+T did not modify B and TNF$\alpha$ secretion (Table

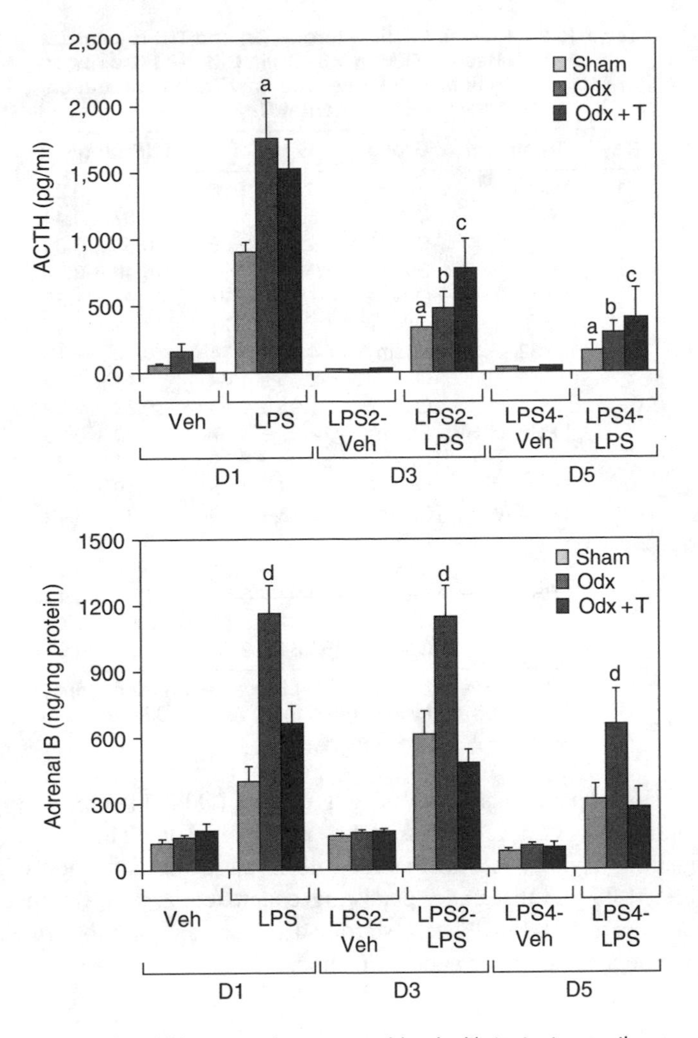


**Figure 16.6** Effects of orchidectomy alone or combined with testosterone therapy on plasma ACTH levels (upper) and adrenal gland glucocorticoid content (lower) 2 hours after the last Veh or LPS (2 mg/Kg) injection. Mean ± SEM (n = 6–11 animals per group). LPS values on D1, D3 and D5 were significantly ($P < 0.02$) higher vs. respective controls. [a]$P < 0.05$ vs. LPS D1 shams; [b]$P < 0.05$ vs. LPS D1 Odx; [c]$P < 0.05$ vs. LPS D1 Odx+T; [d]$P < 0.05$ LPS shams and Odx+T on same day. (Spinedi E et al. *Front Horm Res* 29:91–107, 2002.)

16.2) 2 hours after the first, third, and fifth LPS injections when compared to intact (sham) mice similarly treated. Finally, Figure 16.6 (lower panel) shows that adrenal B content was not modified by Odx alone or combined with T therapy in animals killed 2 hours after Veh administration, with or without previous LPS treatment. Interestingly, the LPS-induced increase in adrenal B content, already observed in sham mice, was significantly enhanced by Odx 2 hours after the first (LPS, D1), third (LPS2-LPS, D3), and fifth (LPS4-LPS, D5) endotoxin administrations compared to the respective sham-operated animals (treated with LPS in a similar fashion).

**Table 16.2 Plasma Corticosterone (B) and TNFα Values (Mean ± SEM; n = 6–11) in Different Groups of Male Mice 2 h after the Last Treatment and on Different Experimental Days**

| Day | Treatment | Group | B (μg/dl) | TNFα (ng/ml) |
|---|---|---|---|---|
| 1 | Veh | Sham | 5.34 ± 1.35 | 0.88 ± 0.23 |
| | | Odx | 6.08 ± 1.58 | 0.37 ± 0.44 |
| | | Odx+T | 5.97 ± 2.38 | 0.55 ± 0.36 |
| | LPS | Sham | 80.31 ± 5.01 | 16.46 ± 3.57 |
| | | Odx | 87.10 ± 8.71 | 11.91 ± 1.85 |
| | | Odx+T | 87.80 ± 8.71 | 11.81 ± 4.41 |
| 3 | LPS2–Veh | Sham | 4.93 ± 1.74 | 0.53 ± 0.11 |
| | | Odx | 2.83 ± 1.04 | 0.44 ± 0.06 |
| | | Odx+T | 1.96 ± 0.88 | 0.39 ± 0.09 |
| | LPS2–LPS | Sham | 72.83 ± 4.37 | 1.73 ± 0.28 |
| | | Odx | 86.09 ± 6.79 | 2.47 ± 0.31 |
| | | Odx+T | 73.24 ± 7.45 | 2.18 ± 0.80 |
| 5 | LPS4–Veh | Sham | 2.13 ± 0.97 | 0.41 ± 0.13 |
| | | Odx | 3.35 ± 1.22 | 0.70 ± 0.08 |
| | | Odx+T | 3.73 ± 1.41 | 0.54 ± 0.11 |
| | LPS4–LPS | Sham | 30.97 ± 5.51 | 1.79 ± 0.46 |
| | | Odx | 37.31 ± 8.81 | 2.55 ± 0.40 |
| | | Odx+T | 36.64 ± 9.75 | 1.91 ± 0.65 |

*Note:* LPS, LPS2–LPS, and LPS4–LPS values are significantly ($P < 0.05$ or less) higher than Veh, LPS2–Veh, and LPS4–Veh, respectively.

Figure 16.6 (lower panel) also shows that T therapy (Odx+T) completely abolished this effect induced by Odx, regardless of the experimental day. These findings suggest that endogenous TNFα has a mediatory role in the acute activation of HPA axis function after LPS and that under persisting endotoxemia, both immune and HPA functions are decreased. Finally, testosterone displayed an inhibitory role on adrenal glucocorticoidogenesis during endotoxic shock.

## METABOLIC EFFECTS OF FOOD RESTRICTION OVERDEVELOPMENT AND THEIR CONSEQUENCES ON THE BODY'S DEFENSE MECHANISMS DURING ADULTHOOD: EVIDENCES SUSTAINING THAT THE REINSERTION OF UNDERNOURISHED INDIVIDUALS IN A PROGRAM OF NORMAL EATING OVERRIDES UNDESIRABLE METABOLIC-NEUROENDOCRINE-IMMUNE DYSFUNCTION

Figure 16.7 shows the results of changes in body weights in well-nourished (WN) and undernourished (UN) (see Table 16.3) female rats. As depicted, restriction of food intake drastically reduced weight gain and induces a significant delay in the vaginal opening day (mean ± SEM: 38.3 ± 1.8 in UN rats vs. 29.9 ± 0.7 in WN animals, $P < 0.05$; n = 20-24 animals per group). In 60-day-old rats, this model of food restriction produced a significant hypoglycemia and hypotriglyceridemia in basal condition (see Table 16.4); however, the food restriction protocol induced no

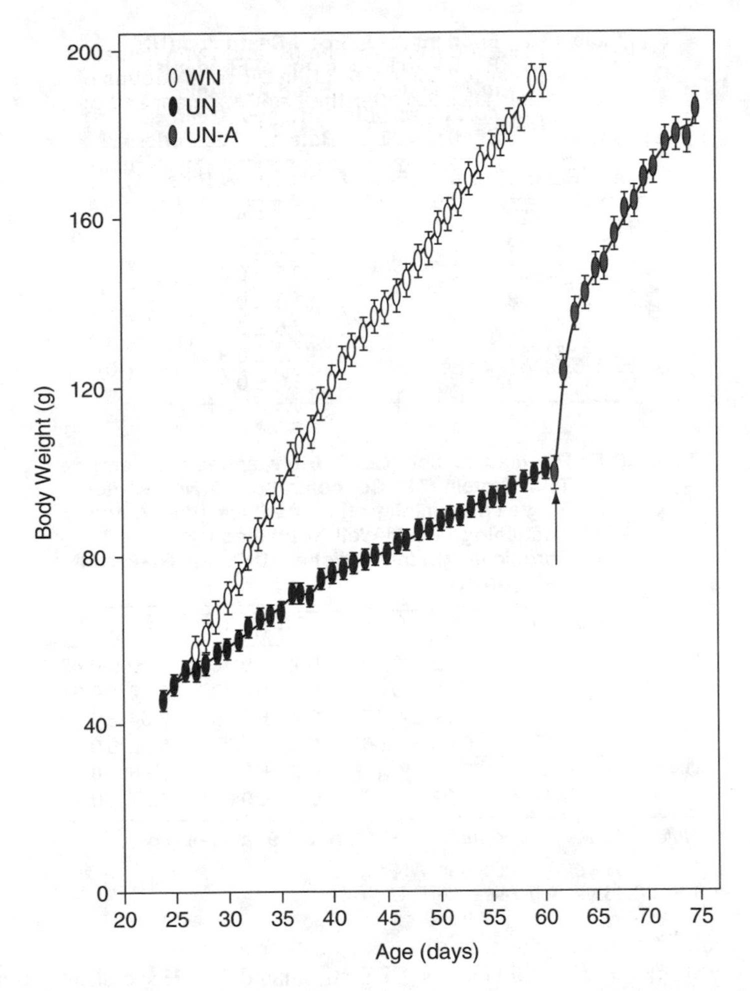


**Figure 16.7** Changes of body weights in well nourished (WN, open circles), chronically undernourished (UN, closed circles), and 15-day re-fed undernourished (UN-R, shaded circles) rats over development. The arrow indicates the day when animals with food restriction were switched to food *ad libitum*. Values are the mean (± SEM) of 20–24 animals per group. (*Neuroimmunomodulation* 7:92–98, 2000.)

significant decrease in either circulating total proteins (TP) (see Table 16.4) or anterior pituitary (AP) ACTH concentrations (in μg ACTH/mg of soluble protein: 70.17 ± 22.36 and 82.39 ± 13.75 in WN and UN rats, respectively). However, basal adrenal gland (AG) corticosterone (B) concentrations (in ng B/mg of soluble protein) were significantly ($P < 0.05$) higher in UN (393.01 ± 44.81) than in WN (75.05 ± 7.61) rats. Single i.p. LPS (180 μg/Kg BW) administration in WN rats induced a significant ($P < 0.05$ vs. vehicle; Veh) decrease in plasma glucose (GLU) without affecting circulating triglyceride (TG) and TP; in UN animals, it did not change any of these parameters (Table 16.4).

Figure 16.8 (upper panel) shows plasma ACTH values 2 hours after a single injection of either Veh or LPS. Circulating ACTH after VEH administration was

**Table 16.3 Food Intake in Food-Restricted (UN) Rats, as a Percentage of Food Intake in Animals with Food Available *ad libitum*, at Different Ages (Mean ± SEM, n = 20–24 Rats per Age Period)**

| Age Period (Days) | Food Intake in UN Rats (%) |
|---|---|
| 24–27 | 85.5 ± 2.7 |
| 28–29 | 74.7 ± 4.4 |
| 30–36 | 63.5 ± 0.8 |
| 37–43 | 51.3 ± 0.7 |
| 44–50 | 47.5 ± 0.7 |
| 51–56 | 43.2 ± 0.8 |
| 57–60 | 43.3 ± 0.9 |

**Table 16.4 Plasma Glucose (GLU), Triglycerides (TG), and Total Protein (TP) Concentrations 2 Hours after Single i.p. Administration of Vehicle (Veh) Alone or Containing LPS in Well Nourished (WN), Chronically Undernourished (UN), and Re-Fed UN Rats (UN-R)**

| Group | Condition | GLU (g/l) | TG (g/dl) | TP (g/dl) |
|---|---|---|---|---|
| WN | Veh | 1.12 ± 0.19 | 0.79 ± 0.10 | 7.00 ± 0.63 |
| | LPS | 0.83 ± 0.07[a] | 0.89 ± 0.12 | 6.22 ± 0.31 |
| UN | Veh | 0.79 ± 0.07[a] | 0.42 ± 0.05[a] | 6.04 ± 0.26 |
| | LPS | 0.66 ± 0.06 | 0.37 ± 0.07 | 5.32 ± 0.29 |
| UN-R | LPS | 0.71 ± 0.14[b] | 0.71 ± 0.03 | 5.96 ± 0.88 |
| | Veh | 1.22 ± 0.12 | 0.78 ± 0.08 | 5.87 ± 0.77 |

*Note:* Values are the mean ± SEM, n = 6–9 rats per group.

[a] P < 0.05 vs. Veh-values in WN rats

[b] P < 0.05 vs. Veh-values in R-UN rats

statistically similar in WN and UN rats. LPS-stimulated ACTH secretion was several fold higher (P < 0.05) vs. the respective Veh-values; however, ACTH released as a consequence of LPS was significantly (P < 0.05) higher in WN than in UN rats. Figure 16.8 (lower panel) shows the adrenal gland function in basal condition (Veh) and after LPS treatment. As depicted, basal plasma B values were already significantly (P < 0.05) higher in UN than in WN animals. LPS administration significantly (P < 0.05) enhanced plasma B concentrations over the respective Veh-values; however, when the adrenal response was compared to the respective baseline, this response in WN was much higher (approximately 13-fold) than in UN (approximately 2-fold) animals.

Figure 16.9 (upper panel) shows TNFα values in plasma after Veh or LPS treatment in WN and UN rats. Food restriction failed to induce any change in basal (Veh-injected rats) plasma TNFα levels. The results of the immune response to LPS challenge indicate that in both groups (WN and UN) of rats, there was a significant (P < 0.05) increase in plasma TNFα levels vs. the respective baseline values; however, peripheral TNFα values were significantly (P < 0.05) higher in WN than in UN animals. Figure 16.9 (lower panel) shows the results of circulating LEP in

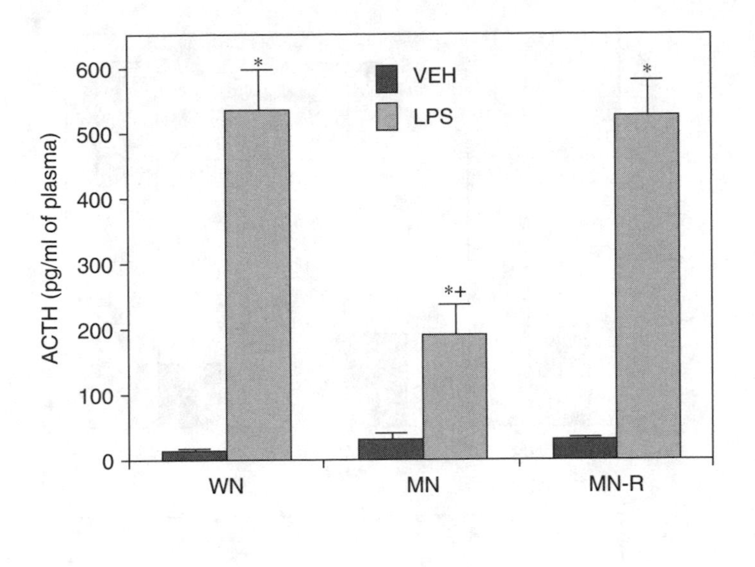


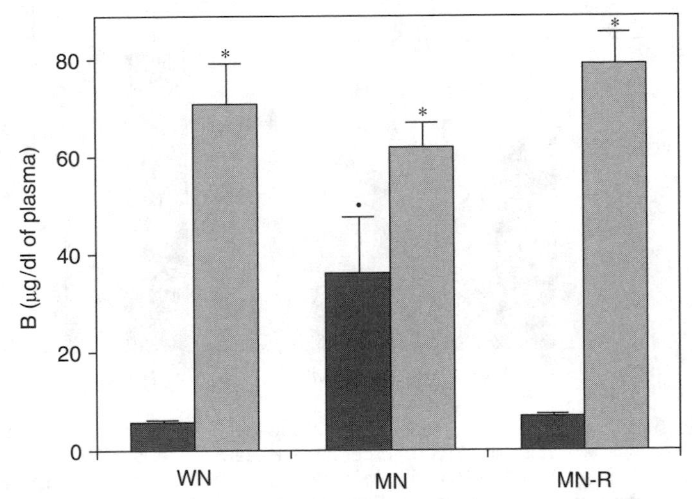


**Figure 16.8** Plasma ACTH (upper panel) and corticosterone (lower panel) in basal (VEH) condition and 2 hours after i.p. administration of endotoxin (LPS) in well nourished (WN), chronically undernourished (UN), and 15-day re-fed undernourished (UN-R) rats. Bars represent the mean (± SEM) values of 6–9 rats per group per condition. *P < 0.05 vs. the respective VEH-values; +P < 0.05 vs. LPS-values in WN and UN-R groups; °P < 0.05 vs. VEH-values in WN and UN-R groups. (*Neuroimmunomodulation* 7:92–98, 2000.)

both basal (Veh) and LPS conditions. Clearly, food-restricted animals showed significantly ($P < 0.05$) lower basal plasma LEP levels than their WN counterparts. Finally, basal plasma LEP levels were unchanged by endotoxin treatment, regardless of the group.

Figure 16.7 shows also that 15 days of *ad libitum* food intake in UN rats, now UN-re-fed (UN-R), was necessary for the recovery of the body weight observed in 60-day-old WN rats. Throughout that period of time, and when analyzed on day 75

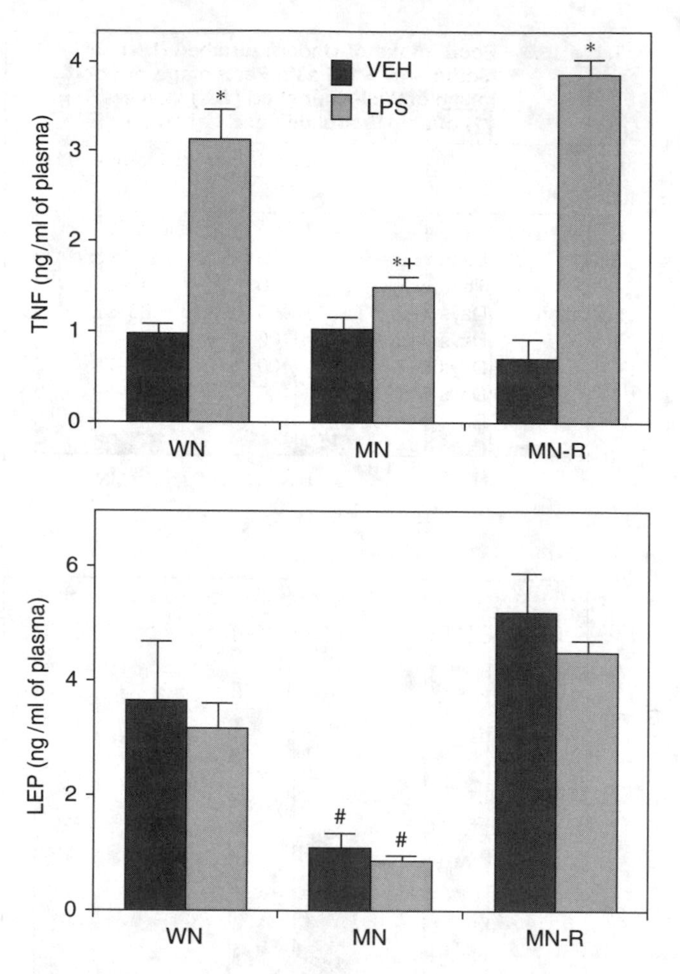


**Figure 16.9** Tumor necrosis factor-α (upper panel) and leptin (lower panel) plasma concentrations in well nourished (WN), chronically undernourished (UN), and 15-day re-fed undernourished (UN-R) rats before (Veh) and 2 hours after i.p. LPS administration. Bars represent the mean (± SEM) values of 6–9 rats per group per condition. *P < 0.05 vs. the respective VEH-values; +P < 0.05 vs. LPS-values in WN and UN-R groups; #P < 0.05 vs. the respective values in WN and UN-R groups. (*NeuroImmunoModulation* 7:92–98, 2000.)

of age, the basal hypoglycemia, hypotriglyceridemia, hypercorticosteronemia, and hypoleptinemia, induced as a consequence of chronic food restriction, were all fully reversed (see Table 16.4, Figure 16.8, and Figure 16.9). UN-R rats also restored their basal AG B concentration to normal WN values (81.48 ± 5.35 ng/mg of protein). In addition, the neuroendocrine, immune, and adipocyte responses found 2 hours after LPS administration in UN-R animals resembled those observed in WN rats (Figure 16.8 and Figure 16.9), thus indicating that full reversion of diminished body weights to WN values is able to restore normal neuroendocrine, immune, and adipocyte functions (both in basal conditions and after LPS treatment).

**Table 16.5 Food Intake of Undernourished (UN) Mothers (n = 19) as a Percentage of Food Intake of Well-Nourished (WN) Mothers (n = 11) during Pregnancy and Lactation**

| Period | | Food Intake of Mothers, % | |
|---|---|---|---|
| | | WN | UN |
| Pregnancy | First week | 100 | 60–72 |
| | Second week | 100 | 47–63 |
| | Third week | 100 | 47–66 |
| Lactation | Days 1–3 | 100 | 83–97 |
| | Days 4–5 | 100 | 69–75 |
| | Days 6–7 | 100 | 66–70 |
| | Days 8–10 | 100 | 55–59 |
| | Days 11–13 | 100 | 44–53 |
| | Days 14–15 | 100 | 44–49 |
| | Days 16–17 | 100 | 48–51 |
| | Days 18–20 | 100 | 50–53 |

## IMPACT OF UNDERNUTRITION IN MOTHERS AND OFFSPRING ON METABOLIC-NEUROENDOCRINE-IMMUNE FUNCTION

### Influence of Food Restriction on Body Weight Changes in Pregnant Rats and Their Offspring

Changes in body weights of mothers as a consequence of a particular food restriction protocol (see Table 16.5) are shown in Figure 16.10. An arrest in body weight gain in pregnant rats was present, though not significantly different vs. their counterparts, at the end of the first week of pregnancy; however, food restriction significantly ($p < 0.05$) reduced body weight gain by the end of both the second and third weeks of pregnancy. Nevertheless, body weights of UN mothers on the day before partum were significantly ($p < 0.05$) higher than the respective day zero values. Restriction of food intake continued after offspring's birth (see Figure 16.11, upper panel, and Table 16.5); however, we did not record body weights of mothers because, in preliminary experiments, we found a high rate of litter mortality as a consequence of the additional stress induced by handling. The number of total weaned pups per mother was significantly ($p < 0.05$) reduced in UN rats (7.68 ± 0.51 vs. 10.36 ± 0.72 in WN mothers), however, not as a consequence in the reduction of the number of weaned males per mother (4.91 ± 0.49 and 3.95 ± 0.33 in WN and UN mothers, respectively). Figure 16.12 (upper panel) shows the results of body weights on different days of age in the four groups established. When day and status of the mother (WN and UN) were compared in rats eating *ad libitum* after weaning (WN-WN and UN-WN groups), body weights were significantly ($p < 0.05$) lower in rats born from UN than from WN mothers up to day 49; similarly, but in groups submitted to the food restriction protocol (WN-UN and UN-UN groups), body weights were significantly ($p < 0.05$) lower in rats born from UN than from WN mothers up to day 47. On the other hand, when comparing the effect of food

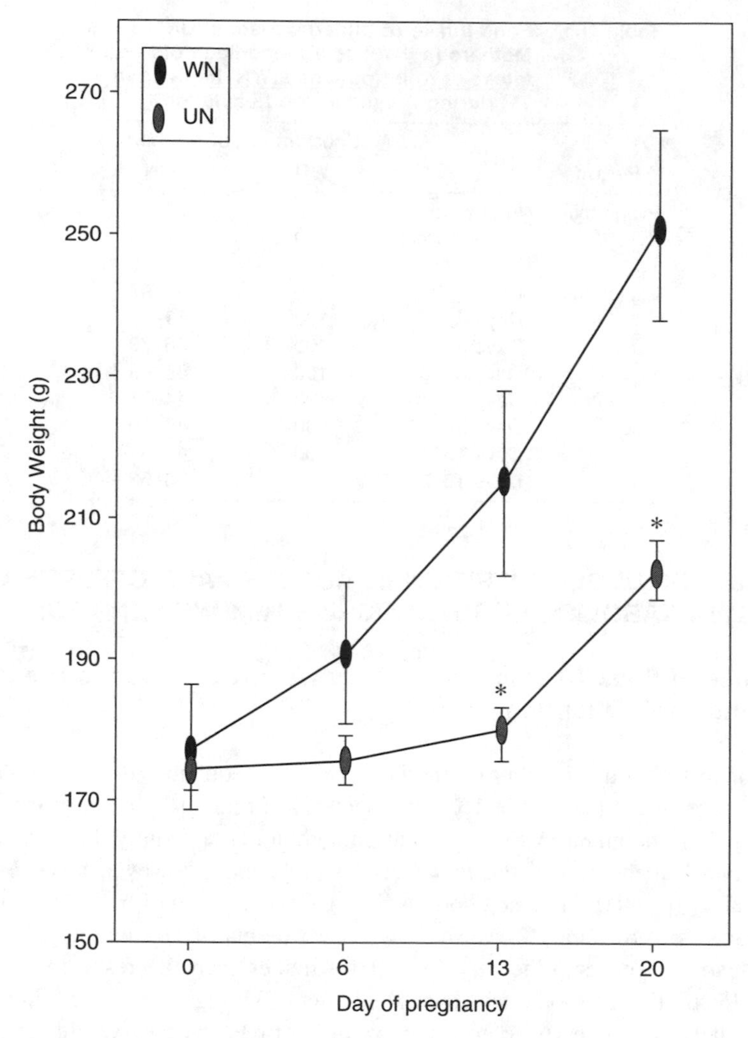


**Figure 16.10** Body weight values of well nourished (WN) and undernourished (UN) mothers at different days of pregnancy. Values are the mean ± SEM (n = 11 and 19 for WN and UN rats, respectively). *$p < 0.05$ vs. the respective values for WN mothers. (Chisari AN et al. *NeuroImmunoModulation* 9:41–48, 2001a.)

restriction on weaned male pups from the same mother group (WN-UN vs. WN-WN and UN-UN vs. UN-WN), data analysis indicates that body weights were significantly ($p < 0.05$) different from day 25 of age.

As depicted in Figure 16.12 (lower left panel), male pups weaned from UN mothers displayed a significantly ($p < 0.05$) lower body weight than their normal counterparts. On day 59 of age (Figure 16.12, lower right panel), no significant differences in body weights were found between *ad libitum* fed male rats born from both WN and UN mothers (WN-WN and UN-WN groups), nor between food-restricted males born from both groups of mothers (WN-UN and UN-UN rats). However, body weights of food-restricted males (WN-UN and UN-UN groups) were

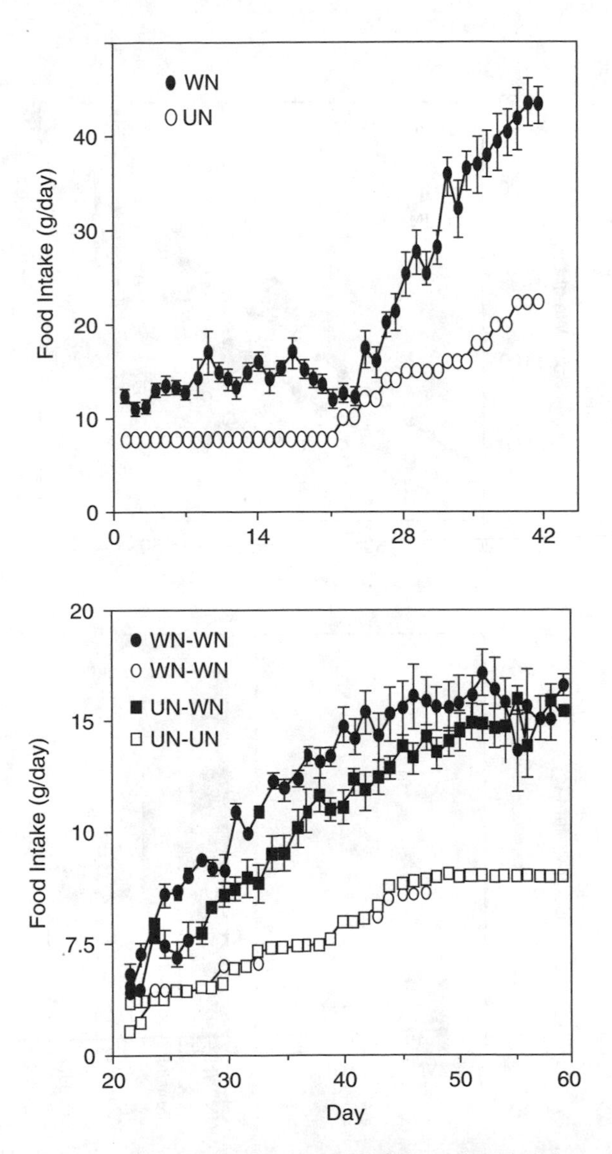


**Figure 16.11** Daily food intake of well-nourished (WN) (n =12) and undernourished (UN) (n = 20) mothers on different days of gestation (days 1 to 21) and lactation (days 21 to 41) (upper panel). Food intake of male rats from several groups: well-nourished mothers during gestation and lactation, well nourished from weaning up to day 59 (WN-WN), WN-Undernourished from day 22 up to day 59 (WN-UN), Undernourished mothers during gestation and lactation-well nourished from weaning up to day 59 (UN-WN) and UN-UN (n = 13–17 rats per group) (lower panel). (Chisari AN et al. *Endocrine* 14:375–382, 2001.)

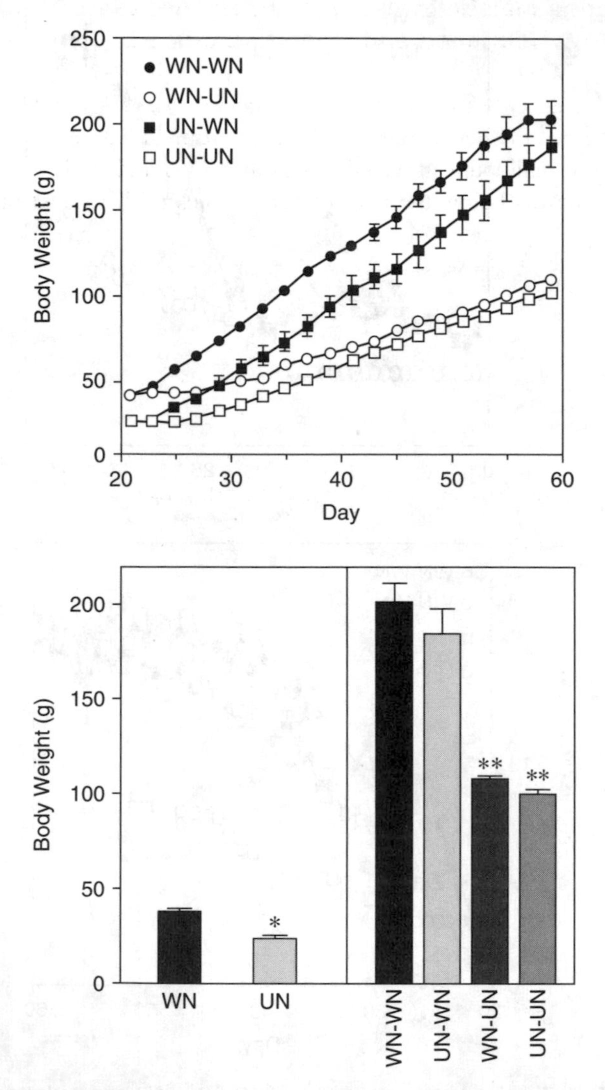


**Figure 16.12** Changes of body weights in several experimental groups, recorded day by day, after weaning (day 21) and up to day 59 of age (upper panel; n = 12–17 rats per group). Body weights of male rats from different groups on days 21 (lower left panel) and 59 (lower right panel) of age. *$p < 0.05$ vs. WN values; **$p < 0.05$ vs. WN-WN and UN-WN values. (Chisari AN et al. *Endocrine* 14:375–382, 2001.)

significantly ($p < 0.01$) lower than those of rats born from both groups of mothers (WN and UN) but fed *ad libitum* until day 59 (Figure 16.12, lower right panel).

## Neuroendocrine-Metabolic and -Immune Responses in 21-Day-Old Rats Born from Well-Nourished and Undernourished Mothers

Basal plasma GLU, ACTH, and glucocorticoid concentrations were not modified by undernutrition of mothers (Figure 16.13A-C). Administration i.p. of insulin (INS) (1 IU/Kg BW) to male pups at weaning was able to significantly ($p < 0.05$) reduce plasma glucose levels below the respective baselines (Veh values) in WN and UN

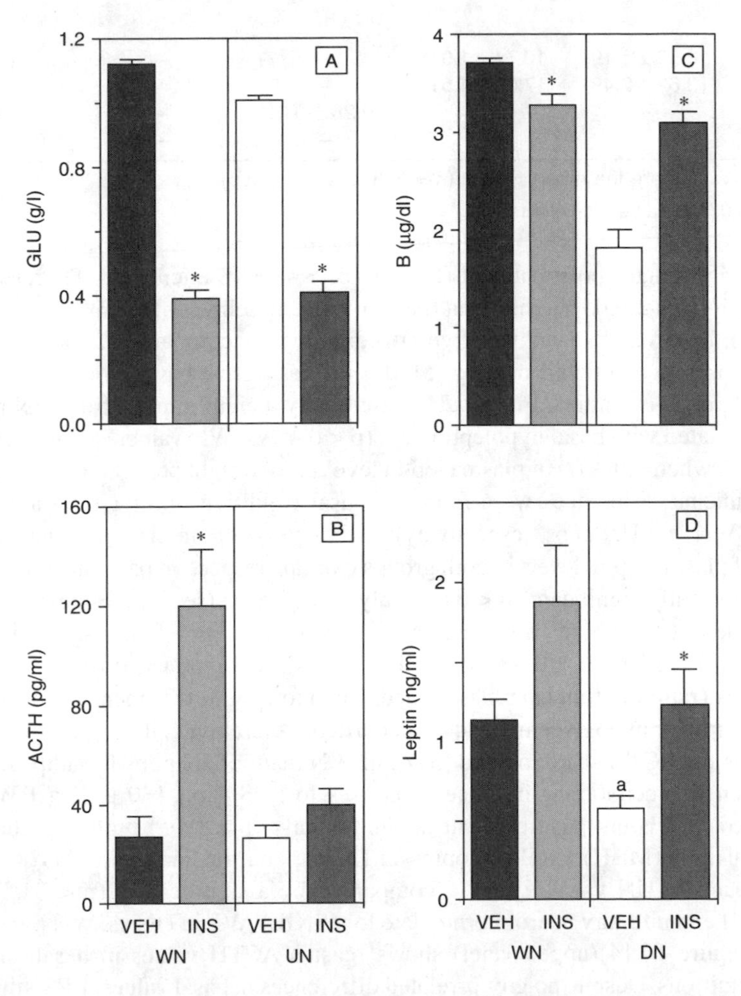


**Figure 16.13** Plasma GLU (A), ACTH (B), glucocorticoid (C), and leptin (D) levels, 45 minutes after Veh or INS i.p. administration, in 21-day-old male pups of well nourished (WN) and undernourished (UN) mothers (n = 6–9 rats per group). *p < 0.05 vs. respective Veh values; [a]p < 0.05 vs. Veh values in WN pups. (Chisari AN et al. *Endocrine* 14:375–382, 2001.)

**Table 16.6 Medial Basal Hypothalamo (MBH) and Median Eminence (ME) Neuropeptide (CRH and AVP; ng per Tissue), AP ACTH (μg per Tissue), and AG B (μg per Gland) Contents in 21-Day-Old WN and UN Rats Killed in Basal Conditions**

| MBH | ME | AP | NIL | AG | |
|---|---|---|---|---|---|
| | | | **WN** | | |
| CRH | 11.65 ± 1.17 | 15.23 ± 3.23 | — | — | — |
| AVP | 6.05 ± 0.98 | 4.97 ± 1.23 | — | 0.14 ± 0.02 | — |
| ACTH | — | — | 0.34 ± 0.02 | — | — |
| B | — | — | — | — | 0.15 ± 0.01 |
| | | | **UN** | | |
| CRH | 16.53 ± 1.19[a] | 10.74 ± 1.01[a] | — | — | — |
| AVP | 13.62 ± 2.49[a] | 17.09 ± 4.51[a] | — | 0.09 ± 0.01[a] | — |
| ACTH | — | — | 0.26 ± 0.01[a] | — | — |
| B | — | — | — | — | 0.08 ± 0.01[a] |

*Note:* Values are the mean ± SEM (n = 6–8 rats per group).

[a] $p < 0.05$ vs. values for WN rats.

pups at 45 minutes posttreatment (Figure 16.13A). INS-elicited ACTH release was significantly ($p < 0.05$) higher than the respective baseline only in WN pups (Figure 16.13B); however, INS-induced hypoglycemia was accompanied by an increase in circulating glucocorticoid over ($p < 0.05$) the respective baseline (Figure 16.13C) in both groups examined. Finally, decreased body weight in pups from UN mothers was correlated with basal hypoleptinemia ($p < 0.05$ vs. WN values) (Figure 16.13D). However, when ratios (r) of plasma leptin level:body weight per 100, were compared, no significant differences were found in basal condition (4.39 ± 0.65 and 3.66 ± 0.51 in WN and UN pups, respectively). Although INS administration increased the mean of plasma leptin levels in both groups over the respective baseline, this increase was statistically significant ($p < 0.05$) only in UN pups (see also Figure 16.13D) [r plasma leptin level:body weight: 6.08 ± 1.73, in WN ($p > 0.05$ vs. baseline) and 6.86 ± 1.31, in UN ($p < 0.05$ vs. baseline) pups]. In summary, different metabolic responses (ratio of stimulated:basal conditions) to insulin treatment were similar in both groups for hypoglycemia and glucocorticoid secreting cells, higher in WN than UN pups for ACTH secretion and lower in WN than in UN pups for adipose tissue.

When neuroendocrine-immune responses to LPS (i.p., 130 μg/Kg BW) were evaluated at 2 hours posttreatment, a significantly ($p < 0.05$) higher medial basal hypothalamus (MBH) CRH, vasopressin (AVP), and median eminence (ME) AVP were found in UN vs. WN rats, accompanied by a significantly ($p < 0.05$) lower ME CRH and pituitary neurointermediate lobe (NIL) AVP in UN vs. WN pups (Table 16.6). Figure 16.14 (upper panel) shows plasma ACTH values in basal and post-LPS conditions. Despite no group-related differences in basal values, LPS-stimulated ACTH secretion was several-fold higher ($P < 0.05$) than basal levels in WN pups, while UN pups displayed no ACTH response. It is important to note that basal AP ACTH content was significantly ($P < 0.05$) lower in UN than in WN pups (Table 16.6). Figure 16.14 (middle panel) shows the characteristically high (statistically

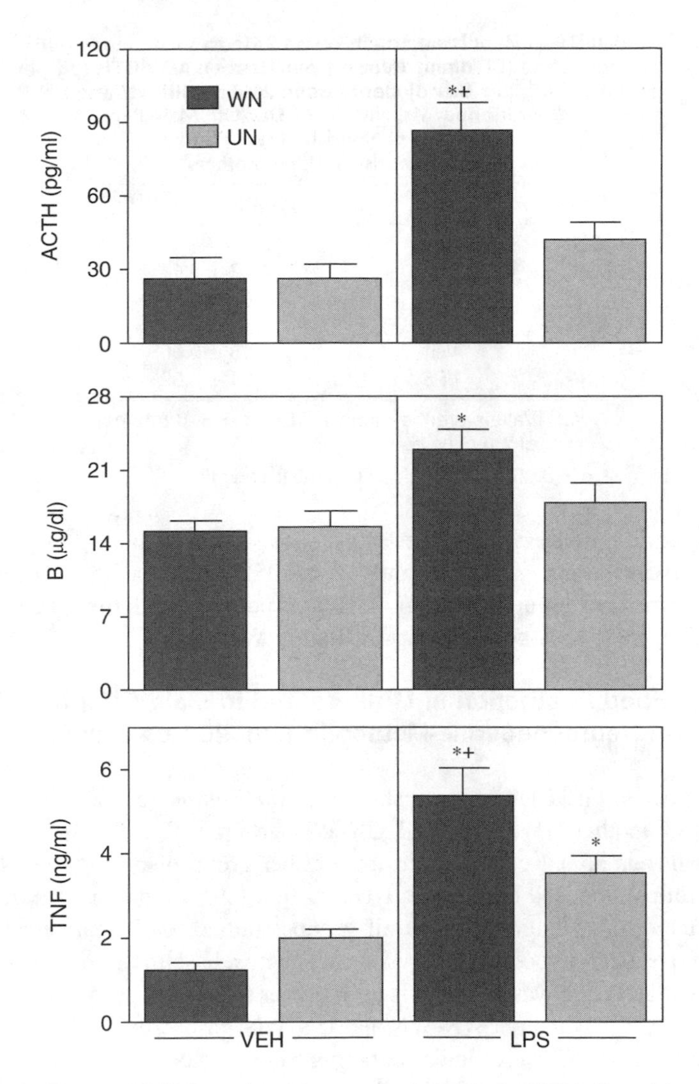


**Figure 16.14** Plasma ACTH (upper), corticosterone (middle), and TNFα (lower) levels, 2 hours after i.p. injection of vehicle (Veh) alone or containing LPS (130 μg/kg body weight), in different groups of male pups at weaning. Values are the mean ± SEM (n = 6–8 rats per group-condition). $^{*}p < 0.05$ vs. the respective Veh values; $^{+}P < 0.05$ vs. UN group-values in similar conditions. (Chisari AN et al. *Endocrine* 14:375–382, 2001.)

similar in the two groups) basal corticosterone plasma levels in both groups of 21-day-old pups and, similarly to findings for the ACTH responses, LPS-elicited adrenal response in WN but not in UN male pups. As for basal AG glucocorticoid, a significantly ($P < 0.05$) lower B content in UN than in WN counterparts (see also Table 16.6) was found. Basal plasma leptin levels concord with body weights and LPS treatment failed to modify plasma leptin levels in both groups (Table 16.7). Finally, Figure 16.14 (lower panel) shows plasma TNFα concentrations in basal and

**Table 16.7 Plasma Leptin Levels 2 Hours after Intraperitoneal Administration of Vehicle (Veh) Alone or Containing LPS (130 μg/kg Body Weight), in 21-Day-Old Male Pups from Well-Nourished (WN) and Undernourished (UN) Mothers**

| Group | | Leptin (ng/ml) |
|---|---|---|
| WN | | |
| | Veh | 1.12 ± 0.14 |
| | LPS | 1.03 ± 0.13 |
| UN | | |
| | Veh | 0.58 ± 0.06[a] |
| | LPS | 0.54 ± 0.09[a] |

*Note:* Values are the mean ± SEM (n = 6–8 rats per group-condition).

[a] $p < 0.05$ vs. WN values in similar condition.

post-LPS conditions. Basal cytokine levels were similar in both groups and LPS-induced cytokine output was significantly ($p < 0.05$) higher than the respective basal values, also for both groups; however, TNFα response to endotoxin administration was significantly ($p < 0.05$) lower in UN than in WN pups.

## Effects of Food Restriction in Mothers and in Male Offspring after Weaning on Neuroendocrine-Metabolic Function on Day 60 of Age

Basal plasma GLU levels were similar in *ad libitum* fed rats after weaning, regardless of mothers (WN-WN and UN-WN groups). Conversely, chronic food restriction in rats after weaning, born from either group of mothers (WN-UN and UN-UN groups), induced basal hypoglycemia ($p < 0.05$ vs. baselines in rats fed *ad libitum* after birth) (Figure 15A). In all groups studied, INS treatment induced a significant ($p < 0.05$) hypoglycemia (vs. the respective baseline). The pattern of INS-induced hypoglycemia was characterized by lower ($p < 0.05$) plasma GLU levels in WN-UN and UN-UN than in WN-WN and UN-WN, respectively. However, plasma glucose levels post-INS were similar in both groups of rats fed *ad libitum* after birth (WN-WN and UN-WN) and in both groups of undernourished rats after birth (WN-UN and UN-UN). We found no impact of undernutrition in mothers on basal TG levels and that chronic undernutrition of weaned rats induced basal hypotriglyceridemia (Table 16.8). Although INS treatment failed to modify basal plasma TG levels in WN-WN, UN-WN, and UN-UN groups, it significantly ($p < 0.05$) reduced (vs. baseline values) triglyceridemia in WN-UN rats (Table 16.8). Neither protocol of food restriction modified basal TP levels, and INS administration significantly ($p < 0.05$ vs. the respective basal) reduced this fuel only in UN-WN rats (Table 16.8).

Figure 16.15B shows the results of basal and post-INS plasma ACTH levels in several groups. Only severe undernutrition (UN-UN rats) induced a significant ($p < 0.05$) enhancement of basal ACTH secretion. INS-elicited ACTH secretion was significant ($p < 0.05$ vs. the respective baseline) in all groups studied. However, INS-stimulated ACTH release (ratio stimulated:basal) was graded and followed a

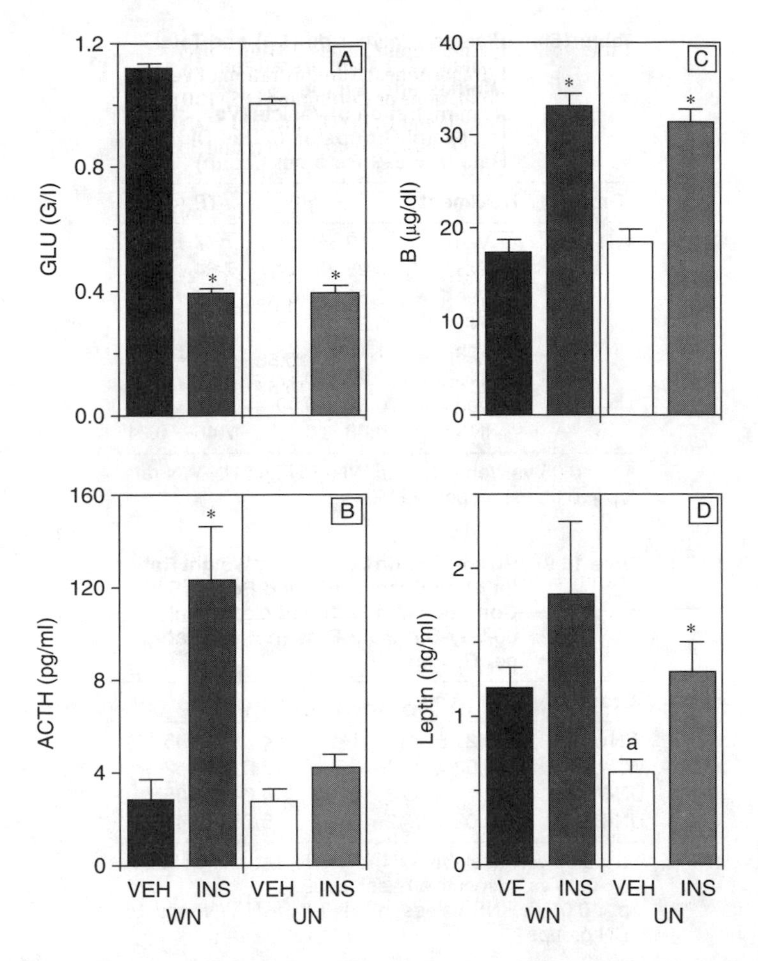


**Figure 16.15** Circulating GLU (A), ACTH (B), glucocorticoid (C), and leptin (D) concentrations in basal (VEH) and postinsulin (INS) conditions, in 60-day-old male rats of different groups (n = 6–9 rats per group). *, $p < 0.05$ vs. respective VEH values; [a]p < 0.05 vs. VEH values in WN-WN rats; [b]p < 0.05 vs. INS values in WN-WN rats; [c]p < 0.05 vs. INS values in UN-WN rats; [d]p < 0.05 vs. VEH values in UN-WN rats; [e]p < 0.05 vs. INS values in WN-UN rats; [f]p < 0.05 vs. INS values in WN-WN rats. (Chisari AN et al. *Endocrine* 14:375–382, 2001.)

group order: WN-WN>UN-WN>WN-UN>UN-UN. Figure 16.15C shows circulating glucocorticoid levels, in basal and post-INS conditions, in different groups. Rats eating *ad libitum* after weaning, regardless of food intake of their mothers, displayed low basal plasma B levels (WN-WN and UN-WN groups). Chronically undernourished rats (WN-UN and UN-UN groups) showed significantly ($p < 0.05$) higher basal B levels than their WN counterparts, although basal plasma B levels were significantly ($p < 0.05$) higher in UN-UN than in WN-UN rats. INS treatment enhanced ($p < 0.05$) plasma B levels (over the respective baseline) in all groups studied. Adrenal response (stimulated:basal) to INS treatment was similarly group-ordered: WN-WN=UN-WN>WN-UN=UN-UN.

**Table 16.8 Plasma Triglyceride (TG) and Total Protein (TP) Concentrations, 45 Minutes after Single i.p. Administration of Vehicle (Veh) or INS, in Several Groups of 60-Day-Old Male Rats (n = 6–9 Rats per Group)**

| Group | Treatment | TG (g/l) | TP (g/dl) |
|---|---|---|---|
| WN-WN | Veh | 1.19 ± 0.11 | 7.77 ± 0.47 |
| | INS | 0.98 ± 0.14 | 7.64 ± 0.48 |
| WN-UN | Veh | 0.61 ± 0.09[a] | 6.66 ± 0.57 |
| | INS | 0.28 ± 0.05[b] | 6.54 ± 0.32 |
| UN-WN | Veh | 1.18 ± 0.23 | 7.51 ± 0.43 |
| | INS | 1.29 ± 0.17 | 5.95 ± 0.21[b] |
| UN-UN | Veh | 0.58 ± 0.09[a] | 6.81 ± 0.75 |
| | INS | 0.48 ± 0.03 | 7.46 ± 0.31 |

[a] $p < 0.05$ vs. Veh values in WN-WN and UN-WN rats.
[b] $p < 0.05$ vs. respective Veh values.

**Table 16.9 Plasma Leptin Level:Body Weight Ratios (in%), in Basal (Veh) and Post-INS Conditions, in Different Groups of 60-Day-Old Male Rats (n = 6-9 Rats per Group)**

| Group | Veh | INS |
|---|---|---|
| WN-WN | 3.21 ± 0.18 | 4.11 ± 0.45 |
| WN-UN | 5.02 ± 0.44[a] | 7.21 ± 0.52[b,c] |
| UN-WN | 2.71 ± 0.22 | 4.64 ± 0.35[b] |
| UN-UN | 3.02 ± 0.42 | 5.28 ± 0.41[b] |

[a] $p < 0.05$ vs. basal WN-WN, UN-WN, and UN-UN values
[b] $p < 0.05$ vs. respective basal values
[c] $p < 0.05$ vs. INS values in WN-WN, UN-WN, and UN-UN groups

Finally, basal plasma leptin levels were significantly ($p < 0.05$ vs. WN-WN values) lower in all groups of rats submitted to either food restriction protocol (Figure 16.15D), although the lowest basal plasma leptin levels were found in UN-UN rats. INS treatment induced a significant ($p < 0.05$ vs. the respective baseline) leptin secretion in all groups, the adipocyte response (stimulated:basal) following this order: UN-UN>UN-WN=WN-UN>WN-WN. When basal plasma leptin level:body weight ratios were expressed, this parameter was significantly ($p < 0.05$) higher in WN-UN rats than in the remaining groups (Table 16.9). Ratios increased, over the respective baselines, post-INS; however, these increases were statistically significant ($p < 0.05$) only in rats submitted to either food restriction protocol (Table 16.9).

## *In Vitro* Effects of Leptin on Adrenal Function in Normal and Undernourished Male Offspring

Figure 16.16 shows the results of basal and ACTH-stimulated corticosterone secretion by isolated total AG cells from 21-day-old WN and UN pups, incubated

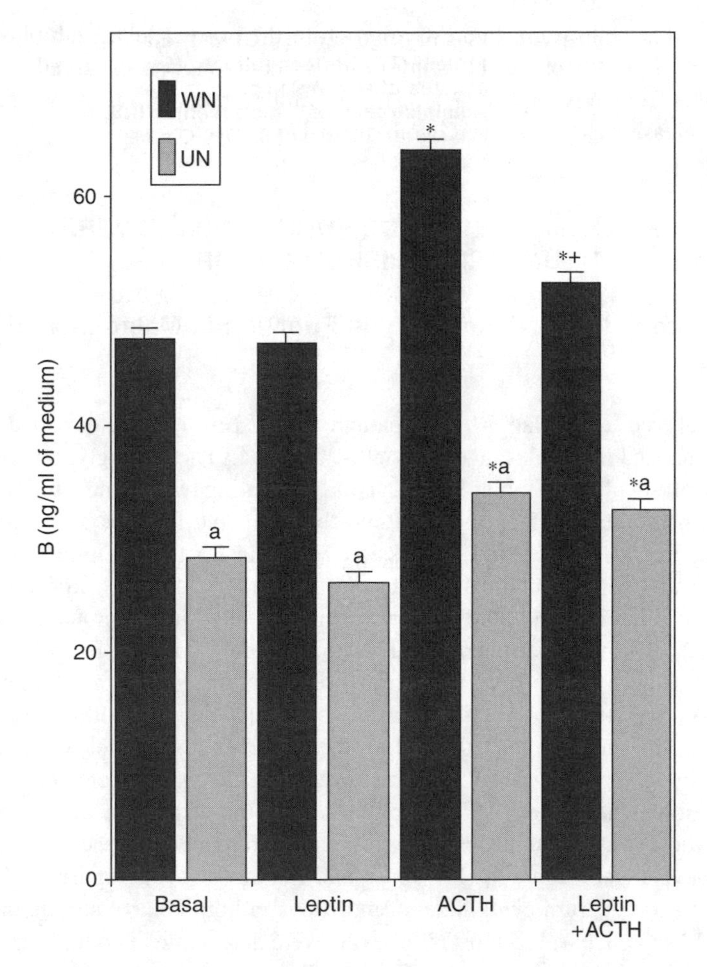


**Figure 16.16** Effects of leptin (100 nM) on spontaneous (Basal) and ACTH (22 pM)-induced corticosterone (B) release by dispersed total adrenal gland cells, from different (WN and UN pups) donors, incubated *in vitro*. Each value represents the mean of 3 different experiments (n = 6–7 tubes per condition in each experiment). [a]p < 0.05 vs. the respective values in cells from WN donors; *p < 0.05 vs. the respective Basal values; [+]p < 0.05 vs. ACTH values in the absence of leptin, in cells from WN donors. (Chisari AN et al. *Endocrine* 14:375–382, 2001.)

*in vitro*. Spontaneous (basal) glucocorticoid output was significantly ($p < 0.05$) higher from WN than from UN adrenal cells. These cell functions were related to the amount of glucocorticoid contained in the AGs of each group. When cells were stimulated with ACTH, a significant ($p < 0.05$) increase in B secretion vs. the respective baseline was found in both cell-groups; however, while a 35% (approximately) increase in B output over the baseline was found after ACTH incubation with cells from WN pups, only an 11% (approximately) increase in B secretion over the baseline was found in cells from the UN group stimulated with a similar concentration of ACTH. When the effect of exogenous leptin was investigated, the results indicated that the adipocyte product failed to modify spontaneous glucocorticoid output in either cell group. Interestingly, while the normal leptin inhibitory effect (Pralong et al. 1998) on ACTH-stimulated B secretion was observed in isolated adrenal cells from WN pups, a lack of leptin inhibition on ACTH-stimulated glucocorticoid release was found in AG cells from UN donors.

## IMPACT ON THE IMMUNE FUNCTION IN SEVERELY UNDERNOURISHED CHILDREN

### Basal Neuroendocrine-Immune Axis Function in Malnourished Children

We analyzed a population of outpatients consisting of malnourished children (four females and five males; age in months: 23.8 ± 4.9) accordingly to national and international standards (Table 16.10). This population was contrasted with age-matched (in months: 17.5 ± 3.9), healthy children (five females and four males). Informed parental consent and permission from the La Plata Children's Hospital Ethical Committee was obtained to carry out this study. Children with fever, obvious infection, or other known pathologies were not included in the protocol.

Figure 16.17 shows peripheral ACTH (upper panel), cortisol (F) (middle panel), and TNFα (lower panel) levels in basal condition, in control and malnourished children. As depicted, the mean basal ACTH and F levels were within normal range (10 to 55 pg/ml and 260 to 720 nM for ACTH and F, respectively) in the control group. Although malnourished children showed a mean peripheral level of these hormones somewhat above (ACTH) and below (F) the respective values for control children, there was no statistically significant difference in these two peripheral hormone levels (see Figure 16.17, upper and middle panels). Figure 16.17 (lower panel) also shows serum cytokine concentrations in both control and malnourished children. As shown, basal serum TNFα levels were detectable in both groups studied, and malnourished children have significantly ($P < 0.05$) higher serum cytokine levels than the control group.

### *In Vitro* Mitogen-Induced Cytokine Production by Peripheral Blood Leukocytes (PBL)

Figure 16.18 shows PBL TNFα production after incubation of cells with several concentrations of PHA (upper panel) and Con A (lower panel). PBL incubated in the presence of the lower concentration of either mitogen released an amount of TNFα into the incubation medium similar to that observed after incubation of cells of different groups with medium alone (mitogen concentration zero). A clear concentration-related effect of PHA and Con A on TNFα output was found in cells from control and malnourished children. However, as shown in Figure 16.18, the middle and the highest concentrations of PHA (upper panel) and Con A (lower panel) induced a higher ($P < 0.05$) TNFα secretion by cells from healthy than from malnourished children.

**Table 16.10 Nutritional and Laboratory Parameters in Malnourished and Healthy Children**

| Patient Number | Age (Months) | BW (Kg) | Height (Meters) | Z Score (w/a I.S.)⁺ | Gomez Mod. (w/a N.S.)⁺ | TNFα (U/ml) | ACTH (pg/ml) | F (Cortisol) (nM) |
|---|---|---|---|---|---|---|---|---|
| 1 | 19 | 7.85 | 0.73 | –3 | II | 3.6 | 12 | 427 |
| 2 | 15 | 8.02 | 0.73 | –2 | I | 3.6 | 65 | 248 |
| 3 | 17 | 8.25 | 0.73 | –2 | I | 2.7 | 152 | 323 |
| 4 | 14 | 6.95 | 0.70 | –2 | II | 4.0 | 20 | 192 |
| 5 | 18 | 9.00 | 0.76 | –2 | I | 4.9 | 56 | 53 |
| 6 | 60 | 13.50 | 0.98 | –3 | II | 5.1 | 270 | 52 |
| 7 | 9 | 5.40 | 0.61 | –3 | II | 3.0 | 25 | 282 |
| 8 | 32 | 9.55 | 0.79 | –2 | II | 5.5 | 5 | 57 |
| 9 | 8.5 | 6.95 | 0.68 | –2 | I | 0.6 | 18 | 79 |
| Ctrs. 17.5 ± 3.9 | NA | NA | 0.005 | | NA | 1.8 ±0.3 | 51 ±19 | 249 ±74 |

*Note:* NA = normal for age; * = Weight for age international standards. Ctrs. = healthy children. Values for Ctrs. children are expressed as the mean ± SEM.

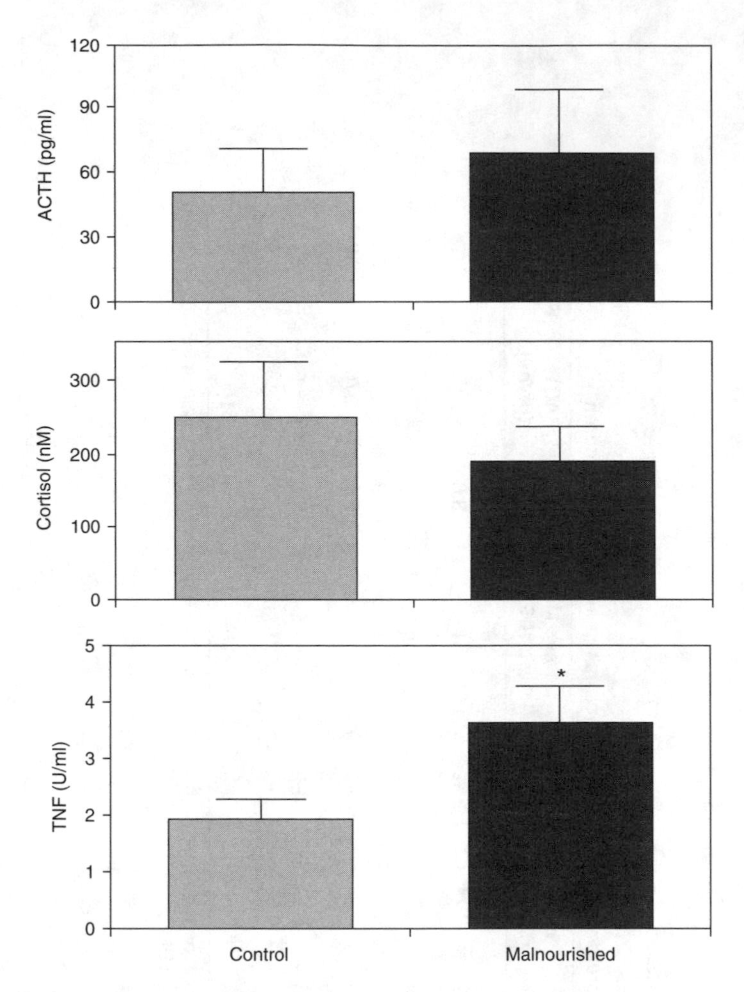


**Figure 16.17** Basal circulating levels of ACTH (upper panel), cortisol (middle panel) and TNFα (lower panel) in normal control and malnourished children. Values are the mean ± SEM (n = 9 children per group). *P < 0.05 vs. control values. (Giovambattista A et al. *Medicina* 60:339–342, 2000b.)

## *In Vitro* Mitogen-Induced PBL Proliferation

Figure 16.19 shows proliferation of PBL from control and malnourished children induced by PHA (upper panel) and Con A (lower panel). Results are expressed in cpm (Thy) incorporated by PBL as a consequence of the effect of different concentrations of mitogens. It must be pointed out that this experiment was performed by adding $H^3$-thymidine in the same wells containing cells from which an aliquot was taken for the measurement of TNFα release; thereafter, incorporated radioactivity was estimated at 4 hours post-incubation in the presence of Thy. Figure 16.19 shows that neither mitogen was able to induce any significant cell proliferation vs. cells incubated in the absence of mitogen, at the two lower concentrations used in the control group; however, a significant ($P < 0.05$ vs. mitogen concentration zero)

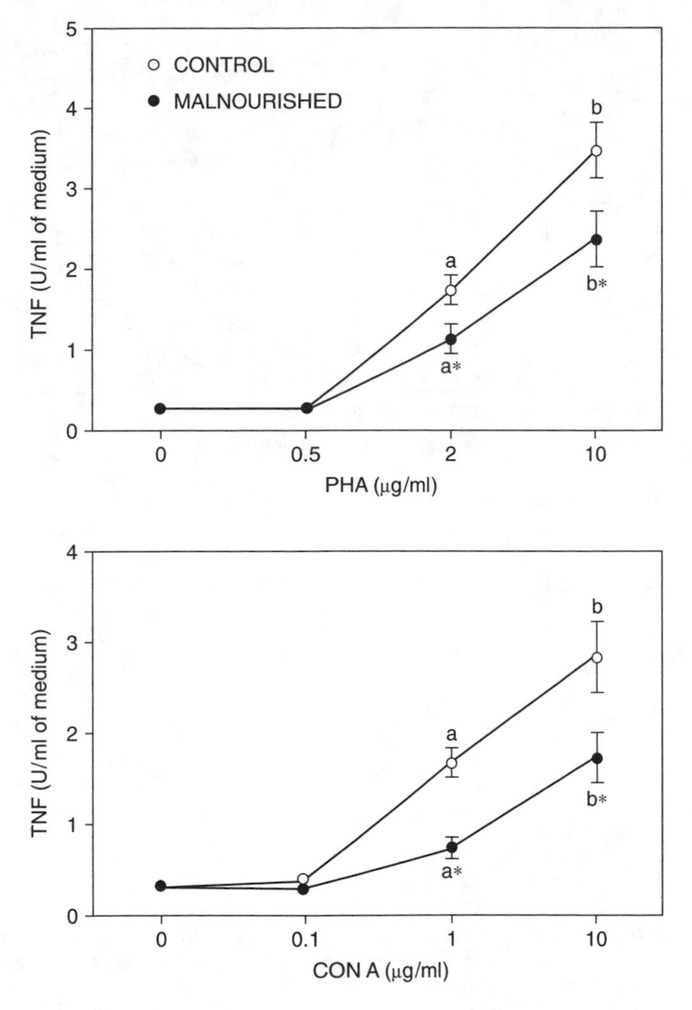


**Figure 16.18** TNFα output into the medium by peripheral blood leukocytes from normal control (open circles) and malnourished (closed circles) children, incubated *in vitro* in the absence (concentration zero) or presence of several concentrations of phytohemaglutinin (PHA) (upper panel) and concavalin A (CON A) (lower panel). Values are the mean ± SEM, n = 12 wells per point. [a]$P < 0.05$ vs. concentration zero values; [b]$P < 0.05$ vs. values found with the middle concentration of either mitogen; *$P < 0.05$ vs. respective control values. (Giovambattista A et al. *Medicina* 60:339–342, 2000b.)

increase in radioactivity incorporated was found with the highest mitogen concentration used in control cells, regardless of mitogen. In cells of marasmic children, neither mitogen was able to enhance, over that induced in the absence of mitogen, radioactivity incorporated when used at the lower concentration; however, conversely to findings for control cells, a proliferative effect in a concentration-dependent fashion was already observed with the middle concentration as well as with the highest mitogen concentrations employed, regardless of mitogen. Figure 16.19 shows that the middle and the highest concentrations of PHA (upper panel) and Con A

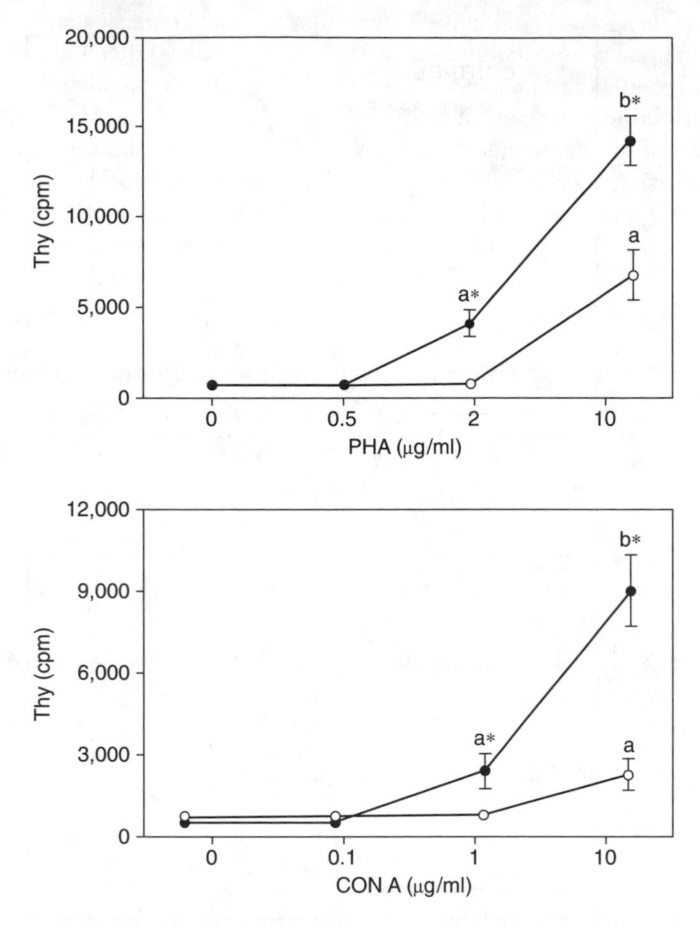


**Figure 16.19** Radioactivity (methyl-$^3$H-thymidine) incorporated by PBL, from normal control (open circles) and malnourished (closed circles) children, stimulated or not (mitogen concentration zero) with various concentrations of PHA (upper panel) and CON A (lower panel). Values are the mean ± SEM, n = 12 wells per point. [a]P < 0.05 vs. concentration zero values; [b]P < 0.05 vs. values found with the middle concentration of either mitogen; *P < 0.05 vs. respective control values. (Giovambattista A et al. *Medicina* 60:339–342, 2000b.)

(lower panel) induced a significantly (P < 0.05) higher incorporation of radioactivity by cells from the malnourished than by those of the control group. These observations further indicate a normal ability of cells from both experimental groups to proliferate, thus indicating satisfactory culture condition in each experiment.

## DISCUSSION OF RESULTS

### Sexual Dimorphism of Neuroendocrine-Immune Interactions

Acute inflammation represents a threat to the integrity of the organism, which requires metabolic changes such as increased secretion of glucocorticoids (Munck et al. 1984) for survival after injury. The results presented indicate that sex hormones have an important modulatory role in HPA axis response after inflammation and further suggest that these molecules blunt the effect of inflammation. In conclusion, all the data strongly support, in addition to gender difference (Rivier 1994), a sex steroid basis for neuroendocrine-immunological sexual dimorphism. Because inflammation and other purely neuroendocrine stressors probably stimulate different subtypes of hypothalamic CRH neurosecretory terminals, it could be extrapolated that these neuronal subpopulations undergo a different process of maturation during development and that such sex-related characteristics run also during adulthood and persist up to middle age.

## Effects of Food Restriction over Development: Evidences Sustaining that the Metabolic-Neuroendocrine-Immune Dysfunction Can Be Appropriately Restored

We have shown in this chapter that a different behavior of WN and UN individuals in basal condition and during the acute phase response of endotoxic shock takes place. In fact, chronic food restriction significantly delayed vaginal opening day (Gruaz et al. 1998), drastically arrested body weight gain, and induced a clear hypoglycemia and hypotriglyceridemia without affecting circulating proteins in basal condition. In our food-restricted rat model, the reduction in food intake varied, depending on the day, to a maximum of approximately, 57%, 24 hours before the experimental day. It has been reported (Picarel-Blanchot et al. 1995) that a 35% restriction in food intake, for 4 weeks and during early life, did not induce hypoglycemia; however, in our experimental design, evolution in body weight was arrested at 50% (approximately) in contrast to the 30% reduction in body weight reported in the above-mentioned model (Picarel-Blanchot et al. 1995). Regarding the effect of reduced food intake on plasma triglyceride levels, we and others found (Stokkan et al. 1991) a decrease in circulating levels of these lipids in chronically undernourished animals. These results are consistent with the individual's adaptive mechanisms that preserve the activity of many metabolic functions as a consequence of life-long food restriction. Dietary protein deficiency, but not food intake restriction, enhances HPA axis activity due to an increase in the drive of ACTH synthesis and secretion (Jacobson et al. 1997a). Conversely, food-restricted rats showed higher basal plasma and adrenal B concentrations than well nourished animals. Increased basal glucocorticoid secretion has previously been reported in both malnourished experimental animals (Jacobson et al. 1997a; Carsia et al. 1988) and fasting animals (Ahima et al. 1996); this fact appears to be associated with a lack of biologically active leptin (Stephens et al. 1995) and/or low plasma leptin levels (Carsia et al. 1988), since administration of leptin to fasting (Ahima et al. 1996) or ob/ob (Stephens et al. 1995) mice fully reversed basal hypercorticosteronemia.

HPA axis response after endotoxin challenge is known to be due mainly to released cytokines from stimulated peripheral immune cells (Dinarello and Mier 1987), which in turn stimulate different levels of the HPA axis (Berkenbosh et al. 1987; Bernton et al. 1987). ACTH secretion in plasma of undernourished rats was impaired when compared with well-nourished animals after single LPS treatment, despite similar basal plasma ACTH concentrations in both groups. Basal plasma glucocorticoid concentrations in undernourished rats were significantly higher than the respective levels found in well-nourished controls, raising the possibility that the impaired ACTH response to LPS in UN rats could be due to enhanced basal plasma glucocorticoid levels. Although LPS-stimulated B levels were similar in both groups, the adrenal response in undernourished animals started from a higher basal level than that of well-nourished controls, thus indicating that the adrenal gland is able to mount an appropriate response during inflammatory stress conditions. However, because we determined total B plasma concentrations, we should not discard the possibility of changes in the free hormone fraction in, sensitivity to corticosteroid feedback or in the metabolic clearance rate of glucocorticoid in undernourished animals.

Although there is a similar basal immune function in both groups, a clear dysfunction of the immune response to LPS was found in food-restricted rats. LPS-stimulated cytokine secretion in undernourished rats, though significantly higher than the respective baseline, was several-fold lower than TNFα output in plasma of well-nourished animals. Whether moderately high basal plasma corticosterone levels are responsible (Besedovsky et al. 1986) for decreased immune response to LPS in undernourished rats remains open for research. This observation of a reduced TNFα release in plasma after LPS administration could also be responsible, at least in part, for the diminished AP output of ACTH, since TNFα, as well as other released cytokines during the acute phase response of endotoxemia, are able to stimulate hypothalamo-corticotrope unit function (Berkenbosh et al. 1987; Bernton et al. 1987).

Regarding adipocyte function, our observations indicate that animals submitted to chronic food restriction showed significantly lower circulating leptin levels than well-nourished controls. In our experimental design, single LPS i.p. administration failed to modify plasma leptin levels in both well-nourished and undernourished rats 2 hours after treatment. Leptin secretion seems to be stimulated by LPS administration (Grunfeld et al. 1996); however, we (Chautard et al. 1999) and others (Yu et al. 1997) agree that this effect occurs only when evaluated 2 hours after single LPS i.v. administration in rats, or 3 hours after LPS i.p. injection in mice, respectively.

When several metabolic functions were evaluated after re-feeding (UN rats were fed *ad libitum* for 15 days), we found that basal hypoglycemia and hypotriglyceridemia as well as AG B concentrations were fully restored. In addition, hypoglycemia and the immune-neuroendocrine response induced by LPS challenge were also restored to normal (similar to those found in WN rats). It is important to point out that, after re-feeding of UN rats, a significant increase in both body weight and circulating leptin were found. It has been suggested that leptin may be a crucial factor in the induction of puberty as the animal nears the so-called critical weight (Yu et al. 1997). The present results strongly suggest that re-feeding of chronically undernourished animals not only allows maturation of the reproductive function (Gruaz et al. 1993; 1998) but also of immune-neuroendocrine and adipocyte activities. However, whether normal plasma leptin levels only or any other mechanisms are responsible for such effects should be investigated.

## Undernutrition in Mothers and Their Offspring

Another part of this chapter is devoted to studies of severely undernourished mothers during pregnancy who displayed an arrest in body weight gain without any modification of the normal length of gestation. This observation concords with earlier results showing that severe food restriction in mothers during the second or third, but not the first, week of gestation induced a significant arrest in body weight gain (Alvarez et al. 1997). Interestingly, when restriction in food intake of mothers continued throughout lactation, the total number of weaned pups per undernourished dam was significantly lower (vs. well nourished dam) due to a decrease in the number of female, but not male, pups. We also observed decreased body weight in 21-day-old male pups from undernourished mothers, which correlated well with low basal circulating levels of leptin (Frederich et al. 1995). An indicative fact for clear adaptation of these pups to basal condition was that no changes were noticed in the basal plasma levels of glucose, ACTH, and glucocorticoid; however, alteration in the circadian rhythm of glucocorticoid secretion due to undernutrition should not be excluded (Xu et al. 1999). Although insulin-induced hypoglycemia and glucocorticoid output were similar in both groups of male pups (from WN and UN mothers), ACTH plasma levels 45 minutes post-INS were significantly lower in 21-day-old UN than in age-matched WN male pups. This could be due to changes in the maturation of the hypothalamo-pituitary unit, resulting, as described in UN female pups at weaning (Chisari et al. 1999), in a shifting to the left in the response of this unit as a consequence of maternal stress (Aird et al. 1997a) due to long-term undernutrition. A significant stimulatory effect of insulin administration on plasma leptin levels (Cusin et al. 1995) was found only in 21-day-old UN pups. It should be noted that body weight and basal circulating leptin levels were lower in UN than in WN pups; however, insulin had an effect (Saladin et al. 1995) in UN pups, mimicking plasma leptin levels in pups born from normally fed mothers.

In the second set of experiments we determined that body weights and plasma levels of glucose, triglyceride, total protein, ACTH, and glucocorticoid were similar in 59-day-old male (and female, data not shown) rats eating *ad libitum* after weaning, regardless of the nutritional status of their mothers during gestation and lactation. Contrary to reports on maternal protein deficiency (Wilson and Hughes 1997), during gestation and lactation, we found that food restriction in mothers during the same critical periods had no impact on the recovery of appropriate body weight of their 59-day-old products when fed *ad libitum* after weaning. However, when mothers were fed with a protein-deficient diet during gestation only, offspring recovered normal body weight at adulthood (Dahri et al. 1991). These observations, taken together, could indicate that postnatal nutrition seems to be the critical period determining adult normal body weight, and nutrient qualitative changes in food composition, rather than the amount of food consumption, could be the most important factor predicting normal body weight at adult life. Interestingly, mild basal hypoleptinemia was found in products born from UN mothers but who ate *ad libitum* after being weaned and up to 60 days of age. This observation led us to speculate that while good nutrition after weaning would be an important determinant of normal body weight during adulthood, undernutrition during the fetal-postnatal period could produce a severe impact upon the hypothalamic-peripheral circuit responsible for control in feeding behavior (Kalra et al. 1999; Woods et al. 1998).

Although starting from different body weights, a similar arrest in body weight gain, from weaning up to day 59 of age, occurred in long-term undernourished male rats born from either group of mothers. Indeed, 60-day-old male products from both groups of mothers (WN-UN and UN-UN) share several characteristics in basal condition, such as hypoglycemia, hypotriglyceridemia, somewhat enhanced plasma ACTH levels, hypercorticosteronemia, and hypoleptinemia. We have shown that chronically undernourished weaned female rats born from WN mothers were hypoglycemic and hypotriglyceridemic, with normal total protein concentration in plasma, in basal conditions (Giovambattista et al. 2000a). Although no changes in basal plasma glucose levels after 35% restriction in food intake during one month have been reported (Picarel-Blanchot et al. 1995), it was accompanied by a 30% reduction in body weight in contrast to the 50% reduction in body weight induced by our design. Regarding long-term undernutrition-induced hypotriglyceridemia, our findings concord with previous results indicating that chronic undernutrition develops, as a compensatory factor for the allostatic state (McEwen 1998), a decrease in plasma levels of these lipids (Stokkan et al. 1991) to further preserve other metabolic pathways, such as protein metabolism (Giovambattista et al. 2000a).

When several responses to INS administration were studied in different groups of 60-day-old well-nourished rats, we found that, on the one hand, male products from both groups of mothers (WN and UN) attained similar decrease in plasma glucose levels, several-fold increase in plasma ACTH and B values, and enhancement in the concentration of plasma leptin. These results clearly indicate that undernutrition of mothers had a serious deleterious impact on both neuroendocrine (HPA axis) and adipocyte responses which cannot be overridden by the recovery of normal body weight. In fact, due to similar insulin-induced hypoglycemia, lower pituitary, adrenal, and adipose tissue responses in UN-WN than in WN-WN have taken place in order to counteract allostasis (McEwen 1998). On the other hand, INS-treated male products, born from WN and UN mothers and long-term undernourished after weaning, displayed similar hypoglycemia, although more pronounced than products that ate *ad libitum*; significant ACTH responses, though lower than products that ate *ad libitum*, and marked glucocorticoid release in plasma; and an enhancement in circulating leptin levels. However, it should be noted that long-term food restriction in products born from UN, but not from WN, mothers resulted in an enhancement in basal plasma ACTH levels. Again, it is known that protein malnutrition (Jacobson et al. 1997) but not chronic undernutrition (Giovambattista et al. 2000a) in rats, born from WN mothers, induces an increase in the drive of ACTH synthesis. Our data indicate that long-term food restriction had an impact on pituitary function in male offspring born from UN but not from WN mothers. It has been reported that the hypothalamic expression of CRH, an anorectic signal (Richard 1993), the physiological regulator of HPA axis function (Vale et al. 1981), is reduced, probably as a compensatory mechanism, after both food restriction and deprivation (Brady et al. 1990). The precise mechanisms involved in the CRH-dependent anorectic effects remain unclear. It has been postulated that food deprivation in lean rats induces a decrease in $CRH_2$-R mRNA expression in the ventromedial nucleus of the hypothalamus (VMH) (Timofeeva and Richard 1997), thus pointing to an effect on the CRH regulatory activity of energetic balance (Bale et al. 2000) rather than HPA axis function (Vale et al. 1981; Pazzoli et al. 1996). However, there seems to be a cooperative involvement of both CRH-$R_1$ and CRH-$R_2$ on feeding behavior (Bradbury et al. 2000). In addition, a compensatory role of the hypothalamic vasopressinergic system on pituitary-adrenal activity (Antoni 1993; Guillon et al. 1995) could be developed to maintain appropriate ACTH and glucocorticoid release after chronic undernutrition; in fact, selective enhancement of the central vasopressinergic pathway in $CRH_1$-R-deficient mice has been recently postulated as a compensatory mechanism to maintain HPA axis function (Muller et al. 2000).

A relation between high-fat diet in mothers and HPA axis function of their products has been described (Tannenbaum et al. 1997; Brindley et al. 1981). It has also been reported that increased dietary fat content modifies both milk lipid levels of lactating mothers (Nicholas and Hartmann 1991) and circulating leptin levels of their products (Trottier et al. 1998). It was further suggested that both the rise in 10-day-old pups and the decrease in 35-day-old offspring of circulating leptin levels could be—at least partially—responsible for decreased and increased, respectively, HPA axis responses to ether-induced stress (Trottier et al. 1998).

In addition, results showed in the present studies indicate that severe undernutrition of mothers during gestation and lactation induces in their male offspring at weaning:

1. adaptive mechanisms to maintain normal basal carbohydrate metabolism, corticotrope-adrenocortical activity, and TNF$\alpha$ secretion
2. reduced body weight and basal circulating leptin level
3. a full and partial abolishment of HPA axis and cytokine responses, respectively, to endotoxin administration
4. *in vitro* adrenocortical dysfunction and leptin-resistance

Although significant body weight gain arrest occurred in undernourished mothers, as stated, the normal rat delivery-time was not modified by the reduction in food intake. However, the development of adaptive mechanisms in male pups from UN mothers was limited. Body weight and plasma leptin concentration were significantly reduced in 21-day-old pups from UN dams; nevertheless, UN pups showed normal basal carbohydrate metabolism. This fact tallies with reports of 4-week undernutrition during early life that did not induce changes in basal glucose levels (Picarel-Blanchot et al. 1995). Conversely, chronically undernourished rats, beginning at weaning and evaluated at 60 days of age, as shown, did develop basal hypoglycemia.

As for HPA axis function, we found similar basal plasma ACTH and glucocorticoid concentrations in both groups. Although significant ACTH and glucocorticoid responses to endotoxic shock was developed by pups from WN mothers, LPS-induced corticotrope and adrenal responses were absent in weaned male offspring from UN dams. It should be noted that *in vivo* basal HPA axis function was similar in UN and WN pups, even though AP ACTH and AG B contents were lower in UN than in WN rats; thus, an increased output of both CRH from the ME and AVP of magnocellular origin could have developed as compensatory mechanisms to assure normal basal HPA axis function in UN rats. These results agree fully with recent data showing that undernutrition of mothers induces HPA axis dysfunction during development in sheep (Hawkins et al. 2000). Our study also provides evidence of significant changes in the ontogeny of the hypothalamic CRH system (Halasz et al. 1997) as a result of maternal UN. It has been reported that, in normal male rat offspring, hypothalamic CRH mRNA increases persistently after birth up to 21 days of age (Aird et al. 1997b). Our findings indicate that UN in mothers significantly enhanced, over WN values, MBH CRH in 21-day-old male pups. This observation finds support in the data reported by Heiman et al. by the fact that long-acting low leptin plasma levels could enhance hypothalamic CRH-ergic activity (Heiman et al. 1997). We also found that hypothalamic AVP content was several fold higher in UN than in WN rats. Both facts could probably be integrated since, as mentioned above, they compensate mechanisms regulating the normal basal circulating glucocorticoid level (Kiss et al. 1988). As expected, the diminished body weight of pups from UN dams was accompanied by hypoleptinemia (Mantzoros and Moschos 1998), and the stimulatory effect of low concentrations of ACTH on the adrenal gland resulted from this reduced inhibitory leptin effect (Pralong et al. 1998). This fact is also strongly supported by the observation of *in vitro* adrenal leptin-resistance in the UN rat. *In vitro* results indicate a potentially impaired adrenal function since both spontaneous and ACTH-stimulated glucocorticoid release is lower in cells from UN than from WN pups. This would indicate an endogenous disadvantage for UN pups to muster appropriate glucocorticoid secretion in inflammatory and other stress conditions. Since *in vitro* basal corticosterone secretion was diminished, perhaps in a direct relationship with decreased adrenal gland corticosterone content, and because basal plasma circulating levels of glucocorticoid are similar to those of WN animals, the hypothesis for the development of several compensatory mechanisms for regulating normal basal HPA axis function in UN pups could be strongly sustained (Kiss et al. 1988; Spinedi et al. 1997). It is important to mention that the development of adrenal leptin-resistance occurred even with hypoleptinemia. Low circulating leptin levels would be expected to induce an up-regulation of leptin receptors; however, our results indicate that from the functional point of view, adrenal leptin receptors are nonresponsive in UN pups. Therefore, the mechanisms whereby this phenomenon takes place remain open for investigation. However, it should be accepted that this pathophysiological change at the adrenal level is relevant for the survival of the UN pup in unfavorable conditions.

The neuroendocrine dysfunction of 21-day-old rats from UN mothers occurred simultaneously with a TNFα hyporesponsiveness post-LPS, despite unchanged basal plasma cytokine and glucocorticoid levels. Decreased immune function as a result of severe protein calorie malnutrition and undernutrition is an accepted fact (Doherty et al. 1993). It is known that macrophage-derived TNFα, among other cytokines, is a key factor to trigger HPA axis activity during the acute phase to endotoxic shock (Berkenbosch et al. 1987); therefore, it is plausible to speculate that decreased peripheral mononuclear cell function after endotoxin administration could be additionally contributing to lower HPA axis stimulation during endotoxemia (Berkenbosch et al. 1987; Bernton et al. 1987). However, changes in the sensitivity of HPA axis response to cytokine stimulation, as a consequence of UN, must be not ruled out.

## Neuroendocrine-Immune Function in Malnourished Children

Malnutrition in children under 4 years of age has been associated with higher risk of mortality. Synergism between wasting and infection is a good predictor of children's mortality as reported in field studies (Fauzi et al. 1997; Zaman et al. 1997). High prevalence of abnormalities in both inflammatory and immune responses in malnourished children has been previously stressed (Scrimshaw and SanGiovann 1997). Plasma cytokine levels in malnourished children have been described as higher in malnourished than in control children (Sauerwein et al. 1997). Even more, the cascade of the inflammatory response seems to be activated *in vivo* in malnourished but not clinically infected children; however, the acute phase response of malnourished children has been reported to be impaired when they are infected or after vaccination (Doherty et al. 1993). In previous studies (Doherty et al. 1994; Bhaskaram and Sivakumar 1986), *in vitro* production of proinflammatory cytokines was found decreased in severely malnourished children. Nevertheless, the relationship between high basal circulating levels and diminished *ex vivo* production of proinflammatory cytokines remained unexplored for several years.

We studied a population of chronically malnourished children from the point of view of their neuroendocrine-immune interactions. We found that there was no difference in basal HPA axis activity, probably due to adaptive mechanisms developed by these organisms under negative energetic balance. Undernourished children displayed higher basal plasma cytokine (TNFα) levels than healthy children. Acting in cooperation, cytokines released after injury are crucial for the development of an appropriate host's response and, in some clinical situations (Waage et al. 1987; Sullivan et al. 1992), their production is potentially related to the final outcome. Previous studies suggested that the inflammatory process may be abnormal in malnourished organisms (Doherty et al. 1993; Doherty et al. 1994; Sauerwein et al. 1997; Bhaskaram and Sivakumar 1986). It has been observed that malnourished war prisoners (Dekaris et al. 1993) and HIV patients (Arnalich et al. 1997), without concomitant disease, display higher circulating TNFα concentrations than their respective counterparts. Additionally, it has been demonstrated that malnourished children have elevated basal production of IL-6 and TNFα-soluble-receptor, irrespective of evident infection (Sullivan et al. 1992). Our data, combined with those observations, suggest a basal activated inflammatory cascade in malnourished patients. Data for blood cells from malnourished children indicated that they develop a significant tolerance to mitogen-stimulated TNFα output *in vitro*, regardless of mitogen. Interestingly, low output of cytokine by cells from marasmic children, after mitogen stimulation, was accompanied by a higher incorporation of radioactivity into these cells than in cells of normal control children, thus indicating that total blood cells of malnourished children have either impaired TNFα synthesis or secretion, or both. Interestingly, *in vivo* production of the acute phase reactants protein-C and serum amyloid A after either infection or DPT vaccination is impaired in malnourished children, although corrected after nutritional recovery (Doherty et al. 1993). This phenomenon could be related to the high circulating levels, in comparison with those found in normal control children, of the cytokine that could induce a negative feedback mechanism on either or both above-mentioned functions of blood cytokine-producing cells. However, changes in the metabolic clearance rate of circulating TNFα, as a consequence of malnutrition, are also possible.

It is thought that cytokines are synthesized *de novo* rather than secreted from intracellular stores. The synthesis of these polypetides, released by various peripheral immune cells in response to inflammatory stimuli, appears to be controlled at both transcriptional and translational steps (Michie et al. 1988). Although marasmic children showed no significant clinical sign of acute inflammation/infection at admission, we have no data certifying absence of any previous inflammatory disorder or known infection that could be responsible for the increased basal plasma cytokine levels. However, despite the higher TNFα plasma concentration in marasmic patients, we found no significant differences vs. control patients in basal HPA axis activity. Although cytokine or endotoxin administration in humans has been reported to increase plasma glucocorticoid levels, it occurred only when injected at pyrogenic doses (Starnes et al. 1988), a condition clearly different from that of our experimental design. In fact, the present study excluded patients when fever or other symptoms of an inflammatory process was detected.

It is well recognized that suppression of food intake and anorexia commonly occur during infection, and that these effects are due to endogenous cytokines (Dinarello 1984; Cerami et al. 1985). It has been shown that infection-induced anorexia in mice can be overridden by force feeding; conversely, the force feeding of the animals induced a deleterious effect on the organism, since mortality increased and survival time significantly shortened (Murray and Murray 1979). Thus, it has been suggested that the loss of appetite could be relevant as a host's defense mechanism for nutrient redistribution, thus contributing to "nutritional immunity" during injury/infection (Murray and Murray 1977a,b). Although the marasmic patients in our study were not in an acute phase of any known infection, it is conceivable that the increase in plasma TNFα concentrations could be related to the malnutrition of these children, or if they were in the recovering phase of any infection, due to their peripheral blood cells still secreting the cytokine. In either case, however, peripheral blood immune cells from this group have developed a tolerance to further release of this cytokine when stimulated *in vitro* with different mitogens. It is very interesting that the reduced cytokine output was not correlated with the rate of peripheral blood cell proliferation after mitogen stimulation, regardless of the mitogen used. In fact, we found an inverse relationship between cytokine output and blood cell proliferation. This fact could be explained, at least in part, by the indirect effects on TNFα secretion induced by other immune cell-derived cytokines, such as interferon gamma (INF-γ). It has been demonstrated that human TNFα secretion *in vitro* could be induced in the absence of endotoxin by INF-γ(Debets et al. 1988); thus, since enhanced production of INF-γ in these children cannot be excluded, persistent stimulation of the latter mentioned cytokine on immune cells could have produced a depletion of TNFα stores, and as a consequence, a decrease in immune cell response to further mitogen stimulation. However, it is also possible that a persistent, slightly increased peripheral level of TNFα in marasmic children could induce a negative feedback mechanism on its own production (synthesis or output). The development of tachyphylaxis to mitogen-stimulated cytokine production by blood cells of marasmic children tallies with other reports indicating tachyphylaxis of some mechanisms stimulated by INF-γ, such as fever among others, whereas fatigue and anorexia tended to be cumulative (Sherwin et al. 1982). It must be pointed out that the plasma glucocorticoid level of malnourished children was not increased; thus, a negative feedback effect of cortisol on immune cells (Dinarello and Mier 1987) does not seem to be responsible for the "*in vitro*" tolerance of blood cells to secreting cytokine after different mitogen stimuli.

Tolerance of the immune function to further mitogen stimuli seems to occur in these children; however, there is some controversy over the importance of this "*ex vivo*" dysfunction in the literature. Genetic background has been suggested to differentiate patients with a graded probability of risk for fatal outcome of meningococcal disease. A very recent study (Westendorp et al. 1997) reports a direct relationship between low capacity of TNFα and high capacity of IL-10 release by whole blood cells in the relatives of patients suffering meningococcal disease. Conversely, other data (Rothe et al. 1993; Fernades and Baldwin 1995) support the hypothesis that patients with the genotype associated with high TNFα production during sepsis (including meningococcal disease) directly correlate with a poorer outcome of the infected patients. However, it is conceivable that the timing of the tests could be essential for the interpretation of this phenomenon. The finding of a low TNFα secretion coinciding with high IL-10 output could be explained because IL-10 is mainly produced by TNFα stimulus and IL-10 is known to be able to reduce TNFα production by a negative feedback mechanism (Gazzinelli et al. 1996). Thus, some mechanisms different from the above mentioned situation could occur in our marasmic patients since a significantly higher level of TNFα was present in their circulation than in control children, while *ex vivo* total blood cells produced less of this cytokine in response to different mitogens, and their cells proliferated significantly more than control cells. It may be speculated, however, that continuous exposition to multiple infectious agents could produce a similar tolerance in the immune response, even in the absence of clinical infection. This altered chronic basal immune activity, accompanied by decreased cytokine response to a novel injury, may account for a dysfunction in the inflammatory response in malnourished children, and thus participate to the severity of some infections in the malnourished host.

## SUMMARY AND CONCLUSIONS

In summary, the sexual dimorphism in the response of the HPA axis to immune signals (Spinedi et al. 1992b,c) may represent an important factor in the understanding of reciprocal interactions between the two systems in hosts of different genders and in both physiological and physiopathological conditions.

Although decreased immune response *ex vivo* (McCarter et al. 1998) and increased resting plasma glucocorticoid levels (Jacobson et al. 1997a,b) have been reported as a consequence of protein-calorie malnutrition, a major implication of the current studies is that restricted food intake after a normal pregnancy of weaned pups is able to induce at adulthood an impaired inflammatory response during endotoxemia. This provides evidence for an *in vivo* reduced TNFα output during the acute phase of the inflammatory response, which may account for the severity of some injuries/infections in the undernourished/malnourished host (Fauzi et al. 1997). However, most important is the demonstration that the neuroendocrine-immune-metabolic dysfunction can be rapidly overridden after discontinuation of metabolically unfavorable conditions. Additionally, our results add valuable information on the body's defense mechanisms necessary for survival in negative energy balance conditions (Sapolsky et al. 2000). It is important to observe that, in contrast with what occurs in products from normally nourished mothers, the deleterious effect of mothers' undernourishment, namely during gestation and lactation, appears to be irreversible, even after the recovery of a normal body weight of the offspring. This clearly suggests a disadvantage of offspring derived from undernourished mothers for survival of infection/injury or any other stress situation (Munck et al. 1984). Finally, our data indicate an adaptation of the basal HPA axis activity in severely malnourished children, while enhanced plasma TNFα level could be a consequence of the mechanism(s) associated with the development of undernourished children (Murray and Murray 1977a,b). The clinical significance of the tachyphylaxis of the immune system developed by malnourished children remains unknown; however, it could be speculated that this adaptation of the immune function in undernourished children might play an important role in the organism's defense mechanisms triggered during acute phase response to novel injury/infection (Doherty et al. 1993; Fauzi et al. 1997).

## ACKNOWLEDGMENTS

The studies presented in this chapter were supported by grants PICT 5-5191/99 (Argentina) and FNSR 32-064107.00 (Switzerland). Authors are indebted to the technical assistance of collaborators from IMBICE and Children's Hospital (La Plata, Argentina), and CHUV (Lausanne, Switzerland). The excellent editorial assistance of Susan Hale Rogers is deeply recognized.

## REFERENCES

Ahima RS, Prabakaran D, Mantzoros C, Qu D, Lowell B, Maratos-Flier E, and Flier JS. 1996. Role of leptin in the neuroendocrine response to fasting. *Nature* 382:25–252.

Aird F, Halasaz I, and Redei E. 1997. Ontogeny of hypothalamic corticotropin-releasing factor and anterior pituitary pro-opiomelanocortin expression in male and female offspring of alcohol-exposed and adrenalectomized dams. *Alcohol Clin Exp Res* 21:1560–1566.

Alvarez C, Martin MA, Goya L, Bertin E, Portha B, and Pascual-Leone AM. 1997. Contrasted impact of maternal rat food restriction on the fetal endocrine pancreas. *Endocrinol* 138:2267–2273.

Antoni FA. Vasopressinergic control of pituitary adrenocorticotropin secretion comes of age. 1993. *Front Neuroendocrinol* 14:76–122.

Arnalich F, Martinez P, and Heranz A. 1997. Altered concentrations of appetite regulators may contribute to the development and maintenance of HIV-associated wasting. *AIDS* 11:1129–1134.

Bale TL, Contarino A, Smith GW, Chan R, Gold LH, Sawchenko PE, Koob GF, Vale WW, and Lee K-F. 2000. Mice deficient for corticotropin-releasing hormone receptor-2 display anxiety-like behaviour and are hypersensitive to stress. *Nat Genet* 24:410–414.

Bateman A, Singh A, Kral T, and Solomon S. 1989. The immune-hypothalamic-pituitary-adrenal axis. *Endocr Rev* 10:92–112.

Berkenbosch F, van Oers J, Del Rey A, Tilders F, and Besedovsky H. 1987. Corticotropin-releasing factor-producing neurons in the rat activated by interleukin-1. *Science* 238:254–256.

Bernton EW, Beach JE, Holaday JW, Smallridge RC, and Fein HG. 1987. Release of multiple hormones by a direct action of interleukin-1 on pituitary cells. *Science* 238:519–521.

Besedovsky H, Del Rey A, Sorkin E, and Dinarello CA. 1986. Immunoregulatory feedback between interleukin-1 and glucocorticoid hormones. 233:652–654.

Bhaskaram P and Sivakumar B. 1986. Interleukin-1 in malnutrition. *Arch Dis Childhood* 61:182–185.

Bingaman EW, Magnuson DJ, Gray TS, and Handa RJ. 1994. Androgen inhibits the increases in hypothalamic corticotropin-releasing hormone (CRH) and CRH-immunoreactivity following gonadectomy. *Neuroendocrinol* 59:228–234.

Bradbury MJ, McBurnie MI, Denton DA, Lee K-F, and Vale WW. 2000. Modulation of urocortin-induced hypophagia and weight loss by corticotropin-releasing factor receptor 1 deficiency in mice. *Endocrinol* 141:2715–2724.

Brady LS, Smith MA, Gold PW, and Herkenham M. 1990. Altered expression of hypothalamic neuropeptide mRNAs in food-restricted and food-deprived rats. *Neuroendocrinol* 52:441–447.

Brindley DN, Cooling J, Glenny HP, Burditt SL, and McKechie IS. 1981. Effects of chronic modification of dietary fat and carbohydrate on the insulin, corticosterone, and metabolic responses of rats fed acutely with glucose, fructose or ethanol. *Biochem J* 200:275–284.

Bronson FH. 1986. Food-restricted, prepubertal, female rats: rapid recovery of luteinizing hormone pulsing with excess food, and full recovery of pubertal development with gonadotropin-releasing-hormone. *Endocrinol* 118.24832487.

Carsia RV, Weber H, and Lauterio TJ. 1988. Protein malnutrition in the domestic fowl induces alterations in adrenocortical function. *Endocrinol* 122.673–680.

Cerami A, Ikeda Y, LeTrang N, Hotez PJ, and Beutler B. 1985. Weight loss associated with an endotoxin-induced mediator from peritoneal macrophages. the role of cachectin (tumor necrosis factor). *Immunol Lett* 11:173–177.

Chautard T, Spinedi E, Voirol M-J, Pralong FP, and Gaillard RC. 1999. Role of glucocorticoids in the response of the hypothalamo-corticotrope, immune and adipose systems to repeated endotoxin administration. *Neuroendocrinol* 69:360–369.

Chisari AN, Giovambattista A, Suescun M, and Spinedi E. 1999. Effects of maternal food restriction on the hypothalamo-pituitary-adrenal axis response to insulin administration in female pups at weaning. Proceedings from the 81st Annual Meeting of the Endocrine Society, San Diego, CA. p. 363 (Abstract).

Christeff N, Auclair MC, Dehennin L, Thobie N, Benassayag C, Carli A, and Nunez EA. 1992. Effects of aromatase inhibitor 4-hydroxy-androstenedione on the endotoxin-induced changes in steroid hormone levels in male rats. *Life Sci* 50:1459–1468.

Cusin I, Sainsbury A, Doyle P, Rohner-Jeanrenaud F, and Jeanrenaud B. 1995. The ob gene and insulin. A relationship leading to clues to the understanding of obesity. *Diabetes* 44:1467–1470.

Dahri S, Snoeck A, Reusens-Billen B, Remacle C, and Hoet JJ. 1991. Islet function in offspring of mothers on low-protein diet during gestation. *Diabetes* 40 (Suppl 2), 115–120.

Debets JMH, van der Linden CJ, Spronken EM, and Buurman WA. 1988. T cell-mediated production of tumor necrosis factor-α by monocytes. *Scand J Immunol* 27:601–608.

Dekaris D, Sabiocello A, and Mazuran R. 1993. Multiple changes of immunologic parameters in prisioners of war. Assessments after release from camp Manjaca, Bosnia. *JAMA* 270:595–599.

Dinarello CA. 1984. Interleukin-1. *Rev Infect Dis* 6:51–95.

Dinarello CA and Mier JW. 1987. Lymphokines. *N Engl J Med* 317:940–945.

Dinarello CA. 1988. The biology of interleukin-1. *FASEB J* 2:108–115.

Doherty JF, Golden MHN, Griffin GE, and McAdam KPWJ. 1989. Febrile response in malnutrition. *West Ind Med J* 38:209–212.

Doherty JF, Golden MHN, Raynes JG, Griffin GE, and McAdam KPWJ. 1993. Acute-phase protein response is impaired in severely malnourished children. *Clin Sci* 84:169–175.

Doherty JF, Golden MHN, Remick DG, and Griffin GE. 1994. Production of interleukin-6 and tumor necrosis factor-α *in vitro* is reduced in whole blood of severely malnourished children. *Clin Sci* 86:347–351.

Erbach GT and Bahr JM. 1991. Enhancement of *in vivo* humoral immunity by estrogen: permissive effect of a thymic factor. *Endocrinol* 128:1352–1358.

Evans GE and Zuckerman SH. 1991. Glucocorticoid-dependent and -independent mechanisms involved in lipopolysaccharide tolerance. *Eur J Immunol* 21:1973–1979.

Faggioni R, Fantuzzi G, Gabay C, Moser A, Dinarello CA, Feingold KR, and Grunfeld C. 1999. Leptin deficiency enhances sensitivity to endotoxin-induced lethality. *Am J Physiol* 276:R136–R142.

Faggioni R, Feingold KR, and Grunfeld C. 2001. Leptin regulation of the immune response and the immunodeficiency of malnutrition. *FASEB J* 15:2565–2571.

Fantino M and Wieteska L. 1993. Evidence for a direct central anorectic effect of tumor-necrosis-factor-alpha in the rat. *Physiol Behav* 53:477–483.

Fauci AS.. Immunosuppressive and anti-inflammatory effects of glucocorticoids. 1979. In: Baxter JD and Rosseau GG (Eds), *Glucocorticoid Hormone Action*. New York: Springer-Verlag, pp 449–465.

Fauzi W, Herrera G, Spiegelman D, El Amin A, Nestel P, and Mohamed K. 1997. A prospective study of malnutrition in relation to child mortality in the Sudan. *Am J Clin Nutr* 65:1002–1009.

Fernandes DM and Baldwin CI. Interleukin-10 downregulates protective immunity to Brucella abortus. 1995. *Infect Immun* 63:1130–1133.

Finck BN, Kelley KW, Dantzer R, and Johnson RW. 1998. In vivo and *in vitro* evidence for the involvement of tumor necrosis factor-α in the induction of leptin by lipopolysaccharide. *Endocrinol* 139:2278–2283.

Frederich RC, Hamann A, Anderson S, Lollman B, Lowell BB, and Flier JS. 1995. Leptin levels reflect body lipid content in mice: evidence for diet-induced resistance to leptin action. *Nat Med* 1:1311–1314.

Gazzinelli RT, Wysocka M, and Hieny S. 1996. In the absence of endogenous IL-10, mice acutely infected with Toxoplasma gondii succumb to a lethal immune response dependent on CD4 T cells and accompanied by overproduction of IL-12, INF-γ and TNF-α. *J Immunol* 157:798–805.

Giovambattista A, Chisari AN, Corro L, Gaillard RC, and Spinedi E. 2000a. Metabolic, neuroendocrine and immune functions in basal conditions and during the acute-phase response to endotoxic shock in undernourished rats. *Neuroimmunomod* 7:92–98.

Girardin E, Grau GE, Dayer J-M, Roux-Lombard P, and Lambert P-H. 1988. Tumor necrosis factor and interleukin-1 in the serum of children with severe infectious purpura. *N Engl J Med* 319:397–400.

Glass AR, Young RA, and Anderson J. 1986. Decreased serum 3,5,3'-triiodothyronine ($T_3$) and abnormal serum binding of $T_3$ in calorie-deficient rats: adaptation after chronic underfeeding. *Endocrinol* 118:2464–2469.

Graff RJ, Lappe MA, and Snell GD. 1969. The influence of the gonads and adrenal glands on the immune response to skin grafts. *Transplant* 5:105–109.

Grossman CJ. 1984. Regulation of the immune system by sex steroids. *Endocr Rev* 5:435–455.

Grunfeld C, Zhao C, Fuller J, Pollock A, Moser A, Friedman J, and Feingold K. 1996. Endotoxin and cytokines induce expression of leptin, the ob gene product, in hamsters. *J Clin Invest* 97:2152–2157.

Gruaz NM, Pierroz DD, Rohner-Jeanrenaud F, Sizonenko PC, and Aubert ML. 1993. Evidence that neuropeptide Y could represent a neuroendocrine inhibitor of sexual maturation in unfavorable metabolic conditions in the rat. *Endocrinol* 133:1891–1894.

Gruaz NM, Lalaoui M, Pierroz DD, Englaro P, Sizonenko PC, Blum WF, and Aubert ML. 1998. Chronic administration of leptin into the lateral ventricle induces sexual maturation in severely food-restricted female rats. *J Neuroendocrinol* 10:627–633.

Guillon G, Trueba M, Jourbet D, Grazzini E, Chouinard L, Cote M, Payet MD, Manzoni O, Barberis C, Robert M, and Gallo-Payet N. 1995. Vasopressin stimulates steroid secretion in human adrenal glands: comparison with angiotensin-II effects. *Endocrinol* 136:1285–1295.

Halasz I, Rittenhouse PA, Zorrilla EP, and Redei E. 1997. Sexually dimorphic effects of maternal adrenalectomy on hypothalamic corticotrophin-releasing factor, glucocorticoid receptor and anterior pituitary POMC mRNA levels in rat neonates. *Develop Brain Res* 100:198–204.

Hawkins P, Steyn C, McGarrigle HH, Saito T, Ozaki T, Stratford LL, Noakes DE, and Hanson MA. 2000. Effect of maternal nutrient restriction in early gestation on responses of the hypothalamic-pituitary-adrenal axis to acute isocapnic hypoxaemia in late gestation fetal sheep. *Exp Physiol* 85:85–96.

Heiman ML, Ahima RS, Craft LS, Schoner B, Stephens TW, and Flier JS. 1997. Leptin inhibition of the hypothalamic-pituitary-adrenal axis response to stress. *Endocrinol* 138:3859–3863.

Jacobson L, Zurakowski D, and Majzoub JA. 1997. Protein malnutrition increases plasma adrenocorticotropin and anterior pituitary proopiomelanocortin messenger ribonucleic acid in the rat. *Endocrinol* 138:1048–1057.

Jacobson L, Zurakowski D, and Majzoub JA. 1997. Protein malnutrition increases plasma adrenocorticotropin and anterior pituitary pro-opiomelanocortin messenger ribonucleic acid in the rat. *Endocrinol* 138:1048–1057.

Kalra SP, Dube MG, Pu S, Xu B, Horvath TL, and Kalra PS. 1999. Interacting appetite-regulating pathways in hypothalamic regulation of body weight. *Endocr Rev* 20:68–100.

Kauffman CA, Jones PG, and Kluger MJ. 1986. Fever and malnutrition: endogenous pyrogen/interleukin-1 in malnourished patients. *Am J Clin Nutr* 44:449–452.

Kiss JZ, Van Eekelen JAM, Reul JMHM, Westphal HM, and de Kloet E. 1988. Glucocorticoid receptor in magnocellular neurosecretory cells. *Endocrinol* 122:444–449.

Lord GM, Matarese G, Howard JK, Baker RJ, Bloom SR, and Lechler RI. 1998. Leptin modulates the T-cell immune response and reverses starvation-induced immunosuppression. *Nature* 394:897–901.

Mantzoros CS and Moschos SJ. 1998. Leptin: in search of role(s) in human physiology and pathophysiology. *Clin Endocrinol* 49:551–567.

McCarter MD, Naama HA, Shou J, Kwi LX, Evoy DA, Calvano SE, and Daly JM. 1998. Altered macrophage intracellular signaling induced by protein-calorie malnutrition. *Cell Immunol* 183:131–136.

McEwen BS. 1998. Stress, adaptation, and disease. Allostasis and allostatic load. *Ann NY Acad Sci* 840:33–44.

Michalek SM, Moore RN, McGhee JR, Rosentreich DL, and Mergenhagen SE. 1980. The primary role of limphoreticular cells in the mediation of host responses to bacterial endotoxin. *J Infect Dis* 41:55–63.

Michie HR, Manogue KB, Sprigss DR, Revhaug A, O'Dwyer S, Dinarello CA, Cerami A, Wolff SM, and Wilmore DW. 1988. Detection of circulating tumor necrosis factor after endotoxin administration. *N Engl J Med* 318:1481–1486.

Muller MB, Landgraf R, Preil J, Sillaber I, Kresse AE, Keck ME, Zimmermann S, Holsboer F, and Wurst W. 2000. Selective activation of the hypothalamic vasopressinergic system in mice deficient for the corticotropin-releasing hormone receptor 1 is dependent on glucocorticoids. *Endocrinol* 141, 4262–4269.

Munck A, Guyre PM, and Holbrook NJ. 1984. Physiological functions of glucocorticoids in stress and their relation to pharmacological actions. *Endocr Rev* 5:25–44.

Murray MJ and Murray AB. 1977b. Suppression of infection by famine and its activation by refeeding—a paradox. *Perspect Biol Med* 20:471–483.

Murray MJ and Murray AB. 1979. Anorexia of infection as a mechanism of host defense. *Am J Clin Nutr* 32:593–596.

Murray MJ and Murray AB. 1977a. Hypothesis-starvation suppression and refeeding activation of infection. *Lancet* ii:123–125.

Nicholas KR and Hartmann PE. 1991. Milk secretion in the rat: progressive changes in milk composition during lactation and weaning and the effect of diet. *Comp Biochem Physiol* 98A:535–542.

Pazzoli G, Bilezikjian LM, Perrin MH, Blount AL, and Vale WW. 1996. Corticotropin-releasing factor (CRF) and glucocorticoids modulate the expression of type 1 CRF receptor messenger ribonucleic acid in rat anterior pituitary cell cultures. *Endocrinol* 137:65–71.

Picarel-Blanchot F, Alvarez C, Bailbe D, Pascual-Leone AM, and Portha B. 1995. Changes in insulin action and insulin secretion in the rat after dietary restriction early in life: influence of food restriction versus low-protein food restriction. *Metabolism* 44:1519–1526.

Pimstone B. 1976. Endocrine function in protein-calorie malnutrition. *Clin Endocrinol* (Oxf) 5:79–95.

Plata-Salaman CR and Borkoski JP. 1994. Chemokines/intercrines and central regulation of feeding. *Am J Physiol* 266:R1711–R1715.

Pralong FP, Roduit R, Waeber G, Castillo E, Mosimann F, Thorens B, and Gaillard RC. 1998. Leptin inhibits directly glucocorticoid secretion by normal human and rat adrenal gland. *Endocrinol* 139:4264–4268.

Richard D. 1993. Involvement of corticotropin-releasing factor in the control of food intake and energy expenditure. *Ann NY Acad Sci* 697:155–172.

Rivest S and Rivier C. 1995. The role of corticotropin-releasing factor and interleukin-1 in the regulation of neurons controlling reproductive functions. *Endocr Rev* 2:177–199.

Rivier C. 1994. Stimulatory effect of interleukin-1β on the hypothalamic-pituitary-adrenal axis of the rat: influence of age, gender and circulating sex steroids. *J Endocrinol* 140:365–372.

Rothe J, Lesslauer W, Lotscher H, Lang Y, Koebel P, Kontgen F, Althage A, Zinkernagel R, Steinmetz M, and Bluethman H. 1993. Mice lacking the tumor necrosis factor receptor 1 are resistant to TNF-mediated toxicity but highly susceptible to infection *Listeria monocytogenes*. *Nature* 364:798–802.

Roubinian JR, Talal N, Siiteri PK, and Sadakin JA. 1979. Sex hormone modulation of autoimmunity in NZB/NZW mice. *Arthritis Rheum* 22:1162–1168.

Saladin R, De Vos P, Guerre-Millo M, Leturque A, Girard J, Staels B, and Auwerx J. 1995. Transient increase in obese gene expression after food intake or insulin administration. *Nature* 377:527–529.

Sanchez-Margalet V and Martin-Romero C. 2001. Human leptin signaling in human peripheral blood mononuclear cells: activation of the JAK-STAT pathway. *Cell Immunol* 211.30–36.

Sapolsky R, Rivier C, Yamamoto G, Plotsky P, and Vale W. 1987. Interleukin-1 stimulates the secretion of hypothalamic corticotropin-releasing factor. *Science* 238:522–524.

Sapolsky RM, Romero LM, and Munck AU. 2000. How do glucocorticoids influence stress responses? Integrating permissive, suppressive, stimulatory, and preparative actions. *Endocr Rev* 21:55–89.

Sauerwein R, Mulder J, and Mulder L. 1997. Inflammatory mediators in children with protein-energy malnutrition. *Am J Clin Nutr* 65:1534–1539.

Scrimshaw N and SanGiovann J. 1997. Synergism of nutrition, infection, and immunity: an overview. *Am J Clin Nutr* 66:464S–477S.

Seeman TE and Robbins RJ. 1994. Aging and hypothalamic-pituitary-adrenal response to challenge in humans. *Endocr Rev* 15:233–260.

Sherwin SA, Knost JA, Fein S, Abrams PG, Foon KA, Ochs JJ, Schoenberger C, Maluish AE, and Oldham RK. 1982. A multiple-dose phase I trial of recombinant leukocyte A interferon in cancer patients. *J Am Med Assoc* 248:2461–2466.

Smith SR, Bledsoe T, and Chhetri MK. 1975. Cortisol metabolism in pituitary-adrenal axis in adults with protein-calorie malnutrition. *J Clin Endocr Metab* 40:43–52.

Spangelo BL, Judd AM, Call GB, Zumwalt J, and Gorospe WC. 1995. Role of the cytokines in the hypothalamic-pituitary-adrenal and gonadal axes. *NeuroImmunoModulation* 2:299–312.

Spinedi E, Hadid R, Daneva T, and Gaillard RC. 1992a. Cytokines stimulate the CRH but not the vasopressin neuronal system: evidence for a median eminence site of interleukin-6 action. *Neuroendocrinol* 56:46–53.

Spinedi E, Suescun MO, Hadid R, Daneva T, and Gaillard RC. 1992b. Effects of gonadectomy and sex hormone therapy on the endotoxin-stimulated hypothalamo-pituitary-adrenal axis: evidence for a neuroendocrine-immune sexual dimorphism. *Endocrinol* 131:2430–2436.

Spinedi E, Suescun MO, Salas M, Daneva T, Hadid R, and Gaillard RC. 1992c. Sexual dimorphism in the hypothalamo-pituitary-adrenal axis response to different stress stimuli in mice. *Neuroendocrinol* (Life Sci Adv) 11:139–145.

Spinedi E, Hadid R, and Gaillard RC. 1997. Increased vasopressinergic activity as a possible compensatory mechanism for normal hypothalamic-pituitary-adrenal axis response to stress in BALB/c nude mice. *Neuroendocrinol* 66:287–293.

Spinedi E and Gaillard RC. 1998. A regulatory loop between the hypothalamo-pituitary-adrenal axis and circulating leptin: a physiological role of ACTH. *Endocrinol* 139:4016–4020.

Starnes Jr HF, Warren RS, Jeevanandam M, Gabrilove JL, Larchian W, Oettgen HF, and Brennan MF. 1988. Tumor necrosis factor and the acute metabolic response to tissue injury in man. *J Clin Invest* 82:1321–1325.

Sterling P and Eyer J. 1988. Allostasis: a new paradigm to explain arousal pathology. In: Fischer J and Reason J (Eds), *Handbook of Life Stress, Cognition and Health*. London: Wiley, pp 629–649.

Stephens TW, Basinski M, Bristow PK, Bue-Valleskey JM, Burgett SG, Craft L, Hale J, Hoffman H, Hsiung HM, Kriauciunas A, MacKellar W, Rosteck Jr PR, Schoner B, Smith D, Tinsley FC, Zhang X-Y, and Heiman M. 1995. The role of neuropeptide Y in the antiobesity action of *obese* gene product. *Nature* 377:530–532.

Sthoeger ZM, Chiorazzi N, and Lahita RG. 1988. Regulation of the immune response by sex hormones. I. *In vivo* effects of estradiol and testosterone on pokeweed mitogen-induced human-B cell differentiation. *J Immunol* 141:91–97.

Stokkan KA, Reiter RJ, Vaughan MK, Nonaka NO, and Lerchl A. 1991. Endocrine and metabolic effects of life-long food restriction in rats. *Acta Endocrinol* (Copenh) 125:93–100.

Sullivan J, Kilpatrick L, Costarino A, Chi Li S, and Harris M. 1992. Correlation of plasma cytokine elevations with mortality rate in children with sepsis. *J Pediatr* 120:510–515.

Suttmann U, Selberg O, Gallati H, Ockenga J, Deicher H, and Muller MJ. 1994. Tumor necrosis factor receptor levels are linked to the acute-phase response and malnutrition in human-immunodeficiency-virus-infected patients. *Clin Sci* 86:461–467.

Tannenbaum BM, Tannenbaum GS, Brindley DN, Dallman MF, McArthur MD, and Meaney MJ. 1997. High-fatfeeding impairs both basal and stress-induced hypothalamic-pituitary-adrenal responsiveness in the rat. *Am J Physiol* 273:E1168–E1177.

Timofeeva E and Richard D. 1997. Functional activation of CRH neurons and expression of the genes encoding CRH and its receptors in food-deprived lean (fa/?) and obese (fa/fa) Zucker rats. *Neuroendocrinol* 66:327–340.

Tracey KJ, Fong Y, Hesse DG, Manogue KR, Lee AT, Kuo GC, Lowry SF, and Cerami A. 1987. Anti-cachectin/TNF monoclonal antibodies prevent septic shock during lethal bacteraemia. *Nature* 330:662–664.

Trottier G, Koski KG, Brun T, Toufexis DJ, Richard D, and Walker C-D. 1998. Increased fat intake during gestation modifies hypothalamic-pituitary-adrenal responsiveness in developing pups: a possible role of leptin. *Endocrinol* 139:3704–3711.

Vale W, Spiess J, Rivier C, and Rivier J. 1981. Characterization of a 41-residue ovine hypothalamic peptide that stimulates the secretion of corticotropin and β-endorphin. *Science* 213:1394–1397.

Vamvakopoulos NC and Chrousos GP. 1993. Evidence for a direct estrogenic regulation of human corticotropin-releasing hormone gene expression. *J Clin Invest* 92:1896–1902.

Waage A, Halstensen A, and Espevik T. 1987. Association between tumor necrosis factor in serum and fatal outcome in patients with meningococcal disease. *Lancet* 1:355–357.

Westendorp RGJ, Langermans JAM, Huizinga TWJ, Elouali AH, Verweij CL, Boomsma DI, and Vandenbrouke JP. 1997. Genetic influence on cytokine production and fatal meningococcal disease. *Lancet* 349:170–173.

Wilson MR and Hughes SJ. 1997. The effect of maternal protein deficiency during pregnancy and lactation on glucose tolerance and pancreatic function in adult rat offspring. *J Endocrinol* 154:177–185.

Woods SC, Seeley RJ, Porte Jr D, and Schwartz MW. 1998. Signals that regulate food intake and energy homeostasis. *Science* 280:1378–1382.

Xu B, Kalra PS, Farmerie WG, and Kalra SP. 1999. Daily changes in hypothalamic gene expression of neuropeptide Y, galanin, pro-opiomelanocortin, and adipocyte leptin gene expression and secretion: effects of food restriction. *Endocrinol* 140:2868–2875.

Yu WH, Kimura M, Walczewska A, Karanth S, and McCann SM. 1997. Role of leptin in hypothalamic-pituitary function. *Proc Natl Acad Sci USA* 94:1023–1028.

Zaman K, Baqui A, Yunus M, Sack R, Chowdhury H, and Black R. 1997. Malnutrition cell-mediated immune deficiency and acute upper respiratory infection in rural Bangladeshi children. *Acta Paediatr* 86:923–927.

Zhang Y, Proenca R, Maffei M, Barone M, Leopold L, and Friedman JM. 1994. Positional cloning of the mouse obese gene and its human homologue. *Nature* 372:425–432.

# Index

## A



## B

## C



## D

## E



## F



## G



## H

## I

## K



## L



## M

## N



## O

## P

## U



## V



## W



## X



## Z